CWS

Certified Wireless Security Professional

Study Guide CWSP-205

Second Edition

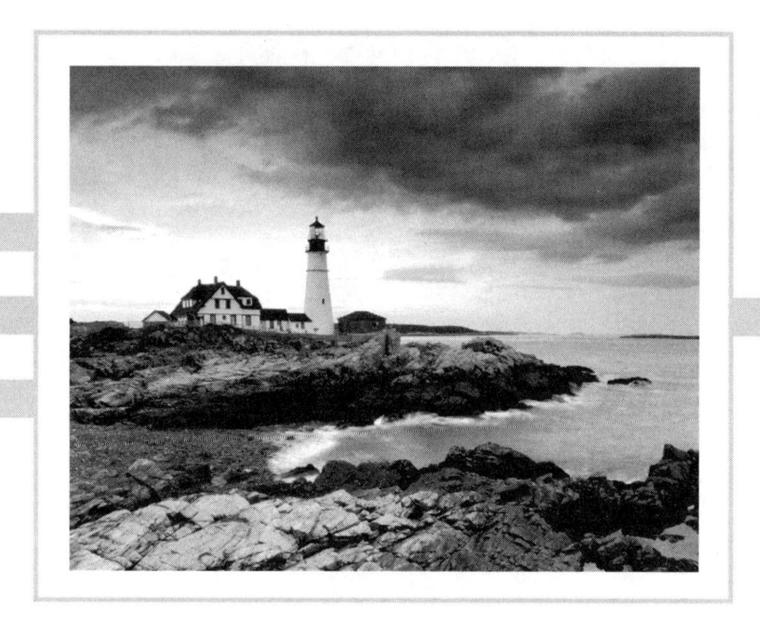

CWSP®

Certified Wireless Security Professional

Study Guide CWSP-205

Second Edition

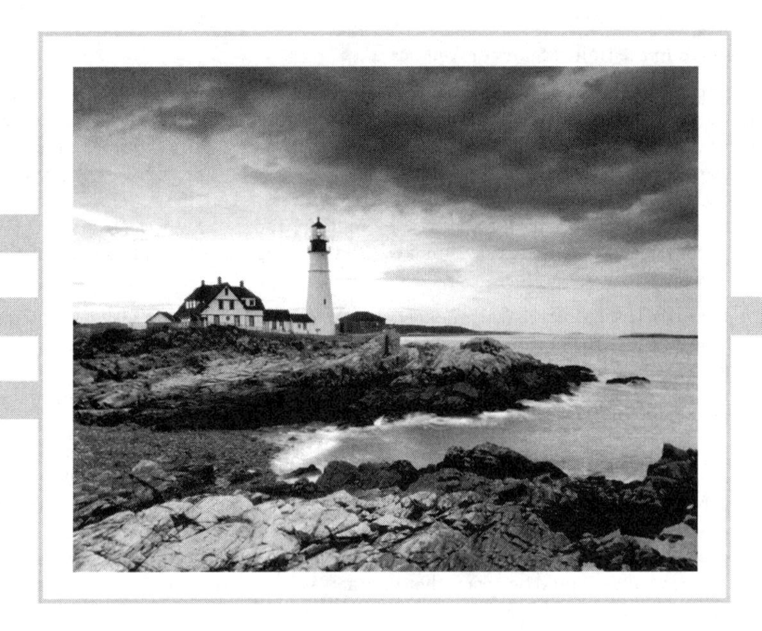

David D. Coleman

David A. Westcott

Bryan Harkins

Executive Editor: Jim Minatel
Development Editor: Kim Wimpsett
Technical Editors: Chris Lyttle and Ben Wilson
Production Editor: Dassi Zeidel
Copy Editor: Liz Welch
Editorial Manager: Mary Beth Wakefield
Production Manager: Kathleen Wisor
Book Designers: Judy Fung and Bill Gibson
Proofreader: Rebecca Rider
Indexer: Ted Laux
Project Coordinator, Cover: Brent Savage
Cover Designer: Wiley
Cover Image: ©Getty Images, Inc./Jeremy Woodhouse

Published simultaneously in Canada

ISBN: 978-1-119-21108-2
ISBN: 978-1-119-24413-4 (ebk.)
ISBN: 978-1-119-21109-9 (ebk.)

Manufactured in the United States of America

Library of Congress Control Number: 9781119211082

10 9 8 7 6 5 4 3 2 1

We dedicate this book to the knowledgeable and competent wireless consultants, designers, and installers, and those who are working diligently to become one. You are the front lines of the industry, explaining the technology to customers, including trying to make them understand that more power and more APs often does not mean better WLAN performance. Wireless networking is a shared medium and a shared community, and we are honored to be part of it and to be able to contribute.

Acknowledgments

When we wrote the first edition of the *CWSP Study Guide*, David Coleman's children, Carolina and Brantley, were just entering college. Carolina now holds a master's degree in public policy from the University of Southern California (USC). Brantley graduated from Boston University and is currently working toward his Ph.D. in biochemistry at the University of Washington. David would like to thank his now adult children for years of support and for making their dad very proud. David would also like to thank his mother Marjorie Barnes, stepfather William Barnes, and brother Rob Coleman, for many years of support and encouragement.

David Coleman would also like to thank the entire Aerohive Networks Knowledge Services department. Additionally, David sends many thanks to Matthew Gast, Paul Levasseur, Abby Strong, Gregor Vucajnk, and all of his co-workers at Aerohive Networks (`www.aerohive.com`). It has been a honor working with you to help build something special.

David Westcott would like to thank Janie for her love and support. I know that my travel and book-writing schedule is difficult to deal with. I say it all of the time and I will continue to say it: "thank you" and "I love you" for your support and for everything that you do for me.

Bryan Harkins would like to thank his wife, Ronda, and his two daughters, Chrystan and Catelynn, and their families, including his three granddaughters, Kaylee, Mikynlee, and Lorali, for allowing him the ability to work with constant travel and the time away from them it has taken to create this book. He would also like to thank his parents for always being there and his brother Chris for getting him involved with IT in the first place. Additionally, he would like to thank Ed Walton, Jeff Manning, and Kent Woodruff for the chance to build something great at Cradlepoint and the team there for their assistance in doing so.

Writing *CWSP: Certified Wireless Security Professional Study Guide* has once again been an adventure. We would like to thank the following individuals for their support and contributions during the entire process.

We must first thank Sybex acquisitions editor Jim Minatel for reaching out to us and encouraging us to write this second edition of our wireless security book. We would also like to thank our development editor, Kim Wimpsett, who has been a pleasure to work with. We also need to send special thanks to our editorial manager, Mary Beth Wakefield; our production editors, Rebecca Anderson and Dassi Zeidel; and Liz Welch, our copyeditor.

We also need to give a big shout-out to our technical editor, Chris Lyttle, CWNE #156. We have personally known Chris for many years. His Wi-Fi background and knowledge were invaluable to providing the amazing technical editing that this book deserved. We encourage you to follow Chris on his blog `www.wifikiwi.com` or on Twitter: `@wifikiwi`. And of course, we offer many thanks to our technical proofreader, Ben Wilson. Ben has accumulated years of Wi-Fi experience working for three major WLAN vendors. We encourage you to follow Ben on Twitter: `@AirNetworkBen`. We would also like to thank Shawn Jackman for his contributions to the first edition of the *CWSP Study Guide*.

We also need to thank Keith Parsons, CWNE #3, and his team at wirelessLAN Professionals. Keith has built a worldwide community of WLAN experts that share knowledge. You can learn more about the wirelessLAN Professionals conferences at www.wlanpros.com. You can also follow Keith on Twitter: @KeithRParsons.

We would also like to thank the CWNP program (www.cwnp.com). All CWNP employees, past and present, should be proud of the internationally renowned wireless certification program that sets the education standard within the enterprise Wi-Fi industry. It has been a pleasure working with all of you the past 16 years.

Finally, we would like to thank Lee Badman for writing his very gracious forward for this book. Lee is also a Wi-Fi expert and he maintains a blog at wirednot.wordpress.com. We encourage you to follow Lee's Wi-Fi question-of-the-day on Twitter via #WIFIQ. You can also follow Lee on Twitter: @wirednot.

About the Authors

David D. Coleman is the Senior Mobility Leader for Aerohive Networks, www.aerohive.com. David collaborates with the Aerohive Knowledge Services team and travels the world for WLAN training sessions and speaking events. He has instructed IT professionals from around the globe in WLAN design, security, administration, and troubleshooting. David has written multiple books, blogs, and white papers about wireless networking, and he is considered an authority on 802.11 technology. Prior to working at Aerohive, he specialized in corporate and government Wi-Fi training and consulting. In the past he has provided WLAN training for numerous private corporations, the US military, and other federal and state government agencies. When he is not traveling, David resides in both Atlanta, Georgia and Seattle, Washington. David is CWNE #4, and he can be reached via email at mistermultipath@gmail.com. Please follow David on Twitter: @mistermultipath.

David Westcott is an independent consultant and technical trainer with over 31 years of experience. David has been a certified trainer for over 23 years, and he specializes in wireless networking, wireless management and monitoring, and network access control. He has provided training to thousands of students at government agencies, corporations, and universities in over 30 countries around the world. David was an adjunct faculty member for Boston University's Corporate Education Center for over 10 years. David has written seven books as well as numerous white papers, and he has developed many courses on wired and wireless networking technologies and networking security.

David was a member of the original CWNE Roundtable. David is CWNE #7 and has earned certifications from many companies, including Cisco, Aruba, Microsoft, Ekahau, EC-Council, CompTIA, and Novell. David lives in Concord, Massachusetts with his wife Janie, his stepdaughters Jennifer and Samantha, and his granddaughter Savannah. David can be reached via email at david@westcott-consulting.com. Please follow David on Twitter: @davidwestcott.

Bryan Harkins has over 30 years experience in the IT field. He has been involved in areas ranging from customer support and sales to network security and design. He has developed custom curriculum for government agencies and Fortune 500 companies alike and delivers both public and private wireless security classes around the world. Previously, Bryan worked as the senior global enablement leader for Aerohive Networks and as the training and courseware development manager for Motorola AirDefense (now Zebra). Currently, Bryan is the Director of Cradlepoint University, where he oversees the training department of Cradlepoint, www.Cradlepoint.com. Bryan also serves on the Board of Advisors for 802Secure, www.802secure.com.

Bryan has presented at multiple industry conferences, including IP Expo, Secure World Expo, Armed Forces Communications and Electronics Association (AFCEA) events, and Microsoft Broad Reach events. He holds a degree in aviation from Georgia State University. He is also a member of the CWNE Roundtable as well as a member of the CWNE Advisory Board. Bryan is CWNE #44, and he can be followed on Twitter: @80211University.

Contents at a Glance

Contents

Table of Exercises

Foreword

Though wireless security options haven't changed significantly since the introduction of 802.11i, the world in which they function certainly has. We are living in strange times for wireless networking. Though our WLAN standards are bringing ever-faster connectivity and more networked devices are coming without Ethernet ports, today's Wi-Fi practitioner operates in a hyper-nuanced security landscape. The media has no shortage of gloom and doom to report on network data breaches, yet many of today's wireless clients are delivered with outdated or limited security capabilities. Where client devices are capable of supporting robust security, users may well opt for ease of use over security. In other situations, WLAN professionals might find themselves being asked to provide an expensive and complicated multitiered security strategy in an environment where there's very little to really protect. Today's CWSPs need be savvy in not only their range of security solutions and analysis tools, but also in how to choose the right option (or combination of options) for complicated situations with diverse user groups and WLAN client devices.

For those just embarking on a wireless career, or for seasoned profressionals trying to broaden their knowledge base, I applaud you for choosing this text. From captive portals to VPN, and MDM solutions to WIPS, the authors give you a knowledge base foundationon which you can build an operational career. David Coleman, Bryan Harkins, and David Westcott bring you decades of wireless security knowledge that spans the gamut from wardriving to Hotspot 2.0. CWSP helps you understand the strengths and disadvantages of any security option you're likely to be faced with in today's real world. It doesn't matter whether you're a one-person company servicing the SMB market or if you support a giant corporate WLAN, you'll do well for yourself and your clients by learning what CWSP has to offer. BYOD, IoT, legacy WLAN concerns—it's all here.

As a long-time wireless professional, I can promise you that there are no shortcuts to building high-quality networks. Good networks support operational goals, and good wireless experts help to make sure those goals are clearly defined and understood before they can be matched with the right solution. When it comes to WLAN security, there are no silver bullets or one-size-fits-all solutions. Thankfully, you're in good hands with David, Bryan, and David as you learn how to think about the broad topic of WLAN security. Best of luck to you.

Lee Badman
CWNA, CWSP, CWDP
Network Architect

Introduction

If you have purchased this book or if you are even thinking about purchasing this book, you probably have some interest in taking the CWSP® (Certified Wireless Security Professional) certification exam or in learning what the CWSP certification exam is about. The authors would like to congratulate you on this first step, and we hope that our book can help you on your journey. Wireless local area networking (WLAN) is currently one of the hottest technologies on the market. Security is an important and mandatory aspect of 802.11 wireless technology. As with many fast-growing technologies, the demand for knowledgeable people is often greater than the supply. The CWSP certification is one way to prove that you have the knowledge and skills to secure 802.11 wireless networks successfully. This study guide is written with that goal in mind.

This book is designed to teach you about WLAN security so that you have the knowledge needed not only to pass the CWSP certification test, but also to be able to design, install, and support wireless networks. We have included review questions at the end of each chapter to help you test your knowledge and prepare for the exam. Extra training resources such as lab materials and presentations are available for download from the book's online resource area, which can be accessed at `www.wiley.com/go/sybextestprep`.

Before we tell you about the certification process and its requirements, we must mention that this information may have changed by the time you are taking your test. We recommend that you visit `www.cwnp.com` as you prepare to study for your test to check out the current objectives and requirements.

Don't just study the questions and answers! The questions on the actual exam will be different from the practice questions included in this book. The exam is designed to test your knowledge of a concept or objective, so use this book to learn the objectives behind the questions.

About CWSP® and CWNP®

If you have ever prepared to take a certification test for a technology with which you are unfamiliar, you know that you are not only studying to learn a different technology, but you are also probably learning about an industry with which you are unfamiliar. Read on and we will tell you about the CWNP Program. *CWNP* is an abbreviation for *Certified Wireless Network Professional.* There is no CWNP test. The CWNP Program develops courseware and certification exams for wireless LAN technologies in the computer networking industry. The CWNP Program certification path is vendor-neutral.

The objective of the CWNP Program is to certify people on wireless networking, not on a specific vendor's product. Yes, at times the authors of this book and the creators of the certification will talk about or even demonstrate how to use a specific product; however, the goal is the overall understanding of wireless technology, not the product itself. If you learned to drive a car, you physically had to sit and practice in one. When you think back and reminisce, you probably do not tell anyone that you learned to drive a Ford; you probably say you learned to drive using a Ford.

There are seven wireless certifications offered by the CWNP Program:

CWTS™: Certified Wireless Technology Specialist The CWTS certification is an entry-level certification for sales professionals, project managers, and networkers who are new to enterprise Wi-Fi. This certification is geared specifically toward both WLAN sales and support staff for the enterprise WLAN industry. The CWTS certification exam (PW0-071) verifies that sales and support staffs are specialists in WLAN technology and have all the fundamental knowledge, tools, and terminology to sell and support WLAN technologies more effectively.

CWNA®: Certified Wireless Network Administrator The CWNA certification is a foundation-level Wi-Fi certification; however, it is not considered an entry-level technology certification. Individuals taking this exam (CWNA-106) typically have a solid grasp on network basics such as the OSI model, IP addressing, PC hardware, and network operating systems. Many candidates already hold other industry-recognized certifications, such as the CompTIA Network+ or Cisco CCNA, and are looking for the CWNA certification to enhance or complement existing skills.

CWSP®: Certified Wireless Security Professional The CWSP certification exam (CWSP-205) is focused on standards-based wireless security protocols, security policy, and secure wireless network design. This certification introduces candidates to many of the technologies and techniques that intruders use to compromise wireless networks and that administrators use to protect wireless networks. With recent advances in wireless security, WLANs can be secured beyond their wired counterparts.

CWAP®: Certified Wireless Analyst Professional The CWAP certification exam (CWAP-402) is a professional-level career certification for networkers who are already CWNA certified and have a thorough understanding of RF technologies and applications of 802.11 networks. This certification provides an in-depth look at 802.11 operations and prepares WLAN professionals to be able to perform, interpret, and understand wireless packet and spectrum analysis.

CWDP®: Certified Wireless Design Professional The CWDP certification exam (CWDP-302) is a professional-level career certification for networkers who are already CWNA certified and have a thorough understanding of RF technologies and applications of 802.11 networks. This certification prepares WLAN professionals to properly design wireless LANs for different applications to perform optimally in different environments.

CWNE®: Certified Wireless Network Expert The CWNE certification is the highest-level certification in the CWNP program. By successfully completing the CWNE requirements, you will have demonstrated that you have the most advanced skills available in today's wireless LAN market. The CWNE certification requires CWNA, CWAP, CWDP, and CWAP certifications. To earn the CWNE certification, a rigorous application must be submitted and approved by CWNP's review team.

CWNT®: Certified Wireless Network Trainer Certified Wireless Network Trainers are qualified instructors certified by the CWNP program to deliver CWNP training courses to IT professionals. CWNTs are technical and instructional experts in wireless technologies, products, and solutions. To ensure a superior learning experience for our customers, CWNP Education Partners are required to use CWNTs when delivering training using official CWNP courseware. More information about becoming a CWNT is available on the CWNP website.

How to Become a CWSP

To become a CWSP, you must do the following three things:

- Agree that you have read and will abide by the terms and conditions of the CWNP Confidentiality Agreement.
- Pass the CWNA certification exam.
- Pass the CWSP certification exam.

The CWNA certification is a prerequisite for the CWSP certification. If you have purchased this book, there is a good chance that you have already passed the CWNA exam and are now ready to move to the next level of certification and plan to study and pass the CWSP exam.

A copy of the CWNP Confidentiality Agreement can be found online at the CWNP website.

When you sit to take any CWNP exam, you will be required to accept this confidentiality agreement before you can continue with the exam. Once you have agreed, you will be able to continue.

The information for the CWNA exam is as follows:

- Exam Name: Certified Wireless Network Administrator
- Exam Number: CWNA-106
- Cost: $175.00 (in U.S. dollars)
- Duration: 90 minutes
- Questions: 60
- Question Types: Multiple choice/multiple answer
- Passing Score: 70% (80% for instructors)
- Available Languages: English
- Availability: Register at Pearson VUE (`www.vue.com/cwnp`)

The information for the CWSP exam is as follows:

- Exam Name: Certified Wireless Security Professional
- Exam Number: CWSP-205

- Cost: $225.00 (in U.S. dollars)
- Duration: 90 minutes
- Questions: 60
- Question Types: Multiple choice/multiple answer
- Passing Score: 70% (80% for instructors)
- Available Languages: English
- Availability: Register at Pearson VUE (www.vue.com/cwnp)

When you schedule the exam, you will receive instructions regarding appointment and cancellation procedures, ID requirements, and information about the testing center location. In addition, you will receive a registration and payment confirmation letter. Exams can be scheduled weeks in advance or, in some cases, even as late as the same day.

After you have successfully completed the CWSP certification requirements, the CWNP Program will award you the CWSP certification, which is good for three years. To recertify, you will need to pass the current CWSP-205 exam or earn the CWNE certification. If the information you provided the testing center with is correct, you will receive an email from CWNP recognizing your accomplishment and providing you with a CWNP certification number. After you earn any CWNP certification, you can purchase a certification kit from the CWNP website.

Who Should Read This Book?

If you want to acquire a solid foundation in WLAN security and your goal is to prepare for the exam, this book is for you. You will find clear explanations of the concepts you need to grasp and plenty of help to achieve the high level of professional competency you need in order to succeed.

If you want to become certified as a CWSP, this book is definitely what you need. However, if you just want to attempt to pass the exam without really understanding WLAN security, this study guide is not for you. It is written for people who want to acquire hands-on skills and in-depth knowledge of wireless networking security.

How to Use This Book

We have included several testing features in the book and via the publisher's website www.wiley.com/go/sybextestprep.

These tools will help you retain vital exam content as well as prepare you to sit for the actual exam:

Before You Begin At the beginning of the book (right after this introduction) is an assessment test you can use to check your readiness for the exam. Take this test before you start reading the book; it will help you determine the areas in which you may need to brush up. The answers to the assessment test appear on a separate page after the last question of the test. Each answer includes an explanation and a note telling you the chapter in which the material appears.

Chapter Review Questions To test your knowledge as you progress through the book, there are review questions at the end of each chapter. As you finish each chapter, answer the review questions and then check your answers; the correct answers appear in Appendix A at the end of the book. You can go back and reread the section that deals with each question you answered wrong to ensure that you answer correctly the next time you are tested on the material.

Interactive Online Learning Environment and Test Bank The interactive online learning environment that accompanies *CWSP: Certified Wireless Security Professional Study Guide* provides a test bank with study tools to help you prepare for the certification exam—and increase your chances of passing it the first time! The test bank includes the following:

Sample Tests All of the questions in this book are provided: the assessment test, which you will find at the end of this introduction, and the chapter tests that include the review questions at the end of each chapter. In addition, there are two practice exams. Use these questions to test your knowledge of the study guide material. The online test bank runs on multiple devices.

Flashcards Questions are provided in digital flashcard format (a question followed by a single correct answer). You can use the flashcards to reinforce your learning and provide last-minute test prep before the exam.

Go to `www.wiley.com/go/sybextestprep` to register and gain access to this interactive online learning environment and test bank with study tools.

Hands-on Exercises Several chapters in this book have exercises that use files that are also provided on the Sybex website. These hands-on exercises will provide you with a broader learning experience by providing hands-on experience and step-by-step problem solving. To get these files go to `www.sybex.com` and search for the book by title or ISBN.

Exam Objectives

The CWSP-205 exam, covering the 2015 objectives, will certify that the successful candidate understands the security weaknesses inherent in WLANs, the solutions available to address those weaknesses, and the steps necessary to implement a secure and manageable WLAN in an enterprise environment. Exam CWSP-205 is required to earn the CWSP certification.

The skills and knowledge measured by this examination are derived from a survey of wireless networking experts from around the world. The results of this survey were used in weighing the subject areas and ensuring that the weighting is representative of the relative importance of the content.

The following chart provides the breakdown of the weight of each section of the exam.

Wireless LAN Security Subject Area	% of Exam
Wireless Network Attacks and Threat Assessment	20%
Security Policy	5%
Wireless LAN Security Design and Architecture	50%
Monitoring and Management	25%
Total	100%

1.0 Wireless Network Attacks and Threat Assessment – 20%

1.1 Describe general network attacks common to wired and wireless networks, including DoS, phishing, protocol weaknesses and configuration error exploits.

1.2 Recognize common attacks and describe their impact on WLANs, including PHY and MAC DoS, hijacking, unauthorized protocol analysis and eavesdropping, social engineering, man-in-the-middle, authentication and encryption cracks and rogue hardware.

1.3 Execute the preventative measures required for common vulnerabilities on wireless infrastructure devices, including weak/default passwords on wireless infrastructure equipment and misconfiguration of wireless infrastructure devices by administrative staff.

1.4 Describe and perform risk analysis and risk mitigation procedures, including asset management, risk ratings, loss expectancy calculations and risk management planning.

1.5 Explain and demonstrate the security vulnerabilities associated with public access or other unsecured wireless networks, including the use of a WLAN for spam transmission, malware injection, information theft, peer-to-peer attacks and Internet attacks.

2.0 Security Policy – 5%

2.1 Explain the purpose and goals of security policies including password policies, acceptable use policies, WLAN access policies, personal device policies, device management (APs, infrastructure devices and clients) and security awareness training for users and administrators.

2.2 Summarize the security policy criteria related to wireless public access network use including user risks related to unsecured access and provider liability.

2.3 Describe how devices and technology used from outside an organization can impact the security of the corporate network including topics like BYOD, social networking and general MDM practices.

3.0 Wireless LAN Security Design and Architecture – 50%

3.1 Describe how wireless network security solutions may vary for different wireless network implementations including small businesses, home offices, large enterprises, public networks and remote access.

3.2 Understand and explain 802.11 Authentication and Key Management (AKM) components and processes including encryption keys, handshakes and pre-shared key management.

3.3 Define and differentiate among the 802.11-defined secure networks, including pre-RSNA security, Transition Security Networks (TSN) and Robust Security Networks (RSN) and explain the relationship of these networks to terms including RSNA, WPA and WPA2.

3.4 Identify the purpose and characteristics of IEEE 802.1X and EAP and the processes used including EAP types (PEAP, EAP-TLS, EAP-TTLS, EAP-FAST and EAP-SIM), AAA servers (RADIUS) and certificate management.

3.5 Recognize and understand the common uses of VPNs in wireless networks, including remote APs, VPN client access, WLAN controllers and cloud architectures.

3.6 Describe centrally-managed client-side security applications, including VPN client software and policies, personal firewall software, mobile device management (MDM) and wireless client utility software.

3.7 Describe and demonstrate the use of secure infrastructure management protocols, including HTTPS, SNMP, secure FTP protocols, SCP and SSH.

3.8 Explain the role, importance, and limiting factors of VLANs and network segmentation in an 802.11 WLAN infrastructure.

3.9 Understand additional security features in WLAN infrastructure and access devices, including management frame protection, Role-Based Access Control (RBAC), Fast BSS transition (preauthentication and OKC), physical security methods and Network Access Control (NAC).

3.10 Explain the purpose, methodology, features, and configuration of guest access networks and BYOD support, including segmentation, guest management, captive portal authentication and device management.

4.0 Monitoring, Management, and Tracking – 25%

4.1 Explain the importance of ongoing WLAN monitoring and the necessary tools and processes used as well as the importance of WLAN security audits and compliance reports.

4.2 Understand how to use protocol and spectrum analyzers to effectively evaluate secure wireless networks including 802.1X authentication troubleshooting, location of rogue security devices and identification of non-compliant devices.

4.3 Understand the common features and components of a Wireless Intrusion Prevention Systems (WIPS) and how they are used in relation to performance, protocol, spectrum and security analysis.

4.4 Describe the different types of WLAN management systems and their features, including network discovery, configuration management, firmware management, audit management, policy enforcement, rogue detection, network monitoring, user monitoring, event alarms and event notifications.

4.5 Describe and implement compliance monitoring, enforcement, and reporting. Topics include industry requirements, such as PCI-DSS and HIPAA, and general government regulations.

CWSP Terminology

In addition to the preceding objectives, the following security specialty terms should be clearly understood by CWSP-205 exam candidates:

802.11r
802.11w
802.1X
Access Control List (ACL)
Access Point (AP)
Advanced Encryption Standard (AES)
Alarms
Asymmetric Encryption
Authentication
Authentication and Key Management (AKM)
Authentication Header (AH)
Authentication Server
Authentication, Authorization, and Accounting (AAA)
Authenticator
Authorization
Availability
Bring Your Own Device (BYOD)
Certificate Authority (CA)
Compliance
Confidentiality
Counter-Mode/CBC Mac Protocol (CCMP)
Denial of Service (DoS)
Discovery
Distributed DoS (DDoS)
EAP Flexible Authentication via Secure Tunneling (EAP-FAST)
EAP Subscriber Identity Module (EAP-SIM)
EAP Transport Layer Security (EAP-TLS)
EAP Tunneled TLS (EAP-TTLS)
Eavesdropping
Encapsulated Security Payload (ESP)
Encryption
Evil Twin
Extensible Authentication Protocol (EAP)
Fast Basic Service Set (BSS) Transition
File Transfer Protocol (FTP)
Firewall
Firmware
Hashing
Health Insurance Portability and Accountability Act (HIPAA)
Hijacking
Hypertext Transfer Protocol over SSL (HTTPS)
Infrastructure
Integrity
Interference
Internet Protocol (IP)
Intrusion Detection System (IDS)
IP Security (IPSec)
Lightweight EAP (LEAP)
Location-Based Access Control (LBAC)
MAC Filter
Malware
Man-in-the-middle
Medium Access Control (MAC)
Mobile Device Management (MDM)
Network Access Control (NAC)
Notifications
Opportunistic Key Caching (OKC)
Payment Card Industry (PCI) Data Security Standard (DSS)
Peer-to-Peer
Phishing
Physical Layer (PHY)
Policy
Pre-authentication
Private Key
Protected EAP (PEAP)
Protocol analysis
Public Key
Public Key Infrastructure (PKI)

RADIUS (Remote Authentication Dial-In User Service)
Risk
Rivest Cipher 4 (RC4)
Robust Security Network (RSN)
Rogue
Role-Based Access Control (RBAC)
Secure Copy (SCP)
Secure FTP (SFTP)
Secure Shell (SSH)
Secure Sockets Layer (SSL)
Service Level Agreement (SLA)
Simple Network Management Protocol (SNMP)
Social Engineering
Spam
Spectrum analysis
Supplicant
Symmetric Encryption
Temporal Key Integrity Protocol (TKIP)
TACACS/TACACS+
Threat
Transition Security Network (TSN)
Virtual Local Area Network (VLAN)
Virtual Private Network (VPN)
Vulnerability
War Driving
Wi-Fi Protected Access (WPA)
Wi-Fi Protected Access v2 (WPA2)
Wi-Fi Protected Setup (WPS)
Wired Equivalent Privacy (WEP)
Wireless Intrusion Prevention System (WISP)
Wireless Local Area Network (WLAN)

Tips for Taking the CWSP Exam

Here are some general tips for taking your exam successfully:

- Bring two forms of ID with you. One must be a photo ID, such as a driver's license. The other can be a major credit card or a passport. Both forms must include a signature.
- Arrive early at the exam center so you can relax and review your study materials, particularly tables and lists of exam-related information.
- Read the questions carefully. Do not be tempted to jump to an early conclusion. Make sure you know exactly what the question is asking.
- Many of the questions will be real-world scenarios. Scenario questions usually take longer to read and often have many distracters. There may be several correct answers to the scenario questions; however, you will be asked to choose the correct answer that best fits the presented scenario.
- All questions will be multiple-choice with a single correct answer.
- Do not spend too much time on one question. This is a form-based test; however, you cannot move backward through the exam. You must answer the current question before you can move to the next question, and once you have moved to the next question, you cannot go back and change your answer to a previous question.
- Keep track of your time. Since this is a 90-minute test consisting of 60 questions, you have an average of 90 seconds to answer each question. You can spend as much or as

little time on any one question, but when the 90 minutes is up, the test is over. Check your progress. After 45 minutes, you should have answered at least 30 questions. If you have not, do not panic. You will simply need to answer the remaining questions at a faster pace. If on average you can answer each of the remaining 30 questions 4 seconds quicker, you will recover 2 minutes. Again, do not panic; just pace yourself.

- For the latest pricing on the exams and updates to the registration procedures, visit CWNP's website at www.cwnp.com.

Assessment Test

1. At which layer of the OSI model does 802.11 technology operate? (Choose all that apply.)
   - **A.** Session
   - **B.** Network
   - **C.** Physical
   - **D.** Presentation
   - **E.** Transport
2. PSK authentication using ARC4 encryption is mandatory in which of the following? (Choose all that apply.)
   - **A.** WPA-Personal
   - **B.** WPA Enterprise
   - **C.** WPA-2 SOHO
   - **D.** WPA-2 Enterprise
   - **E.** WPA2-Personal
3. 802.11 pre-RSNA security defines which wireless security solution?
   - **A.** Dynamic WEP
   - **B.** 802.1X/EAP
   - **C.** 128-bit static WEP
   - **D.** Temporal Key Integrity Protocol
   - **E.** CCMP/AES
4. Which one of the following technologies can be used to provide the access security needed to expand outside of the organization's network?
   - **A.** SAM
   - **B.** Auth
   - **C.** SAML
   - **D.** CRM
   - **E.** OAuth
5. Which of the following is a self-service process for an employee to provision a BYOD device to connect to the secure corporate network?
   - **A.** Captive Portal
   - **B.** 802.1X/EAP Configurator
   - **C.** MDM
   - **D.** Over-the-air management
   - **E.** Onboarding

6. Which of the following encryption methods uses asymmetric communications?
   - A. WEP
   - B. TKIP
   - C. Public key cryptography
   - D. CCMP

7. For an 802.1X/EAP solution to work properly with a WLAN, which two components must both support the same type of encryption?
   - A. Supplicant and authenticator
   - B. Authorizer and authenticator
   - C. Authenticator and authentication server
   - D. Supplicant and authentication server

8. Which of these types of EAP do not use tunneled authentication? (Choose all that apply.)
   - A. EAP-LEAP
   - B. EAP-PEAPv0 (EAP-MSCHAPv2)
   - C. EAP-PEAPv1 (EAP-GTC)
   - D. EAP-FAST
   - E. EAP-TLS (normal mode)
   - F. EAP-MD5

9. What type of WLAN security is depicted by this graphic?

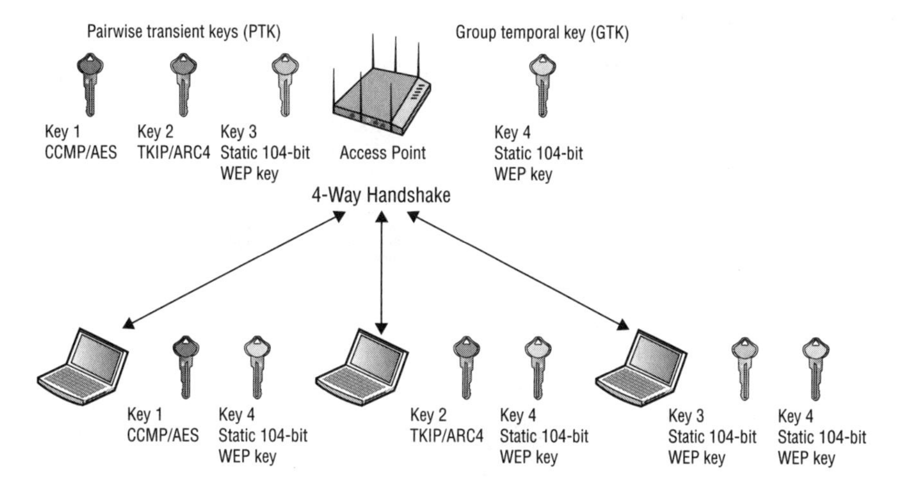

   - A. RSN
   - B. TSN
   - C. VPN
   - D. WPS
   - E. WMM

10. The 802.11-2012 standard defines authentication and key management (AKM) services. Which of these keys are part of the key hierarchy defined by AKM? (Choose all that apply.)

    A. MSK

    B. GTK

    C. PMK

    D. ACK

    E. ATK

11. Which of these Wi-Fi Alliance security certifications are intended for use only in a home office environment? (Choose all that apply.)

    A. WPA-Personal

    B. WPA-Enterprise

    C. WPA2-Personal

    D. WPA2-Enterprise

    E. WPS

12. Which of these fast secure roaming (FSR) methods requires an authenticator and supplicant to establish an entire 802.1X/EAP exchange prior to the creation of dynamic encryption keys when a supplicant is roaming?

    A. PMK caching

    B. Opportunistic key caching

    C. Fast BSS transition

    D. Preauthentication

13. What is the main WLAN security risk shown in this graphic?

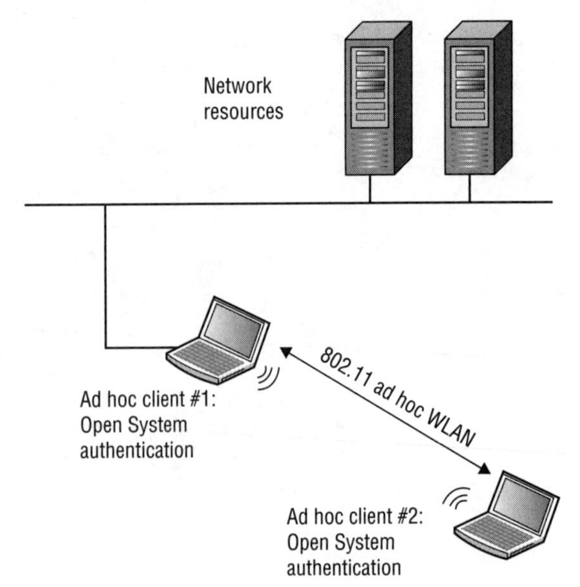

A. The ad hoc clients are not using encryption.
B. The ad hoc clients are using weak authentication.
C. The ad hoc clients are not communicating through an access point.
D. The ad hoc client #1 Ethernet card is connected to an 802.3 wired network.

14. Which components of 802.11 medium contention can be compromised by a DoS attack? (Choose all that apply.)
    A. Physical carrier sense
    B. Interframe spacing
    C. Virtual carrier sense
    D. Random backoff timer

15. After viewing this graphic, determine which type of WLAN attack tool could be used to create this Layer 1 denial of service to the WLAN.

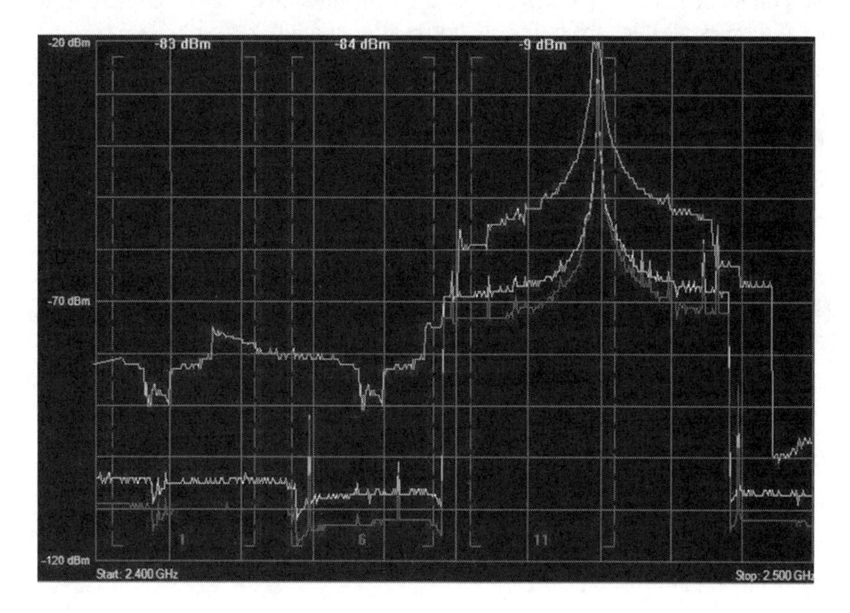

    A. All-band hopping jammer
    B. Wide-band jammer
    C. Narrow-band jammer
    D. Queensland software utility
    E. Packet generator

16. Bill is designing a WLAN that will use an integrated WIPS with dedicated full-time sensors. The WLAN predictive modeling software solution that Bill is using has recommended a ratio of one dedicated sensor for every six access points. Bill needs to make sure that the entire building can be monitored at all times, and he is also concerned about the accuracy of location tracking of rogue devices. What considerations should Bill give to sensor placement in order to properly meet his objectives? (Choose all that apply.)

    A. Installing the sensors in a straight line

    B. Installing the sensors in a staggered arrangement

    C. Installing sensors around the building perimeter

    D. Increasing the transmit power

    E. Installing more sensors

17. Which of these WIDS/WIPS software modules allows an organization to monitor WLAN statistics on hidden nodes, excessive Layer 2 retransmissions, excessive wired to wireless traffic, and excessive client roaming? (Choose all that apply.)

    A. Spectrum analysis

    B. Protocol analysis

    C. Forensic analysis

    D. Signature analysis

    E. Performance analysis

18. When deploying 802.1X/EAP security, which IETF standard RADIUS attribute can be used to encapsulate up to 255 custom RADIUS attributes?

    A. (11) Filter-id

    B. (26) Vendor-Specific

    C. (79) EAP-Message

    D. (80) Message-Authenticator

    E. (97) Frame-Encapsulator

19. Identify the protocols that are normally used to manage WLAN infrastructure devices securely. (Choose all that apply.)

    A. HTTPS

    B. Telnet

    C. SSH2

    D. TLS

    E. IPsec

    F. CCMP/AES

20. What type of WLAN security policy defines WLAN security auditing requirements and policy violation report procedures?

    A. Functional policy

    B. General policy

    C. Protocol policy

    D. Performance policy

Answers to Assessment Test

1. C. The IEEE 802.11-2012 standard only defines communication mechanisms at the Physical layer and MAC sublayer of the Data-Link layer of the OSI model. For more information, see Chapter 1.

2. A. The security used in SOHO environments is preshared key (PSK) authentication; however, WPA-2 defines CCMP/AES encryption. The Wi-Fi Alliance WPA-Personal and WPA2-Personal certifications both use the PSK authentication method; however, WPA-Personal specifies TKIP/ARC4 encryption and WPA2-Personal specifies CCMP/AES. WLAN vendors have many names for PSK authentication, including WPA/WPA2-Passphrase, WPA/WPA2-PSK, and WPA/WPA2-Preshared Key. For more information, see Chapter 2.

3. C. The original 802.11 standard ratified in 1997 defined the use of a 64-bit or 128-bit static encryption solution called Wired Equivalent Privacy (WEP). WEP is considered pre-RSNA security. Dynamic WEP was never defined under any wireless security standard. The use of 802.1X/EAP, TKIP/ARC4, and CCMP/AES are all defined under the current 802.11-2012 standard for robust network security (RSN). For more information, see Chapter 2.

4. C. Two technologies, Security Assertion Markup Language (SAML) and open standard for authorization (OAuth), can be used to provide the access security needed to expand outside of the organizations network. SAML provides a secure method of exchanging user security information between your organization and an external service provider, such as a third-party cloud-based customer relationship management (CRM) platform. OAuth is different from SAML because it is an authorization standard and not an authentication standard. For more information, see Chapter 10.

5. E. The main purpose of onboarding solutions is to provide an inexpensive and simple way to provision employee personal WLAN devices onto a secure corporate SSID. For more information, see Chapter 10.

6. C. WEP, TKIP, and CCMP use symmetric algorithms. WEP and TKIP use the ARC4 algorithm. CCMP uses the AES cipher. Public key cryptography is based on asymmetric communications. For more information, see Chapter 3.

7. A. An 802.1X/EAP solution requires that both the supplicant and the authentication server support the same type of EAP. The authenticator must be configured for 802.1X/EAP authentication, but it does not care which EAP type passes through. The authenticator and the supplicant must support the same type of encryption. The 802.1X/EAP process provides the seeding material for the 4-Way Handshake process that is used to create dynamic encryption keys. For more information, see Chapter 4.

8. A, E, F. Tunneled authentication is used to protect the exchange of client credentials between the supplicant and the AS within an encrypted TLS tunnel. All flavors of EAP-PEAP use tunneled authentication. EAP-TTLS and EAP-FAST also use tunneled authentication. While EAP-TLS is highly secure, it rarely uses tunneled authentication. Although rarely supported, an optional privacy mode does exist for EAP-TLS, which can be used to

establish a TLS tunnel. EAP-MD5 and EAP-LEAP do not use tunneled authentication. For more information, see Chapter 4.

9. B. A transition security network (TSN) supports RSN-defined security as well as legacy security such as WEP within the same BSS. Within a TSN, some client stations will use RSNA security using TKIP/ARC4 or CCMP/AES for encrypting unicast traffic. However, some legacy stations might use static WEP keys for unicast encryption. All of the clients will use WEP encryption for the broadcast and multicast traffic. Because all the stations share a single group encryption key for broadcast and multicast traffic, the lowest common denominator must be used for the group cipher. For more information, see Chapter 5.

10. A, B, C. AKM services defines the creation of encryption keys. Some of the encryption keys are derived from the authentication process, some of the keys are master keys, and some are the final keys that are used to encrypt/decrypt 802.11 data frames. The keys include the master session key (MSK), group master key (GMK), pairwise master key (PMK), group temporal key (GTK), and pairwise transient key (PTK). For more information, see Chapter 5.

11. A, C, E. WPA/WPA2-Enterprise solutions use 802.1X/EAP methods for authentication in enterprise environments. Most SOHO wireless networks are secured with WPA/WPA2-Personal mechanisms. WPA-Personal and WPA2-Personal both use the PSK authentication methods. PSK authentication is sometimes used in the enterprise, but is not recommended due to known weaknesses. Wi-Fi Protected Setup (WPS) defines simplified and automatic WPA and WPA2 security configurations for home and small-business users. Users can easily configure a network with security protection by using a personal identification number (PIN) or a button located on the access point and the client device. WPS is intended only for SOHO environments and is not meant to be used in the enterprise. For more information, see Chapter 6.

12. D. The 802.11-2012 standard defines two fast secure roaming mechanisms called preauthentication and PMK caching. Most WLAN vendors currently use an enhanced method of FSR called opportunistic key caching. The 802.11r-2008 amendment defines more complex Fast BSS transition (FT) methods of FSR. PMK caching, opportunistic key caching (OKC), and fast BSS transition (FT) all allow for 802.1X/EAP authentication to be skipped when roaming. Preauthentication still requires another 802.1X/EAP exchange through the original AP prior to the client roaming to a new target AP. For more information, see Chapter 7.

13. D. Probably the most overlooked rogue device is the ad hoc wireless network. The technical term for an 802.11 ad hoc WLAN is an independent basic service set (IBSS). The radio cards that make up an IBBS network consist solely of client stations, and no access point is deployed. The more common name for an IBSS is an ad hoc wireless network. An Ethernet connection and a Wi-Fi card can be bridged together—an intruder might access the ad hoc wireless network and then potentially route their way to the Ethernet connection and get onto the wired network. For more information, see Chapter 12.

14. A, C. 802.11 uses a medium contention process called Carrier Sense Multiple Access with Collision Avoidance (CSMA/CA). To ensure that only one radio card is transmitting on the half-duplex RF medium, CSMA/CA uses four checks and balances. The four

checks and balances are virtual carrier sense, physical carrier sense, the random backoff timer, and interframe spacing. Virtual carrier sense uses a timer mechanism known as the network allocation vector (NAV) timer. Physical carrier sense uses a mechanism called the clear channel assessment (CCA) to determine whether the medium is busy before transmitting. Virtual carrier sense is susceptible to a DoS attack when an attacker manipulates the duration value of 802.11 frames. Physical carrier sense is susceptible to DoS when there is a continuous transmitter on the frequency channel. For more information, see Chapter 12.

15. C. A Layer 1 DoS attack can be accomplished using a wide-band jamming device or narrow-band jamming device. A wide-band jammer transmits a signal that raises the noise floor for most of the entire frequency band and therefore disrupts communications across multiple channels. The graphic shows a spectrum analyzer view of the narrow-band jammer that is disrupting service on several channels but not the entire frequency band. For much less money, an attacker could also use the Queensland Attack to disrupt an 802.11 WLAN. A major chipset manufacturer of 802.11b radio cards produced a software utility that placed the radios in a continuous transmit state for testing purposes. This utility can also be used for malicious purposes and can send out a constant RF signal much like a narrowband signal generator. For more information, see Chapter 13.

16. B, C, E. Every WLAN vendor has their own sensor deployment recommendations and guidelines; however, a ratio of one sensor for every three to five access points is highly recommended. Full-time sensors are often placed strategically at the intersection points of three AP coverage cells. A common mistake is placing the sensors in a straight line as opposed to staggered sensor arrangement, which will ensure a wider area of monitoring. Another common sensor placement recommendation is to arrange sensors around the perimeter of the building. Perimeter placement increases the effectiveness of triangulation and also helps to detect WLAN devices that might be outside the building. Some of the better WLAN predictive modeling software solutions will also create models for recommended sensor placement. For more information, see Chapter 14.

17. B, E. Although the main purpose of an enterprise WIDS/WIPS is security monitoring, information collected by the WIPS can also be used for performance analysis. Since everything WLAN devices transmit is visible to the sensors, the Layer 2 information gathered can be used to determine the performance level of a WLAN, including capacity and latency. The Layer 2 information can also be gathered using standard protocol analysis. For more information, see Chapter 14.

18. B. RADIUS vendor-specific attributes (VSAs) are derived from the IETF attribute (26) Vendor-Specific. This attribute allows a vendor to create any additional 255 attributes however they wish. Data that is not defined in standard IETF RADIUS attributes can be encapsulated in the (26) Vendor-Specific attribute. For more information, see Chapter 9.

19. A, C. Secure Shell, or SSH, is typically used as the secure alternative to Telnet. SSH2 implements authentication and encryption using public key cryptography of all network traffic traversing between a host and a WLAN infrastructure device. HTTPS is essentially an SSL session that uses the HTTP protocol and is implemented on network devices for management via a graphical user interface (GUI). For more information, see Chapter 8.

20. B. When establishing a wireless security policy, you must first define a general policy. A general wireless security policy establishes why a wireless security policy is needed for an organization. General policy defines a statement of authority and the applicable audience. General policy also defines threat analysis and risk assessments. General policy defines internal auditing procedures as well as the need for independent outside audits. WLAN security policy should be enforced, and clear definitions are needed to properly respond to policy violations. For more information, see Chapter 15.

CWSP®

Certified Wireless Security Professional

Study Guide CWSP-205

Second Edition

Chapter 1

WLAN Security Overview

IN THIS CHAPTER, YOU WILL LEARN ABOUT THE FOLLOWING:

✓ **Standards organizations**

- International Organization for Standardization (ISO)
- Institute of Electrical and Electronics Engineers (IEEE)
- Internet Engineering Task Force (IETF)
- Wi-Fi Alliance

✓ **802.11 networking basics**

✓ **802.11 security basics**

- Data privacy
- Authentication, authorization, accounting (AAA)
- Segmentation
- Monitoring
- Policy

✓ **802.11 security history**

- 802.11i security amendment and WPA certifications
- Robust Security Network
- The future of 802.11 security

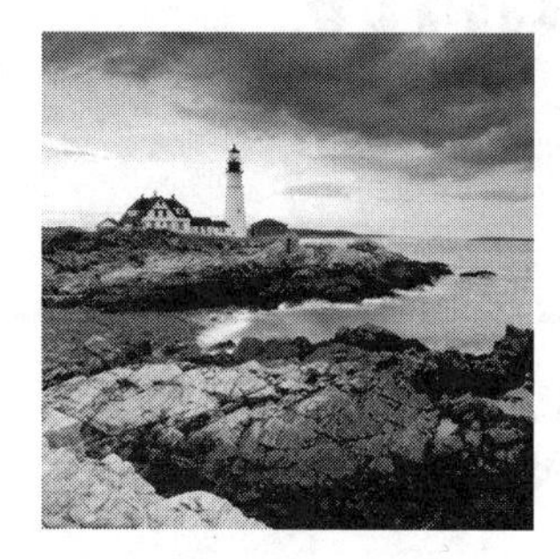

The 802.11-2012 standard defines *wireless local area network (WLAN)* technology, including all Layer 2 security mechanisms. To better understand WLAN security, you need to have a general appreciation of computer security and the components that are used to provide computer security. Security should never be taken lightly for wired or wireless networks. Since the early days of Wi-Fi communications, there has been a concern about the ability to transmit data securely over a wireless medium and properly protect wired network resources. This concern is as valid now as it was in 1997 when 802.11 was introduced. The difference between then and now is that the technologies and standards for Wi-Fi communications are much more secure and easier to implement. In addition to the standards providing better WLAN security, the people who are installing and managing these networks are much more knowledgeable about the design and implementation of secure wireless networks.

In 2004, the 802.11i amendment was ratified by the IEEE, defining stronger encryption and better authentication methods. The 802.11i amendment, which is now part of the 802.11-2012 standard, fully defines a robust security network (RSN), which is discussed later in this chapter. If proper encryption and authentication solutions are deployed, a wireless network can be as secure as, if not more secure than, the wired segments of a network.

Before you learn about the various wireless security methods, techniques, and tools, it is important to learn some of the basic terms and concepts of encryption and computer security. WLAN security is based on many of the same concepts and principles as hard-wired systems, with the main difference being the natural reduced security of the unbounded medium (RF waves) that is used in wireless communications. Because data is transmitted freely and openly in the air, proper protection is needed to ensure data privacy. Thus strong encryption is needed.

The function of most wireless networks is to provide a portal into some other network infrastructure, such as an 802.3 Ethernet backbone. The wireless portal must be protected, and therefore an authentication solution is needed to ensure that only authorized users can pass through the portal via a wireless access point. After users have been authorized to pass through the wireless portal, virtual local area networks (VLANs) and identity-based mechanisms are needed to restrict access, additionally, to network resources. 802.11 wireless networks can be further protected with continuous monitoring by networking accounting and a wireless intrusion detection system. All of these security components should also be cemented with policy enforcement.

In this chapter we will explore the basic terminology of WLAN security. We will discuss the organizations that create the standards, certifications, and recommendations that help guide and direct wireless security. In addition, you will learn about these wireless security standards and certifications.

Standards Organizations

Each of the standards organizations discussed in this chapter helps guide a different aspect of security that is used in wireless networking.

The International Organization for Standardization (ISO) created the Open Systems Interconnection (OSI) model, which is an architectural model for data communications.

The Institute of Electrical and Electronics Engineers (IEEE) creates standards for compatibility and coexistence between networking equipment, not just wireless networking equipment. However, in this book we are concerned primarily with its role in wireless networking and more specifically wireless security.

The Internet Engineering Task Force (IETF) is responsible for creating Internet standards. Many of these standards are integrated into the wireless networking and security protocols and standards.

The Wi-Fi Alliance performs certification testing to make sure wireless networking equipment conforms to interoperable WLAN communication guidelines, which are similar to the IEEE 802.11-2012 standard.

You will look at each of these organizations in the following sections.

International Organization for Standardization (ISO)

The *International Organization for Standardization*, or *ISO*, is a global, nongovernmental organization that identifies business, government, and society needs and develops standards in partnership with the sectors that will put them to use. The ISO is responsible for the creation of the Open Systems Interconnection (OSI) model, which has been a standard reference for data communications between computers since the late 1970s.

Why Is It ISO and Not IOS?

ISO is not a mistyped acronym. It is a word derived from the Greek word *isos*, meaning *equal*. Because acronyms can be different from country to country, based on varying translations, the ISO decided to use a word instead of an acronym for its name. With this in mind, it is easy to see why a standards organization would give itself a name that means *equal*.

The OSI model is the cornerstone of data communications. Becoming familiar with it is one of the most important and fundamental tasks a person in the networking industry can undertake.

The layers of the OSI model are as follows:

OSI Model

Layer	OSI Model	
Layer 7	Application	
Layer 6	Presentation	
Layer 5	Session	
Layer 4	Transport	
Layer 3	Network	
Layer 2	Data-Link	LLC / MAC
Layer 1	Physical	

The IEEE 802.11-2012 standard defines communication mechanisms only at the Physical layer and the MAC sublayer of the Data-Link layer of the OSI model. By design, the 802.11 standard does not address the upper layers of the OSI model, although there are interactions between the 802.11 MAC layer and the upper layers for parameters such as quality of service (QoS).

You should have a working knowledge of the OSI model for both this book and the CWSP exam. Make sure you understand the seven layers of the OSI model and how communication takes place at the different layers. If you are not comfortable with the concepts of the OSI model, spend some time reviewing it on the Internet or from a good networking fundamentals book prior to taking the CWSP exam. More information about the ISO can be found at `www.iso.org`.

Institute of Electrical and Electronics Engineers (IEEE)

The Institute of Electrical and Electronics Engineers, commonly known as the IEEE, is a global professional society with more than 400,000 members. The IEEE's mission is to "foster technological innovation and excellence for the benefit of humanity." To networking professionals, that means creating the standards that we use to communicate.

The IEEE is probably best known for its LAN standards, the IEEE 802 project. IEEE projects are subdivided into working groups to develop standards that address specific problems or needs. For instance, the IEEE 802.3 working group was responsible for the creation of a standard for Ethernet, and the IEEE 802.11 working group was responsible for creating the WLAN standard. The numbers are assigned as the groups are formed, so the 11 assigned to the wireless group indicates that it was the 11th working group formed under the IEEE 802 project. IEEE 802.11, more commonly referred to as Wi-Fi, is a standard technology for providing local area network (LAN) communications using radio frequencies (RF). The IEEE designates the 802.11-2012 standard as the most current guideline to provide operational parameters for WLANs.

As the need arises to revise existing standards created by the working groups, task groups are formed. These task groups are assigned a sequential single letter (multiple letters are assigned if all single letters have been used) that is added to the end of the standard number (for example, 802.11g, 802.11i, and 802.3at). Some letters are not assigned. For example, o and l are not assigned to prevent confusion with the numbers 0 and 1. Other letters may not be assigned to task groups to prevent confusion with other standards. For example, 802.11x has not been assigned because it can be easily confused with the 802.1X standard and because 802.11x has become a common casual reference to the 802.11 family of standards.

More information about the IEEE can be found at `www.ieee.org`.

It is important to remember that the IEEE standards, like many other standards, are written documents describing how technical processes and equipment should function. Unfortunately, this often allows for different interpretations when the standard is being implemented, so it is common for early products to be incompatible between vendors, as was the case with early 802.11 products.

The CWSP exam is based on the most recently published version of the standard, 802.11-2012. The 802.11-2012 standard can be downloaded from `http://standards.ieee.org/getieee802/802.11.html`.

Internet Engineering Task Force (IETF)

The Internet Engineering Task Force, commonly known as the IETF, is an international community of people in the networking industry whose goal is to make the Internet work better. The mission of the IETF, as defined by the organization in a document known as RFC 3935, is "to produce high quality, relevant technical and engineering documents that influence the way people design, use, and manage the Internet in such a way as to make the Internet work better. These documents include protocol standards, best current practices, and informational documents of various kinds." The IETF has no membership fees, and anyone may register for and attend an IETF meeting.

The IETF is one of five main groups that are part of the Internet Society (ISOC). The ISOC groups include the following:

- Internet Engineering Task Force (IETF)
- Internet Architecture Board (IAB)

- Internet Corporation for Assigned Names and Numbers (ICANN)
- Internet Engineering Steering Group (IESG)
- Internet Research Task Force (IRTF)

The IETF is broken into eight subject matter areas: Applications, General, Internet, Operations and Management, Real-Time Applications and Infrastructure, Routing, Security, and Transport. Figure 1.1 shows the hierarchy of the ISOC and a breakdown of the IETF subject matter areas.

FIGURE 1.1 ISOC hierarchy

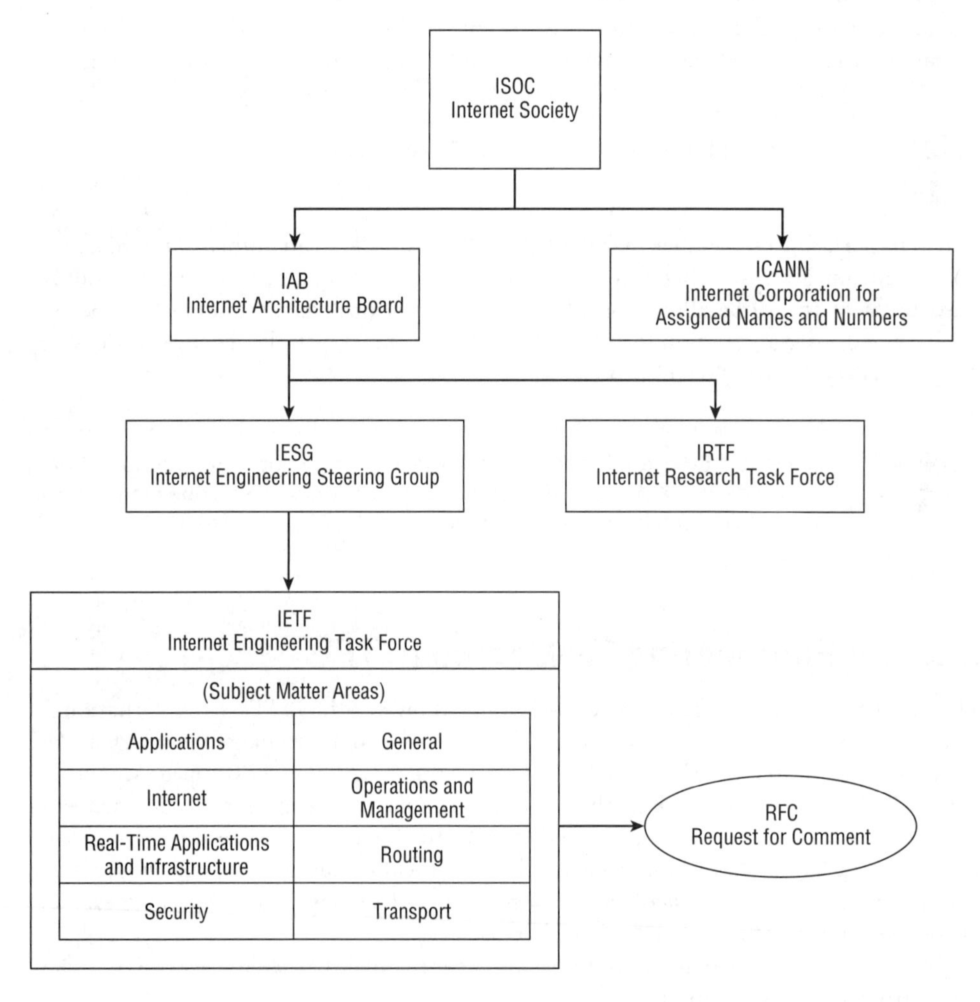

The IESG provides technical management of the activities of the IETF and the Internet standards process. The IETF is made up of a large number of groups, each addressing specific topics. An IETF working group (WG) is created by the IESG and is given a specific charter or topic to address. There is no formal voting process for the working groups. Decisions in working groups are made by rough consensus, or basically a general sense of agreement among the working groups.

The results of a working group are usually the creation of a document known as a Request for Comment (RFC). Contrary to its name, an RFC is not actually a request for comment, but a statement or definition. Most RFCs describe network protocols, services, or policies and may evolve into an Internet standard. RFCs are numbered sequentially, and once a number is assigned it is never reused. RFCs may be updated or supplemented by higher numbered RFCs. As an example, Mobile IPv4 is described in RFC 3344 and updated in RFC 4721. In 2012, RFC 5944 made RFC 3344 obsolete. At the top of the RFC document, it states whether it is updated by another RFC and also if it makes any other RFCs obsolete.

Not all RFCs are standards. Each RFC is given a status, relative to its relationship with the Internet standardization process: Informational, Experimental, Standards Track, or Historic. If it is a Standards Track RFC, it could be a Proposed Standard, Draft Standard, or Internet Standard. When an RFC becomes a standard, it still keeps its RFC number, but it is also given an "STD xxxx" label. The relationship between the STD numbers and the RFC numbers is not one to one. STD numbers identify protocols whereas RFC numbers identify documents.

Many of the protocol standards, best current practices, and informational documents produced by the IETF affect WLAN security. In Chapter 4, "802.1X/EAP Authentication," you will learn about the many varieties of the Extensible Authentication Protocol (EAP) that are defined by the IETF RFC 3748.

More information about the IETF can be found at `www.ietf.org`.

Wi-Fi Alliance

The Wi-Fi Alliance is a global, nonprofit industry association of about 600 member companies devoted to promoting the growth of WLANs. One of the primary tasks of the Wi-Fi Alliance is to market the Wi-Fi brand and raise consumer awareness of new 802.11 technologies as they become available. Because of the Wi-Fi Alliance's overwhelming marketing success, the majority of the worldwide Wi-Fi users are likely to recognize the Wi-Fi logo seen in Figure 1.2.

FIGURE 1.2 Wi-Fi Alliance logo

The Wi-Fi Alliance's main task is to ensure the interoperability of WLAN products by providing certification testing. During the early days of the 802.11 standard, the Wi-Fi Alliance further defined some of the ambiguous standards requirements and provided a set of guidelines to ensure compatibility between different vendors. This is still done to help simplify the complexity of the standards and to ensure compatibility. As seen in Figure 1.3, products that pass the Wi-Fi certification process receive a Wi-Fi Interoperability Certificate that provides detailed information about the individual product's Wi-Fi certifications.

FIGURE 1.3 Wi-Fi Interoperability Certificate

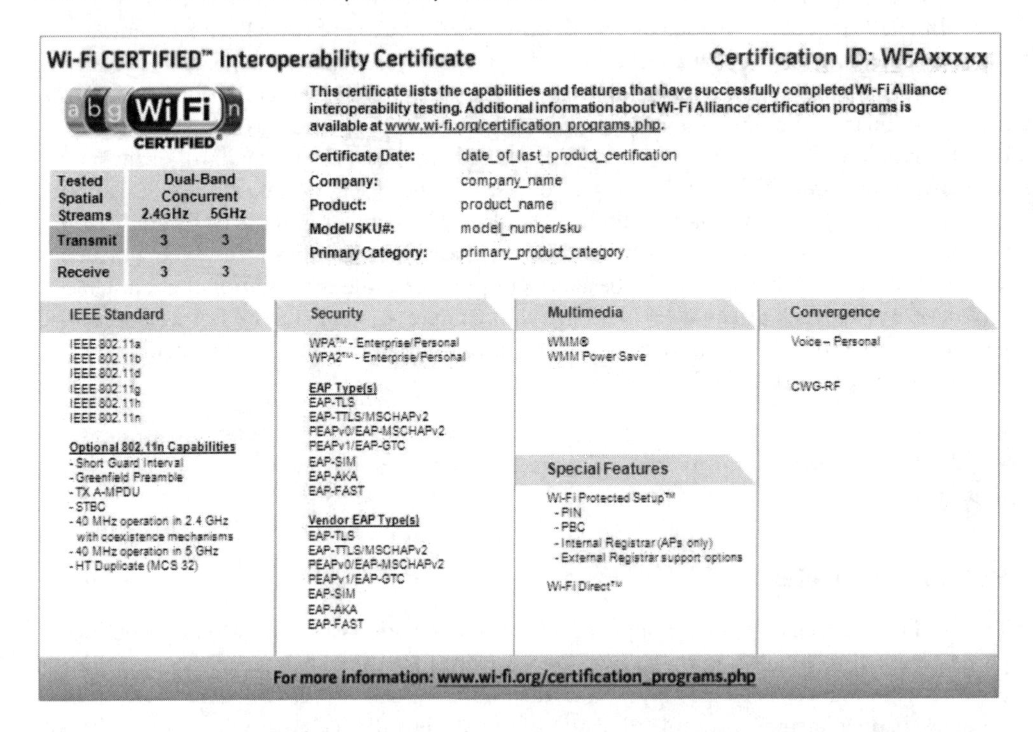

The Wi-Fi Alliance, originally named the Wireless Ethernet Compatibility Alliance (WECA), was founded in August 1999. The name was changed to the Wi-Fi Alliance in October 2002.

The Wi-Fi Alliance has certified more than 25,000 Wi-Fi products for interoperability since testing began in April 2000. Multiple Wi-Fi CERTIFIED programs exist that cover basic connectivity, security, quality of service (QoS), and more. Testing of vendor Wi-Fi products is performed in independent authorized test laboratories worldwide. A listing of these testing laboratories can be found on the Wi-Fi Alliance's website. The guidelines for interoperability for each Wi-Fi CERTIFIED program are usually based on key components and functions that are defined in the IEEE 802.11-2012 standard and various 802.11 amendments. In fact, many of the same engineers who belong to 802.11 task groups are also contributing members of the Wi-Fi Alliance. However, it is important to understand that the IEEE and the Wi-Fi Alliance are two separate organizations. The IEEE 802.11 task group defines the WLAN standards, and the Wi-Fi Alliance defines interoperability certification programs. The Wi-Fi CERTIFIED programs include the following:

Core Technology & Security The core technology and security program certifies 802.11a, b, g, n, and/or ac interoperability to ensure that the essential wireless data transmission works as expected. Each device is tested according to its capabilities. Table 1.1 lists the five different core Wi-Fi transmission technologies along with the frequencies and maximum data rate that each is capable of.

TABLE 1.1 Five generations of Wi-Fi

Wi-Fi technology	Frequency band	Maximum data rate
802.11a	5 GHz	54 Mbps
802.11b	2.4 GHz	11 Mbps
802.11g	2.4 GHz	54 Mbps
802.11n	2.4 GHz, 5 GHz, 2.4 or 5 GHz (selectable), or 2.4 and 5 GHz (concurrent)	450 Mbps
802.11ac	5 GHz	1.3 Gbps

Each certified product is required to support one frequency band as a minimum, but it can support both. The CWSP exam will not use the terms 802.11 a/b/g/n/ac; however, the a/b/g/n/ac terminology is commonplace within the industry because of the Wi-Fi Alliance certifications.

Although 802.11n specifies data rates of up to 600 Mbps and 802.11ac specifies data rates of up to 6.93 Gbps, as of this writing, equipment to support these maximum data rates had not been developed yet. Therefore, the Wi-Fi certification tests do not test up to the maximum 802.11n or 802.11ac specified data rates.

In addition to having the required transmission capabilities, each device must support *robust security network (RSN)* capabilities, security mechanisms that were originally defined in the IEEE 802.11i amendment. Devices must support Wi-Fi Protected Access (WPA) and Wi-Fi Protected Access 2 (WPA2) security mechanisms for personal (WPA2-Personal) or enterprise (WPA2-Enterprise) environments. Additionally, enterprise devices must support *Extensible Authentication Protocol (EAP)*, which is used to validate the identity of the wireless device or user. In 2012, support for Protected Management Frames extended WPA2 protection to unicast and multicast management action frames.

Wi-Fi Multimedia Wi-Fi Multimedia (WMM) is based on the QoS mechanisms that were originally defined in the IEEE 802.11e amendment. WMM enables Wi-Fi networks to prioritize traffic generated by different applications. In a network where WMM is supported by both the access point and the client device, traffic generated by time-sensitive applications such as voice or video can be prioritized for transmission on the half-duplex RF medium. WMM certification is mandatory for all core certified devices that support 802.11n. WMM certification is optional for core certified devices that support 802.11 a, b, or g.

WMM Power Save WMM Power Save (WMM-PS) helps conserve battery power for devices using Wi-Fi radios by managing the time the client device spends in sleep mode. Conserving battery life is critical for handheld devices such as barcode scanners and voice over Wi-Fi (VoWiFi) phones. To take advantage of power-saving capabilities, both the device and the access point must support WMM Power Save.

Wi-Fi Protected Setup Wi-Fi Protected Setup defines simplified and automatic WPA and WPA2 security configurations for home and small-business users. Users can easily configure a network with security protection by using a personal identification number (PIN) or a button located on the access point and the client device. This technology is defined in the Wi-Fi Simple Configuration Technical Specification.

Wi-Fi Direct Wi-Fi Direct enables Wi-Fi devices to connect directly without the use of an access point, making it easier to print, share, sync, and display. Wi-Fi Direct is ideal for mobile phones, cameras, printers, PCs, and gaming devices needing to establish a one-to-one connection, or even for connecting a small group of devices. Wi-Fi Direct is simple to configure (in some cases as easy as pressing a button), provides the same performance and range as other Wi-Fi CERTIFIED devices, and is secured using WPA2 security. This technology is defined in the Wi-Fi Peer-to-Peer Services Technical Specification.

Converged Wireless Group-RF Profile Converged Wireless Group-RF Profile (CWG-RF) was developed jointly by the Wi-Fi Alliance and the Cellular Telecommunications and Internet Association (CTIA), now known as The Wireless Association. CWG-RF defines performance metrics for Wi-Fi and cellular radios in a converged handset to help ensure that both technologies perform well in the presence of the other. All CTIA-certified handsets now include this certification.

Voice Personal Voice Personal offers enhanced support for voice applications in residential and small-business Wi-Fi networks. These networks include one access point, mixed voice and data traffic from multiple devices (such as phones, PCs, printers, and other consumer electronic devices), and support for up to four concurrent phone calls. Both the access point and the client device must be certified to achieve performance matching the certification metrics.

Voice Enterprise Voice Enterprise offers enhanced support for voice applications in enterprise Wi-Fi networks. Enterprise-grade voice equipment must provide consistently good voice quality under all network load conditions and coexist with data traffic. Both access point and client devices must support prioritization using WMM, with voice traffic being placed in the highest-priority queue (Access Category Voice, AC_VO). Voice Enterprise equipment must also support seamless roaming between access points (APs), WPA2-Enterprise security, optimization of power through the WMM-Power Save mechanism, and traffic management through WMM-Admission Control.

Tunneled Direct Link Setup Tunneled Direct Link Setup (TDLS) enables devices to establish secure links directly with other devices after they have joined a traditional Wi-Fi network. This will allow consumer devices such as TVs, gaming devices, smartphones, cameras, and printers to communicate quickly, easily, and securely between each other.

Passpoint Passpoint is designed to revolutionize the end-user experience when connecting to Wi-Fi hotspots. This is done by allowing security identity module (SIM) and non-SIM mobiles devices to automatically identify a Wi-Fi network and connect to it, automatically authenticating the user to the network using Extensible Authentication Protocol (EAP), and providing secure transmission using WPA2-Enterprise encryption. Passpoint is also known as Hotspot 2.0. Passpoint has also been specified by the Wireless Broadband Alliance and the GSMA Terminal Steering Group.

WMM-Admission Control WMM-Admission Control allows Wi-Fi networks to manage network traffic based on channel conditions, network traffic load, and type of traffic (voice, video, best effort data, or background data). The access point allows only the traffic that it can support to connect to the network, based on the available network resources. This allows users to confidently know that, when the connection is established, the resources will be there to maintain it.

IBSS with Wi-Fi Protected Setup IBSS with Wi-Fi Protected Setup provides easy configuration and strong security for ad hoc (peer-to-peer) Wi-Fi networks. This is designed for mobile products and devices that have a limited user interface, such as smartphones, cameras, and media players. Features include easy push button or PIN setup, task-oriented short-term connections, and dynamic networks that can be established anywhere.

Miracast Miracast seamlessly integrates the display of streaming video content between devices. Wireless links are used to replace wired connections. Devices are designed to identify and connect with each other, manage their connections, and optimize the transmission of video content. It provides wired levels of capabilities but the portability of Wi-Fi. Miracast provides 802.11n performance, ad hoc connections via Wi-Fi Direct, and WPA2 security. This technology is defined in the Wi-Fi Display Technical Specification.

Wi-Fi Aware Wi-Fi Aware provides a real-time and energy-efficient discovery mechanism for Wi-Fi devices to discover other devices and services within its proximity. It is designed as an enabling technology for personalized social, local, and mobile applications and services, and is optimized to work well even in crowded environments.

As 802.11 technologies evolve, new Wi-Fi CERTIFIED programs will be detailed by the Wi-Fi Alliance.

Wi-Fi Alliance and Wi-Fi CERTIFIED

Learn more about the Wi-Fi Alliance at `www.wi-fi.org`. The Wi-Fi Alliance website contains many articles, FAQs, and white papers describing the organization along with additional information about the certification programs. The Wi-Fi Alliance technical white papers are recommended extra reading when preparing for the CWSP exam. The Wi-Fi Alliance white papers can be accessed at `www.wi-fi.org`.

802.11 Networking Basics

In addition to understanding the OSI model and basic networking concepts, you must broaden your understanding of many other networking technologies in order to design, deploy, and administer an 802.11 wireless network properly. For instance, when administering an Ethernet network, you typically need a comprehension of TCP/IP, bridging, switching, and routing. The skills to manage an Ethernet network will also aid you as a WLAN administrator, because most 802.11 wireless networks act as "portals" into wired networks. The IEEE defines the 802.11 communications at the Physical layer and the MAC sublayer of the Data-Link layer.

To understand the 802.11 technology completely, you need to have a clear concept of how wireless technology works at the Physical layer of the OSI model, and at the heart of the Physical layer is *radio frequency (RF)* communications. A clear concept of how wireless works at the second layer of the OSI model is also needed. The 802.11 *Data-Link layer* is divided into two sublayers. The upper portion is the IEEE 802.2 *Logical Link Control (LLC)* sublayer, which is identical for all 802-based networks, although not used by all of them. The bottom portion of the Data-Link layer is the *Media Access Control (MAC) sublayer*, which is identical for all 802.11-based networks. The 802.11-2012 standard defines operations at the MAC sublayer.

Because the main focus of this study guide is WLAN security, it is beyond the scope of this book to discuss general 802.11 networking topics in great detail. For a broad overview of 802.11 technology, we suggest *CWNA: Certified Wireless Network Administrator Official Study Guide: (Exam CWNA-106)*, by David D. Coleman and David A. Westcott (Sybex, 2014).

If you have ever taken a networking class or read a book about network design, you have probably heard the terms *core*, *distribution*, and *access* when referring to networking architecture. Proper network design is imperative no matter what type of network topology is used. The core of the network is the high-speed backbone or the superhighway of the network. The goal of the core is to carry large amounts of information between key data centers or distribution areas, just as superhighways connect cities and metropolitan areas.

The core layer does not route traffic or manipulate packets but rather performs high-speed switching. Redundant solutions are usually designed at the core layer to ensure fast and reliable delivery of packets. The distribution layer of the network routes or directs traffic toward the smaller clusters of nodes or neighborhoods of the network.

The distribution layer routes traffic between virtual LANs (VLANs) and subnets. The distribution layer is akin to the state and county roads that provide medium travel speeds and distribute the traffic within a city or metropolitan area.

The access layer of the network is responsible for a delivery of the traffic directly to the end user or end node. The access layer mimics the local roads and neighborhood streets that are used to reach your final address; thus the speed of delivery in this layer is slower than at the core and distribution layers. (Remember that speed is a relative concept.) The access layer ensures the final delivery of packets to the end user.

Because of traffic load and throughput demands, speed and throughput capabilities increase as data moves from the access layer to the core layer. Additional speed and throughput tend to also mean higher cost.

Just as it would not be practical to build a superhighway so that traffic could travel between your neighborhood and the local school, it would not be practical or efficient to build a two-lane road as the main thoroughfare to connect two large cities such as New York and Boston. These same principles apply to network design. Each of the network layers—core, distribution, and access—are designed to provide a specific function and capability to the network. It is important to understand how wireless networking fits into this network design model.

Wireless networking can be implemented as either point-to-point or point-to-multipoint solutions. Most wireless networks are used to provide network access to the individual client stations and are designed as point-to-multipoint networks. This type of implementation is designed and installed on the access layer, providing connectivity to the end user. 802.11 wireless networking is most often implemented at the access layer. In Chapter 8, "WLAN Security Infrastructure," you will learn about the evolution of WLAN architecture from the early days of autonomous APs to the many hybrid architectures that exist today. Security design and deployment considerations are unique between autonomous, centralized, and distributed WLAN architecture. Various components of these architectures exist between the core/distribution and access layers.

Wireless bridge links are generally used to provide connectivity between buildings in the same way that county or state roads provide distribution of traffic between neighborhoods. The purpose of wireless bridging is to connect two separate, wired networks wirelessly. Routing data traffic between networks is usually associated with the distribution layer. Wireless bridge links cannot typically meet the speed or distance requirements of the core layer, but they can be very effective at the distribution layer. An 802.11 bridge link is an example of wireless technology being implemented at the distribution layer.

Throughout this study guide, you will learn how to provide proper 802.11 wireless security integration at the access, distribution, and core layers of network design.

802.11 Security Basics

When you are securing a wireless 802.11 network, five major components are typically required:

- Data privacy
- Authentication, authorization, and accounting (AAA)
- Segmentation
- Monitoring
- Policy

Because data is transmitted freely and openly in the air, proper protection is needed to ensure *data privacy*, so strong encryption is needed. The function of most wireless networks is to provide a portal into some other network infrastructure, such as an 802.3 Ethernet backbone. The wireless portal must be protected, and therefore an authentication solution is needed to ensure that only authorized users can pass through the portal via a wireless AP. After users have been authorized to pass through the wireless portal, VLANs and identity-based mechanisms are needed to further restrict access to network resources. Interim accounting along with RADIUS change of authorization (CoA) can be used to monitor the user's connection and authorization, and dynamically change the user's access to network resources. 802.11 wireless networks can be further protected with continuous monitoring by a wireless intrusion detection system. All of these security components should also be cemented with policy enforcement. If properly implemented, these five components of 802.11 security discussed throughout this book will lay a solid foundation for protecting your WLAN.

Data Privacy

The history of secure communications is as old as the history of communications itself. Sharing ideas and thoughts with specific people but not others is a natural human desire. The task of sharing a thought is easy when the person you want to share it with is nearby. A quiet, private conversation is often all that is needed to do this. Sharing an idea in a secure way starts to become a problem when you need to send your message over a longer distance. Whether you have to yell the message to the other person from a distance, send it to the other person via a courier, or write it down on a piece of paper and send it, all of these methods run a risk of being intercepted. Because of these types of risks, methods of encrypting or encoding messages were created. The goal of encrypting a message, even if it is overheard or intercepted, is for it to be legible only to the person who created the message and the person for whom the message is intended—or at least, that is the intent.

802.11 wireless networks operate in license-free frequency bands, and all data transmissions travel in the open air. Protecting data privacy in a wired network is much easier because physical access to the wired medium is more restricted. However, physical access to wireless transmissions is available to anyone in listening range. Therefore, using cipher encryption technologies to obscure information is mandatory to provide proper data privacy in wireless networks. A *cipher* is an algorithm used to perform encryption.

Along with the desire by some to keep things a secret, there is often an equal desire by others to reveal these secrets. The techniques needed to encrypt and decrypt information form the science known as *cryptology*. As you would expect, the science of cryptology uses specific words and phrases, many of which you will learn about in this section.

The term *cryptology* is derived from the Greek language and translates to mean "hidden word." The goal of cryptology is to take a piece of information, often referred to as *plaintext*, and, using a process or algorithm, also referred to as a *key* or *cipher*, transform the plaintext into encrypted text, also known as *ciphertext*. The process of encrypting plaintext is shown in Figure 1.4. This ciphertext could then only be decrypted, converted back into plaintext, by someone who knows the key or cipher. The process of decrypting the ciphertext is shown in Figure 1.5.

FIGURE 1.4 The encryption process

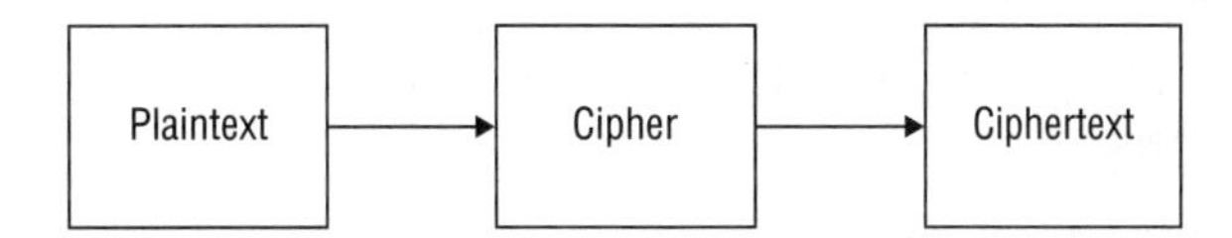

FIGURE 1.5 The decryption process

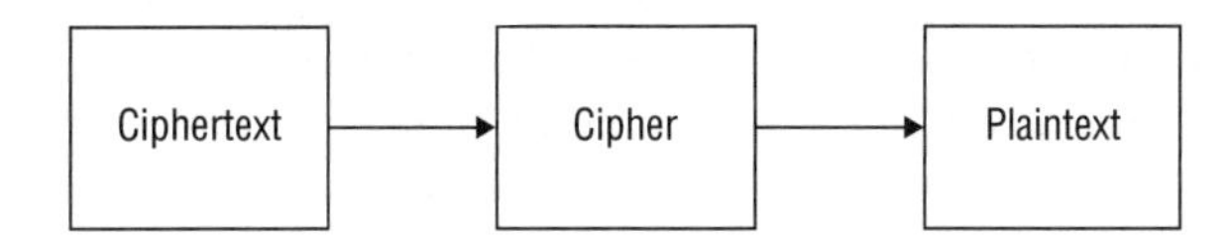

When a plaintext message is encrypted, the encrypted message is referred to as ciphertext. A detailed discussion about encryption methods used for WLAN security can be found in Chapter 3, "Encryption Ciphers and Methods."

When discussing cryptology, it is important to use the term *cipher* instead of *code*, as a code is simply a way of representing information in a different way. For example, the American Standard Code for Information Interchange (ASCII) represents information as letters, numbers, and other characters. Morse code uses dots and dashes to represent letters. Like a code, a cipher also represents information in a different way. However, a cipher uses a secret technique of representing the information in a different way, with this technique known only to a select few.

The science of concealing the plaintext and then revealing it is known as *cryptography*. In the computer and networking industries, the process of converting plaintext into ciphertext is commonly referred to as *encryption*, and the process of converting ciphertext back to plaintext is commonly referred to as *decryption*. The science of decrypting the ciphertext without knowledge of the key or cipher is known as *cryptanalysis*. If cryptanalysis is successful in decrypting the ciphertext, then that key or cipher is considered to be broken, cracked, or not secure.

The term *steganography* is also derived from the Greek language and is translated as "concealed writing." Unlike cryptography, where the goal is to make the message unreadable to someone without access to the key or cipher, steganography strives to hide the fact that there is a message. This is often referred to as "security through obscurity" or "hiding a message in plain sight." A classic example of using steganography to hide a message is to write a document with the first letter of each sentence or word as the hidden message (see Figure 1.6).

FIGURE 1.6 Steganography example

Sent Message: Here is my obscure message. Saturday evening nobody dared make outbursts, not even Yolanda.

Here **i**s **m**y **o**bscure **m**essage. **S**aturday **e**vening **n**obody **d**ared **m**ake **o**utbursts, **n**ot **e**ven **Y**olanda.

Hidden Message: Hi mom send money

Steganography is often used for digital watermarking, which embeds an artist or photographer's information in an image so that ownership can be proven in case someone tries to use the image without permission. Steganography is useful for hiding a message where one would not be expected, which is why it is not used in environments like wireless networking where data communications is the key objective.

Authentication, Authorization, Accounting (AAA)

Authentication, authorization, and accounting (AAA) is a common computer security concept that defines the protection of network resources.

Authentication is the verification of user identity and credentials. Users must identify themselves and present credentials, such as usernames and passwords or digital certificates. More secure authentication systems use multifactor authentication, which requires at least two sets of different credentials to be presented.

Authorization determines if the device or user is authorized to have access to network resources. This can include identifying whether you can have access based on the type of device you are using (laptop, tablet, or phone), time-of-day restrictions, or location. Before authorization can be determined, proper authentication must occur.

Accounting is tracking the use of network resources by users and devices. It is an important aspect of network security, used to keep a historical trail of who used what resource, when, where, and how. A record is kept of user identity, which resource was accessed, and at what time. Keeping an accounting trail is often a requirement of many industry regulations, such as the payment card industry (PCI).

Remember that the usual purpose of an 802.11 wireless network is to act as a portal into an 802.3 wired network. It is therefore necessary to protect that portal with very strong authentication methods so that only legitimate users with the proper credentials will be authorized to access network resources.

Segmentation

Although it is of the utmost importance to secure an enterprise wireless network by utilizing both strong encryption and an AAA solution, an equally important aspect of wireless security is segmentation. Segmentation is the chosen method of separating user traffic within a network. Prior to the introduction of stronger authentication and encryption techniques, wireless was viewed as an untrusted network segment. Therefore, before the ratification of the 802.11i security amendment, the entire wireless segment of a network was commonly treated as the untrusted segment and the wired 802.3 network was considered the trusted segment.

Now that better security solutions exist, properly secured WLANs are more seamlessly and securely integrated into the wired infrastructure. It is still important to separate users and devices into proper groups, much like what is done on any traditional network. Once authorized onto network resources, users and devices can be further restricted as to what resources may be accessed and where they can go. Segmentation can be achieved through a variety of means, including firewalls, routers, VPNs, VLANs, and encapsulation or tunneling techniques; such as generic routing encapsulation (GRE). The most common wireless segmentation strategy used in 802.11 enterprise WLANs is segmentation using virtual LANs (VLANs). Encapsulation is heavily used to transport data between infrastructure devices in enterprise WLAN environments. Segmentation is also intertwined with role-based access control (RBAC), which is discussed in Chapter 9, "RADIUS and LDAP."

Monitoring

After you have designed and installed your wireless network, it is important to monitor it. In addition to monitoring it to make sure that it is performing up to your expectations and those of your users, it may be necessary to constantly monitor it for attacks and intrusions. Similar to a business placing a video camera on the outside of its building to monitor the traffic going in and out of a locked door, it is important for the wireless network administrator to monitor the wireless traffic of a secured network. To monitor potentially malicious wireless activity on your network, you should install a wireless intrusion detection system (WIDS). WLAN security monitoring can be an integrated solution or an overlay solution. Network management solutions that include WLAN security monitoring can be cloud based or run on a private data center server.

In addition to a WIDS, you could implement a wireless intrusion prevention system (WIPS). Both WIDSs and WIPSs have the ability to classify valid and invalid devices on the network. WIPSs can also mitigate attacks from rogue access points and rogue clients by performing attacks against the rogue devices, effectively disabling their ability to communicate with your network. WIPSs will be discussed in detail in Chapter 14, "Wireless Security Monitoring."

Policy

Securing a wireless network and monitoring for threats are absolute necessities, but both are worthless unless proper security policies are in place. What good is an 802.1X/EAP solution if the end users share their passwords? Why purchase an intrusion detection system if a policy has not been established for dealing with rogue access points? WLAN security policies must be clearly defined and enforced to solidify the effectiveness of all WLAN security components.

In most countries, mandated regulations exist for protecting and securing data communications within all government agencies. In the United States, the National Institute of Standards and Technology (NIST) maintains the *Federal Information Processing Standards (FIPS)*. Of special interest to wireless security is the FIPS 140-2 standard, which defines security requirements for cryptography modules. Additionally, other legislation and regulations exist for protecting information and communications in certain industries such as healthcare and banking. WLAN policy enforcement is needed to meet compliance mandates set forth by these regulations. Policies and compliance regulations will be discussed in greater detail in Chapter 15, "Wireless Security Policies."

802.11 Security History

From 1997 to 2004, not much was defined in terms of security in the original IEEE 802.11 standard. Two key components of any wireless security solution are data privacy (encryption) and authentication (identity verification). For seven years, the only defined method of encryption in an 802.11 network was the use of 64-bit static encryption called *Wired Equivalent Privacy (WEP)*.

WEP encryption has long been cracked and is not considered an acceptable means of providing data privacy. The original 802.11 standard defined two methods of authentication. The default method is *Open System authentication*, which verifies the identity of everyone regardless. Another defined method is called *Shared Key authentication*, which opens up a whole new can of worms and potential security risks. Outdated 802.11 security mechanisms will be discussed in detail in Chapter 2, "Legacy 802.11 Security."

802.11i Security Amendment and WPA Certifications

The 802.11i amendment, which was ratified and published as IEEE Std. 802.11i-2004, defined stronger encryption and better authentication methods. The 802.11i amendment defined a *robust security network (RSN)*. The intended goal of an RSN was to hide the data flying through the air better while at the same time placing a bigger guard at the front door. The 802.11i security amendment was without a doubt one of the most important enhancements to the original 802.11 standard because of the seriousness of properly protecting a wireless network. The major security enhancements addressed in the 802.11i amendment are as follows:

Enhanced Data Privacy Confidentiality needs have been addressed in 802.11i with the use of a stronger encryption method called *Counter Mode with Cipher Block Chaining Message Authentication Code Protocol (CCMP),* which uses the *Advanced Encryption Standard (AES)* algorithm. The encryption method is often abbreviated as CCMP/AES, AES CCMP, or often just CCMP. The 802.11i supplement also defines an optional encryption method known as *Temporal Key Integrity Protocol (TKIP),* which uses the ARC4 stream cipher algorithm and is basically an enhancement of WEP encryption. It should be noted that TKIP is essentially being phased out. The IEEE and Wi-Fi Alliance mandate the use of only CCMP encryption for 802.11n and 802.11ac data rates.

Enhanced Authentication 802.11i defines two methods of authentication using either an IEEE 802.1X authorization framework or *preshared keys (PSKs).* An 802.1X solution requires the use of an *Extensible Authentication Protocol (EAP),* although the 802.11i amendment does not specify what EAP method to use.

In 2004, the 802.11i security amendment was ratified and is now part of the 802.11-2012 standard. All aspects of the 802.11i ratified security amendment can now be found in clause 11 of the 802.11-2012 standard.

The current 802.11-2012 standard defines an enterprise authentication method as well as a method of authentication for home use. The current standard requires the use of an 802.1X/EAP authentication method in the enterprise and the use of a preshared key, technically a passphrase in a small office/home office (SOHO) environment. The 802.11-2012 standard also requires the use of strong, dynamic encryption-key generation methods. CCMP/AES encryption is the default encryption method, whereas TKIP/ARC4 is an optional encryption method.

Prior to the IEEE ratification of the 802.11i amendment, the Wi-Fi Alliance introduced the *Wi-Fi Protected Access (WPA)* certification as a snapshot of the not-yet-released 802.11i amendment, supporting only TKIP/ARC4 dynamic encryption-key generation. 802.1X/EAP authentication was required in the enterprise, and passphrase authentication was required in a SOHO environment.

After 802.11i was ratified by the IEEE, the Wi-Fi Alliance introduced the *Wi-Fi Protected Access 2 (WPA2)* certification. WPA2 is a more complete implementation of the 802.11i amendment and supports both CCMP/AES and TKIP/ARC4 dynamic encryption-key generation. 802.1X/EAP authentication is required in the enterprise, and passphrase authentication is required in a SOHO environment. WPA version 1 was considered a preview of 802.11i, whereas WPA version 2 is considered more of a mirror of the 802.11i security amendment. Once again, you should understand that all aspects of the 802.11i ratified security amendment are now defined as part of the 802.11-2012 standard. Table 1.2 compares the various security standards and certifications.

TABLE 1.2 Security standards and certifications

802.11 standard	Wi-Fi Alliance certification	Authentication method	Encryption method	Cipher	Key generation
802.11 legacy	No Certification	Open System or Shared Key	WEP	ARC4	Static
	WPA-Personal	WPA Passphrase (also known as WPA PSK and WPA Pre-Shared Key)	TKIP	ARC4	Dynamic
	WPA-Enterprise	802.1X/EAP	TKIP	ARC4	Dynamic
802.11-2012	WPA2-Personal	WPA2 Passphrase (also known as WPA2 PSK and WPA2 Pre-Shared Key)	CCMP (mandatory)	AES (mandatory)	Dynamic
			TKIP (optional)	ARC4 (optional)	Dynamic
802.11-2012	WPA2-Enterprise	802.1X/EAP	CCMP (mandatory)	AES (mandatory)	Dynamic
			TKIP (optional)	ARC4 (optional)	Dynamic

Robust Security Network (RSN)

The 802.11-2012 standard defines what is known as a *robust security network (RSN)* and *robust security network associations (RSNAs)*. Two stations (STAs) must establish a procedure to authenticate and associate with each other as well as create dynamic encryption keys through a process known as the 4-Way Handshake. This association between two stations is referred to as an RSNA. In other words, any two radios must share dynamic encryption keys that are unique between those two radios. CCMP/AES encryption is the mandated encryption method, whereas TKIP/ARC4 is an optional encryption method. An RSN is a network that allows for the creation of only RSNAs. The 802.11-2012 standard does allow for the creation of pre-robust security network associations (pre-RSNAs) as well as RSNAs. In other words, legacy security measures can be supported in the same *basic service set (BSS)* along with RSN-security-defined mechanisms. A *transition security network (TSN)* supports RSN-defined security as well as legacy security such as WEP within the same BSS.

Robust network security, the 4-Way Handshake, and dynamic encryption are discussed in more detail in Chapter 5, "802.11 Layer 2 Dynamic Encryption Key Generation."

Where Else Can I Learn More about 802.11 Security and the Wi-Fi Industry?

Reading this book from cover to cover is a great way to start understanding WLAN security. Because of the rapidly changing nature of 802.11 WLAN technologies, the authors of this book would like to recommend these additional resources:

Wi-Fi Alliance As mentioned earlier in this chapter, the Wi-Fi Alliance is the marketing voice of the Wi-Fi industry and maintains all the industry's certifications. The knowledge center section of the Wi-Fi Alliance website, `www.wi-fi.org`, is an excellent resource.

CWNP The Certified Wireless Networking Professional program maintains learning resources such as user forums and a WLAN white paper database. The website `www.cwnp.com` is also the best source of information about all the vendor-neutral CWNP wireless networking certifications.

WLAN Vendor Websites Although the CWSP exam and this book take a vendor-neutral approach to 802.11 security, the various WLAN vendor websites are often an excellent resource for information about specific Wi-Fi security solutions. Many of the major WLAN vendors are mentioned throughout this book, and a complete listing of most of the major WLAN vendor websites can be found in this book's appendix.

Wi-Fi Blogs In recent years, numerous personal blogs about the subject of Wi-Fi have sprung up all over the Internet. One great example is the Revolution Wi-Fi blog written by CWNE #84, Andrew von Nagy.

`www.revolutionwifi.net`

The technical editor of this book, Chris Lyttle, who is CWNE #156, is the author of another great blog called WiFi Kiwi.

`www.wifikiwi.com`

Summary

This chapter explained the roles and responsibilities of four key organizations involved with wireless security and networking:

- ISO
- IEEE

- IETF
- Wi-Fi Alliance

To provide a basic understanding of the relationship between networking fundamentals and 802.11 technologies, we discussed these concepts:

- OSI model
- Core, distribution, and access

To provide a basic knowledge of data privacy, we introduced some of the basic components of security:

- Cryptology
- Cryptography
- Cryptanalysis
- Steganography
- Plaintext
- Key
- Cipher
- Ciphertext

Five major components that are typically required to secure an 802.11 network were also discussed:

- Data privacy
- Authentication, authorization, and accounting (AAA)
- Segmentation
- Monitoring
- Policy

To provide an initial foundation, we reviewed the 802.11 security history, including:

- 802.11 legacy, WEP
- WPA-Personal
- WPA-Enterprise
- 802.11-2012 (RSN) - WPA2-Personal
- 802.11-2012 (RSN) - WPA2-Enterprise

Exam Essentials

Know the four industry organizations. Understand the roles and responsibilities of the ISO, the IEEE, the IETF, and the Wi-Fi Alliance.

Understand data privacy, AAA, segmentation, monitoring, and policy. Know the five major components typically required to secure a wireless network.

Understand cryptology, cryptography, cryptanalysis, steganography, plaintext, cipher, and ciphertext. Know the definition of each of these security terms, how they relate to each other, and the differences between them.

Know the history of 802.11 security. Know the history of wireless security and the differences between the different IEEE standards and Wi-Fi Alliance certifications.

Review Questions

1. The IEEE 802.11-2012 standard mandates this encryption for robust security network associations and the optional use of which other encryption?

   A. WEP, AES

   B. IPsec, AES

   C. MPPE, TKIP

   D. TKIP, WEP

   E. CCMP, TKIP

2. What wireless security solutions are defined by Wi-Fi Protected Access? (Choose all that apply.)

   A. Passphrase authentication

   B. LEAP

   C. TKIP/ARC4

   D. Dynamic WEP

   E. CCMP/AES

3. Which wireless security standards and certifications call for the use of CCMP/AES encryption? (Choose all that apply.)

   A. WPA

   B. 802.11-2012

   C. 802.1X

   D. WPA2

   E. 802.11 legacy

4. A robust security network (RSN) requires the use of which security mechanisms? (Choose all that apply.)

   A. 802.11x

   B. WEP

   C. IPsec

   D. CCMP/AES

   E. CKIP

   F. 802.1X

5. The Wi-Fi Alliance is responsible for which of the following certification programs? (Choose all that apply.)

   A. WPA2

   B. WEP

   C. 802.11-2012

D. WMM

E. PSK

6. Which sublayer of the OSI model's Data-Link layer is used for communication between 802.11 radios?

A. LLC

B. WPA

C. MAC

D. FSK

7. What encryption methods are defined by the IEEE 802.11-2012 standard? (Choose all that apply.)

A. 3DES

B. WPA-2

C. SSL

D. TKIP

E. CCMP

F. WEP

8. Which organization is responsible for the creation of documents known as Requests for Comments?

A. IEEE

B. ISO

C. IETF

D. Wi-Fi Alliance

E. RFC Consortium

9. Which of the following is not a standard or amendment created by the IEEE? (Choose all that apply.)

A. 802.11X

B. 802.1x

C. 802.3af

D. 802.11N

E. 802.11g

10. TKIP can be used with which of the following? (Choose all that apply.)

A. WEP

B. WPA-Personal

C. WPA-Enterprise

D. WPA-2 Personal

E. WPA-2 Enterprise

F. 802.11-2012 (RSN)

11. Which of the following is simply a means of representing information in a different way?
    - A. Cryptography
    - B. Steganography
    - C. Encryption
    - D. Cipher
    - E. Code

12. What wireless security components are mandatory under WPA version 2? (Choose all that apply.)
    - A. 802.1X/EAP
    - B. PEAP
    - C. TKIP/ARC4
    - D. Dynamic WEP
    - E. CCMP/AES

13. An AP advertising an 802.1X/EAP employee SSID along with a guest SSID is considered to be operating what type of network device?
    - A. Core
    - B. Distribution
    - C. Access
    - D. Network layer
    - E. Session layer

14. The science of concealing plaintext and then revealing it is known as ________, and the science of decrypting the ciphertext without knowledge of the key or cipher is known as ________.
    - A. encryption, decryption
    - B. cryptanalysis, cryptology
    - C. cryptology, cryptanalysis
    - D. cryptography, cryptanalysis
    - E. cryptography, steganography

15. What is the chronological order in which the following security standards and certifications were defined?
    1. 802.11-2012
    2. 802.11i
    3. WEP
    4. WPA-2
    5. WPA
       - A. 3, 5, 2, 4, 1
       - B. 3, 2, 5, 4, 1

C. 3, 5, 2, 1, 4
D. 1, 3, 2, 5, 4
E. 1, 3, 5, 4, 2

16. The 802.11 legacy standard defines which wireless security solution?

A. Dynamic WEP
B. 802.1X/EAP
C. 64-bit static WEP
D. Temporal Key Integrity Protocol
E. CCMP/AES

17. These qualifications for interoperability are usually based on key components and functions that are defined in the IEEE 802.11-2012 standard and various 802.11 amendments.

A. Request for Comments
B. Wi-Fi Alliance
C. Federal Information Processing Standards
D. Internet Engineering Task Force
E. Wi-Fi CERTIFIED

18. Which of the following can be used with a wireless network to segment or restrict access to parts of the network? (Choose all that apply.)

A. VLANs
B. WPA-2
C. Firewall
D. 802.11i
E. RBAC

19. 802.1X/EAP is mandatory in which of the following? (Choose all that apply.)

A. WPA SOHO
B. WPA Enterprise
C. WPA-2 SOHO
D. WPA-2 Enterprise
E. WPA2-PSK

20. Monitoring potentially malicious wireless activity on the network is handled by ________, whereas intrusion remediation and mitigation is handled by ______________.

A. WIDS, WIDS
B. WIPS, WIPS
C. WIDS, FIPS
D. WIPS, FIPS
E. WIDS, WIPS

Legacy 802.11 Security

IN THIS CHAPTER, YOU WILL LEARN ABOUT THE FOLLOWING:

✓ **Authentication**

- Open System authentication
- Shared Key authentication

✓ **Wired Equivalent Privacy (WEP) encryption**

✓ **Temporal Key Integrity Protocol (TKIP) Encryption**

✓ **Virtual Private Networks (VPNs)**

- Point-to-Point Tunneling Protocol (PPTP)
- Internet Protocol Security (IPsec)
- Secure Sockets Layer (SSL)
- VPN Configuration Complexity
- VPN Scalability

✓ **MAC Filters**

✓ **SSID Segmentation**

✓ **SSID Cloaking**

Many changes to the security mechanisms of the 802.11 standard have taken place since its ratification in 1997. Three pre-RSNA or legacy security mechanisms exist: Open System authentication, Shared Key authentication, and WEP encryption. These pre-RSNA security mechanisms are currently defined in clause 11.2 of the 802.11-2012 standard. It should be noted that the current 802.11-2012 standard also defines the more current *robust security network (RSN)* operations that are meant to replace legacy 802.11 security. Even though these legacy security mechanisms have been superseded and should be avoided, they are still integrated into many 802.11 devices to provide backward compatibility with existing equipment. It is important to understand these security methods and to understand why Open System authentication is still valid and why Shared Key authentication and WEP encryption should be avoided.

If this chapter was strictly about legacy 802.11 security as defined by the standard, then it would be a very short chapter indeed, since Open System authentication, Shared Key authentication, and WEP encryption are the only legacy security methods originally defined by the IEEE. So why is it there is more to this chapter than just that? Well, two types of standards exist in the world of technology; *de jure* standards and *de facto* standards. Essentially de jure (Latin for "concerning law") standards are typically defined and ratified by a standards body, such as the IEEE, whereas de facto (Latin for "concerning fact") standards are established by practice or usage.

Over the years, different nonstandard security solutions have been implemented to enhance the wireless network security or to make up for shortcomings in the standard. Some of these, such as VPN over wireless, provided solutions to overcome flaws that arose in the standard. Others, such as MAC filtering, SSID segmentation, and SSID cloaking, provided enhancements or additional capabilities that were not in the standard. Although all of these may still have their place in some environments, for the most part, they should be avoided as the newer security mechanisms are capable of providing a faster and more secure wireless network.

Authentication

Authentication is the first of two steps required to connect to the 802.11 basic service set. Both authentication and association must occur, in that order, before an 802.11 client can pass traffic through the access point to another device on the network. Authentication is a process that is often misunderstood. When many people hear authentication, they think of what is commonly referred to as network authentication—entering a username and password in order to get access to the network. In this chapter, we are referring to 802.11 authentication that occurs at Layer 2 of the OSI model. When an 802.3 device needs to communicate with other devices, the first step is to plug the Ethernet cable into a wall jack. When this cable is plugged in, the client creates a physical link to the wired switch and is then able to start transmitting frames. When an 802.11 device needs to communicate, it must first authenticate with the access point. This authentication is not much more of a task than plugging the Ethernet cable into the wall jack. The 802.11 authentication merely establishes an initial connection between the client and the access point. The 802.11-2012 standard specifies two different methods of legacy authentication: Open System authentication and Shared Key authentication. These legacy authentication methods were not so much an authentication of user or device identity, but more of an authentication of capability. Prior to the 802.11 standard, all other wireless technology was proprietary. Think of these authentication methods as verification between the two radios that they are both valid 802.11 devices.

Open System Authentication

Open System authentication is the only pre-RSNA security mechanism that has not been deprecated. Open System authentication is the simpler of the two authentication methods. It provides authentication without performing any type of client verification. It is essentially an exchange of hellos between the client and the access point. It is considered a null authentication because there is no exchange or verification of identity between the devices. It is assumed that the devices already have all of the appropriate information to authenticate to the network. In other words, every station (STA) is validated during Open System authentication.

Within a basic service set (BSS), Open System authentication occurs with an exchange of frames between the client station and the access point station. Open System authentication utilizes a two-message authentication transaction sequence. The first message asserts identity and requests authentication. The second message returns the authentication result. If the result is "successful," the STAs will be declared mutually authenticated. Open System authentication is also used by STAs in an independent basic service set (IBSS), which is more commonly known as an ad hoc WLAN.

Open System authentication occurs after a client STA knows about the existence of an access point (AP) by either passive or active scanning. The client STA can passively find out about the parameters of the BSS from the AP's beacon management frame or extract the same information during the active probing process from the AP's probe response frame. An Open System authentication frame exchange process then begins with the goal of eventually joining the BSS. As shown in Figure 2.1, the client STA must first become authenticated before exchanging two more association frames. Once Open System authentication and association occurs, the client STA establishes a Layer 2 connection to the AP and is a member of the BSS.

FIGURE 2.1 Open System authentication

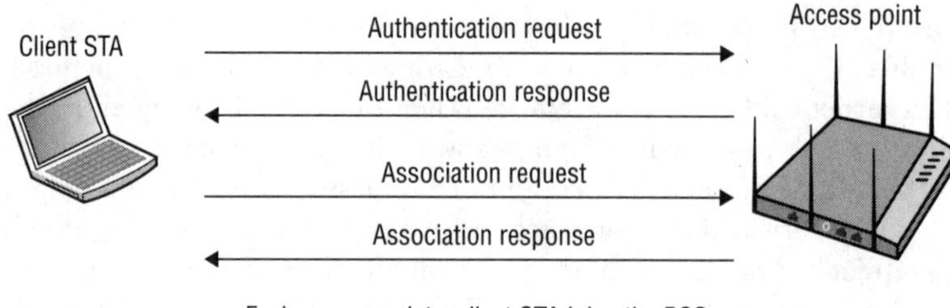

WEP encryption is optional with Open System authentication. For data privacy, Wired Equivalent Privacy (WEP) encryption can be used with Open System authentication, but WEP is used only to encrypt the Layers 3–7 MAC Service Data Unit (MSDU) payload of 802.11 data frames and only after the client station is authenticated and associated. In other words, WEP is not used as part of the Open System authentication process, but WEP encryption can be used to provide data privacy after authentication and association occur. So, if Open System authentication is so simple and basic—providing no verification of identity—then why is it still used when security is so important? The answer to this question is simple. It does not need to be secure, because other more advanced overlay security authentication methods such as 802.1X/EAP are now being implemented. As you can see in Figure 2.2, Open System authentication and association between the client STA and AP still occurs prior to the 802.1X/EAP authentication exchange between the client STA and a RADIUS server. In Exercise 2.1, you will look at a packet capture containing Open System authentication frames.

FIGURE 2.2 Open System and 802.1X/EAP authentication

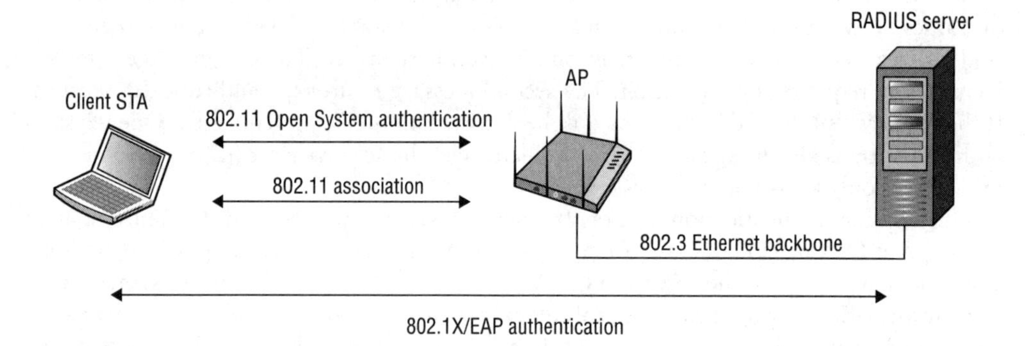

The 802.11-2012 standard now defines more advanced authentication methods. A detailed discussion about 802.1X/EAP can be found in Chapter 4, "802.1X/EAP Authentication."

Shared Key Authentication

Shared Key authentication uses WEP to authenticate client stations and requires that a static WEP key be configured on both the client STA and the access point. In addition to WEP being mandatory, authentication will not work if the static WEP keys do not match. The authentication process is similar to Open System authentication but includes a challenge and response between the AP station and client station within the BSS. Shared Key authentication can also be used between two STAs in an IBSS.

Shared Key authentication is a four-way authentication frame exchange, as shown in Figure 2.3. The client station sends an authentication request to the AP, and the AP sends a cleartext challenge to the client station in an authentication response. The client station then encrypts the cleartext challenge and sends it back to the AP in the body of another authentication request frame. The AP decrypts the station's response and compares it to the challenge text. If they match, the AP will respond by sending a fourth and final authentication frame to the station, confirming successful authentication. If they do not match, the AP will respond negatively. If the AP cannot decrypt the challenge, it will also respond negatively. If Shared Key authentication is successful, the same static WEP key that was used during the Shared Key authentication process will also be used to encrypt the 802.11 data frames.

FIGURE 2.3 Shared Key authentication exchange

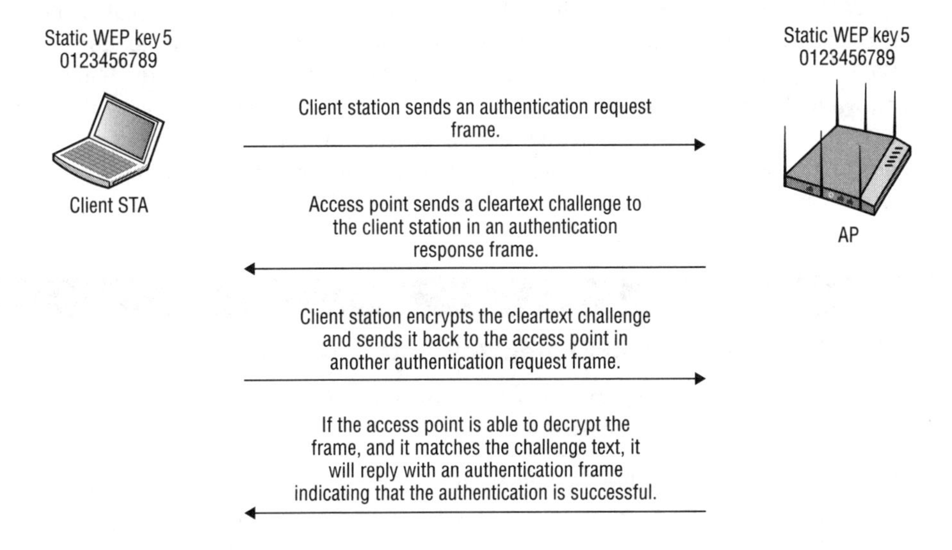

Although it might seem that Shared Key authentication is a more secure solution than Open System authentication, in reality Shared Key could be the bigger security risk. Anyone who captures the cleartext challenge phrase and then captures the encrypted challenge phrase in the response frame could potentially derive the static WEP key. If the static WEP key is compromised, a whole new can of worms has been opened because now all the data frames can be decrypted.

Do not confuse the Shared Key authentication with Preshared Key (PSK) authentication. Shared Key authentication is a legacy method defined as a pre-RSNA security method. The 802.11-2012 standard defines robust security that requires either 802.1X/EAP authentication or PSK authentication. PSK authentication methods are discussed in greater detail in Chapters 3 and 4.

In Exercise 2.1, you will look at packet captures containing encrypted and decrypted Shared Key authentication frames. Since it is our own network, and we know the WEP key, we can look at the decrypted authentication frames.

Which Legacy Authentication Method Is Better?

On the surface, Shared Key authentication may appear to be the more secure solution because the static WEP key is also being used for credentials during the authentication process. Open System authentication, on the other hand, does not require credentials. However, if the static WEP key is compromised from the Shared Key authentication process, all 802.11 data frames encrypted with that same static WEP key are at risk. In reality, static WEP encryption should never be used because WEP can easily be cracked using hacker tools such as Aircrack-ng. Much better dynamic encryption methods such as CCMP/AES are widely available. It is becoming rare that WEP is the only encryption option available, but it does still happen. In these instances, using simple Open System authentication together with WEP encryption instead of Shared Key authentication is probably the better choice. Also be aware that it may be possible for end users to change the configuration settings on a client station from Open System to Shared Key, especially when they are having trouble configuring their wireless client software utility. Improperly configured legacy authentication settings are something to watch for when troubleshooting legacy client problems.

EXERCISE 2.1

Viewing Open System and Shared Key Authentication Frames

In this exercise, you will use a protocol analyzer to view the 802.11 frame exchanges used to authenticate a client to the access point.

1. To perform this exercise, first download the following files from the book's online resource area, which can be accessed at `www.sybex.com/go/cwsp2e`.

   `OPEN_SYSTEM_AUTHENTICATION.PCAP`

   `SHARED_KEY_AUTHENTICATION_ENCRYPTED.PCAP`

   `SHARED_KEY_AUTHENTICATION_DECRYPTED.PCAP`

2. After the files are downloaded, you will need packet analysis software to open these files. If you do not already have a packet analyzer installed on your computer, you can download Wireshark from `www.wireshark.org`.

3. Using the packet analyzer, open the `OPEN_SYSTEM_AUTHENTICATION.PCAP` file. Most packet analyzers display a list of capture frames in the upper section of the screen, with each frame numbered sequentially in the first column.
4. Click on packet 1. Typically in the lower section of the screen is the Packet Details pane. This section contains annotated details about the selected frame. In this pane, locate and expand the IEEE 802.11 Wireless LAN Management Frame section. Then expand the Fixed Parameters section, and note that Auth Algorithm is set to Open System and that Auth Seq Num is set to 1. This indicates that it is an Open System authentication request frame.
5. Click on packet 3. In the Packet Details pane, locate and expand the IEEE 802.11 Wireless LAN Management Frame section. Then expand the Fixed Parameters section, and you can see that Auth Algorithm is set to Open System and that Auth Seq Num is set to 2. This indicates that it is an Open System authentication reply frame. Also note that Status Code indicates that the authentication was successful.
6. Using the packet analyzer, open the `SHARED_KEY_AUTHENTICATION_ENCRYPTED.PCAP` file.
7. The two packets of most interest are packets 3 and 5. In packet 3, the access point is responding to the authentication request and it includes the unencrypted challenge text. Click on packet 3. In the Packet Details pane, locate and expand the IEEE 802.11 Wireless LAN Management Frame section. In the Tagged Parameters section, you can see the challenge text. Expand the challenge text section and note that the length of the text is 128 bytes. Also note the text that is being used.
8. Click on packet 5. In the Packet Details pane, locate and expand the Data section. In this section you can see 136 bytes of encrypted data. In the IEEE 802.11 Authentication section above the Data section, note the WEP parameters.
9. Using the packet analyzer, open the `SHARED_KEY_AUTHENTICATION_DECRYPTED.PCAP` file.
10. This file is a decrypted version of the file that you were just examining. Repeat steps 7 and 8 using the decrypted file. Packets 3 and 5 contain the challenge text and the response; however, you will not see any WEP references in packet 5, since the packet is decrypted. Also note that the challenge text and response text are identical.

Wired Equivalent Privacy (WEP) Encryption

Wired Equivalent Privacy (WEP) is a Layer 2 encryption method that uses the ARC4 streaming cipher. Because WEP encryption occurs at Layer 2, the information that is being protected is the upper Layers of 3–7. The payload of an 802.11 data frame is called the *MAC Service Data Unit (MSDU)*. The MSDU contains data from the LLC and Layers 3–7. A simple definition of the MSDU is that it is the data payload that contains the IP packet plus some LLC data. WEP and other Layer 2 encryption methods encrypt the MSDU payload of an 802.11 data frame. The original 802.11 standard defined both 64-bit WEP and 128-bit WEP as supported encryption methods. The three main intended goals of WEP encryption include confidentiality, access control, and data integrity. The primary goal of confidentiality was to provide data privacy by encrypting the data before transmission. WEP also provides access control, which is basically a crude form of authorization. Client stations that do not have the same matching static WEP key as an access point are refused access to network resources. A data integrity checksum, known as the Integrity Check Value (ICV), is computed on data before encryption and used to prevent data from being modified. The current 802.11-2012 standard still defines WEP as a legacy encryption method for pre-RSNA security.

If You Thought WEP Used RC4 Encryption, You Are Right and Wrong!

RC4 is also known as ARC4 or ARCFOUR. ARC4 is short for Alleged RC4. RC4 was created in 1987 by Ron Rivest of RSA Security. It is known as either "Rivest Cipher 4" or "Ron's Code 4." RC4 was initially a trade secret; however, in 1994 a description of it was leaked onto the Internet. Comparison testing confirmed that the leaked code was genuine. RSA has never officially released the algorithm, and the name "RC4" is trademarked, hence the reference to it as ARCFOUR or ARC4.

Although both 64-bit and 128-bit WEP were defined in 1997 in the original IEEE 802.11 standard, the U.S. government initially allowed the export of only 64-bit technology. After the U.S. government loosened export restrictions on key size, radio card manufacturers began to produce equipment that supported 128-bit WEP encryption. The 802.11-2012 standard refers to the 64-bit version as WEP-40 and the 128-bit version as WEP-104. As shown in Figure 2.4, 64-bit WEP uses a secret 40-bit static key, which is combined with a 24-bit number selected by the card's device drivers. This 24-bit number, known as the *initialization vector (IV)*, is sent in cleartext and a new IV is created for every frame. Although the IV is said to be new for every frame, there are only 16,777,216 different IV combinations; therefore, over time, you are forced to reuse the IV values. The standard also does not define what algorithm to use to create the IV. The effective key strength of combining the IV with the 40-bit static key is 64-bit encryption. 128-bit WEP encryption uses a 104-bit secret static key that is also combined with a 24-bit IV.

FIGURE 2.4 Static WEP encryption key and initialization vector

64-bit WEP	24-bit IV	40-bit static key
128-bit WEP	24-bit IV	104-bit static key

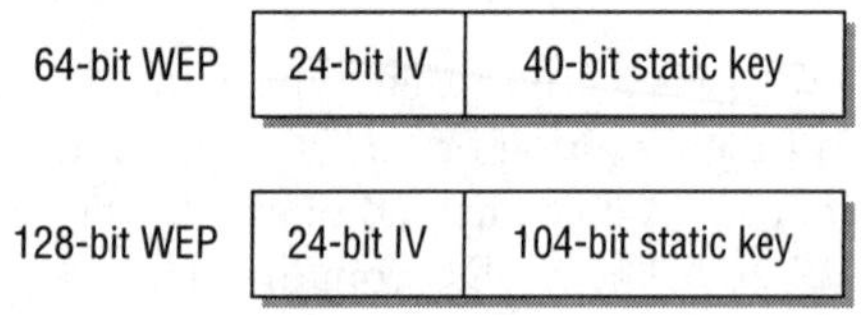

A static WEP key can usually be entered as hex characters (0–9 and A–F) or ASCII characters. The static key must match on both the access point and the client device. A 40-bit static key consists of 10 hex characters or 5 ASCII characters, whereas a 104-bit static key consists of 26 hex characters or 13 ASCII characters. Not all client stations or access points support both hex and ASCII. Some clients and access points support the use of up to four separate static WEP keys, from which a user can choose one as the default transmission key (Figure 2.5 shows an example of one such client).

FIGURE 2.5 WEP transmission key

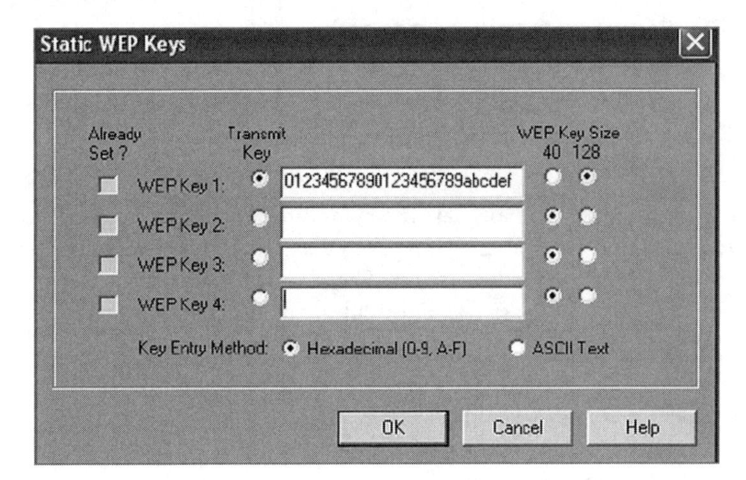

The transmission key is the static key that is used to encrypt data by the transmitting radio. A client or access point may use one key to encrypt outbound traffic and a different key to decrypt received traffic. However, each key used must match exactly on both sides of a link for encryption/decryption to work properly. When a device creates a WEP-encrypted frame, a key identifier is added to the IV field indicating which of the four possible static keys was used to encrypt the data and which key will be used to decrypt the data. As an example, if a transmitting device uses key 3 to encrypt the data, the receiving device will use key 3 to decrypt the data. If the receiving device does not have key 3 defined, or does not have the same WEP key entered in key 3, the data will not be decrypted.

How does WEP work? WEP runs a *cyclic redundancy check (CRC)* on the plaintext data that is to be encrypted and then appends the *Integrity Check Value (ICV)* to the end of the plaintext data. The ICV is used for data integrity and should not be confused with the initialization vector (IV). A 24-bit cleartext IV is generated and combined with the static secret key. WEP then uses both the static key and the IV as seeding material through a pseudorandom algorithm that generates random bits of data known as a *keystream.* These pseudorandom bits are equal in length to the plaintext data that is to be encrypted. The pseudorandom bits in the keystream are then combined with the plaintext data bits by using a Boolean XOR process. The end result is the WEP *ciphertext*, which is the encrypted data. The encrypted data is then prefixed with the cleartext IV, along with the number of which of the four keys was used for encryption. Figure 2.6 illustrates this process.

FIGURE 2.6 WEP encryption process

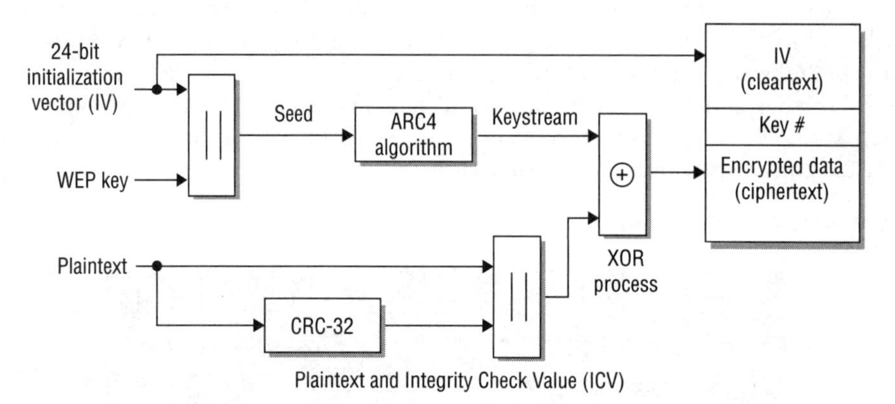

To decrypt a frame, WEP first extracts the IV and the key identifier, recognizing which key to use. WEP then uses the static key and the IV as seeding material through the pseudorandom algorithm to generate the keystream. The keystream is combined with the ciphertext using a Boolean XOR process. The end result is the decryption of the ciphertext and the creation of the plaintext data. WEP runs a CRC on the plaintext data and compares it to the decrypted ICV from the ciphertext. If the two are bit-wise identical, the frame is considered valid. Figure 2.7 illustrates this process.

FIGURE 2.7 WEP encryption process

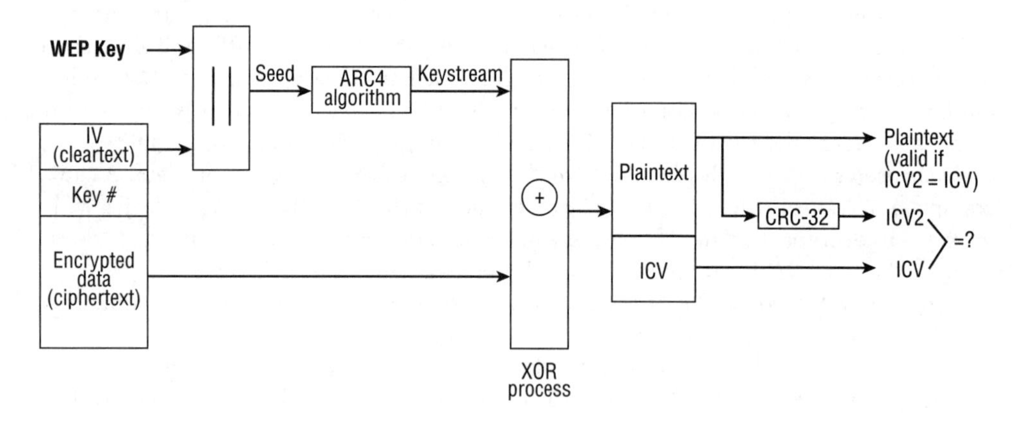

Unfortunately, WEP has quite a few weaknesses, which is why it has been deprecated. Its weaknesses include the following four main attacks:

IV Collisions Attack Because the 24-bit initialization vector is in cleartext and a new IV is generated for every frame, all 16 million IVs will eventually be used and will be forced to repeat themselves in a busy WEP-encrypted network. Because of the limited size of the IV space, IV collisions occur and an attacker can recover the secret key much easier when IV collisions occur in wireless networks.

Weak Key Attack Because of the ARC4 key-scheduling algorithm, weak IV keys are generated. An attacker can recover the secret key much easier by recovering the known weak IV keys.

Reinjection Attack Hacker tools exist that implement a packet reinjection attack to accelerate the collection of weak IVs on a network with little traffic.

Bit-Flipping Attack The ICV data integrity check is considered weak. WEP-encrypted packets can be tampered with. Current WEP cracking tools may use a combination of the first three mentioned attacks and can crack WEP in less than 5 minutes. After an attacker has compromised the static WEP key, any data frame can be decrypted with the newly discovered key. As defined by the 802.11-2012 standard, WEP encryption is considered a legacy pre-RSNA security method. Although WEP encryption has indeed been cracked and is viewed as unacceptable in the enterprise, it is still better than using no encryption at all. If legacy WEP is the only encryption solution available, it should be deployed. Many legacy devices that only support static WEP are still deployed in enterprise environments.

Are All WEP Keys Static?

As defined by the 802.11-2012 standard, WEP uses either a 40-bit static key or a 104-bit static key. These keys are combined with a 24-bit IV to bring the effective key strength to 64-bit and 128-bit. Before the Wi-Fi Alliance created the Wi-Fi Protected Access (WPA) security certification, various WLAN vendors offered a dynamic key generation security solution using WEP encryption. Dynamic WEP was never defined as any part of the 802.11 security and was intended only as a stop-gap dynamic encryption solution until the stronger encryption methods of TKIP/ARC4 or CCMP/AES became available. Dynamic WEP is discussed in Chapter 5, "802.11 Layer 2 Dynamic Encryption Key Generation."

EXERCISE 2.2

Viewing Encrypted MSDU Payload of 802.11 Data Frames

In this exercise, you will use a protocol analyzer to view an FTP file transfer of a text file. You will look at the MAC Service Data Unit (MSDU) payload of 802.11 data frames.

1. To perform this exercise, first download the following files from the book's online resource area, which can be accessed at `www.sybex.com/go/cwsp2e`.

   `NON_ENCRYPTED_MSDU.PCAP`

   `ENCRYPTED_MSDU.PCAP`

EXERCISE 2.2 *(continued)*

2. After the files are downloaded, you will need packet analysis software to open these files. If you do not already have a packet analyzer installed on your computer, you can download Wireshark from `www.wireshark.org`.
3. Using the packet analyzer, open the `NON_ENCRYPTED_MSDU.PCAP` file. Most packet analyzers display a list of capture frames in the upper section of the screen, with each frame numbered sequentially in the first column.
4. Click on packet 14. Typically the Packet Details pane appears in the lower section of the screen. This section contains annotated details about the selected frame. In this window, locate and expand the Transmission Control Protocol section. In this section notice that the source port is 20 and the application is ftp-data. Further down in this section and in the hex view window, you can read the text file. The cleartext data is not encrypted and contains multiple instances of the text "These are the times that try men's souls."
5. Using the packet analyzer, open the `ENCRYPTED_MSDU.PCAP` file. You can now view an MSDU payload of an 802.11 data frame that is encrypted.
6. Click on packet 15. In the Packet Details pane, locate and expand the Data section. Notice the 636 bytes of encrypted data. All the 3–7 information of the MSDU is now encrypted and cannot be seen. Select the IEEE 802.11 Data section and expand the WEP Parameters subsection, and you can see the initialization vector (IV) and Integrity Check Value (ICV).
7. If you are interested in decrypting the file, you can do so by entering the WEP key into the packet analyzer software. If you are using Wireshark, click the Wireshark-Preferences menu. In the window that appears, expand the Protocols section and then scroll down and select the IEEE 802.11 protocol. Then select Enable Decryption. Next to Decryption Key, click on the Edit button and add a key type of WEP with the key value of 55:55:55:55:55 (which is the key that was used when this transmission was performed). After accepting these changes, notice that the packet capture immediately appears different. All of the encrypted packets are decrypted, with the contents fully visible. If you are using packet analyzer software other than Wireshark, you will need to check the user's guide for that software to determine how to enter the WEP key.

TKIP

Temporal Key Integrity Protocol (TKIP) is a security protocol that was created to replace WEP. After WEP encryption was broken, 802.11 networks were left without a reliable security solution. The IEEE 802.11i security task group first defined TKIP to provide a stronger security solution without requiring users to replace their legacy equipment. Most legacy 802.11 radios could implement TKIP with a firmware upgrade, but not all legacy APs and STAs were upgradeable. The intent of TKIP was to provide a better temporary security solution until WLAN vendors could provide hardware that supported CCMP/AES encryption.

In April 2002, the Wi-Fi Alliance introduced the *Wi-Fi Protected Access (WPA)* certification, which requires the use of TKIP encryption. The IEEE 802.11-2012 standard defines two RSNA data confidentiality and integrity protocols: TKIP and CCMP, with TKIP support optional. Although TKIP is defined in the IEEE 802.11-2012 standard, due to security risks, it has been deprecated. TKIP still remains as an optional security protocol to provide support for older legacy devices.

TKIP is an enhancement of WEP. Like WEP, TKIP uses the ARC4 algorithm for performing its encryption and decryption processes. The TKIP enhancements were also intended to address the many known weaknesses of WEP. TKIP modifies WEP as follows:

Temporal Keys TKIP uses dynamically created encryption keys as opposed to the static keys. Any two radios use a 4-Way Handshake process to create dynamic unicast keys that are unique to those two radios. Static keys are susceptible to social engineering attacks. Dynamic encryption key generation is designed to defeat social engineering attacks.

Sequencing TKIP uses a per-MPDU *TKIP sequence counter (TSC)* to sequence the MPDUs it sends. An 802.11 station drops all MPDUs that are received out of order. Sequencing is designed to defeat replay and reinjection attacks that are used against WEP.

Key Mixing TKIP uses a complex two-phase cryptographic mixing process to create stronger seeding material for the ARC4 cipher. The key mixing process is designed to defeat the known IV collisions and weak-key attacks used against WEP.

Enhanced Data Integrity TKIP uses a stronger data integrity check known as the *Message Integrity Code (MIC)*. The MIC is sometimes also referred to as the Message Integrity Check. The MIC is designed to defeat bit-flipping and forgery attacks that are used against WEP.

TKIP Countermeasures Because of the design constraints of the TKIP MIC, it is still possible for an adversary to compromise message integrity; therefore, TKIP also implements countermeasures. The countermeasures bound the probability of a successful forgery and the amount of information an attacker can learn about a key.

Figure 2.8 shows the TKIP encryption and data integrity process. You will find it helpful to refer to this figure as you read about the steps TKIP performs.

TKIP starts with a 128-bit temporal key. An often-asked question is "Where does the 128-bit temporal key come from?" The answer is that the 128-bit temporal key is a dynamically generated key that comes from a 4-Way Handshake creation process. The 128-bit temporal key can either be a *pairwise transient key (PTK)* used to encrypt unicast traffic or a *group temporal key (GTK)* used to encrypt broadcast and multicast traffic. The creation of these dynamic temporal keys is discussed in great detail in Chapter 5.

FIGURE 2.8 TKIP encryption and data integrity process

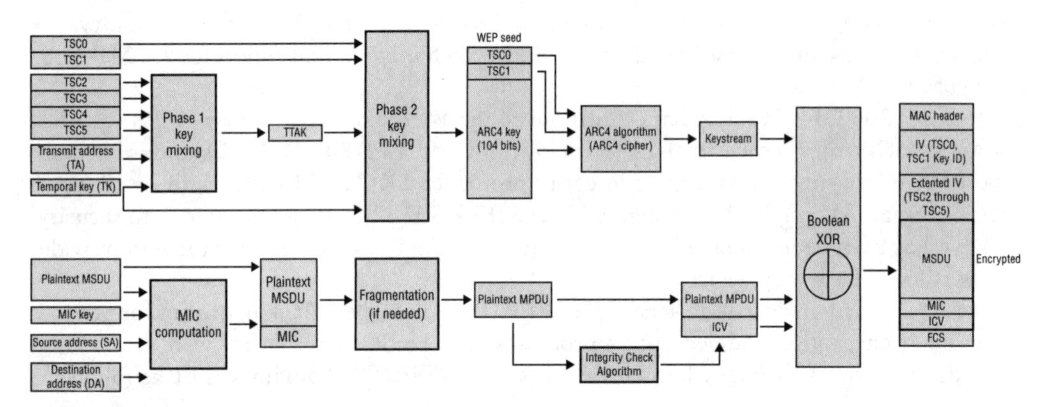

After the appropriate 128-bit temporal key (pairwise or group) is created, the two-phase key-mixing process begins. A 48-bit TKIP sequence counter (TSC) is generated and broken into 6 octets labeled TSC0 (least significant octet) through TSC5 (most significant octet). Phase 1 key mixing combines the 128-bit temporal key (TK) with the TSC2 through TSC5 octets of the TSC as well as the *transmit address (TA)*. The TA is the MAC address of the transmitting 802.11 radio. The output of the Phase 1 key mixing is the creation of the *TKIP-mixed transmit address and key (TTAK)*.

After the TTAK is generated, the Phase 2 key mixing can begin. Phase 2 key mixing combines the TTAK with the TSC0 and TSC1 octets of the TSC with the 128-bit TK. The output of the Phase 2 key mixing is referred to as the WEP seed. This WEP seed is then run through the ARC4 algorithm, and the keystream is created. The WEP seed is represented as a WEP IV and 104-bit WEP key when fed into the ARC4 algorithm. You may often hear TKIP referenced as using a 48-bit IV. During the Phase 2 key mixing process, TKIP encodes the TSC value from the sender as a WEP IV and an extended IV. The encoding of the 48-bit TSC effectively creates a 48-bit IV.

The two-phase key-mixing process can be summarized as follows:

- TTAK = Phase 1 (TK, TA, TSC)
- WEP seed = Phase 2 (TTAK, TK, TSC)

TKIP uses a stronger data integrity check known as the Message Integrity Code (MIC) to mitigate known forgery attacks against WEP. The MIC is often referred to by its nickname of Michael. The MIC can be used to defeat bit-flipping attacks, fragmentation attacks, redirection, and impersonation attacks. The MIC is computed using the destination address (DA), source address (SA), MSDU priority, and the entire unencrypted MSDU plaintext data. After the MIC is generated, it is appended to the end of the MSDU payload. The MIC is 8 octets in size and is labeled individually as M0 through M7. The MIC contains only 20 bits of effective security strength, making it somewhat vulnerable to brute-force attacks. Because the MIC only provides weak protection against active attacks, the 802.11-2012 standard defines *TKIP countermeasures* procedures. The TKIP countermeasures include the following:

Logging MIC failure would indicate an active attack that should be logged. MIC failure events can then be followed up by a system administrator.

60-Second Shutdown If two MIC failures occur within 60 seconds of each other, the STA or AP must disable all reception of TKIP frames for 60 seconds. This shutdown method theoretically provides a risk of a denial-of-service (DoS) attack. In reality, there are much easier ways to perform DoS attacks.

New Temporal Keys As an additional security feature, the PTK and (in the case of the Authenticator) the GTK should be changed.

The TKIP MIC augments, but does not replace, the WEP ICV. Because the TKIP MIC is still considered weak, TKIP protects the MIC with encryption, which makes TKIP MIC forgeries more difficult. The WEP ICV helps prevent false detection of MIC failures that would cause the countermeasures to be invoked.

After the MIC is created and appended to the plaintext MSDU, the 802.11 MAC performs its normal processing on this MSDU. If fragmentation is enabled, it is possible that this could be broken up into one or more MPDUs. It is even possible for the MIC to wind up split between two MPDUs. To keep things simple, we will assume that only one MPDU is created. An integrity check is performed on the plaintext MPDU, and the WEP ICV is then appended to the MPDU. A Boolean XOR is then performed on the keystream and the MPDU/ICV to generate the encrypted payload. A *frame check sequence (FCS)* is calculated over all the fields of the header and entire frame body. The resulting 32-bit CRC is then placed in the FCS field.

Before verifying the MIC, a receiving 802.11 STA will check the FCS, ICV, and TSC of all MPDUs. Any MPDU that has an invalid FCS, an incorrect ICV, or a TSC value that is less than or equal to the TSC replay counter is dropped before checking the MIC. This avoids unnecessary MIC failure events. Checking the TSC before the MIC makes countermeasure-based DoS attacks harder to perform. Checking the TSC also protects against replay/injection attacks. Although considered weak for data integrity protection, the ICV also offers some error detection. If the MPDU is corrupted by multipath interference or collisions, the FCS fails and the entire MPDU must be retransmitted. After the FCS, ICV, and TSC are checked, the MIC is used for verification of data integrity.

Can WEP and TKIP Still Be Used?

The 802.11n and higher amendments do not permit the use of WEP encryption or TKIP encryption for the High Throughput (HT) and Very High Throughput (VHT) data rates. HT data rates were introduced with 802.11n, and VHT data rates were introduced with 802.11ac. The Wi-Fi Alliance will certify 802.11n and 802.11ac radios to use CCMP encryption only for the higher data rates. For backward compatibility, newer radios can still support TKIP and WEP for the slower data rates defined by legacy 802.11a/b/g.

EXERCISE 2.3

TKIP-Encrypted Frames

In this exercise, you will use a protocol analyzer to view 802.11 data frames encrypted with TKIP.

1. To perform this exercise, first download the following file from the book's online resource area, which can be accessed at `www.sybex.com/go/cwsp2e`.

   `TKIP_FRAMES.PCAP`

2. After the file is downloaded, you will need packet analysis software to open this file. If you do not already have a packet analyzer installed on your computer, you can download Wireshark from `www.wireshark.org`.
3. Click on one of the 802.11 TKIP data packets. Typically the Packet Details pane appears in the lower section of the screen. This section contains annotated details about the selected frame. Scroll down to look at the different sections of the frame and the different fields. You will also see a TKIP Parameters section.
4. Click on one of the beacons with the source address of 00:1A:1E:94:4C:31. Locate and expand the IEEE 802.11 Wireless LAN Management Frame section. In this section, expand the Tagged Parameters section, and then expand Tag ➢ Vendor Specific ➢ Microsoft ➢ WPA Information Element. In this section, you can see TKIP listed as the cipher suite type.

Virtual Private Networks (VPNs)

Although the 802.11-2012 standard clearly defines Layer 2 security solutions, the use of upper-layer *virtual private network (VPN)* solutions can also be deployed with WLANs. VPNs are typically no longer recommended to provide security for WLAN client access in the enterprise due to the extra overhead from VPN encryption and the sometimes complex configuration. Furthermore, because faster, more secure Layer 2 solutions are now available for securing WLAN client access, use of Layer 3 VPNs for WLAN client access is outdated.

VPNs still have their place in Wi-Fi security and should definitely be used for remote access. They are also often used in wireless bridging environments and for branch office connectivity. Use of VPN technology is mandatory for remote access. Your end users will take their laptops off site and will most likely use public access Wi-Fi hot spots. Since there is no security at most hot spots, a VPN solution is a necessity. The VPN user will need to bring the security to the hot spot in order to provide a secure connection. It is imperative that users implement a VPN solution coupled with a personal firewall whenever accessing any public access Wi-Fi networks.

The two major types of VPN topologies are router-to-router or client/server based. Router-to-router VPNs are used to protect communications between two separate networks. Client/server VPNs are used to protect client communication to and from a network. VPNs have several major characteristics. They provide encryption, encapsulation, authentication, and data integrity. A simple definition of a Layer 3 VPN is that it is a router that also provides data privacy with encryption. VPNs use secure tunneling, which is the process of encapsulating one IP packet within another IP packet. The first packet is encapsulated inside the outer packet. The original destination and source IP address of the first packet is encrypted along with the data payload of the first packet. VPN tunneling therefore protects your original Layer 3 addresses and also protects the data payload of the original packet. Layer 3 VPNs use Layer 3 encryption; therefore, the payload that is being encrypted is the Layer 4 to 7 information. The IP addresses of the second packet are seen in cleartext and are used for communications between the tunnel end points. The destination and source IP addresses of the outer second packet will point to the virtual IP addresses of the VPN server and VPN client software.

Although no longer a recommended practice, VPNs were often used for WLAN security for client access because a VPN server was already deployed within the wired infrastructure. Since the VPN server already existed, the company only needed to install the WLAN on a network with a firewall between the WLAN and the corporate network. As mentioned earlier, a Layer 3 VPN has both routing and encryption capabilities. The encryption component of the VPN is used to provide data privacy. The Layers 4–7 payload of an 802.11 data frame will be encrypted across the wireless and wired medium. The VPN client software on the WLAN client station encrypts and decrypts at one end of the tunnel and the VPN server encrypts and decrypts at the other end of the tunnel. A firewall sits between the WLAN client and the VPN server. As shown in Figure 2.9, the firewall is configured to allow only VPN traffic to pass through the firewall. Once the VPN traffic passes through the firewall, the VPN server terminates it. If the traffic is a VPN client attempting to connect, and if the client credentials are valid, the client will be authenticated and assigned an IP address on the internal network. A VPN tunnel is created between the client and the VPN server. When the VPN tunnel passes through the firewall, the tunnel terminates at the VPN server, and the traffic is decrypted and routed to network resources. The VPN provides data privacy and the firewall is used together with the VPN to segment the WLAN from network resources.

The major protocols used in Layer 3 VPN technologies are *Point-to-Point Tunneling Protocol (PPTP)* and *Internet Protocol Security (IPsec)*. Unlike 802.1X/EAP solutions, an IP address is needed before a VPN tunnel can be established. A downside to using a VPN solution is that access points are potentially open to attack because a potential attacker can get both a Layer 2 and a Layer 3 connection before the VPN tunnel is established. WEP, which encrypts at Layer 2, was often used together with VPN security to protect the Layer 3 information. The problem with this strategy is that a double-encryption solution was being used, which created overhead that negatively affected the throughput and performance of the WLAN. Furthermore, as mentioned earlier in this chapter, WEP encryption can be cracked. A better solution would be 802.1X/EAP, which requires that all security credentials and transactions be completed before any Layer 3 connectivity is even possible.

FIGURE 2.9 VPN and WLAN client access security

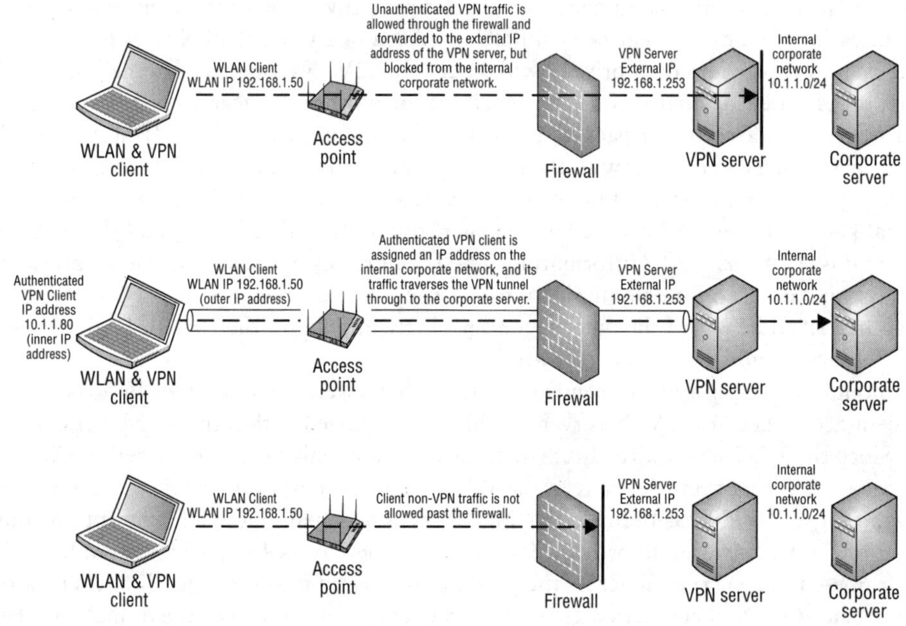

Point-to-Point Tunneling Protocol (PPTP)

Point-to-Point Tunneling Protocol (PPTP) does not provide encryption or confidentiality, relying on the tunneled protocol to provide it. PPTP typically uses 128-bit *Microsoft Point-to-Point Encryption (MPPE)*, which uses the ARC4 algorithm. PPTP encryption is considered adequate but not strong. PPTP uses MS-CHAP version 2 for user authentication. Unfortunately, MS-CHAP version 2 can be compromised with offline dictionary attacks, using auditing software such as *Asleap*. PPTP VPNs are considered to be outdated VPN technology and have been replaced with IPSec and SSL VPNs.

Layer 2 Tunneling Protocol (L2TP)

As its name implies, Layer 2 Tunneling Protocol (L2TP) is a tunneling protocol and is used to create VPNs. Like PPTP, L2TP does not provide encryption or confidentiality, relying on the tunneled protocol to provide it. IPsec is often used with L2TP to provide confidentiality, authentication, and integrity. When the two protocols are used together, they are typically referred to as L2TP/IPsec. The L2TP packets from one network are transported over IP to another network. IPsec provides a secure channel or connection between the two systems, and L2TP provides the tunnel.

Internet Protocol Security (IPsec)

Internet Protocol Security (IPsec) is a bundle of different security protocols, hashing techniques, and encryption algorithms. IPsec is flexible and allows vendors and users to select the different protocols that they want to use. IPsec VPNs use stronger encryption methods and more secure methods of authentication than PPTP. IPsec uses this assortment of protocols to perform the functions that are often linked to it. IPsec can be configured for transport mode or tunnel mode. Transport mode only encrypts the payload of the message and is typically used for host-to-host communications. In tunnel mode, the payload, header, and routing information are all encrypted, and it is typically used for gateway to gateway communications. IPsec VPNs use public and private key cryptography to establish a connection.

Internet Security Association and Key Management Protocol (ISAKMP) is a protocol that is used to establish *security associations (SAs)* and cryptographic keys. ISAKMP typically uses the *Internet Key Exchange (IKE and IKEv2)* protocol to set up SAs. SAs are unidirectional, with one SA needing to be created for each direction of the link. IKE uses a *Diffie-Hellman* key exchange to establish a shared session secret. Diffie-Hellman key exchange is a protocol that allows two devices to exchange a secret key across an insecure communications channel, without any prior knowledge. The type of encryption used is determined by the SA.

When configuring IPsec, you can choose *Authentication Header (AH)* or *Encapsulating Security Payload (ESP)*. AH only provides authentication. As part of the authentication, AH guarantees connectionless integrity of the packets along with authentication of their data origin. This helps to protect against replay attacks. ESP is used for both encryption and authentication of the IP packet. ESP provides origin confidentiality, integrity, and authenticity of packets. It does not protect the IP packet header, but it protects the entire inner IP packet, including the header.

IPsec uses *Message Digest 5 (MD5)* or *Secure Hash Algorithm 1 (SHA-1)* hash functions in the IKE authentication. Both of these hash algorithms are cryptographically secure. These hash algorithms have evolved into what is known as *Hashed Message Authentication Codes (HMAC)*, combining additional cryptographic functions to the algorithm. IPsec supports multiple ciphers, including *Data Encryption Standard (DES)*, *Triple DES (3DES)*, and *Advanced Encryption Standard (AES)*. Device authentication is achieved by using either a server-side certificate or a preshared key.

When configuring IPsec, you must define a *transform set*. A transform set is a combination of the security protocols and algorithms that will be used to protect your data. This transform set is used in the configuration of the VPN server.

Secure Sockets Layer (SSL)

VPN technologies do exist that operate at other layers of the OSI model, including *Secure Sockets Layer (SSL)* tunneling. Unlike an IPsec VPN, an SSL VPN does not require the installation and configuration of client software on the end user's computer. A user connects to an SSL VPN server via a web browser. The traffic between the web browser and the SSL VPN server is encrypted with the SSL protocol or Transport Layer Security (TLS). Note that SSL v3.0 has been deprecated and should no longer be used. SSL v3.0 should be replaced by TLS. TLS and SSL encrypt data connections above the Transport layer, using asymmetric cryptography for privacy and a keyed message authentication code for message reliability. Although most IPsec VPN solutions are Network Address Translation (NAT)-transversal, SSL VPNs are often chosen because of issues with NAT or restrictive firewall policies at remote locations.

VPN Configuration Complexity

As stated earlier in this chapter, the installation of a VPN is not the optimal way of securing your WLAN. There are many components that have to be installed and configured. Because most VPNs operate at Layer 3, static routes often have to be configured and advanced routing skills may be required of the administrator. The configuration would consist of a VPN server, which often is sized by the number of simultaneous connections. The VPN server would have to be configured to authenticate all of the WLAN users as they log on, so the VPN server would either need to have an internal user database, or more likely be configured to authenticate against an external database, such as a RADIUS server. If the number of clients on the WLAN grows, you might have to upgrade your VPN server or add an additional one. You may also have a firewall between the VPN server and the WLAN. This may be necessary because the WLAN itself has no security on it, allowing anyone to connect to it. You would need to have VPN software installed and configured on each wireless client, and you would have to train users how to connect to the WLAN and then log on to the VPN server using the VPN dialer.

VPN Scalability

Scaling a VPN secured WLAN compared to scaling an 802.1X/EAP secured WLAN requires more effort and resources. When scaling an 802.1X/EAP network, the addition of new users only requires an account on the authentication server and the configuration of the 802.1X client, which nowadays is often built into the operating system. The addition of new users in a VPN-secured WLAN also requires the addition of new users on the authentication server. Along with adding the user, at some point the VPN server will need to be upgraded or expanded to support the additional users. Also, the VPN client will need to be configured and installed on each of the computers that will be used to connect to the WLAN. Any firewalls that the VPN tunnel traverses will have to be configured so that the proper ports are open to allow the VPN packets to pass through. Also, if the VPN clients will need to roam across Layer 3 boundaries, the VPN tunnel will collapse. A complex MobileIP solution will have to be put in place along with the VPN infrastructure. As the size of the network grows, more resources will be used as new users are added to the network, making it a less flexible and scalable solution for WLAN client access. VPNs are rarely used anymore for WLAN client access as an enterprise solution. IPsec VPNs are mostly used to secure communications over the Internet from remote branch offices to corporate headquarters.

When Should VPNs Be Used with a WLAN?

Although VPN security is an outdated solution for WLAN client access security in the enterprise, many scenarios exist where VPNs can still be used as part of WLAN security. As mentioned earlier, use of VPN technology is mandatory for remote access when end users connect to public access WLANs. Since there is no encryption used at public access WLANs, a VPN solution is needed to provide for data privacy. Router-to-router VPNs are also used between WLAN controllers across wide area network (WAN) links. VPNs are used for secure connectivity between branch offices and the corporate office. VPN tunnels are also often used to secure point-to-point WLAN bridge links used for wireless backhaul between buildings. Finally, split-tunnel VPNs are also used for *remote* solutions across WAN links.

MAC Filters

Every network card has a physical address known as a *media access control (MAC)* address. This address is a 12-character hex number. 802.11 client stations, like all network-enabled devices, each have unique MAC addresses, and 802.11 access points use MAC addresses to direct frame traffic. Most vendors provide MAC filtering capabilities on their access points and WLAN controllers. MAC filters can be configured either to allow or deny traffic from specific MAC addresses.

Most MAC filters apply restrictions that only allow traffic from specific client stations to pass through based on their unique MAC addresses. Any other client stations whose MAC addresses are not on the allowed list will not be able to pass traffic through the virtual port of the access point and onto the distribution system medium. It should be noted, however, that MAC addresses can be *spoofed*, or impersonated, and any amateur hacker can easily bypass any MAC filter by spoofing an allowed client station's address. Many network adapters have the ability to change the MAC address as an option built into the advanced configuration window for the adapter, as shown in Figure 2.10. MAC addresses can also be very easily spoofed from the command line of many operating systems. Entering the new address and re-enabling the network card is all that is needed to change the MAC identity of the computer. Because of spoofing and because of all the administrative work that is involved with setting up MAC filters, MAC filtering is not considered a reliable means of security for wireless enterprise networks. The 802.11 standard does not define MAC filtering, and any implementation of MAC filtering is vendor specific.

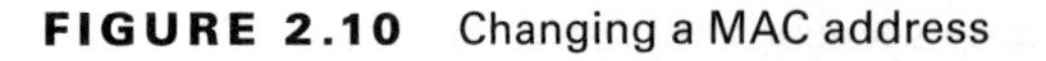

FIGURE 2.10 Changing a MAC address

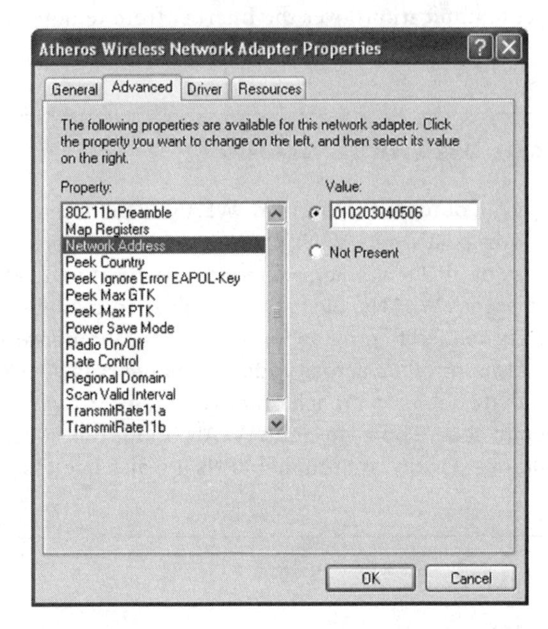

MAC filters are often used as a security measure to protect legacy radios that do not support stronger security. For example, older handheld barcode scanners may use 802.11 radios that support only static WEP. Best practices dictate an extra layer of security by segmenting the handheld devices in a separate VLAN with a MAC filter based on the manufacturer's *organizationally unique identifier (OUI)* address (the first three octets of the MAC address that are manufacturer specific).

SSID Segmentation

Another technique to provide security in a WLAN environment using autonomous access points was through SSID and VLAN segmentation. It was common for companies to create different SSIDs for different types of users. In a WLAN environment using enterprise-class autonomous APs, SSIDs would typically be mapped to individual VLANs, and users could be segmented by the SSID/VLAN pair, all while communicating through a single access point. Each SSID could also be configured with separate security settings. As you can see in Figure 2.11, the AP would have to be configured to support VLAN tagging using a protocol such as 802.1Q, and the AP would be connected to an upstream Layer 2 or Layer 3 switch using a trunked connection. Vendors could often have as many as 16 wireless SSIDs and VLANs with the capability of segmenting the users into separate Layer 3 domains because the VLANs were mapped to unique IP subnets. It should be noted that no more than three or four SSIDs should ever be transmitted by a single radio. A unique set of beacon frames and other management frames is generated for every SSID, creating excessive MAC layer overhead. Do not go overboard and have 8, 10, or even 16 different SSIDs. The MAC Layer overhead generated by using too many SSIDs can negatively affect the throughput and performance of the WLAN.

FIGURE 2.11 SSID/VLAN/Subnet mapping

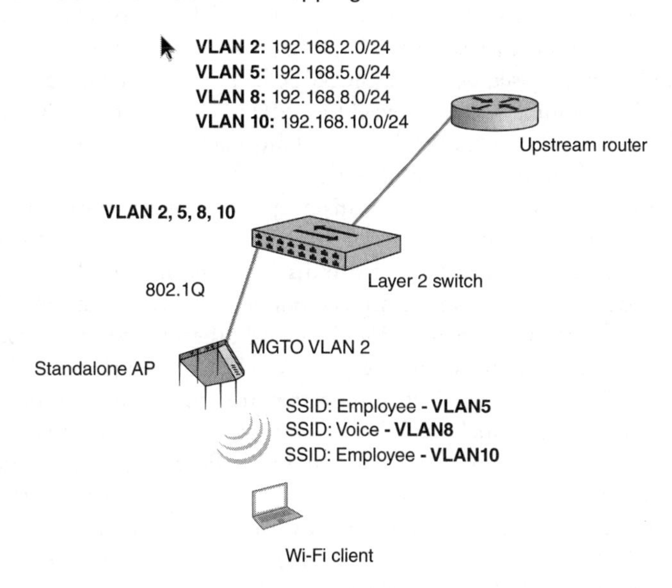

A common strategy is to create a guest, voice, and data VLAN. The SSID mapped to the guest VLAN will have captive portal authentication, and all users are restricted from corporate network resources and routed off to an Internet gateway. The SSID mapped to the voice VLAN might be using a security solution such a WPA2-Personal, and the VoWiFi client phones are routed to a VoIP server that provides QoS services through the VLAN. The SSID mapped to the data VLAN uses a stronger security solution such as WPA2-Enterprise, and the data users are allowed full access to network resources once authenticated.

SSID and VLAN segmentation is highly recommended and is discussed in more detail in Chapter 8, "WLAN Security Infrastructure." In Chapter 12, "Wireless Security Risks," you will also learn that multiple VLANs and access policies for different groups of users and/or devices can be assigned to a single SSID by leveraging RADIUS attributes. Assigning multiple VLANs to a single SSID is a great strategy to cut down on MAC layer overhead and airtime usage.

SSID Cloaking

Remember in *Star Trek* when the Romulans "cloaked" their spaceship but somehow Captain Kirk always found the ship anyway? Well, there is a way to "cloak" your SSID. Access points typically have a setting called *Closed Network* or *Broadcast SSID*. By either enabling a closed network or disabling the broadcast SSID feature, you can hide, or cloak, your wireless network name.

The *service set identifier (SSID)*, which is also often called the *extended service set identifier (ESSID)*, is the logical identifier, or logical name, of a WLAN. The SSID WLAN name is comparable to a Windows workgroup name. The SSID is a configurable setting on all radio cards, including access points and client stations. The SSID can be made up of as many as 32 characters and is case sensitive.

When you implement a closed network, the SSID field in the beacon frame is null (empty), and therefore passive scanning will not reveal the SSID to client stations that are listening to beacons. The idea behind cloaking the SSID is that any client station that does not know the SSID of the WLAN will not be able to associate.

Many wireless client software utilities transmit probe requests with null SSID fields when actively scanning for access points. Additionally, there are many WLAN discovery applications that can be used by individuals to discover and list nearby wireless networks.

These WLAN discovery applications also send out null probe requests actively scanning for access points. When you implement a closed network, the access point responds to null probe requests with probe responses; however, as in the beacon frame, the SSID field is null, and therefore the SSID is hidden to client stations that are using active scanning. Effectively, your wireless network is temporarily invisible, or cloaked. It should be noted that an access point in a closed network will respond to any configured client station that transmits directed probe requests with the properly configured SSID. This ensures that legitimate end users will be able to authenticate and associate to the AP. However, any station that is not configured with the correct SSID will not be able to authenticate or associate. Although implementing a closed network will indeed hide your SSID from WLAN discovery tools, anyone with a WLAN protocol analyzer can capture the frames transmitted by any legitimate end user and discover the SSID, which is transmitted in cleartext. In Exercise 2.4 you will search through a packet capture to see some of the different types of frames that include the SSID, even when it is hidden. In other words, a hidden SSID can usually be found in seconds with a WLAN protocol analyzer. Many wireless professionals will argue that hiding the SSID is a waste of time, whereas others view a closed network as just another layer of security. Cloaking the SSID usually keeps the SSID hidden from most WLAN discovery tools that use null probe requests. However, some of the WLAN discovery tools use alternate methods of discovering an SSID and may be able to display the SSID of a hidden network.

Although you can hide your SSID to cloak the identity of your WLAN from novice hackers (often referred to as *script kiddies*) and nonhackers, do not consider SSID cloaking as a WLAN security solution. The 802.11-2012 standard does not define SSID cloaking, and therefore all implementations of a closed network are vendor specific. As a result, incompatibility can potentially cause connectivity problems. Some wireless clients will not connect to a hidden SSID, even when the SSID is manually entered in the client software. Therefore, be sure to know the capabilities of your devices before implementing a closed network. Cloaking the SSID can also become an administrative and support issue. Requiring end users to configure the SSID in the radio software interface often results in more calls to the help desk because of misconfigured SSIDs. We highly recommended that you never cloak your SSID and instead broadcast your SSID for anyone and everyone to see.

EXERCISE 2.4

Viewing Hidden SSIDs

In this exercise, you will use a protocol analyzer to view a 70-second packet capture. You will search through this packet capture for the name of a hidden SSID. You will see that even though an SSID is hidden, it is contained in many different types of packets and can be easily found.

1. To perform this exercise, first download the following file from the book's online resource area, which can be accessed at `www.sybex.com/go/cwsp2e`.

   `CWSP_HIDDEN2_SSID.PCAP`

2. After the files are downloaded, you will need packet analysis software to open this file. If you do not already have a packet analyzer installed on your computer, you can download Wireshark from `www.wireshark.org`.
3. Using the packet analyzer, open the `CWSP_HIDDEN2_SSID.PCAP` file. Most packet analyzers display a list of capture frames in the upper section of the screen, with each frame numbered sequentially in the first column.
4. Click on packet 1, which is a beacon frame. In the Packet Details pane, expand the IEEE 802.11 Wireless LAN Management Frame section. Then expand the Tagged Parameters section. Notice that the Tag: SSID parameter set displays Broadcast instead of an SSID.
5. Click on packet 58, which is a null probe request from a client station. In this frame you will see that the Tag: SSID parameter set also displays Broadcast. The client is sending a null probe request looking for any and all access points. Now click on packet 59, which is the probe response frame from the AP. In the frame you will again see that the Tag: SSID parameter set displays Broadcast since the AP is cloaking the SSID.
6. Search through the packet capture for the CWSP-Hidden2 SSID. Even though it is hidden, it is still transmitted in many frames. If you are using Wireshark, select Edit ➢ Find Packet from the menu. You want to search for the string `CWSP-Hidden2`. Remember, SSIDs are case sensitive, so you want to make sure that you type it correctly, and you may want to click the Case Sensitive option. If you deselected this option, you do not have to be exact with the entry's letter case.
7. After you have entered the SSID that you are searching for, click the Find button and the program will highlight the next frame that matches your search. Every time you click the Find button, the program will find the next frame that contains the SSID. These are frame exchanges between the AP and legitimate clients that are preconfigured with the SSID.

Real World Scenario

Should Legacy Security Still Be Used?

The simple answer to this question is yes if the situation warrants the use of legacy security. Legacy devices are still being used on many WLANs. Some of these legacy devices that do not support WPA or WPA2 are still deployed in the enterprise. Common examples are older wireless barcode scanners and other handheld or specialty hardware devices that still only support static WEP encryption. The preferred scenario is to replace all of the legacy devices with new devices that support WPA2-Enterprise or at least WPA2-Personal. However, budgetary reasons often prevent an upgrade to newer and better devices. The bottom line is that some security is better than no security. If static WEP encryption is the strongest solution available, it should still be used even though WEP has been cracked.

An often-quoted security model is the concept of the "lowest hanging fruit." If someone walks up to an apple tree and wants an apple to eat, they will usually pick an apple that is at eye level and within arm's reach. They will probably not pick an apple that is hanging from a branch 20 feet above the ground. Think of a WLAN using no security as an apple within arm's reach and think of a WLAN using WEP encryption as an apple that is on a branch 20 feet from ground level. The passing individual can still get a ladder if they really want the apple that is 20 feet in the air, but chances are they will ignore the higher apple and still choose to pick an apple that is within arm's reach. If a hacker is determined to crack WEP, they will do it. However, if WEP is used on legacy devices, the hacker may try to access a WLAN with no security as opposed to the WLAN using WEP.

If you are using legacy security, we highly recommend that you segment the legacy devices in a separate VLAN with a separate SSID. We also suggest restricting access so that the wireless devices are limited to performing only the tasks that they require and limiting access to anywhere else on the network. Some legacy devices do not even support static WEP. Those devices should also be put in a separate VLAN on a separate SSID and a MAC filter should be used for security. Any hacker can get around a MAC filter, but at least the apple is higher on the tree.

Summary

In this chapter, you learned about the de jure and de facto standards that have been used to secure legacy 802.11 networks. We discussed Open System and Shared Key authentication and why Shared Key authentication should no longer be used. We also looked at the encryption and decryption processes of WEP and explained some of its shortcomings that have led to it being deprecated.

It is important to remember that, although VPN technology is widely used and a recommended technology to provide secure communications between networks and between individual clients and networks, it has fallen out of favor for securing WLAN users. The reason for this is because easier and faster security solutions are available that require less configuration and can scale up better. VPN solutions can still provide secure access for a WLAN, but with an ease, speed, and growth penalty when compared to an 802.1X/EAP solution. It is important to remember that over the years, in attempts to provide more security to the WLAN, vendors have included many features with their products to allow users to try to secure their networks. MAC filters and SSID cloaking are two legacy features that are both easily bypassed. Both techniques add a level of inconvenience to the network, to the user, and to a potential intruder, but neither adds any level of additional security.

SSID segmentation is one of the legacy security solutions that, if properly configured, has provided security to the network. By integrating SSIDs with VLANs, the flow of traffic can be controlled and isolated.

Exam Essentials

Understand pre-RSNA authentication methods. Be able to explain the differences between Open System authentication and Shared Key authentication.

Describe the WEP encryption and decryption process. Explain the components and process used to encrypt and decrypt a WEP packet. Identify and describe each of the fields that make up a WEP-encrypted 802.11 data frame.

Understand VPN technology and how it is used in a WLAN. Know the differences between the types of VPN technology that have been used with WLANs. Know the benefits and detriments of each.

Define other non-802.11 WLAN security mechanisms. Explain the other non-802.11 security mechanisms, such as MAC filters, SSID segmentation, and SSID cloaking.

Review Questions

1. Before an 802.11 client STA can pass traffic through the AP, which two of the following must occur? (Choose two answers.)
   - **A.** 802.1X
   - **B.** EAP
   - **C.** Association
   - **D.** Authentication
   - **E.** WEP keys must match
2. Which of the following is contained in a WEP-encrypted frame? (Choose all that apply.)
   - **A.** IV in cleartext format
   - **B.** IV in encrypted format
   - **C.** Key identifier
   - **D.** WEP key in encrypted format
   - **E.** 64-bit initialization vector
3. 128-bit WEP encryption uses a user-provided static key of what size?
   - **A.** 64 bits
   - **B.** 104 bits
   - **C.** 104 bytes
   - **D.** 128 bits
   - **E.** 128 bytes
4. When SSID cloaking is enabled, which of the following occurs? (Choose all that apply.)
   - **A.** The SSID field is set to null in the beacon frame.
   - **B.** The SSID field is set to null in the probe request frame.
   - **C.** The SSID field is set to null in the probe response frame.
   - **D.** The AP stops transmitting beacon frames.
   - **E.** The AP stops responding to probe request frames.
5. Which technologies use the RC4 or ARC4 cipher? (Choose all that apply.)
   - **A.** Static WEP
   - **B.** Dynamic WEP
   - **C.** PPTP
   - **D.** L2TP
   - **E.** MPPE
6. Which of the following is not defined by the 802.11-2012 standard? (Choose all that apply.)
   - **A.** WEP
   - **B.** VPN

C. MAC filtering
D. SSID segmentation
E. SSID cloaking

7. 802.11 pre-RSNA security defines which wireless security solution?
   A. Dynamic WEP
   B. 802.1X/EAP
   C. 64-bit static WEP
   D. Temporal Key Integrity Protocol (TKIP)
   E. CCMP/AES

8. Which of the following have been deprecated in the 802.11-2012 standard? (Choose all that apply.)
   A. Wired Equivalent Privacy
   B. Temporal Key Integrity Protocol
   C. Point-to-Point Tunneling Protocol
   D. Shared Key authentication
   E. Open System authentication

9. Peter is configuring a standalone AP to provide segmentation of three groups of wireless user traffic on the corporate network. Which deployment strategies will reach this goal? (Choose all that apply.)
   A. Create three separate SSIDs, one for each group, and have each SSID linked with a separate VLAN.
   B. Create a trunk for each of the VLANs between the AP and the access layer switch.
   C. Create a single trunk for all of the VLANs between the AP and the access layer switch.
   D. Configure each of the SSIDs with the same encryption keys for easier management and administration.
   E. Configure each of the SSIDs with different encryption keys.
   F. Consider leveraging RADIUS attributes to assign different groups of users and devices to different VLANS tied to the same SSID.

10. Evan has configured a laptop and an AP, each with two WEP keys. WEP key 1 is the same on both devices, and WEP key 2 is the same on both devices. He configured the laptop to use WEP key 1 to encrypt its data. He configured the AP to use WEP key 2 to encrypt its data. Will this configuration work?
    A. No, since there is only one WEP key on each device.
    B. No, since the value of the WEP key must be identical on both the laptop and the AP.
    C. Yes, as long as the value of WEP key 1 is identical on both computers and the value of WEP key 2 is identical on both computers.
    D. Yes. The laptop and AP will only use the first WEP key, so as long as the value of these keys is identical, the configuration will work.
    E. Yes. The laptop and AP will attempt to use each of the WEP keys when decrypting a frame.

11. Laura is attempting to diagnose a WLAN by using a packet analyzer to capture the exchange of frames and packets between a wireless client and the AP. In the process of analyzing the packets, she sees two 802.11 authentication frames, two 802.11 association frames, and DHCP requests and responses, and then she begins to see encrypted data. Which of the following could the client be using? (Choose all that apply.)

- **A.** Open System authentication
- **B.** Shared Key authentication
- **C.** 802.1X/EAP
- **D.** WEP
- **E.** IPsec

12. This graphic shows a packet capture of a successful 802.11 authentication. In which of the following types of client connections could this authentication not occur? (Choose all that apply.)

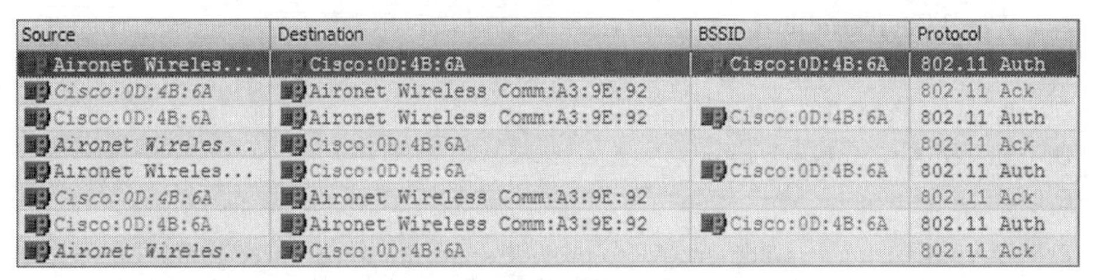

Source	Destination	BSSID	Protocol
Aironet Wireles...	Cisco:0D:4B:6A	Cisco:0D:4B:6A	802.11 Auth
Cisco:0D:4B:6A	Aironet Wireless Comm:A3:9E:92		802.11 Ack
Cisco:0D:4B:6A	Aironet Wireless Comm:A3:9E:92	Cisco:0D:4B:6A	802.11 Auth
Aironet Wireles...	Cisco:0D:4B:6A		802.11 Ack
Aironet Wireles...	Cisco:0D:4B:6A	Cisco:0D:4B:6A	802.11 Auth
Cisco:0D:4B:6A	Aironet Wireless Comm:A3:9E:92		802.11 Ack
Cisco:0D:4B:6A	Aironet Wireless Comm:A3:9E:92	Cisco:0D:4B:6A	802.11 Auth
Aironet Wireles...	Cisco:0D:4B:6A		802.11 Ack

- **A.** 802.1X/EAP
- **B.** WEP with Shared Key authentication
- **C.** WEP with Open System authentication
- **D.** Open System authentication with WEP

13. This graphic shows a WLAN discovery tool screen capture. How many SSIDs are configured with cloaking enabled? (Choose all that apply.)

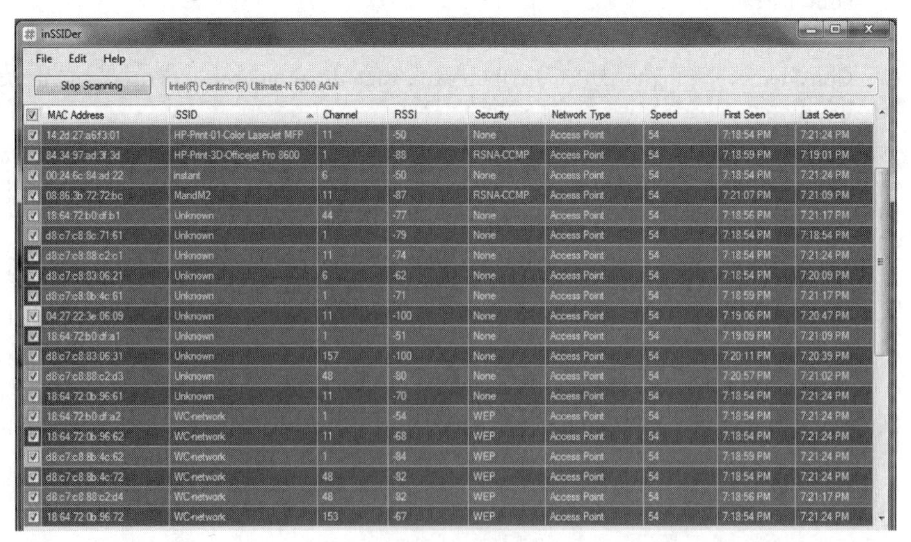

MAC Address	SSID	Channel	RSSI	Security	Network Type	Speed	First Seen	Last Seen
14:2d:27:a6:f3:01	HP-Print-01-Color LaserJet MFP	11	-50	None	Access Point	54	7:18:54 PM	7:21:24 PM
84:34:97:ad:3f:3d	HP-Print-3D-Officejet Pro 8600	1	-88	RSNA-CCMP	Access Point	54	7:18:59 PM	7:19:01 PM
00:24:6c:84:ad:22	instant	6	-50	None	Access Point	54	7:18:54 PM	7:21:24 PM
08:86:3b:72:72:bc	MandM2	11	-87	RSNA-CCMP	Access Point	54	7:21:07 PM	7:21:09 PM
18:64:72:b0:df:b1	Unknown	44	-77	None	Access Point	54	7:18:56 PM	7:21:17 PM
d8:c7:c8:8c:71:61	Unknown	1	-79	None	Access Point	54	7:18:54 PM	7:18:54 PM
d8:c7:c8:88:c2:c1	Unknown	11	-74	None	Access Point	54	7:18:54 PM	7:21:24 PM
d8:c7:c8:83:06:21	Unknown	6	-62	None	Access Point	54	7:18:54 PM	7:20:09 PM
d8:c7:c8:8b:4c:61	Unknown	1	-71	None	Access Point	54	7:18:59 PM	7:21:17 PM
04:27:22:3e:06:09	Unknown	11	-100	None	Access Point	54	7:19:06 PM	7:20:47 PM
18:64:72:b0:df:a1	Unknown	1	-51	None	Access Point	54	7:19:09 PM	7:21:09 PM
d8:c7:c8:83:06:31	Unknown	157	-100	None	Access Point	54	7:20:11 PM	7:20:39 PM
d8:c7:c8:88:c2:d3	Unknown	48	-80	None	Access Point	54	7:20:57 PM	7:21:02 PM
18:64:72:0b:96:61	Unknown	11	-70	None	Access Point	54	7:18:54 PM	7:21:24 PM
18:64:72:b0:df:a2	WC-network	1	-54	WEP	Access Point	54	7:18:54 PM	7:21:24 PM
18:64:72:0b:96:62	WC-network	11	-68	WEP	Access Point	54	7:18:54 PM	7:21:24 PM
d8:c7:c8:8b:4c:62	WC-network	1	-84	WEP	Access Point	54	7:18:59 PM	7:21:24 PM
d8:c7:c8:8b:4c:72	WC-network	48	-82	WEP	Access Point	54	7:18:54 PM	7:21:24 PM
d8:c7:c8:88:c2:d4	WC-network	48	-82	WEP	Access Point	54	7:18:56 PM	7:21:17 PM
18:64:72:0b:96:72	WC-network	153	-67	WEP	Access Point	54	7:18:54 PM	7:21:24 PM

A. None
B. At least ten
C. One
D. Ten
E. Exact number cannot be determined

14. 128-bit WEP encryption uses an IV and a static key. What are the lengths of these?
A. 64 bit IV and 64 bit key
B. 24 bit IV and 104 bit key
C. 28 bit IV and 100 bit key
D. 20 bit IV and 108 bit key
E. None of the above

15. The graphic shows a packet capture of a successful 802.11 authentication. In which of the following types of client connections could this not occur?

Source	Destination	BSSID	Protocol
Aironet Wireles...	00:1A:1E:94:4C:30	00:1A:1E:94:4C:30	802.11 Auth
00:1A:1E:94:4C:30	Aironet Wireless Comm:A3:9...		802.11 Ack
00:1A:1E:94:4C:30	Aironet Wireless Comm:A3:9...	00:1A:1E:94:4C:30	802.11 Auth
Aironet Wireles...	00:1A:1E:94:4C:30		802.11 Ack

A. 802.1X/EAP
B. WEP with Shared Key authentication
C. WEP with Open System authentication
D. Unencrypted with Open System authentication

16. Which hash algorithms can be used in the IKE authentication process? (Choose all that apply.)
A. Diffie-Hellman
B. MS-CHAPv2
C. MD5
D. ISAKMP
E. SHA-1

17. Which of these temporal keys are used by TKIP/ARC4 encryption? (Choose all that apply.)
A. PMK
B. PTK
C. GTK
D. GMK
E. 256-bit, 256-bit

18. Which of these authentication methods is the most secure?
A. Open System authentication with WEP
B. Open System authentication without WEP

C. Shared Key authentication

D. 802.1X/EAP authentication

19. Which of the following specifications are true for an SSID? (Choose all that apply.)

A. Up to 20 characters.

B. Up to 32 characters.

C. Case sensitive.

D. Spaces are allowed.

E. Spaces are not allowed.

20. Which 802.11 Layer 2 protocol is used for authentication in an 802.1X framework?

A. Extensible Authentication Protocol

B. Extended Authentication Protocol

C. MS-CHAP

D. Open System

E. Shared Key

Chapter 3

Encryption Ciphers and Methods

IN THIS CHAPTER, YOU WILL LEARN ABOUT THE FOLLOWING:

✓ **Encryption basics**

- Symmetric and asymmetric algorithms
- Stream and block ciphers
- RC4/ARC4
- RC5
- DES
- 3DES
- AES

✓ **WLAN Encryption methods**

✓ **WEP**

- WEP MPDU

✓ **TKIP**

- TKIP MPDU

✓ **CCMP**

- CCMP MPDU

✓ **WPA/WPA2**

✓ **Future Encryption Methods**

✓ **Proprietary Layer 2 implementations**

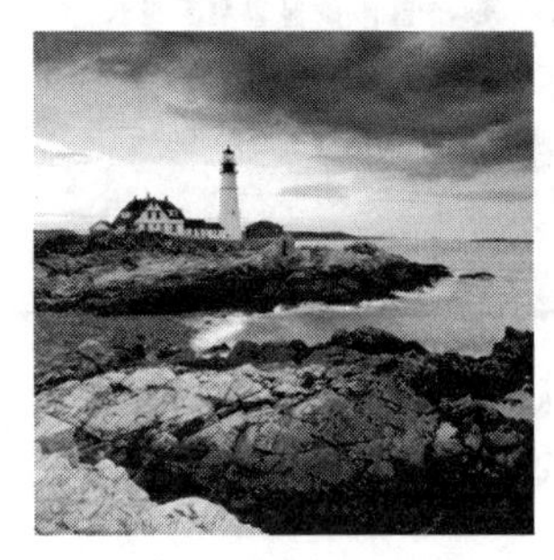

Over the years people have created many ways to secure data for many purposes. In this chapter, you will learn about the various encryption algorithms that are used to secure wireless networks. You will see how encryption ciphers work to create encrypted data from plaintext data. This chapter will also discuss the encryption methods that are part of the 802.11-2012 standard, and the ciphers that they use. When data is encrypted, additional overhead is added to the frames. In this chapter, you will also see the MAC Protocol Data Unit (MPDU) format of Wired Equivalent Privacy (WEP), Temporal Key Integrity Protocol (TKIP), and Counter Mode with Cipher Block Chaining Message Authentication Code Protocol (CCMP) encrypted frames.

Encryption Basics

One of the major concerns with wireless networking has always been the fact that wireless communications use what is referred to as an *unbounded medium*. The wireless signal radiates away from the transmitting device in all directions, unlike a wired signal, which travels along the path of the cable. In other words, the RF physical medium is not limited to a cable and has no set boundaries. Since wireless is unbounded, and the signal can essentially be heard by anyone within listening range, measures need to be taken to secure the transmission so that only the intended recipients can understand the message. Therefore, data privacy should be considered mandatory. All essential data must be encrypted prior to transmission and then decrypted after being received.

Chapter 1, "WLAN Security Overview," introduced the concept of cryptology. To review this concept briefly, the goal of cryptology is to process a piece of information, often referred to as *plaintext*, through an algorithm, often referred as a *cipher*, and transform the plaintext into encrypted text, which is also known as *ciphertext*. This ciphertext can then be decrypted, or converted back into plaintext, only by someone who knows the cipher.

The cipher is the process or algorithm that transforms the plaintext into encrypted text. One of the earliest ciphers, the Caesar cipher, is named after Julius Caesar. The Caesar cipher is a type of substitution cipher in which each letter of the alphabet is replaced by a different letter. When writing a message, Caesar shifted the alphabet by three characters to encrypt and protect his messages. Figure 3.1 shows the plaintext alphabet along with the shifted cipher key. The figure also shows a sample message in both plaintext and its ciphertext.

FIGURE 3.1 Example of the Caesar cipher

```
Plaintext Key:       ABCDEFGHIJKLMNOPQRSTUVWXYZ
Ciphertext Key:      DEFGHIJKLMNOPQRSTUVWXYZABC

Plaintext  Message: CWSP IS A GREAT CERTIFICATION TO EARN
Ciphertext Message: FZVS LV D JUHDW FHUWLILFDWLRQ WR HDUQ
```

Ciphers have come a long way since the Caesar cipher. Present-day ciphers use mathematical calculations along with multiple repetitions or transformation rounds, with each round typically consisting of several processing steps.

Symmetric and Asymmetric Algorithms

Most cipher algorithms can be categorized as either a *symmetric algorithm* or an *asymmetric algorithm*. When using a symmetric algorithm, both the encrypting and decrypting parties share the same key. To ensure the privacy of a symmetric algorithm encrypted communication, the key needs to be kept secret. A potential problem with this is that the key must be shared between two or more parties prior to establishing the secure communications channel. Therefore, it is necessary to have a secure method of sharing the key. WEP, TKIP, and CCMP are encryption methods that all use symmetric algorithms.

Instead of using a single shared key, asymmetric algorithms use a pair of keys. As shown in Figure 3.2, one key is used for encryption and the other is used for decryption. The decryption key is typically kept secret and is known as the *private key*, or "secret key." The encryption key is shared and is referred to as the *public key*. The premise is that if you wanted to send someone an encrypted message, you would obtain and use the public key to encrypt the message. The public key is only good for encrypting the message and cannot be used to decrypt the message. So even though many people could have the public key, none of them would be able to decrypt your message. Only someone with the private key would be able to decrypt it. Public key cryptography builds on the use of asymmetric encryption and digital signatures. Some methods of EAP authentication use digital certificates based on the X.509 standard for a *public key infrastructure (PKI)*.

FIGURE 3.2 Asymmetric keys

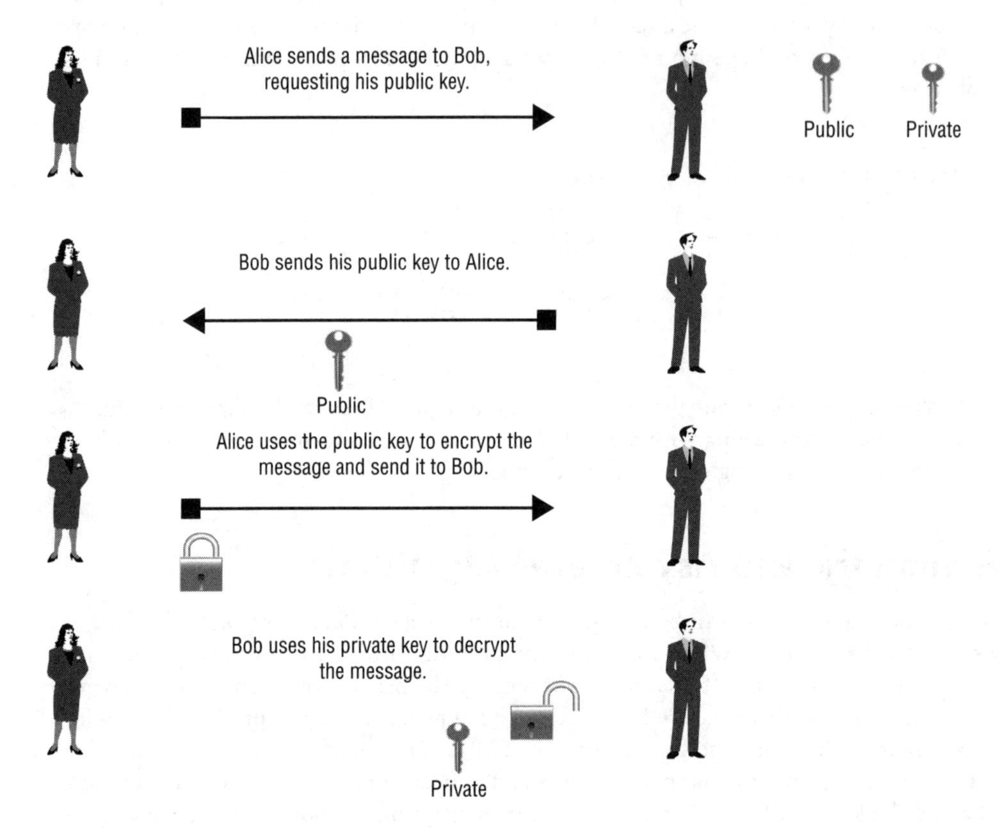

Symmetric algorithms generally require less computer processing power than asymmetric algorithms and, therefore, are typically much faster. However, the problem still exists that, with a symmetric algorithm, the key must be exchanged prior to the establishment of the secure communications, whereas with an asymmetric algorithm, the shared key that is used to decrypt the message never needs to be disclosed. Each method has benefits and drawbacks.

EAP and Digital Certificates

You can find a more detailed explanation about EAP authentication methods in Chapter 4, "802.1X/EAP Authentication." A more detailed discussion about digital certificates and public key infrastructure (PKI) can be found in Chapter 9, "RADIUS and LDAP."

Stream and Block Ciphers

During the cryptographic process, the plaintext needs to be combined with random data bits to create the ciphertext. One common way of performing this task is sequentially, on a bit-by-bit basis. Ciphers that use this technique are known as stream ciphers. A *stream cipher* is a symmetric key cipher where plaintext bits are combined with a pseudorandom cipher bit stream called the *keystream*. The keystream is generated when some sort of seed is used to feed the stream cipher algorithm. For example, WEP encryption uses a static key that feeds the ARC4 stream cipher, which then generates the pseudorandom keystream. The stream cipher then combines the plaintext with the keystream, typically using a Boolean *Exclusive-OR (XOR)* operation. Stream ciphers are often used when the plaintext is not consistently one size, such as the data transmitted on a wireless LAN.

Boolean logic is a mathematical way to compare or combine bits. As shown in Table 3.1, an Exclusive-OR (XOR) will generate a 0 when both of the input values are the same and will generate a 1 when both of the input values are different. When an XOR is used with a shared key, the same key will successfully encrypt and then decrypt the data.

TABLE 3.1 Exclusive-OR (XOR)

Input 1	Input 2	XOR output
0	0	0
0	1	1
1	0	1
1	1	0

Another common method of creating the ciphertext is using a *block cipher*. Unlike stream ciphers, which operate on one bit at a time, a block cipher takes a fixed-length block of plaintext and generates a block of ciphertext of the same length. A block cipher is a symmetric key cipher operating on fixed-length groups of bits, called *blocks*. For example, a block cipher will use a 128-bit block of input of plaintext, and the resulting output would be a 128-bit block of ciphertext. The fixed length of the blocks is referred to as the *block size*, with block sizes often ranging from 64 bits up to 256 bits. Most block ciphers are designed to apply a simpler function repeatedly to the block. Each iterative process or function is referred to as a round. Depending on the specific block cipher, the *round function* could be repeated as many as a few dozen times. In most instances, the greater the number of rounds, the greater the level of security; however, performance will be effected due to the time needed to perform the rounds.

RC4/ARC4

RC4 is a stream cipher that was designed by Ron Rivest of RSA Security in 1987. The RC in RC4 stands for Rivest Cipher or Ron's Code. RC4 was originally a trade secret; however, a description of it was anonymously leaked on the Internet in September 1994. After the code was leaked and confirmed to be genuine, and the algorithm was known, it was clearly no longer a trade secret. Although the algorithm was no longer a trade secret, the name "RC4" was trademarked and could only be used with permission. RSA never released the algorithm, so unofficial versions of it are often referred to as *Arcfour* or *ARC4*, which stands for "Alleged RC4." ARC4 is fast and simple, and it is widely used in protocols such as WEP and Secure Sockets Layer (SSL). Due to weaknesses in the cipher, it is not recommended for use in newer networks.

RC5

RC5 is a symmetric block cipher that was designed by Ron Rivest in 1994, and a U.S. patent was granted for it in May 1997. RC5 allows for a variable block size, a variable key size, and a variable number of rounds. The block size can be set to 32, 64, or 128 bits; the key size can range from 0 bits to 2040 bits; and the number of rounds can range from 0 to 255. A key table is created, with the size of the table varying depending on the number of rounds that will be performed. When the user-provided secret key is entered, a key-expansion routine will expand the user-provided key to fill the key table. This key table is used for both encryption and decryption.

DES

Data Encryption Standard (DES) is a symmetric block cipher that was developed in the early 1970s. In 1976, it was selected by the National Bureau of Standards (NBS), currently known as the National Institute of Standards and Technology (NIST), as part of the official Federal Information Processing Standards (FIPS) for the United States. DES uses a 56-bit symmetric key, and it is now considered to be insecure, primarily due to the key size being too small. Multiple groups have successfully cracked DES using what are known as *brute-force attacks*, sequentially trying every possible key. DES has a 64-bit block size and a 64-bit key; however, 8 bits are used for checking parity and are discarded, so the effective key length is 56 bits. DES performs 16 identical rounds on each block.

The history of DES is very interesting. A quick search on the Internet will reveal that the National Security Agency (NSA) was accused of tampering with it and covertly weakening the algorithm. These accusations were later shown to be false.

3DES

Triple Data Encryption Algorithm (TDEA), also known as *Triple DES (3DES)*, is a symmetric block cipher published in 1998. 3DES uses a key bundle, which is made up of three DES keys (K1, K2, and K3), each with an effective key length of 56 bits. Like DES, each key is made up of 64 bits; however, 8 bits are used for checking parity and are discarded. 3DES is essentially DES run three different times using three keys. Therefore, it performs 48 DES-equivalent rounds on each block. 3DES defines three keying options:

Keying Option 1 All three keys are unique.

Keying Option 2 K1 and K2 are unique, but K3 = K1.

Keying Option 3 All three keys are identical; K1 = K2 = K3.

Keying option 1 is the strongest, because all three keys are unique, giving it an effective key size of 168 bits. Keying option 3 is the weakest, and it is essentially equal to very slow DES. Remember that with a symmetric algorithm, the same key that encrypts the data also decrypts the data. With 3DES, the first pass with K1 encrypts the data, the second pass with K2 actually decrypts the data, and the third pass with K3 encrypts the data again. Keying option 2 provides an effective key size of 112 bits. In 1999, DES was reaffirmed by FIPS for the fourth time, with 3DES the preferred method and single DES permitted only in legacy systems.

AES

Advanced Encryption Standard (AES) is an encryption standard adopted by the U.S. government. AES uses an algorithm that is a symmetric block cipher that supports three key sizes of 128, 192, and 256 bits. These different key lengths are referred to as AES-128, AES-192, and AES-256. AES was announced in November 2001 by NIST as FIPS 197. Although FIPS 197 specifies AES, it is just an algorithm that can be incorporated into many types of security solutions. For example, AES can be used by an IPsec VPN as well as in WPA2 security. It is based on the *Rijndael* algorithm, which was developed by two Belgian cryptographers, Vincent Rijmen and Joan Daemen. AES is used by the 802.11 encryption protocol CCMP, which officially became part of wireless security when 802.11i was ratified in June 2004. Remember that 802.11i is now part of the 802.11-2012 standard.

AES uses a fixed block size of 128 bits, which is actually a 4×4 array of bytes, called a *state*. The number of rounds performed on the block varies depending on the key sizes. AES-128 performs 10 rounds, AES-192 performs 12 rounds, and AES-256 performs 14 rounds. The 802.11ad-2012 amendment defines Very High Throughput (VHT) enhancements using the unlicensed frequency band of 60 GHz. The 802.11ad-2012 amendment standardized on the use of Galois/Counter Mode Protocol (GCMP), which uses AES cryptography. However, GCMP calculations can be run in parallel and are computationally less intensive than the cryptographic operations of CCMP. The extremely high data rates defined by 802.11ad need GCMP because it is more efficient than CCMP.

802.11 WLANs and FIPS Validation

Security is the number one concern when deploying wireless technology in government environments. In most countries, there are mandated regulations on how to protect and secure data communications within all government agencies. In the United States, the *National Institute of Standards and Technologies (NIST)* maintains the *Federal Information Processing Standards (FIPS)*. FIPS 197 defines the Advanced Encryption Standard (AES). However, of special interest to wireless security is the FIPS 140-2 standard, which defines security requirements for cryptography modules. The use of validated cryptographic modules is required by the U.S. government for all unclassified communications. A WLAN infrastructure cannot be deployed in most U.S. government agencies unless the solution has been FIPS 140-2–validated by NIST. WLAN vendors spend a lot of time and money getting their WLAN security solutions FIPS 140-2–validated. Local governments and other nations also recognize the FIPS 140-2 requirements or have similar regulations. Both the FIPS 197 and FIPS 140-2 publications can be downloaded in PDF format from the NIST website at `http://csrc.nist.gov/publications/PubsFIPS.html`.

WLAN Encryption Methods

The 802.11-2012 standard defines three encryption methods that operate at Layer 2 of the OSI model: WEP, TKIP, and CCMP. The information that is being protected by these Layer 2 encryption methods is data found in the upper Layers of 3–7. Layer 2 encryption methods are used to provide data privacy for 802.11 data frames. The technical name for an 802.11 data frame is a *MAC Protocol Data Unit (MPDU)*. The 802.11 data frame, as shown in Figure 3.3, contains a Layer 2 MAC header, a frame body, and a trailer, which is a 32-bit CRC known as the *frame check sequence (FCS)*. The Layer 2 header contains MAC addresses and the duration value. Encapsulated inside the frame body of an 802.11 data frame is an upper-layer payload called the *MAC Service Data Unit (MSDU)*. The MSDU contains data from the Logical Link Control (LLC) and Layers 3–7. A simple definition of the MSDU is that it is the data payload that contains an IP packet plus some LLC data. The 802.11-2012 standard states that the MSDU payload can be anywhere from 0 to 2,304 bytes. The frame body may actually be larger due to encryption overhead.

FIGURE 3.3 802.11 MAC Protocol Data Unit (MPDU)

MAC header	Frame body MSDU 0–2,304 bytes	FCS

MPDU—802.11 data frame

WEP, TKIP, CCMP, and other proprietary Layer 2 encryption methods are used to encrypt the MSDU payload of an 802.11 data frame. Therefore, the information that is being protected is the upper Layers of 3–7, which is more commonly known as the IP packet.

It should be noted that many types of 802.11 frames are either never encrypted, or typically not encrypted. 802.11 management frames only carry a Layer 2 payload in their frame body, so encryption is not necessary for data security purposes. However, wireless denial of service (DoS) attacks often use spoofed or impersonated management frames, which led to the development of 802.11w, also known as management frame protection. 802.11w makes wireless DoS attacks more difficult. 802.11 control frames only have a header and a trailer; therefore, encryption is not necessary. Some 802.11 data frames, such as the null function frame, actually do not have an MSDU payload. Non-data-carrying data frames have a specific function, but they do not require encryption. Only 802.11 data frames with an MSDU payload can be encrypted. As a matter of corporate policy, 802.11 data frames should always be encrypted for data privacy and security purposes.

WEP, TKIP, and CCMP are encryption methods that all use symmetric algorithms. WEP and TKIP use the ARC4 cipher, whereas CCMP uses the AES cipher. The current 802.11-2012 standard defines WEP as a legacy encryption method for pre-RSNA security. TKIP and CCMP are considered to be *robust security network (RSN)* encryption protocols. The 802.11ad-2012 amendment standardized the use of Galois/Counter Mode Protocol (GCMP), which uses AES cryptography. GCMP is also considered an optional encryption method for 802.11ac radios. However, none of the chipset vendors have implemented GCMP in the first several generations of 802.11ac radios. Hardware upgrades for both access points and client radios will be required for GCMP to be used with 802.11ac radios.

How Are 802.11 Encryption Keys Created?

As you have already learned, WEP uses a preconfigured static key. Static keys are always susceptible to social engineering attacks; therefore, TKIP and CCMP use encryption keys that are dynamically generated by the *4-Way Handshake*. How TKIP and CCMP keys are created using the 4-Way Handshake is discussed in great detail in Chapter 5, "802.11 Layer 2 Dynamic Encryption Key Generation."

WEP

Wired Equivalent Privacy (WEP) is a Layer 2 security protocol that uses the ARC4 streaming cipher. The original 802.11 standard defined both 64-bit WEP and 128-bit WEP as supported encryption methods. The current 802.11-2012 standard still defines WEP as a legacy encryption method for pre-RSNA security. The Wi-Fi Alliance has been certifying 802.11 radios using WEP encryption since 2000. You learned about WEP encryption in Chapter 2, "Legacy 802.11 Security."

For a brief review of the WEP encryption process, recall that a 24-bit cleartext *initialization vector (IV)* is randomly generated and combined with the static secret key. As shown in Figure 3.4, the static key and the IV are used as WEP seeding material through the pseudorandom ARC4 algorithm that generates the keystream. The pseudorandom bits in the keystream are then combined with the plaintext data bits by using a Boolean XOR process. The end result is the WEP ciphertext, which is the encrypted data. WEP also runs a *cyclic redundancy check (CRC)* on the plaintext data that is to be encrypted and then appends the *Integrity Check Value (ICV)* to the end of the plaintext data. The ICV is used for data integrity and should not be confused with the IV, which is a part of the seeding material for the ARC4 cipher.

FIGURE 3.4 WEP encryption process

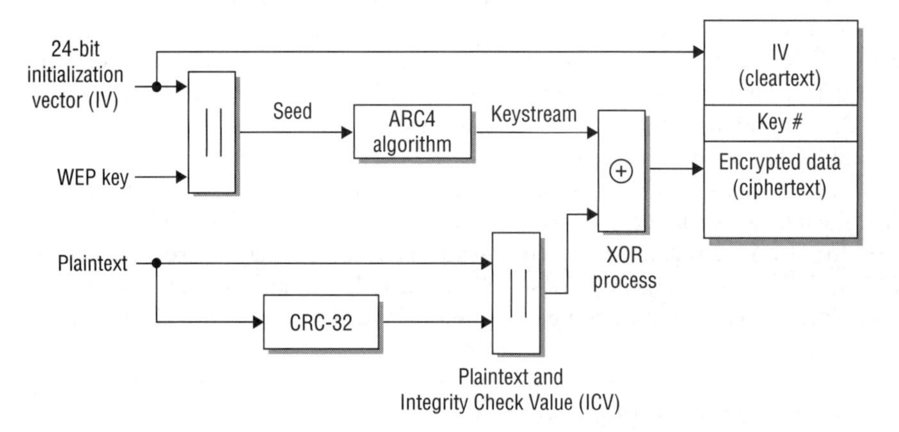

WEP encryption is covered extensively in Chapter 2, therefore, rather than exploring the topic again in great detail, we will direct you back to Chapter 2.

WEP MPDU

The encryption and decryption process for WEP is the same whether you are using WEP-40 or WEP-104. Figure 3.5 shows the WEP frame body, which contains an encrypted MSDU.

To create the WEP-encrypted MSDU, WEP runs a CRC on the plaintext data that is to be encrypted and then appends the ICV to the end of the plaintext data. The ICV adds 32 bits (4 octets) of overhead to an 802.11 data frame. The data and ICV are then encrypted. WEP can be configured with up to four different keys. A Key ID identifies which WEP key was combined with the system-generated 24-bit IV to perform the encryption. This 24-bit IV is combined with the KEY ID and 6 bits of padding to create a 32-bit IV. The IV adds 32 bits (4 octets) of overhead to the frame body of an 802.11 data frame. The IV is not encrypted and is appended to the front of the encrypted MSDU payload.

FIGURE 3.5 WEP MPDU format

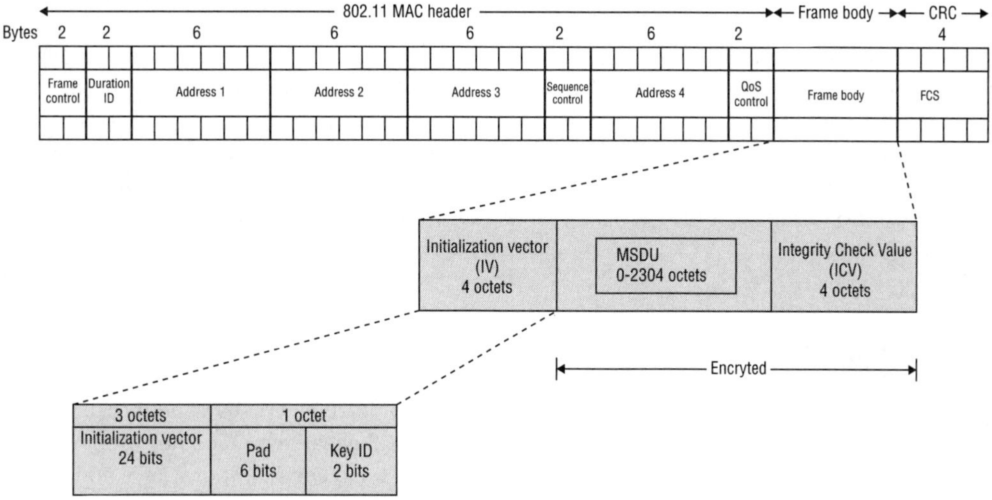

Bits, Bytes, Octets

A *bit* is a binary digit, taking a value of either 0 or 1. Binary digits are a basic unit of communication in digital computing. A byte of information comprises 8 bits. An *octet* is another name for one byte of data. The terminology of octet and byte can generally be used interchangeably.

Remember that WEP encrypts the MSDU upper-layer payload that is encapsulated in the frame body of an MPDU. The MSDU payload has a maximum size of 2,304 bytes. Because the IV adds 4 octets and the ICV also adds 4 octets, when WEP is enabled, the entire size of the body inside an 802.11 data frame is expanded by 8 bytes to a maximum of 2,312 bytes. In other words, WEP encryption adds 8 bytes of overhead to an 802.11 MPDU.

TKIP

Temporal Key Integrity Protocol (TKIP) is a security protocol developed by the Wi-Fi Alliance in 2002. It was designed to replace WEP after its encryption was broken. WPA was designed to replace WEP, however, using the existing WLAN hardware. This meant that WPA was restricted to using the existing ARC4 algorithm that was used by WEP. When TKIP was introduced, most legacy 802.11 radios could implement TKIP with a firmware upgrade, but not all legacy APs and STAs were upgradeable. WPA and TKIP were designed to be around for 5 years, allowing the 802.11i amendment time to become ratified and implemented in the industry. As you have already learned, TKIP has been phased out and is not a supported encryption method for 802.11n or 802.11ac data rates. For a brief review of the TKIP encryption process, recall that TKIP uses a 128-bit temporal key, plus a 48-bit TKIP sequence counter (TSC), and the transmit address (TA) as seeding material, mixed together in the two-phase key-mixing process. This seed is then fed through the ARC4 algorithm to generate the keystream. The keystream is then combined with the plaintext to create the encrypted ciphertext. The Message Integrity Code (MIC) is computed and added to the end of the MSDU payload as the data integrity check. TKIP encryption is covered extensively in Chapter 2. Therefore, rather than exploring the topic again in great detail, we will direct you back to Chapter 2.

 **Real World Scenario**

What Is the Difference Between TKIP and CKIP?

Prior to the ratification of the 802.11i security amendment in 2004, WLAN vendor Cisco Systems offered a prestandard proprietary version of TKIP called the *Cisco Key Integrity Protocol (CKIP)*. Cisco's proprietary enhancement of WEP also used temporal keys and a key-mixing process. A *Cisco Message Integrity Check (CMIC)* used for data integrity was designed to detect forgery attacks. Because of the extra overhead created, the CMIC was optional when CKIP was used as the encryption method. It should be noted that CKIP and CMIC only worked within a Cisco WLAN infrastructure due to the proprietary nature of the protocols.

TKIP MPDU

Figure 3.6 shows the TKIP MPDU. The first 32 bytes are the 802.11 MAC header, which does not change. The encrypted frame body is made up of five key pieces:

- IV/Key ID
- Extended IV
- MSDU payload
- MIC
- ICV

FIGURE 3.6 TKIP MPDU

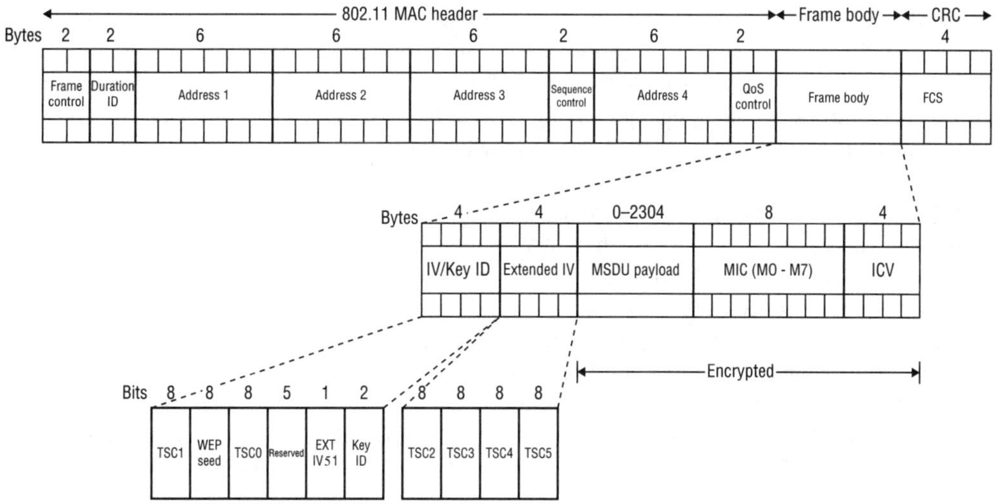

It begins with the IV/Key ID combination. This is 4 octets in size and is similar to the IV/KEY ID that is found in WEP. TSC0 and TSC1, the first 2 octets of the 48-bit TKIP sequence counter (TSC0 and TSC1), make up part of the IV/Key ID. If TKIP is being used, which is the case in this example, the Extended IV field is set to 1, indicating that an extended IV of 4 octets will follow the original IV. The Extended IV is 4 octets and is made up of the other 4 octets of the 48-bit TKIP sequence counter (TSC2 through TSC5). Both the original IV and the Extended IV are not encrypted. The 8 bytes that comprise the IV/Key ID and Extended IV could be considered a TKIP header.

After the original IV and the Extended IV comes the MSDU payload, followed by the 8 MIC octets, which are then followed by the 32-bit Integrity Check Value (ICV) that was calculated on the MPDU. The MSDU upper-layer payload as well as the MIC and ICV are all encrypted. The frame is then completed by adding the 32-bit frame check sequence (FCS) that is calculated over all the fields of the header and frame body.

Because of the extra overhead from the IV (4 bytes), Extended IV (4 bytes), MIC (8 bytes), and ICV (4 bytes), a total of 20 bytes of overhead is added to the frame body of a TKIP encrypted 802.11 data frame. When TKIP is enabled, the entire size of the frame body inside an MPDU is expanded by 20 bytes to a maximum of 2,324 bytes. In other words, TKIP encryption adds 20 bytes of overhead to an 802.11 MPDU.

CCMP

Counter Mode with Cipher-Block Chaining Message Authentication Code Protocol (CCMP) is the security protocol that was created as part of the 802.11i security amendment and was designed to replace TKIP and WEP. CCMP uses the AES block cipher instead of the ARC4 streaming cipher used by WEP and TKIP. As mentioned earlier, the IEEE 802.11-2012 standard defines two RSNA data confidentiality and integrity protocols: TKIP and CCMP, with CCMP support mandatory. CCMP is mandatory for RSN compliance. In September 2004, the Wi-Fi Alliance introduced version 2 of the Wi-Fi Protected Access certification, called *WPA2,* which requires the use of CCMP/AES encryption. Because the AES cipher is processor-intensive, older legacy 802.11 devices that only supported WEP and TKIP in most cases had to be replaced with newer hardware to support CCMP/AES encryption processing.

CCMP is made up of many components that provide different functions. Before going any further in this section, there are numerous acronyms and abbreviations relating to CCMP to which you need to be introduced. These acronyms and abbreviations are commonly used in the wireless industry and in the IEEE 802.11-2012 standard. Since CCMP is made up of many different components, it is common to reference the components individually. *CounterMode* is often represented as *CTR.* The CTR is used to provide data confidentiality. The acronym for *Cipher-Block Chaining* is *CBC.* You should also be familiar with *CBC-MAC,* which is the acronym for *Cipher-Block Chaining Message Authentication Code.* The CBC-MAC is used for authentication and integrity.

The full phrase of Counter Mode with Cipher-Block Chaining Message Authentication Code Protocol is represented by the acronym of CCMP. However, the shorter phrase of CTR with CBC-MAC is also sometimes represented by the CCMP acronym.

Some references to CCMP leave off the letter P and use the term *CCM* when referencing the block cipher and not the actual protocol. CCMP is based on the CCM of the AES encryption algorithm. CCM combines CTR to provide data confidentiality and CBC-MAC for authentication and integrity. In simpler words, the CCM process uses the same key for encrypting the MSDU payload and provides for a cryptographic integrity check. The integrity check is used to provide data integrity for both the MSDU data and portions of the MAC header of the MPDU.

CCM is used with the AES block cipher. Although it is capable of using different key sizes, when implemented as part of the CCMP encryption method AES uses a 128-bit key and encrypts the data in 128-bit blocks.

The inputs used by the CCMP encryption/data integrity process include the following:

Temporal Keys Just like TKIP, CCMP starts with a 128-bit temporal key. The 128-bit temporal key can either be a pairwise transient key (PTK) used to encrypt unicast traffic or a group temporal key (GTK) used to encrypt broadcast and multicast traffic.

Packet Number The 48-bit packet number (PN) is much like a TKIP sequence number. The PN uniquely identifies the frame and is incremented with each frame transmission. This protects CCMP from replay and injection attacks.

Nonce A *nonce* is a random numerical value that is generated one time only. A 104-bit unique nonce is constructed from the packet number (PN), priority data used in QoS, and the transmitter address (TA). Do not confuse this nonce with the nonces used during the 4-Way Handshake process described in Chapter 5.

802.11 data frame (MPDU) The frame body encapsulates the MSDU upper-layer payload that will be encrypted and protected by a MIC. The MPDU header, also known as the MAC header, will not be encrypted but is partially protected by the MIC.

AAD Additional authentication data (AAD) is constructed from portions of the MPDU header. This information is used for data integrity of portions of the MAC header. Receiving stations can then validate the integrity of these MAC header fields.

Figure 3.7 shows the CCMP encryption and data integrity process. It will be helpful to refer to this figure as you read about the steps CCMP performs.

FIGURE 3.7 CCMP encryption and data integrity process

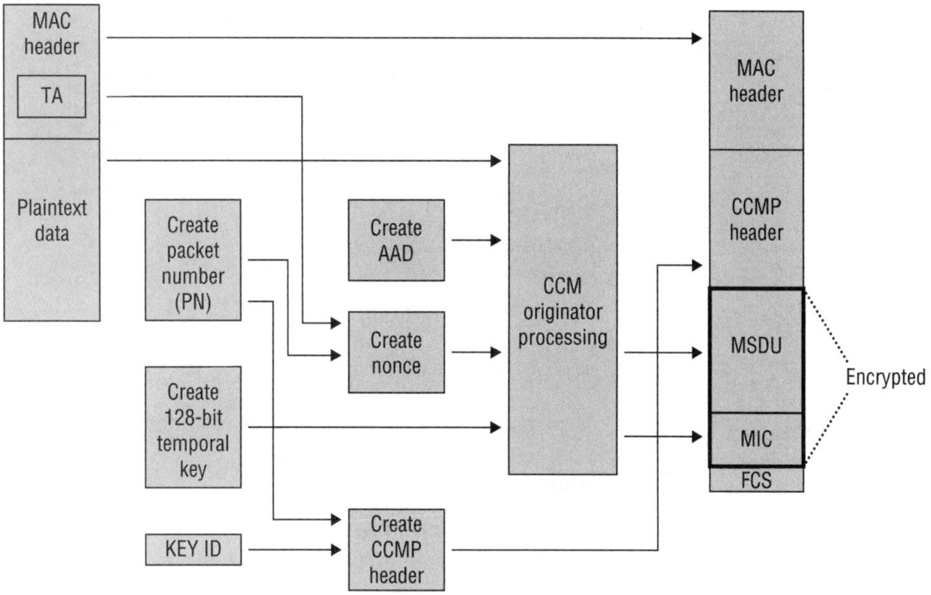

CCMP encrypts the payload of a plaintext MPDU using the following steps:

1. A 48-bit PN is created. Packet numbers increment with each individual MPDU, although they remain the same for retransmissions.
2. As shown in Figure 3.8, certain fields in the MPDU header are used to construct the AAD. The MIC provides integrity protection for these fields in the MAC header as well as for the frame body. All of the MAC addresses, including the BSSID, are protected. Portions of the other fields of the MAC header are also protected. Receiving stations will validate the integrity of these protected portions of the MAC header. For example, the frame type and the distribution bits that are subfields of the Frame Control field are protected. Receiving stations will validate the integrity of these protected portions of the MAC header. The AAD does not include the header Duration field, because the Duration field value can change due to normal IEEE 802.11 operation. For similar reasons, several

subfields in the Frame Control field, the Sequence Control field, and the QoS Control field are masked to 0 and therefore are not protected. For example, the Retry bit and Power Management bits are also masked and are not protected by CCM integrity.

FIGURE 3.8 Additional authentication data (AAD)

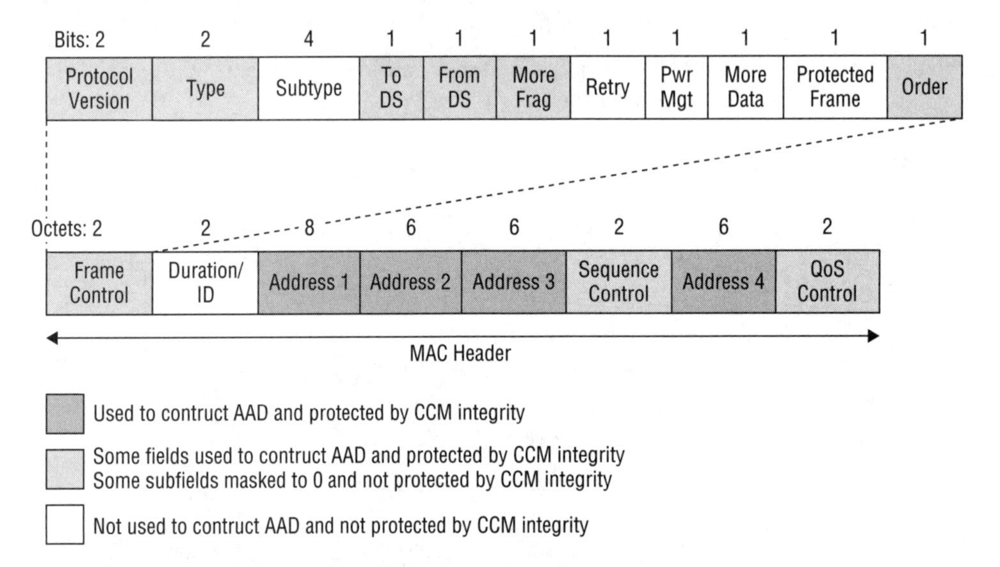

3. A nonce is created from the PN, the TA, and priority data used in QoS.
4. The 8-octet CCMP header is constructed. The CCMP header includes the Key ID and the PN, which is divided into 6 octets. You will notice that the construction of the CCMP header is basically identical to the 8-octet TKIP header.
5. The CCM module, which uses the AES clock cipher, will now be used to create a data integrity check and encrypt the upper-layer data. The 128-bit temporal key, the nonce, the AAD, and the plaintext data are then processed to create an 8-byte MIC. The MSDU payload of the frame body and the MIC are then encrypted in 128-bit blocks. This process is known as CCM originator processing.
6. The original MAC header is appended to the CCMP header, the encrypted MSDU, and the encrypted MIC. A FCS is calculated over all of the fields of the header and entire frame body. The resulting 32-bit CRC is then placed in the FCS field.

CCMP MPDU

Figure 3.9 shows the CCMP MPDU. The first 32 bytes are the 802.11 MAC header, which does not change. The frame body consists of the CCMP header, the MSDU upper-layer payload, and the MIC. The CCMP header includes the Key ID and the PN, which is divided into 6 octets. You will notice that the format of the CCMP header is basically identical to the format of the 8-octet TKIP header (IV/Extended IV). The CCMP header is not encrypted. The MSDU payload and the 8-byte MIC are encrypted.

FIGURE 3.9 CCMP MPDU

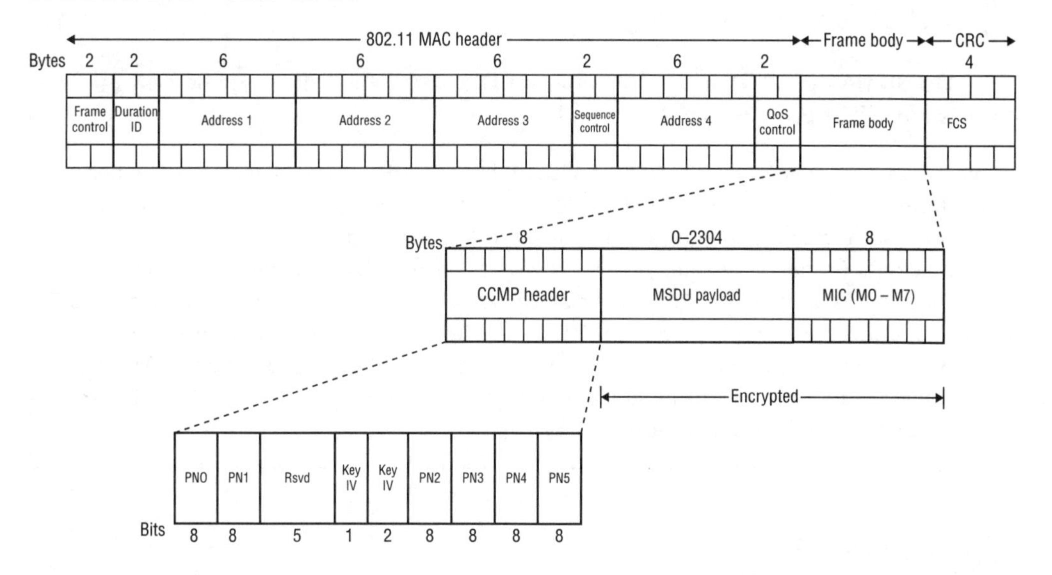

The overhead that results from CCMP encryption includes CCMP header (8 bytes) and the MIC (8 bytes). When CCMP is enabled, the entire size of the frame body inside an MPDU is expanded by 16 bytes to a maximum of 2,320 bytes. In other words, CCMP encryption adds 16 bytes of overhead to an 802.11 MPDU.

EXERCISE 3.1

CCMP Encrypted Frames

In this exercise, you will use a protocol analyzer to view 802.11 data frames encrypted with CCMP.

1. To perform this exercise, first download the following file from the book's online resource area, which can be accessed at `www.sybex.com/go/cwsp2e`:

   `CCMP_FRAMES.PCAP`

2. After the file is downloaded, you will need packet analysis software to open this file. If you do not already have a packet analyzer installed on your computer, you can download Wireshark from `www.wireshark.org`.

3. Click on one of the beacons with the source address of 00:1A:1E:94:4C:32. Typically the Packet Details pane appears in the lower section of the screen. This section contains annotated details about the selected frame. In this window, locate and expand the IEEE 802.11 Wireless LAN Management Frame section. In this section, expand the Tagged Parameters section, and then expand Tag: RSN Information. Here you will see AES (CCM) listed as the cipher suite type.

Real World Scenario

Why Do Some WLAN Protocol Analyzers Display CCMP Encrypted Data Frames as TKIP Encrypted Data Packets?

As you have already learned, the format of the 8-byte CCMP header is basically identical to the format of the 8-byte TKIP header (IV/Extended IV) used by TKIP. Therefore, some protocol analyzers cannot distinguish between TKIP-encrypted data frames and CCMP-encrypted data frames. However, you can always determine which cipher is being used by looking at a field called the *RSN information element*. The RSN information element is found in four different 802.11 management frames: beacon management frames, probe response frames, association request frames, and reassociation request frames. More information about the RSN information element can be found in Chapter 5, "802.11 Layer 2 Dynamic Encryption Key Generation."

WPA/WPA2

Prior to the ratification of the 802.11i amendment, the Wi-Fi Alliance introduced the Wi-Fi Protected Access (WPA) certification. WPA was a snapshot of the not-yet-released 802.11i amendment, but only supported TKIP/ARC4 dynamic encryption key generation. 802.1X/EAP authentication was required in the enterprise, and passphrase authentication was required in a SOHO environment. TKIP was designed as a stopgap measure, with an expected limited lifespan (5 years), until 802.11i was finalized. Numerous publicly announced attacks have shown that there are flaws in TKIP and that it is able to be exploited. The Beck-Tews attack can recover plaintext from an encrypted short packet, recover the MIC key, and inject forged frames. However, the attack has a limitation in that the targets are restricted to WPA implementations that support WMM QoS features. The Ohiagi/Morii attack further enhances the Beck-Tews attack with a man-in-the-middle-approach. It should be noted that these attacks do not recover the encryption key but instead are used to recover the MIC checksum used for packet integrity.

These exploits and others can usually be prevented by changing TKIP settings, such as keying intervals on a WLAN controller or AP; however, the proper solution is to stop using TKIP and upgrade to CCMP with AES. TKIP has provided WLAN data privacy for more than the five years for which it was intended. WLANs should now be protected with CCMP to provide the necessary data privacy and data integrity.

The further migration from TKIP to CCMP can be seen in the IEEE 802.11n amendment, IEEE 802.11ac amendment, and IEEE 802.11-2012 standard, which all state that *High Throughput (HT)* or *Very High Throughput (VHT)* data rates are not allowed to be used if WEP or TKIP is enabled. The Wi-Fi Alliance also began requiring that all HT radios not use TKIP when using HT or VHT data rates. Starting on September 1, 2009, the Wi-Fi Alliance began testing 802.11n APs and client STAs for compliance with this requirement. This compliance was automatically applied to VHT APs and client STAs when

802.11ac was ratified. WLAN vendors still need to provide support for TKIP and WEP, but this support will only be with data rates up to 54 Mbps.

WPA2 is a Wi-Fi Alliance certification that is a mirror of the IEEE 802.11i security amendment. Testing of WPA2 interoperability certification began in September 2004. WPA2 incorporates the AES algorithm in CCMP, providing government-grade security based on the NIST FIPS 140-2 compliant AES encryption algorithm. WPA2 supports 802.1X/EAP authentication or preshared keys, and is backward-compatible with WPA.

Future Encryption Methods

As with any active technology, enhancements and advancement of the technology is commonplace, especially when security is involved. Counter Mode with CBC-MAC Protocol (CCMP) is at the core of the WPA2 certification. CCMP with AES was originally implemented with 54 Mbps WLANs and remained essentially unchanged with the introduction of newer PHY technologies, such as 802.11n and 802.11ac.

Cryptographic suites consist of much more than just the encryption algorithm. CCM defines a specific way of using AES encryption. CCM divides the data that is being encrypted into 16 byte blocks or chunks, and then links or chains them together. The ciphertext output of the first chunk is used as part of the encryption of the second chunk, which is then used as part of the encryption of the third chunk, and so on. This is where the term *cipher block chaining (CBC)* comes from. Additionally, due to technical reasons, CCMP requires that two AES operations be performed for each chunk. The sequential aspect of this process, along with the dual processes, requires a large number of AES operations, leading to a concern as to whether CCMP will be capable of keeping up with the faster WLAN gigabit rates that are being introduced.

Galois/Counter Mode (GCM) is significantly more efficient and faster than CCM. Like CCM, GCM uses the same AES encryption algorithm, however it is applied differently. GCM only needs a single AES operation per block, immediately reducing the encryption process by half. Additionally, GCM does not link or chain the blocks together. Since each block is not dependent on the previous block, they are independent of each other, and can be processed simultaneously using parallel circuits.

Galois/Counter Mode Protocol (GCMP) is not backward-compatible with existing Wi-Fi equipment and thus requires new hardware. GCMP has already been specified in the 802.11ad amendment, and is optional in 802.11ac. It will likely be seen in other technologies in the future.

Another technology, *Suite B*, is a suite of cryptographic algorithms specified by the National Institute of Standards and Technology (NIST) and is approved by the U.S. National Security Agency (NSA) for protecting classified and unclassified National Security Systems. Suite B consists of four encryption algorithms:

- Advanced Encryption Standard (AES) Block Encryptions with key sizes of 128 or 256 bits used with either with Counter Mode (CTR) or Galois/Counter Mode (GCM)
- Elliptic Curve Digital Signature Algorithm (ECDSA)

- Elliptic Curve Diffie-Hellman (ECDH) Key Exchange
- Secure Hash Algorithm (SHA) using SHA-256 and SHA-384

Suite B was announced in 2005. As of this writing, only a couple of WLAN vendors have provided Suite B–capable products.

Proprietary Layer 2 Implementations

In addition to the security and encryption standards, it is worthwhile to mention that there have also been vendor-specific products that have provided authentication and encryption for WLANs. Since these vendor-specific products have been based on industry-standard security and authentication technologies, they are better referred to as proprietary implementations rather than as proprietary solutions. These proprietary implementations required the installation of custom supplicant software on the client, which then used authentication and encryption standards to communicate with the authenticator and authentication server on the network. The proprietary implementations historically provided higher levels of security than was found on a typical 802.11 network. These products were designed and marketed to organizations that needed extremely high levels of security, usually government and military agencies. Most proprietary products were FIPS 140-2 compliant and incorporated standards such as AES 128, 192, and 256.

xSec was a Layer 2 protocol that provided a unified framework for securing both wired and wireless connections. xSec was jointly developed by Aruba Networks and Funk Software, a division of Juniper Networks. xSec implemented a FIPS-compliant mechanism for providing identity-based security. xSec used the AES-CBC-256 with HMAC-SHA1 algorithm, which provided for a 256-bit encryption key and even stronger data integrity. It was based on the IEEE 802.1X framework and supported many versions of EAP. xSec provided an extremely secure connection from the client, through the access point, all the way to an Aruba Networks WLAN controller.

Fortress Technologies was another company that used proprietary client software to enable secure communications between the wireless device and the Fortress Technologies controller or bridge, which was either connected to the wired network or provided wireless backhaul. The Fortress Technologies solution also used proprietary Fortress encryption that could use a 256-bit AES encryption key and stronger data integrity.

Summary

In this chapter, you learned about five different encryption ciphers: RC4, RC5, DES, 3DES, and AES. We discussed how symmetric and asymmetric algorithms function, as well as how stream and block ciphers are used to process plaintext data into ciphertext. We also covered the three different encryption technologies that are used in the 802.11 standard:

- WEP uses a 24-bit IV and a static key as the seed that is fed through the ARC4 algorithm to generate a keystream. The keystream is combined with the plaintext data bits to create the ciphertext. The ICV is calculated and appended as the data integrity check.
- TKIP uses a 128 bit temporal key + 48-bit TSC + TA address as the seeding material mixed together in the two-phase key-mixing process. The seed is then fed through the ARC4 algorithm to generate the keystream that is combined with the plaintext data bits to create the ciphertext. The MIC is also calculated and added as the data integrity check.
- CCMP uses a 128-bit temporal key + the AAD + a nonce as the seed for the AES block cipher. No key mixing is needed due to the strength of the AES algorithm. Data is encrypted in 128-bit blocks. A MIC is also calculated and added for integrity. The CCMP MIC is stronger than the TKIP MIC.

The frame formats of each of these encryption technologies were reviewed. The relationship between the Wi-Fi Alliance certifications (WPA/WPA2) and the 802.11 encryption technologies was also discussed.

A couple of new encryption technologies were also introduced and explained: Galois/Counter Mode Protocol (GCMP) and Suite B.

Exam Essentials

Know the two categories of algorithms. Understand the differences between symmetric and asymmetric algorithms and how they are used when encrypting and decrypting data.

Know the process of how public key and private key encryption works. Understand that the public key is available to anyone and is used to encrypt the data, whereas the private key is kept secret and is used to decrypt the data.

Describe stream and block ciphers. Explain the process used by stream and block ciphers to encrypt the plaintext, along with how a Boolean XOR is incorporated in the process.

Define RC4, ARC4, RC5, DES, 3DES, and AES. Be able to explain the differences and similarities of all six of these algorithms.

Define WEP, TKIP, CCMP, and GCMP. Be able to explain the differences and similarities between these four security protocols. Understand the individual encryption and data integrity processes associated with each security protocol.

Describe the frame MPDU format. Explain the differences and similarities between the MPDU frame formats of the three 802.11 security protocols. Explain why the frame sizes vary and which portions of the frame are actually encrypted.

Explain the relationship between WPA/WPA2 and 802.11. Understand how the 802.11 standard relates to the Wi-Fi Alliance certification. Know which 802.11 security protocols are required for the Wi-Fi Alliance certifications.

Review Questions

1. CCMP/AES encryption adds an extra ________ of overhead to the body of an 802.11 data frame.
   - A. 16 bytes
   - B. 12 bytes
   - C. 20 bytes
   - D. 10 bytes
   - E. None of the above
2. TKIP/ARC4 encryption adds an extra ________ of overhead to the body of an 802.11 MPDU.
   - A. 16 bytes
   - B. 12 bytes
   - C. 20 bytes
   - D. 10 bytes
   - E. None of the above
3. An HT client STA is transmitting to an HT AP using modulation and coding scheme (MCS) #12 that defines 16-QAM modulation, two spatial streams, a 40-MHz bonded channel, and an 800 ns guard interval to achieve a data rate of 162 Mbps. According to the IEEE, which types of encryption should be used by the HT client STA? (Choose all that apply.)
   - A. Static WEP
   - B. Dynamic WEP
   - C. TKIP/ARC4
   - D. CCMP/AES
   - E. All of the above
4. CCMP/AES uses a ________ temporal key and encrypts data in ________ blocks.
   - A. 128-bit, 128-bit
   - B. 128-bit, 192-bit
   - C. 192-bit, 192-bit
   - D. 192-bit, 256-bit
   - E. 256-bit, 256-bit
5. When using an encryption suite that implements an asymmetric algorithm, which of the following statements is true? (Choose all that apply.)
   - A. Both the encrypting and decrypting parties share the same key.
   - B. Asymmetric algorithms generally are faster than symmetric algorithms.
   - C. Asymmetric algorithms generally are slower than symmetric algorithms.
   - D. Public and private encryption keys must be generated.

6. Andy calls the help desk for assistance with sending an encrypted message to Chris. Without knowing what type of security protocol and encryption Andy and Chris are using, which of the answers here could make the following scenario true?

   In order for Andy to send an encrypted message successfully to Chris, Andy is told to enter ____________ on his computer; Chris needs to enter ____________ on his computer. (Choose all that apply.)

   A. A public key, the companion private key

   B. A private key, the companion public key

   C. An asymmetric key, the same asymmetric key

   D. A symmetric key, the same symmetric key

7. The IEEE 802.11-2012 standard states which of the following regarding 802.11n data rates and encryption? (Choose all that apply.)

   A. WEP and TKIP must not be used.

   B. CCMP can only be used.

   C. WEP cannot be used; however, TKIP can be used if also using 802.1X.

   D. Any encryption defined by the standard is allowed.

8. Given that CCMP uses a MIC for data integrity to protect the frame body and portions of the MAC header, what information needs to be constructed to protect certain fields in the MAC header?

   A. Nonce

   B. Extended IV

   C. ICV

   D. AAD

   E. PN

   F. IV

9. Which of the following is a FIPS encryption standard that uses a single 56-bit symmetric key? (Choose all that apply.)

   A. RC4

   B. RC5

   C. DES

   D. 3DES

   E. AES

10. CCMP is an acronym made up of multiple components. Which of the following is an expanded version of this acronym? (Choose all that apply)

    A. Counter Mode with Cipher-Block Chaining Message Authentication Code Protocol

    B. Counter Message with Cipher-Block Chaining Mode Authentication Code Protocol

    C. CTR with CBC-MAC Protocol

D. Counter Mode with CBC-MAC Protocol

E. None of these is accurate.

11. Which of the following is a random numerical value that is generated one time only and is used in cryptographic operations? (Choose all that apply.)

A. Pseudo-random function (PRF)

B. One-time password (OTP)

C. Single sign-on (SSO)

D. Throw-away variable (TV)

E. Nonce

12. Which of the following are Layer 2 encryption methods defined by the 802.11-2012 standard? (Choose all that apply.)

A. WEP

B. WPA

C. WPA2

D. TKIP

E. CCMP

F. GCMP

13. Which of the following encryption methods use symmetric algorithms? (Choose all that apply.)

A. WEP

B. TKIP

C. Public-key cryptography

D. CCMP

14. The Rijndael algorithm was the foundation for which of the following ciphers?

A. TKIP

B. DES

C. AES

D. CCMP

E. 3DES

15. A data integrity check known as Message Integrity Code (MIC) is used by which of the following? (Choose all that apply.)

A. WEP

B. TKIP

C. CCMP

D. AES

E. DES

16. Given that additional authentication data (AAD) is constructed from portions of the MPDU header and that the information is used for data integrity, which fields of the MAC header comprise the AAD? (Choose all that apply.)

    A. Frame Control field
    B. Transmitter address
    C. Sequence Control field
    D. Receiver address
    E. Destination address
    F. BSSID

17. Which of the following are encryption algorithms specified by the IEEE 802.11-2012 standard to be used for data encryption? (Choose all that apply.)

    A. ARC4
    B. RC5
    C. IPsec
    D. DES
    E. 3DES
    F. AES

18. The CCMP header is made up of which of the following pieces? (Choose all that apply.)

    A. PN
    B. TTAK
    C. TSC
    D. Key ID
    E. MIC

19. 3DES has effective key sizes of how many bits? (Choose all that apply.)

    A. 56
    B. 64
    C. 112
    D. 128
    E. 168
    F. 192

20. AES supports three key lengths of 128, 192, and 256. The number of rounds performed for AES-128 is _____, for AES-192 is _____, and for AES-256 is _____. (Choose all that apply.)

    A. 8
    B. 10
    C. 12
    D. 14
    E. 16

Chapter 4

802.1X/EAP Authentication

IN THIS CHAPTER, YOU WILL LEARN ABOUT THE FOLLOWING:

✓ **WLAN authentication overview**

✓ **AAA**

- Authentication
- Authorization
- Accounting

✓ **802.1X**

- Supplicant
- Authenticator
- Authentication server

✓ **Supplicant credentials**

- Usernames and passwords
- Digital certificates
- Protected Access Credentials (PACs)
- One-time passwords
- Smart cards and USB tokens
- Machine authentication

✓ **802.1X/EAP and certificates**

- Server certificates and root CA certificates
- Client certificates

✓ **Shared secret**

✓ **Legacy authentication protocols**

- PAP
- CHAP

- MSCHAP
- MSCHAPv2

✓ **EAP**

- Weak EAP protocols
- EAP-MD5
- EAP-LEAP
- Strong EAP protocols
- EAP-PEAP
- EAP-TTLS
- EAP-TLS
- EAP-FAST
- Miscellaneous EAP protocols
- EAP-SIM
- EAP-AKA

This chapter discusses the key concepts, components, and methods involved in WLAN authentication. You will learn about authentication, authorization, and accounting (AAA); what roles are played in the authentication process; 802.1X; and all of the EAP methods you will encounter in the real world that will assist you in securing your enterprise WLAN. 802.1X/EAP authentication, in particular, can be a difficult topic to grasp. This chapter will describe how EAP authentication works within an 802.1X network access control framework when used in enterprise WLANs. As we introduce each topic, we will provide real-world examples and design principles to help solidify each major concept.

WLAN Authentication Overview

WLAN authentication, in all of its many flavors, is what needs to occur before an individual or a device is allowed to access network resources. Authentication is the verification of users' identity and credentials. Users must identify themselves and present credentials, such as usernames and passwords or digital certificates. Systems with higher levels of authentication use multifactor authentication, which requires at least two sets of different credentials to be presented. In a nutshell, authentication is based on *who* you are. Furthermore, *who* you are may indeed require more than one identifying element, but more on that later. After you successfully authenticate (via whatever method), encryption can then take place, but an important distinction must be made; authentication should be considered mutually exclusive from encryption. In Chapter 5, "802.11 Layer 2 Dynamic Encryption Key Generation," you will learn that the authentication process provides the seeding material to create the necessary encryption keys, but conceptually it is still a separate concept and process. While the encryption process is a by-product of the authentication process, understand that the goals of authentication and encryption are very different. Authentication provides mechanisms for validating user and device identity; encryption provides mechanisms for data privacy and confidentiality.

Since authentication requires proving who you are, you must present *credentials* that fall into three categories:

- Something you know
- Something you have
- Something you are

If multiple credentials are provided for validating identity, greater trust can be assigned to the identity of the party requesting access to the WLAN. Requiring multiple credentials to be presented for user validation is known as *multifactor authentication*. Requiring two sets of credentials to be presented for user validation is often called *two-factor authentication*. Some of the EAP authentication methods you will learn about in this chapter are capable of two-factor authentication.

For example, consider your bank debit card. When you insert your physical card (something you have) into the ATM machine, it then requests your PIN (something you know). Some banks even have a picture of the card owner imprinted on the card. The ATM machine can't verify that, but what about the cashier clerk at your local supermarket? It immediately tells the clerk that when you present the card that it is indeed yours (and not someone else's). The picture on the debit card immediately ties together for the clerk both something you *are* and something you *have*. If the debit card was used along with the PIN to complete a debit transaction and the clerk verifies your face with the picture identity on your card, three forms of credentials will have been used for authentication.

Biometrics, such as fingerprint, retina, and other biological verification methods, can also be used as *something you are* to prove your identity. Although biometrics are not usually employed in WLAN security authentication, biometric devices have become commonplace as built-in components in laptops, tablets, and smart phones. Even facial recognition, using the built-in video cameras in laptop displays, is being discussed as an additional form of authentication.

As these technologies develop, you can rest assured that yet another new authentication standard will emerge. The important thing to remember is that it will be just another authentication method using the principles presented in this chapter.

AAA

While we have already addressed authentication at a conceptual level, who or what is the one performing the authorization? Can we also have a record of this activity? In fact, how about having a record of when this authorization activity occurred, how long the user was on the network, where the user went, and whether the user is either still online or has ended their session? *Authentication, authorization, and accounting (AAA)* is a common computer security concept that defines the protection of network resources.

As we have already discussed, *authentication* is the verification of user identity and credentials. Users must identify themselves and present credentials, such as usernames and passwords or digital certificates. More secure authentication systems use multifactor authentication, which requires at least two sets of different credentials to be presented.

Authorization involves granting access to network resources and services. Before authorization to network resources can be granted, proper authentication must occur.

Accounting is tracking the use of network resources by users and devices. It is an important aspect of network security, used to keep a paper trail of who used what resource, when, and where. A record is kept of user identity, which resource was accessed, and at what time. Keeping an accounting trail is often a requirement of many industry regulations, such as the U.S. Health Insurance Portability and Accountability Act of 1996 (HIPAA).

Authentication

Let's discuss authentication in a more technical manner now. What can we take advantage of in order to prove the identity of someone wanting to gain access to your precious WLAN and, therefore, potentially all the data you consider secure? Answers to that question will vary depending on the level of concern over the access to that network and thus the data that resides on it. Understanding this concept is critical from a design perspective when you are being consulted to help an enterprise decide how to best perform authentication.

The following are common examples of authentication credentials used in enterprise WLANs today:

- Usernames and passwords
- Digital certificates
- Dynamic/one-time passwords (for example, RSA SecurID)
- Smart cards or credentials stored on USB devices
- Machine authentication (based on an embedded machine identity)
- Preshared keys (PSK)

A bank system processing billions of dollars every second would have very different security concerns than a small company that sells, say, car tires. Both may consider their data secure, but there is additional cost and/or complexity in both deploying and operating stronger authentication types such as two- and three-factor authentication.

Two-factor authentication would function just as in our previous example where we are using the bank ATM with a debit card (first factor = something you have) and having to enter our PIN (second factor = something you know). A specific WLAN security example of two-factor authentication would be using a computer (a computer object) that is part of a Microsoft *Active Directory (AD)* and then having to use a Windows AD user account to log in from that computer. The purpose is to ensure access is only granted to your WLAN by being from both a valid and current AD computer and AD user account. For example, if you brought in your laptop from home that isn't known by AD, the entity performing the authorization will not let you on the network regardless of your valid AD user credentials because you are not accessing the network from a valid "enterprise asset."

Real World Scenario

Cost vs. Security

A warehouse company that requires a barcode system (which is only sending quantities and item numbers electronically) to operate over the WLAN has requested your expertise. The network the system will reside on already has access control mechanisms (firewalls and/or ACLs) that prevent access to any other network or device beyond where the barcode devices can communicate.

In your consultation with the IT director of the warehouse company, she has asked for the most secure method possible to be employed for the company's WLAN. She claims to have a large budget, if necessary, but has limited support staff. You also learn that the operators of the barcode devices have limited computer skills and will likely require a lot of training. Therefore, simplicity of security is a major requirement.

Every salesperson with a quota would cringe hearing this, but this is when it's time to balance cost and complexity. A two-factor authentication system would not be a good choice in this case because of the extra burden and cost involved in a high-end security infrastructure. Two-factor authentication systems would provide an unnecessary operational burden to the end users of the system and the technical staff as well.

People will typically opt to have their network secure as long as it does not become overly burdensome. However, at some level, the concern for security takes a front seat over the amount of burden an end user will have to endure to use the WLAN. In one of our previous examples, not only does the bank processing billions of dollars per second have concerns over the direct financial impact to the viability of its core business should somebody have access to manipulate this data, but there are also regulatory requirements such as PCI that dictate the minimum requirements to which the bank must adhere. Furthermore, there may also be trade secrets that could be gleaned from this data that competitors would want to access. Quite literally, a security breach of this network might mean the end of the bank's business. The bank scenario is an example of when multifactor authentication may be required regardless of the additional burden placed on the end users. Later in this chapter, all of the commonly used and available enterprise EAP authentication methods will be discussed in detail.

Authorization

When you log into your email account, you will likely enter a username and a password. Who is authorizing your username and password? This is conceptually the email application server itself. The difference with WLAN security is that there are multiple applications being used via the WLAN portal, so some sort of overlay authorization solution is needed to validate user identity. An 802.11 WLAN normally serves as a portal to preexisting wired network resources, such as a corporate server farm. Authorization is about properly protecting network resources. Authorization allows authenticated users access to network resources, but users who cannot provide the proper authentication credentials will not be authorized. Authorization should be considered as a framework in which proper authentication can occur. Enterprise WLANs should use an 802.1X authorization framework.

In WLANs, a RADIUS server is typically the entity performing the authorization from the WLAN hardware's perspective. *Remote Authentication Dial-in User Service (RADIUS)* is a networking protocol that provides AAA capabilities for computers to connect to and use network services. RADIUS authentication and authorization is defined in IETF RFC 2865. Accounting is defined in IETF RFC 2866. RADIUS servers are sometimes referred to as AAA servers. While a RADIUS server may actually communicate with other systems like Windows Active Directory or an LDAP server, from the WLAN's perspective it is convenient to think of the RADIUS server as the single authorization entity. You will, in fact, find that many WLAN vendors offer RADIUS server capabilities directly in their access points and WLAN controllers. A deeper discussion on RADIUS deployment and scalability will be covered in Chapter 9, "RADIUS and LDAP."

The IEEE 802.11-2012 standard does not dictate the use of a RADIUS server. However, the IEEE 802.11-2012 WLAN standard does dictate the use of the IEEE 802.1X-2004 standard for authentication and port control within an enterprise *robust security network (RSN)*. 802.1X is a *port-based access control* standard that defines the mechanisms necessary to authenticate and authorize devices to network resources. This was a clever and logical way for IEEE 802.11 designers to leverage the use of a preexisting and popular standards-based authorization method and inherit all its strengths and benefits. RADIUS servers are usually one of the main components of an 802.1X authorization framework.

The original IEEE 802.1X-2004 standard has been updated as the IEEE 802.1X-2010 Port-Based Network Access Control standard. The 802.1X-2010 standard can be downloaded from this URL: `http://standards.ieee.org/getieee802/802.1.html`. Clause 11.5.8 of the IEEE 802.11-2012 WLAN standard defines how 802.1X mechanisms are used for authentication and port control within an 802.11 WLAN. These mechanisms will be described in detail in this chapter. Clause 11.5.12 of the IEEE 802.11-2012 WLAN standard defines how 802.11 and 802.1X mechanisms are used together to provide for robust secure key management. Key management and creation is discussed in great detail in Chapter 5.

Accounting

When you purchase a gallon of milk from the supermarket with a check, many accounting records are generated from this transaction. For example, if you wrote the check, it is always wise to write down the check number in your check ledger, as well as the date, merchant, and amount of the check you just wrote. Better yet, much of the manual effort can be eliminated by using a checkbook with carbon copies. Next, the merchant will make their own accounting records, and then the bank, and so on. We can refer to this as the *accounting trail*.

An accounting trail is considered to be the Holy Grail if you are ever audited by the Internal Revenue Service (IRS), the U.S. government agency that ensures proper tax contributions. An IRS agent can follow the entire transaction from beginning to end with a proper accounting trail.

In the case of WLANs, an accounting record is also quite useful in security forensics in attempting to analyze a security breach or even when evaluating normal network activities. An accounting server will typically contain information such as the user account logged in, where it was logged in from (the actual AP or the controller), the time the transaction started, the amount of traffic sent and received by the user, when the user logged off, and so forth. Figure 4.1 shows an example of an accounting record from an accounting server.

FIGURE 4.1 Accounting trail

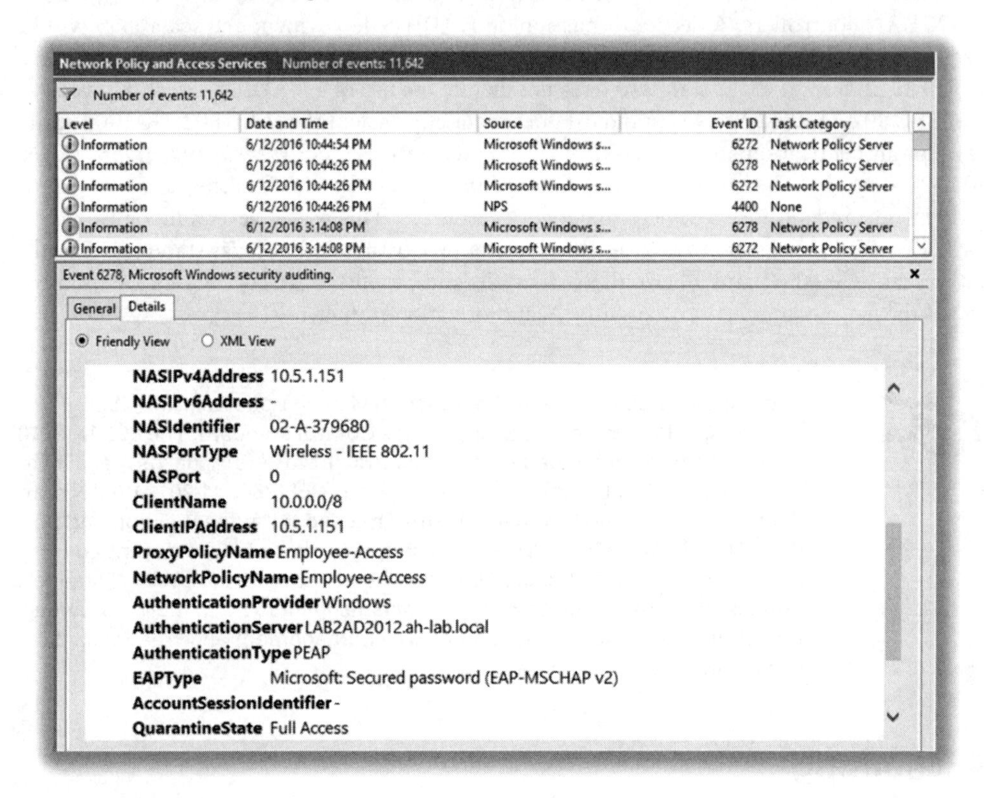

RADIUS accounting records typically contain the following elements at a minimum:

- Date
- Time
- Username
- RADIUS group
- Accounting status type
- Accounting session ID
- Accounting session time
- Service type
- Framed protocol
- Input octets

- Output octets
- Input packets
- Output packets
- Framed IP address
- NAS port
- NAS IP address
- Attribute value pairs

Vendors make use of their own *attribute value pairs (AVPs)* that can be captured by a RADIUS accounting service. Consult the manual for your particular RADIUS accounting system if there is other information not included in the standard reporting that you may want to have included.

Together, authentication, authorization, and accounting are commonly referred to as AAA. You will also find RADIUS servers typically referred to as AAA servers as most RADIUS servers perform the role of all three services. RADIUS accounting is defined in RFC 2866.

From a WLAN hardware device's perspective, the authentication and authorization is typically broken off separately in the device configuration from the accounting server.

802.1X

The IEEE 802.1X-2004 standard is not specifically a wireless standard and is often mistakenly referred to as 802.11x. As mentioned earlier, the 802.1X standard is a port-based access control standard. The 802.1X-2001 standard was originally developed for 802.3 Ethernet networks. Later, 802.1X-2004 provided additional support for 802.11 wireless networks and Fiber Distributed Data Interface (FDDI) networks. The current version of the port-based access control standard, 802.1X-2010, defined further enhancements. 802.1X provides an authorization framework that allows or disallows traffic to pass through a port and thereby access network resources. An 802.1X framework may be implemented in either a wireless or wired environment. The 802.1X authorization framework consists of three main components, each with a specific role. These three 802.1X components work together to make sure only properly validated users and devices are authorized to access network resources. A Layer 2 authentication protocol called *Extensible Authentication Protocol (EAP)* is used within the 802.1X framework to validate users at Layer 2. EAP will be discussed in detail later in this chapter. The three major components of an 802.1X framework are as follows:

Supplicant A host with software that is requesting authentication and access to network resources. Each supplicant has unique authentication credentials that are verified by the authentication server. In a WLAN, the supplicant is often the laptop or wireless handheld device trying to access the network.

Authenticator A device that blocks or allows traffic to pass through its port entity. Authentication traffic is normally allowed to pass through the authenticator, while all other traffic is blocked until the identity of the supplicant has been verified. The authenticator maintains two virtual ports: an *uncontrolled port* and a *controlled port*. The uncontrolled port allows EAP authentication traffic to pass through, whereas the controlled port blocks all other traffic until the supplicant has been authenticated. In a WLAN, the authenticator is usually either an AP or a WLAN controller.

Authentication Server A server that validates the credentials of the supplicant that is requesting access and notifies the authenticator that the supplicant has been authorized. The authentication server will maintain a native database or may proxy query with an external database, such as an LDAP database, to authenticate the supplicant credentials. A RADIUS server normally functions as the authentication server.

You will see this terminology repeatedly over the course of your reading and hands-on work in WLAN security both inside this study guide and throughout industry publications. Each of these 802.1X components will now be discussed in further detail in the sections that follow.

Supplicant

The *supplicant* is the device that will need to be validated by the authentication server before being allowed access to network resources. The supplicant will use an EAP protocol to communicate with the authentication server at Layer 2. The supplicant will not be allowed to communicate at the upper Layers of 3–7 until the supplicant's identity has been validated at Layer 2 by the authentication server. Once again, this EAP authentication process will be described in great detail later in this chapter.

Think of the supplicant as the client software on a Wi-Fi device where the WLAN client security is configured. This isn't to be confused with the *driver* for the 802.11 radio of the device. The supplicant is a software application that performs the 802.1X endpoint services on a client device such as a laptop or smart phone. Fully featured enterprise supplicants can offer support for the wired Ethernet adapter and perhaps multiple 802.11 network adapters if necessary. Different types of supplicant client utility software exist, including

- Integrated OS supplicant
- Vendor-specific supplicant
- Third-party supplicant

The software interface that is most widely used to configure a Wi-Fi radio is usually the integrated operating system Wi-Fi client utilities. The client utilities are where the 802.1X supplicant security settings are defined. Laptop users will most likely use the Wi-Fi NIC configuration interface that is a part of the OS running on the laptop. The client software utilities are different depending on the OS of the laptop being used. The Wi-Fi client utilities also vary between different versions of operating systems, but they all support 802.1X supplicant capabilities. Support for 802.1X/EAP configuration in the supplicant software has improved over the years. For example, the Wi-Fi client utility in Windows 7 is much improved and drastically different from the client utility found in Windows XP. The Windows 8 client utility is different than the Windows 7 client utility. Figure 4.2 shows the most recent supplicant interface found in Windows 10. The Mac OS X 10.6 (Snow Leopard) client utility is different from the Mac OS X 10.11 (El Capitan) client utility. The operating systems of handheld devices usually also include some sort of Wi-Fi client utility that includes the supplicant security settings integrated into the OS of the device.

FIGURE 4.2 Integrated OS supplicant for Windows 10

Vendor-specific software supplicants are sometimes available for use instead of an integrated operating system software interface. SOHO client utilities are usually simplistic in nature and are designed for ease of use for the average home user. The majority of vendor-specific software utilities are for peripheral device WLAN radios such as a USB 802.11 radio. The use of vendor-specific client utilities has decreased dramatically in recent years as the use of peripheral Wi-Fi radios has also declined. Enterprise-grade vendor client utilities provide the software interface for the more expensive enterprise-grade vendor radios.

In the past, third-party 802.1X supplicant client utilities often brought the advantage of supporting many different EAP types, giving a WLAN administrator a wider range of security choices. The main disadvantage of third-party client utilities is that they cost extra money. Because integrated client utilities have improved over the years, the use of third-party Wi-Fi client utilities as supplicants has declined. As shown in Figure 4.3, the supplicant software is where all the 802.1X security settings are defined. Later in this chapter, you will learn about EAP protocols. The supplicant software must support the same EAP protocol as the RADIUS server. The majority of the EAP protocols require the use of X.509 digital certificates, which must be defined within the supplicant configuration. A root certificate from a trusted *certificate authority (CA)* must be designated, as shown in Figure 4.4. Some flavors of EAP also make use of client-side certificates, which must be defined in the supplicant configuration. Certificates and their role within 802.1X/EAP will be discussed in greater detail later in this chapter.

FIGURE 4.3 Supplicant EAP configuration

FIGURE 4.4 Supplicant Root CA certificate

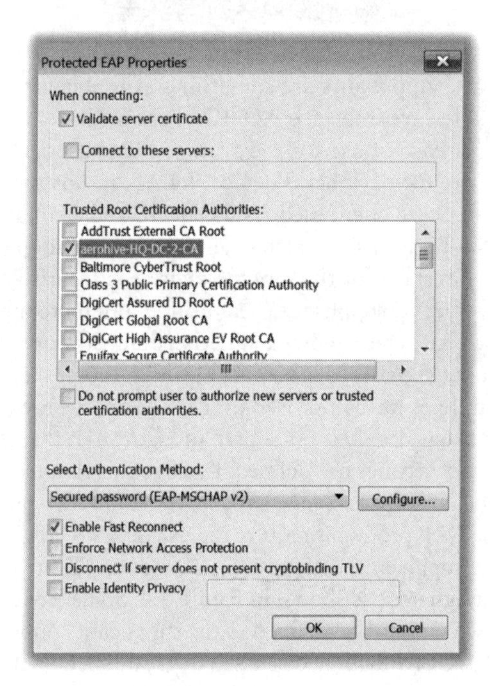

Finally, the user security credentials are also configured within the supplicant software. As shown in Figure 4.5, supplicant software will allow for single sign-on capabilities for domain users in Active Directory. Although username and password credentials are what are normally validated within the 802.1X/EAP process, machine credentials can also be designated for validation with some supplicants. Remember that the role of the supplicant is to request network access within an 802.1X authorization framework. You will learn later in this chapter that the authentication server validates the supplicant credentials using a Layer 2 EAP authentication protocol.

FIGURE 4.5 Supplicant user credentials

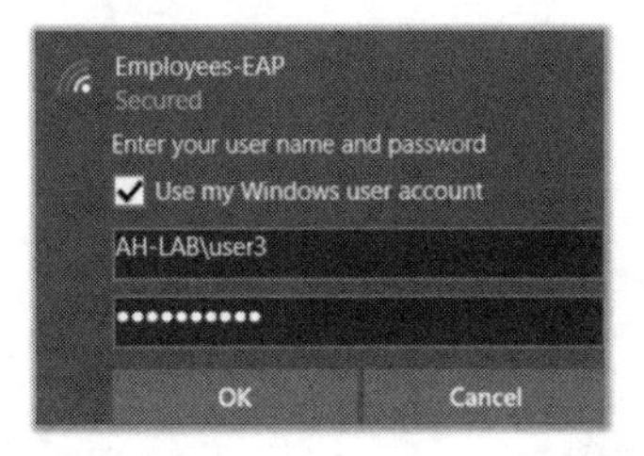

Authenticator

From the context of EAP authentication, the role of the authenticator is quite simple. The authenticator plays the role of the intermediary, passing messages between the supplicant and the authentication server. These messages travel via an EAP authentication protocol. Remember that authenticator is an 802.1X term. Also remember that 802.1X was described as a port-based access control standard. 802.1X essentially blocks traffic until a successful Layer 2 EAP authentication occurs. As mentioned earlier, the authenticator maintains two virtual ports: an uncontrolled port and a controlled port. The uncontrolled port only allows EAP authentication traffic to pass through, while the controlled port blocks all other traffic until the supplicant has been authenticated. EAP will be discussed later in this chapter, so don't worry about how it works at this point.

As shown in Figure 4.6, a standalone access point functions as the authenticator and the authentication server is typically a RADIUS server. Figure 4.7 shows that when an 802.1X security solution is used with a WLAN controller solution, the WLAN controller is the authenticator—and not the controller-based access points. In either case, directory services are often provided by a *Lightweight Directory Access Protocol (LDAP)* database that the RADIUS server queries. Active Directory would be an example of an LDAP database that is queried by a RADIUS server. Note that some WLAN vendors offer solutions where either a standalone AP or a WLAN controller can dual-function as a RADIUS server and perform direct LDAP queries, thus eliminating the need for an external RADIUS server.

FIGURE 4.6 802.1X comparison-standalone vs. controller-based APs

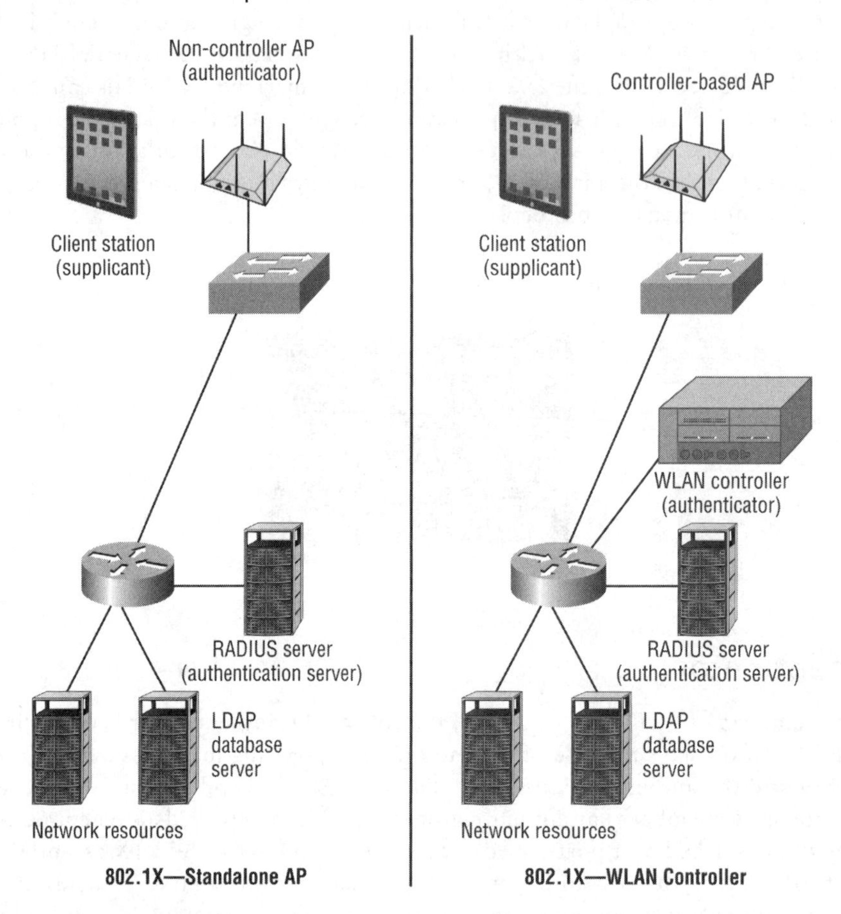

FIGURE 4.7 WLAN bridging and 802.1X

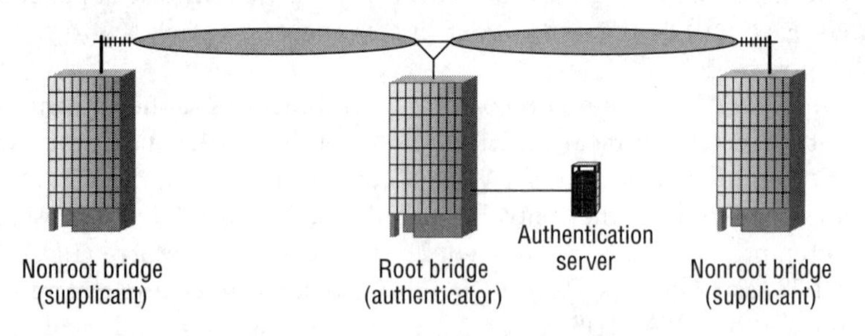

What about using an 802.1X framework with a WLAN bridging solution? As you can see in Figure 4.8, the root bridge would be the authenticator and the nonroot bridge would be the supplicant if 802.1X security is used in a WLAN bridged network.

FIGURE 4.8 Authenticator bouncer analogy

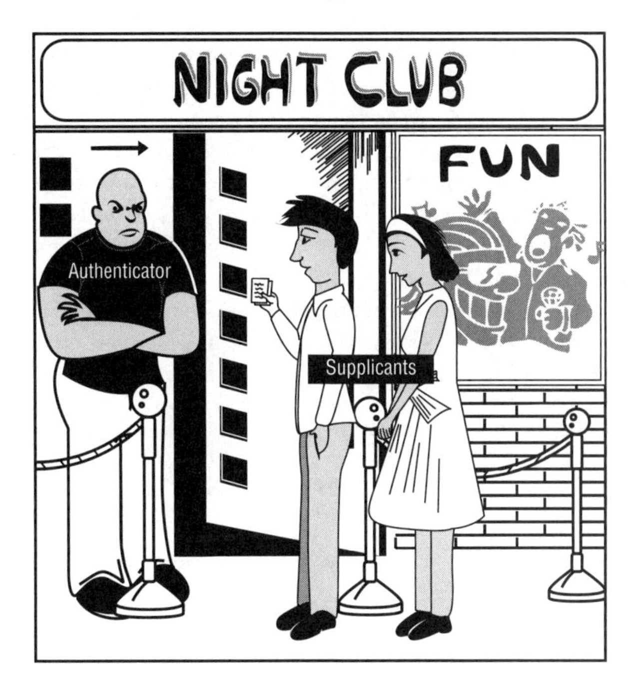

The term *authenticator* is a misnomer because the authenticator does not validate the supplicant's credentials. The authenticator's job is simply to either let traffic pass through or not pass through. As shown in Figure 4.8, a common analogy used to describe the authenticator is a bouncer at an exclusive nightclub that everybody is trying to get into. Picture some big, buff dude with a dark suit and an earpiece to communicate with the person holding the "guest list." The bouncer serves the role of the authenticator. He holds out his hand in a blocking fashion as would-be party-goers request entrance to the club. The would-be party guest (supplicant) shows their ID and the bouncer passes the identity to the guest list holder on the other end of his communication link. The guest list holder or nightclub owner (authentication server) looks up the identity on the list and simply responds with a verbal thumbs up or thumbs down.

If the bouncer gets a thumbs up, he will unblock the entrance for the authorized party guest to enter. Once the guest enters the nightclub, the bouncer will then block the entrance again for all new attempts to enter the nightclub. If the bouncer receives a thumbs down from the nightclub owner, he will kick the posing party guest out of line and send them away.

As mentioned previously, the authenticator is either an AP or WLAN controller in your enterprise WLAN. What the authenticator needs to know is essentially who is going to provide the guest list services—which is the role of the authentication server. Therefore, when configuring a WLAN controller or AP as an authenticator, you would need to be able to point the authenticator in the direction of an authentication server. Typically, the authentication server is a RADIUS server. As shown in Figure 4.9, the authenticator would need to be configured with the RADIUS server's IP address and along with a *shared secret* in order to communicate with the server. A shared secret is only used to validate and encrypt the communication link between the authenticator and the authentication server. The shared secret configured on both the authenticator and authentication server is not used for any part of supplicant validation. The shared secret is only for the authenticator-to-authentication server communication link. The authenticator and authentication server should be configured to use UDP port 1812 for authentication communications and UDP port 1813 for accounting communications. It should also be noted that the authenticator will be configured to "require" EAP authentication, but a specific EAP type is not chosen. Remember that the authenticator is essentially a pass-through device that either allows or disallows traffic to flow through the authenticator's virtual ports.

FIGURE 4.9 Authenticator configuration

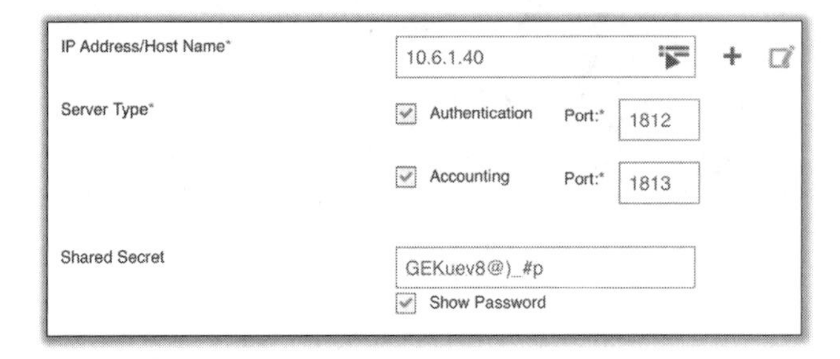

Authentication Server

As mentioned earlier, the *authentication server (AS)* validates the credentials of a supplicant that is requesting access and notifies the authenticator that the supplicant has been authorized. The authentication server will maintain a user database or may proxy with an external user database to authenticate user or device credentials. The authentication server and the supplicant communicate using a Layer 2 EAP authentication protocol. In almost all cases, a RADIUS server functions as the authentication server. The RADIUS server may indeed hold a master native user database, but usually will instead query to a preexisting external database. Any *Lightweight Directory Access Protocol (LDAP)*-compliant database can be queried by the RADIUS authentication server. LDAP is an application protocol for querying and modifying directory services running over TCP/IP networks. Active Directory is the most commonly used external LDAP database, but a RADIUS server can also query LDAP-compliant databases such as Apple Open Directory and Novell eDirectory. As shown in Figure 4.10, typically a RADIUS server performs authentication server duties and the RADIUS server initiates a proxy query to a preestablished LDAP-compliant database, such as Active Directory. This is referred to as *proxy authentication*.

FIGURE 4.10 Proxy authentication

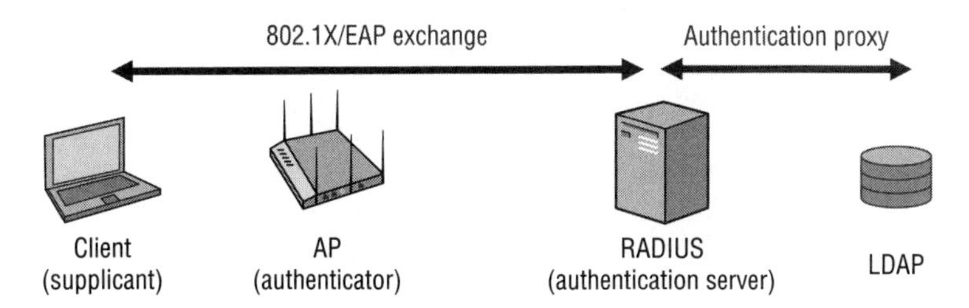

Some WLAN controller vendors allow for direct queries from the authenticator (WLAN controller) to an LDAP database. This method may have design constraints for some situations but may be desirable for some enterprises wanting to increase authentication performance where RADIUS servers aren't adding special features or additional user databases.

Although LDAP and direct AD integration is an option, RADIUS has been around for a very long time in WLAN security and is by far the most common method used for authentication servers. Table 4.1 contains examples of authentication servers.

TABLE 4.1 Examples of authentication servers

Product name	Protocol
Cisco ACS	RADIUS
Juniper Steel Belted RADIUS	RADIUS
Microsoft Network Policy Server (NPS)	RADIUS
Microsoft AD 2003 and higher	Kerberos and LDAP
FreeRADIUS (open source)	RADIUS

When configuring a RADIUS server, you need to be able to point the authentication server back in the direction of the authenticator. Typically the authenticator is a WLAN controller or access point. As shown in Figure 4.11, the AS would need to be configured with the authenticator's IP address and shared secret in order to communicate with the authenticator. A more detailed discussion of RADIUS server configuration can be found in Chapter 9.

FIGURE 4.11 Authentication server configuration

Configuration Gotchas

When configuring authenticators (APs or WLAN controllers) and RADIUS servers, there are usually two configuration problems that people regularly encounter: nonmatching shared secrets and wrong UDP ports. Ensure these values are correct before attempting any supplicant authentication attempts. RADIUS uses UDP ports 1812 for RADIUS authentication and 1813 for RADIUS accounting. These ports were officially assigned by the Internet Assigned Number Authority (IANA). However, prior to IANA allocation of UDP ports 1812 and 1813, the UDP ports of 1645 and 1646 (authentication and accounting, respectively) were used as the default ports by many RADIUS server vendors.

From years of experience, most RADIUS servers used in enterprise WLAN security implementations are implemented quite simply. As mentioned earlier, RADIUS servers usually are only used to proxy a user authentication to some master list of users, which is typically Active Directory. Although there are several features that can be used to enhance security and operational features, most enterprises do not take advantage of them. One example of this is a feature called dynamic VLAN assignment. If a user who is requesting authentication is part of the Accounting group, that user will be assigned to a specific VLAN by passing down a VLAN identifier RADIUS attribute in the RADIUS response message when a user authentication is accepted. If the user requesting authentication is part of the Marketing group or General Staff group, those users can be placed in a different VLAN based on RADIUS attributes. RADIUS attributes carry the specific authentication, authorization information, and configuration details for requests and replies, including VLAN identifiers. RADIUS servers are also often used for dynamic role assignment.

Additional information such as user roles can be sent using *vendor-specific attributes (VSAs)*. Without digressing too much, this capability would allow the WLAN administrator to inherit any currently implemented *role-based access control (RBAC)* mechanisms already implemented on the wired network, dynamically assigning them to the proper VLAN. For example, we generally want to keep the hands of non-accounting employees from accessing accounting applications like paycheck and expense approval systems, so typically most networks already employ some level of segmentation and access control to accomplish this for the wired network.

More information about RADIUS-based VLAN assignment and RADIUS-based role assignment can be found in Chapter 9.

That being said, there is another server that can, in a way, be referred to as a new type of authentication server—NAC. *Network Access Control (NAC)* is a computer network security methodology that integrates endpoint security technology with user authentication. NAC can allow for access decisions to be based on a client's antivirus state and OS patch version. NAC is an upcoming security enhancement gaining a lot of industry attention, and is typically designed to work with RADIUS servers. NAC servers are often being tied into an 802.1X framework design.

The NAC server could interrogate the user's machine when attempting to authenticate, and it could additionally make sure the computer being used has the most recent antivirus definitions update installed. Therefore, not only is the 802.1X framework validating that the user is valid, it is also validating other factors important to controlling virus outbreaks and patching operating system and software security vulnerabilities. If a NAC server determines a computer does not meet certain minimum criteria, it then dynamically assigns the WLAN device to a "quarantine network" VLAN whereby the user only has access to update the virus definitions and download the necessary software patch.

Authentication servers provide the following key features:

Client/server model RADIUS servers receive user connection requests, authenticate the user, and provide any other information required to support the user network connection.

Network security Transactions between the authenticator and the RADIUS server contain sensitive user identity information and are encrypted by the RADIUS protocol.

Flexible authentication mechanisms RADIUS can support a variety of authentication methods that can, in turn, support a wide variety of applications and systems.

Extensible Authentication Protocol This protocol was designed to be flexible and to accommodate new attributes or enhancements easily.

Some vendors may supply features in addition to the ones listed here, but you will only be tested on your understanding of these topics.

Supplicant Credentials

As you have learned, the supplicant is the device that will need to be validated by the authentication server before being allowed access to network resources. The supplicant will use an EAP protocol to present credentials to the authentication server to prove identity. Depending on which type of EAP protocol is used, the supplicant identity credentials can be in many different forms, including

- Usernames and passwords
- Digital certificates
- Protected Access Credentials (PACs)
- Dynamic security token devices
- Smart cards
- USB devices
- Machine authentication

How these supplicant user credentials are used with EAP will be discussed later in this chapter. The following sections will offer a broad overview of the various types of supplicant identity credentials.

Usernames and Passwords

Simple usernames and their respective passwords are the most common form of identity information that is supplied by a supplicant to an authentication server. Many EAP methods use usernames and passwords. What may also be used is a domain name or realm that might tell the RADIUS server what master database to use. This is a clever way to use a single RADIUS infrastructure that proxies the correct user database based on the domain or realm. For example, if you are using Windows Active Directory and have more than one domain where user accounts exist, the supplicant can pass its credentials in a `DOMAIN/username` format. Here's an example:

```
CWNP/Tom Carpenter
```

Most RADIUS servers accept multiple formats of the domain or realm identifier. It can, in some cases, also be accepted in the form of

```
Tom Carpenter@CWNP
```

Basically, the RADIUS server will dictate what forms are acceptable for this identity to be formatted. Be careful, though—some supplicants try to tinker with this format, making the assumption that the RADIUS server might always want to see it in the first form `DOMAIN-NAME/username` even if a user entered it using the second method.

Supplicant Configuration

Remember, supplicants are software. Software designers make certain assumptions when they write their code. Some supplicants will ask you for username, password, and domain. Some will just ask you for username and password, assuming you will actually type `DOMAIN/username` in the Username box.

Furthermore, some supplicants will check to make sure the end user did not perform a configuration error and type the domain name twice. For example, say the desktop engineers embedded the domain name into all suppliants installed on all mobile devices so the user didn't have to type it. What if the user actually typed `DOMAIN/username`? You might find that the EAP method presents `DOMAIN/DOMAIN/username` to the authentication server. There are other cases where login scripts might be configured to precede the username with a domain as well. The bottom line is that you should assume nothing and should always check for proper configurations. Check your authentication server log files for failed authentications if you are having failures, and issues like this will usually quickly reveal themselves.

Digital Certificates

Any Internet user who browses to a website with the `https://` prefix encounters digital certificates. Likewise, every time an Internet user logs into a website or uses online banking, they are using digital certificates.

Most of us know that the web server's certificate is used for encryption—a technology called *Secure Socket Layer (SSL)*. SSL is a cryptographic protocol normally used to provide secure communications over the Internet. SSL uses end-to-end encryption at the Transport layer of the OSI model. What may not be self-evident is that your computer is also authenticating the web server based on the contents of the SSL certificate, as shown in Figure 4.12. In this example, we used the Wells Fargo website. The bank uses a *Public Key Infrastructure (PKI)* digital certificate. Notice that the URL matches exactly the "Issued to" statement in the certificate. You can also see the certificate was issued by Symantec, who provides the PKI management service for the bank's website.

FIGURE 4.12 SSL certificate

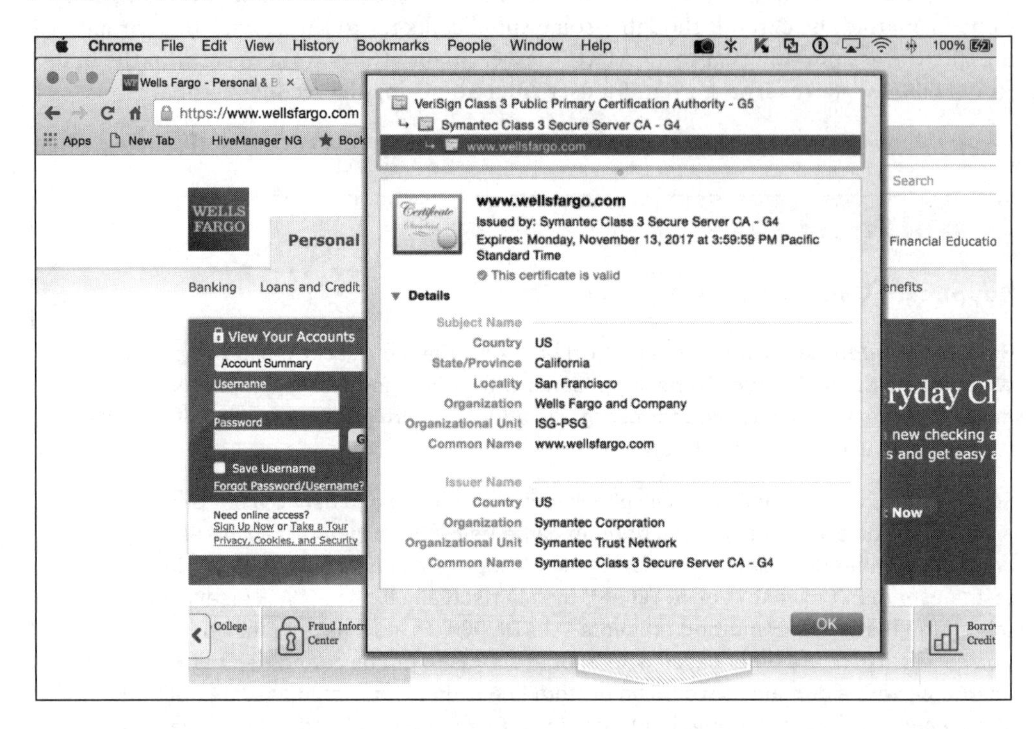

Then by utilizing the X.509 PKI certificate, we were able to verify the identity of the remote computer. In this case, it is the web server for the bank. If you have ever viewed a certificate, you will see that it contains information about the holder, such as

- Common name
- Subject
- Issuer
- Valid dates

EAP authentication protocols can use both server-side and client-side certificates. Later in this chapter, we will discuss how some EAP protocols use server-side certificates to create an encrypted tunnel during the authentication process. If the EAP protocol supports client-side certificates, the implementation of a PKI will be necessary. Unique client certificates will have to be created, issued, and managed for every WLAN user or device. Upon expiration, certificates will also require updating. Because the client-side certificates are being used as credentials to validate the supplicant, the 802.1X authentication server validates the client-side certificate issued by a PKI.

Public Key Infrastructure (PKI)

A PKI is a framework used for creating a secure method for exchanging information based on public key cryptography. A PKI uses hardware, software, people, policies, and procedures to create, manage, store, distribute, and revoke digital certificates. Certificates require a common format and are largely based on the ITU-T X.509 standard.

Protected Access Credentials (PACs)

Another certificate-like credential that can be used by supplicants is called a *Protected Access Credential (PAC)*. EAP-FAST uses PACs as credentials instead of the more standards-based X.509 certificates. EAP-FAST is an EAP protocol that was originally developed by Cisco as a replacement for EAP-LEAP and was designed for ease of deployment and renewal.

Because using client-side certificates requires the implementation of a PKI, the installation, administration, and cost of the PKI is often considered to be undesirable to maintain just for WLAN devices. Cisco knew this when they designed EAP-FAST and designed a PAC file. A PAC file is a close cousin of a digital certificate, but the RADIUS server is the issuing party. We already know the RADIUS server is also the authenticating party that validates the supplicant. While it is not quite the same, as a mental model you can think of an EAP-FAST PAC/RADIUS infrastructure as a mini-PKI. The RADIUS server issues and validates the correct PAC that is used by the supplicant. Each PAC is tied to an individual user identity. EAP-FAST will be discussed in greater detail later in this chapter.

One-Time Passwords

A *security token* is any physical device that is issued to an authorized user of computer services to enhance authentication strength. Some security tokens also incorporate one-time password capabilities. A *one-time password (OTP)* is a password that is only valid for a single login session or transaction. RSA Security is largely considered the inventor of the one-time password and is the dominant market leader in this space at the time of this writing. RSA developed a technology called SecurID that uses a dynamically updated one-time password at fixed intervals—usually 30 or 60 seconds. OTP security tokens can exist in either a hardware or software form factor. A hardware OTP security token device is shown in Figure 4.13.

Each token device is unique and is assigned to an individual user ID. The OTP generated from the token device is designed to be different from other token devices at any given time.

When that user requests an authentication, the token device OTP must be supplied with the user's normal password. If the user's password is correct but the one-time password is wrong, the authentication fails. Therefore, this protects against stolen user identities.

FIGURE 4.13 OTP token device

Courtesy of RSA

The OTP token devices and the OTP authentication server verifying the passwords work off a precise time clock. A mathematical calculation is performed using the token device clock value as well as other token information. The OTP authentication server's clock runs a similar calculation because it is synchronized with the token device. The OTP server already knows the other token device identity for each unique user in order to construct the same calculated result the token device uses.

OTP achieves a two-factor authentication for the user requesting access based on their user identity *and* something that user has in their possession. In theory, using OTP security tokens with 802.1X/EAP WLAN security sounds like a great idea because of the two-factor authentication. Several flavors of the EAP protocol can support OTP tokens. In reality, OTP tokens have not really been deployed extensively with WLANs because of issues during roaming. If a user roams from one AP to another AP and if a full 802.1X/EAP authentication occurs, a new OTP must be entered into the token. Because of the complexity, cost, and potential complications during roaming, OTP security with 802.1X/EAP has not really caught on.

Smart Cards and USB Tokens

Smart cards and USB tokens can also be used for a single-factor authentication. They really haven't gained the mass popularity that was once anticipated, but they do have their niche. As shown in Figure 4.14, a *smart card* is any pocket-sized card with embedded integrated circuits that can process data. A smart card is also often referred to as an *integrated circuit card (ICC)*. The key concept around a smart card or a USB token is to store unique user identity information securely on the integrated circuit chip. Typically, this data cannot be modified or copied. The contents can be nearly any type of information, but it is usually similar to or quite literally a client-side digital certificate. Most smart card technology is based on X.509 certificates. The core problem behind this identity method falls into the same PKI cost and management issues that are needed for client-side certificates on a per-user and perhaps even a per-machine basis. Effectively, a smart card or USB token functions as a client-side certificate and is often used when the EAP-TLS protocol is deployed with 802.1X.

FIGURE 4.14 Smart card

The United States Department of Defense (DoD) uses a Common Access Card (CAC) for identification of active-duty military personnel, reserve personnel, civilian employees, and contractor personnel. The CAC is based on X.509 certificates, with software middleware enabling an operating system to interface with the card via a hardware smart card reader. As shown in Figure 4.15, most smart card readers are external devices; however, internal smart card readers are now being built into laptops, handheld scanners, and even phones. As shown in Figure 4.16, USB security tokens can also function as client-side credentials. However, it should be noted that most government agencies and many corporations do not allow the use of any kind of USB device due to the security risks associated with the USB interface.

FIGURE 4.15 Smart card reader

FIGURE 4.16 USB security token

Courtesy of Aladdin

Machine Authentication

Although 802.1X/EAP is most often used to authenticate and authorize network access for users of a WLAN, computer devices can also be authorized. Computer authentication, more often know as *machine authentication*, is the concept of ensuring the *device* requesting access to the network is authorized in a separate but chained authentication process. Machine authentication is often deployed as an extra layer of security in Active Directory environments. As shown in Figure 4.17, when a Windows-based computer (machine) joins Active Directory, there is a computer account that gets created and a unique password is negotiated between the machine and AD. In the case of Windows, the machine credentials are based on a *System Identifier (SID)* value that is stored on a Windows domain computer after being joined to a Windows domain with Active Directory. The information stored in this SID is unique for each AD machine.

The computer account is used to identify the machine, even when no user is logged in, which can be used to provide the machine access to the network. Usually the machine does not need full access to the entire network and is often very restricted. Machine authentication is usually more about validating through 802.1X/EAP that the computer is authorized on the corporate network. 802.1X/EAP user authentication can then be used to grant further access. As shown in Figure 4.18, a computer might be placed into a unique VLAN after machine authentication and is transitioned to a different VLAN after user authentication. The machine authenticates using the cached machine credentials and the user authenticates when logging into the domain. Machine authentication is often used in enterprise environments where the computer is shared by multiple employees with different user accounts.

In Figure 4.18 the advanced settings show that the default User Or Computer Authentication mode is selected. The following process will occur if the computer is part of a domain. On boot, the computer domain account will be used to authenticate. This will mean the computer appears in the RADIUS server log as authenticating with `HOST\computername` as the username. Once the user logs into the computer, Windows will authenticate again using the username and end the machine authenticated session. If the user logs out, then Windows will again switch to authenticating with the computer domain account. This maintains the security concept of individual user authentication for the system.

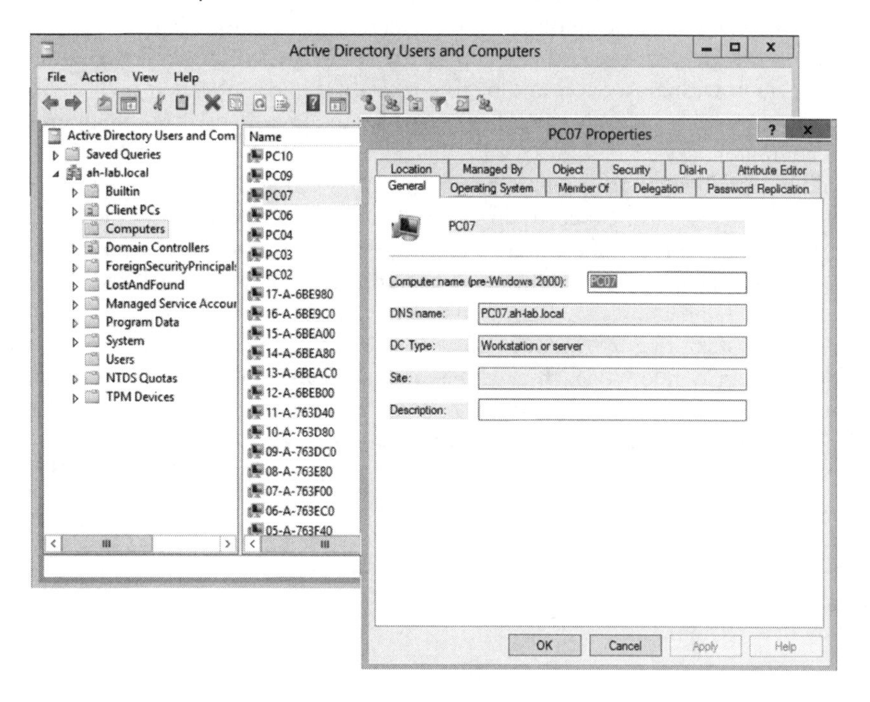

FIGURE 4.17 Computer domain account

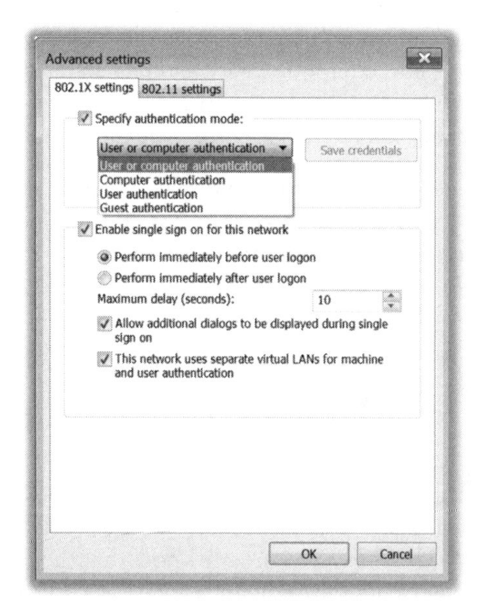

FIGURE 4.18 Supplicant: machine authentication

Many RADIUS servers are designed to note the multiple accounts authenticating from the same MAC and use this to chain the authentication process together to provide multiple authentication factors that are used to evaluate the security level of the device. An example would be a person with a user account logging into a network with their own device being put into a more restricted VLAN than if they had logged in with a corporate approved Windows computer.

It should be noted that machine authentication with 802.1X/EAP and Active Directory was designed for devices using the Windows OS. This is because when a user logs into a Windows system, the context of the system on the network changes and a new 802.1X/EAP authentication occurs. Your options for machine authentication with non-Windows operating systems are limited and complex. Although Macs have three different modes for configuring authentication (System, Login Window, and User), it's not possible to use multiple modes and switch context as in Windows. Machine authentication is also not really an option for mobile devices using iOS and Android OS. Instead, mobile device management (MDM) solutions are often deployed for authorization purposes of company-owned and personal mobile devices. MDM is discussed in greater detail in Chapter 10, "Bring Your Own Device (BYOD) and Guest Access."

Real World Scenario

Machine Authentication

There are several supplicants that can be configured for machine authentication. The computer will associate to the wireless network using 802.1X/EAP, and instead of passing user credentials, it will use a system identifier that Active Directory stores on all AD computer objects. When machine authentication is configured, after a machine boots up—even for the first time—and the user is asked for their user credentials, the machine can be already on the WLAN via these machine credentials. This will allow a never-before-seen user to log in to that computer because it already has communication with Active Directory.

Requirements for using machine credentials are that you must use a supplicant capable of machine authentication and 802.1X/EAP must be employed. Most enterprise supplicants support machine authentication. Furthermore, your RADIUS server needs to be integrated with Active Directory and machine authentication must be enabled.

802.1X/EAP and Certificates

You have already learned that within any 802.1X framework the authentication server validates the supplicant's credentials. The most secure EAP authentication methods incorporate the concept of *mutual authentication*. We already know that the supplicant's identity gets validated, but how about validating the authentication server's identity? In other words, the EAP-protocol allows the supplicant to also validate the authentication server. Mutual authentication validates both the supplicant and the AS and, if implemented properly, will prevent man-in-the-middle attacks. If the authentication server is to be validated by the supplicant, the AS must present credentials to the supplicant. Most EAP protocols use a server-side certificate for the authentication server credentials. A server-side certificate can also be used to provide protection of username/password credentials in a process known as tunneled authentication.

Server Certificates and Root CA Certificates

As shown in Figure 4.19, a server certificate must be created and installed on the authentication server. Additionally, the root Certificate Authority (CA) public certificate that was used to create the server certificate must be installed on the supplicants. Distribution and installation of the root CA certificate to multiple WLAN supplicants is often the biggest challenge when deploying 802.1X/EAP security.

FIGURE 4.19 Server certificate and Root CA certificate

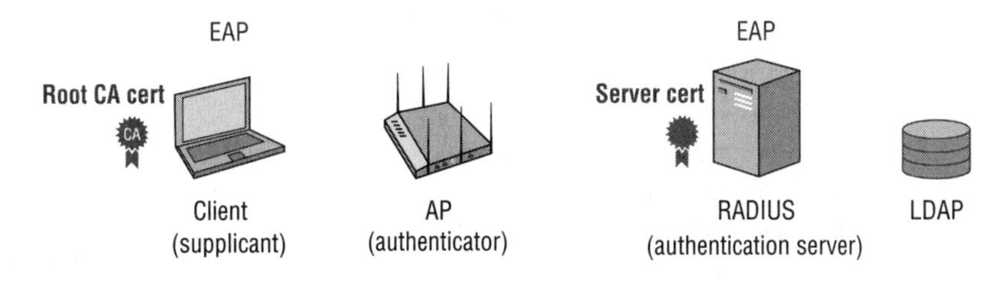

The authentication server certificate, in conjunction with the root CA public certificate, serves two major purposes:

Validates the Authentication Server The AS certificate is first used to validate the identity of the server to the WLAN supplicant. This process is akin to the supplicant saying, "Oh, I know who you are," before the supplicant submits its own sensitive identity information. The server certificate is validated by the root CA certificate that resides on the supplicant. This is possible because the server certificate was created and signed by the root CA.

Creates an Encrypted TLS Tunnel EAP protocols that require a server-side certificate for the authentication server are used to create *Transport Layer Security (TLS)* encryption tunnels. TLS is a cryptographic protocol normally used to provide secure communications at the Transport layer of the OSI model. However, in the case of 802.1X/EAP, TLS technology is leveraged at Layer 2. Similar to a browser-based HTTPS session, the TLS protocol uses end-to-end encryption. Once the supplicant is sure of the identity of the authentication server, the supplicant then uses the certificate to establish an encrypted TLS tunnel. The supplicant identity credentials are then exchanged within the encrypted TLS tunnel. This process is known as *tunneled authentication*. The supplicant identity, you have already learned, can come in many forms. Whatever form of identity that is passed by supplicant, it will be passed within the encrypted TLS tunnel. The TLS tunnel protects the supplicant credentials from offline dictionary attacks and from eavesdropping. This is just like the method employed with e-commerce websites using HTTPS where credit card and personal information is passed securely through an SSL/TLS tunnel.

What Is the Difference between SSL and TLS?

Essentially there is no difference. The original term Secure Sockets Layer (SSL) refers to an earlier version of the Transport Layer Security (TLS) protocol that is used to provide end-to-end encryption. TLS uses symmetric cryptography to provide bidirectional encryption between two computer applications. The keys for this symmetric encryption are generated uniquely for each connection and are based on a shared secret negotiated using a TLS handshake. The identity of the two devices using TLS can be authenticated using public-key cryptography, thus the need for certificates. The term SSL is still often used to refer to TLS. For example, HTTPS is HTTP-within-SSL/TLS. SSL (TLS) establishes an encrypted bidirectional tunnel for data between two hosts. HTTP is an application protocol that is the foundation of data communication on the World Wide Web. When HTTP is communicated via an SSL/TLS tunnel, it is known as HTTPS.

So where does the server certificate get created before it is installed on the RADIUS server? The simple method is to purchase a server certificate from a trusted root Certificate Authority (CA) such as GoDaddy (`www.godaddy.com`) or Verisign (`www.verisign.com`). The server certificates usually cost several hundred dollars a year for each RADIUS server you deploy. The good news is that the root CA certificate already resides on most WLAN devices that can function as supplicants, as seen in Figure 4.20. The major trusted CAs pay a lot of money to have their public root certificates accessible within the various operating systems. The main advantage of purchasing a server certificate from a trusted CA is that there is no need to distribute and install root certificates on WLAN clients because they already are there.

The other option is to create a server certificate signed by an internal private CA such as Microsoft Certificate Services. Much like a public CA, a private CA establishes an internal company trust chain using separate certificates for the root and the servers. Many companies choose this method because they prefer to keep all the security in-house. One downside of using a public CA with 802.1X/EAP is that an attacker can possibly perform a man-in-the middle attack. An attacker can use a rogue AP along with a rogue RADIUS server and a server certificate that was also created from the same public CA. This attack is complex and has many moving parts. But because the chain of trust might actually be compromised, most organizations instead choose to install a server certificate signed by an internal CA on the RADIUS server. The main challenge with using an internal private CA is the distribution and installation of the root CA certificate from the internal CA to all the supplicants.

FIGURE 4.20 Trusted Root CA certificates

AAA Certificate Services
Root certificate authority
Expires: Sunday, December 31, 2028 at 3:59:59 PM Pacific Standard Time
This certificate is valid

Name	Kind	Expires	Keychain
Entrust Root Certification Authority - EC1	certificate	Dec 18, 2037, 7:55:36 AM	System Roots
Entrust Root Certification Authority - G2	certificate	Dec 7, 2030, 9:55:54 AM	System Roots
Entrust.net Certification Authority (2048)	certificate	Dec 24, 2019, 10:20:51 AM	System Roots
Entrust.net Certification Authority (2048)	certificate	Jul 24, 2029, 7:15:12 AM	System Roots
ePKI Root Certification Authority	certificate	Dec 19, 2034, 6:31:27 PM	System Roots
Federal Common Policy CA	certificate	Dec 1, 2030, 8:45:27 AM	System Roots
GeoTrust Global CA	certificate	May 20, 2022, 9:00:00 PM	System Roots
GeoTrust Primary Certification Authority	certificate	Jul 16, 2036, 4:59:59 PM	System Roots
GeoTrust Primary Certification Authority - G2	certificate	Jan 18, 2038, 3:59:59 PM	System Roots
GeoTrust Primary Certification Authority - G3	certificate	Dec 1, 2037, 3:59:59 PM	System Roots
Global Chambersign Root	certificate	Sep 30, 2037, 9:14:18 AM	System Roots
Global Chambersign Root - 2008	certificate	Jul 31, 2038, 5:31:40 AM	System Roots
GlobalSign	certificate	Mar 18, 2029, 3:00:00 AM	System Roots
GlobalSign	certificate	Jan 18, 2038, 7:14:07 PM	System Roots
GlobalSign	certificate	Jan 18, 2038, 7:14:07 PM	System Roots
GlobalSign	certificate	Dec 15, 2021, 12:00:00 AM	System Roots
GlobalSign Root CA	certificate	Jan 28, 2028, 4:00:00 AM	System Roots
Go Daddy Class 2 Certification Authority	certificate	Jun 29, 2034, 10:06:20 AM	System Roots
Go Daddy Root Certificate Authority - G2	certificate	Dec 31, 2037, 3:59:59 PM	System Roots

Real World Scenario

Distributing the Root CA Certificate from an Internal Certificate Authority

Using an internal PKI infrastructure as a private CA first requires the creation of an original root certificate. Server certificates are then created and can be signed to the root CA. However, the recommended design for an internal PKI is to have two to three tiers, so at minimum there is a root CA and an issuing CA with a possible intermediate CA in between those. Security recommendations are that you never use a self-signed root CA to issue certificates and many RADIUS supplicants will reject a certificate from a root CA. The server certificates can then be installed on the RADIUS servers. However, there still has to be a means in which to distribute and install the root certificate to all of the WLAN supplicants. For example, the root certificate must be installed in the *Trusted Root Certification Authorities Store* of a Windows machine. Installing the root certificate onto Windows laptops can be easily automated using a *Group Policy Object (GPO)* if the Windows laptop is part of the Active Directory (AD) domain. However, a GPO cannot be used for Mac OS, iOS, or Android mobile devices, or for personal Windows BYOD devices that are not joined to the AD domain. Manually installing certificates on mobile devices and employee-owned devices is an administrative nightmare. For this reason, mobile device management (MDM) solutions are often deployed. An MDM solution uses an encrypted over-the-air provisioning of certifications during the MDM enrollment process. Instead of a full-blown MDM solution, another option is a *self-service device onboarding* solution.

Several WLAN vendors offer self-service solutions so employees can easily self-install security credentials such as an 802.1X /EAP root CA certificate. Third-party self-service onboarding solutions such as SecureW2 (`www.securew2.com`) are also available. MDM and device onboarding solutions are discussed in greater detail in Chapter 10.

802.1X using EAP protocols that require certificates provides solid WLAN security solution when properly configured at both the client and server side. Trusted root certificate authorities are being updated all of the time and distributed with regular OS service patches, or if you're using Windows Active Directory, can be passed via a domain Group Policy automatically. The list of trust root authorities, sometimes referred to as a Certified Trust List (CTL), keeps on growing. Put simply, a client should be configured to be *skeptical* of the RADIUS server identity. Most supplicants allow this in the simple form of a *Validate Server Certificate* checkbox, just as with Windows supplicants, as shown in Figure 4.21.

FIGURE 4.21 Server certificate validation

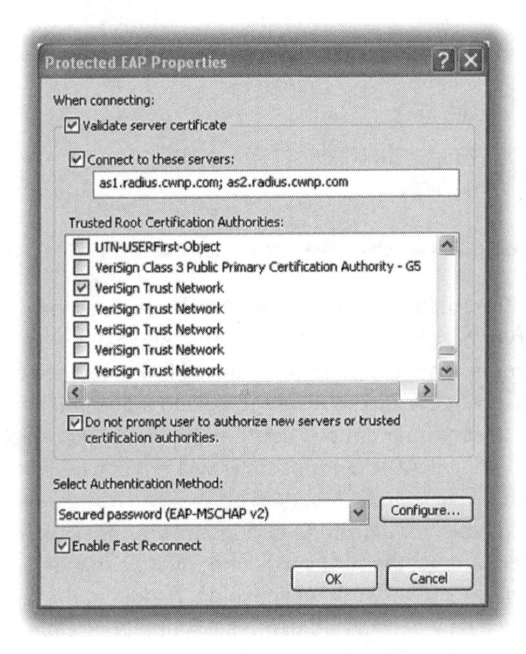

Once this is checked, known trusted root authorities and imported computer certificates are listed. In some supplicants, when you click this checkbox with default options, you are essentially saying that *any* server with a certificate issued from *any* one of these trusted root authorities is fine with you. This leaves a potentially large security hole for a hacker to perform a man-in-the-middle attack. All an attacker would have to do is purchase a valid, current certificate from any one of those sources, and it would pass the supplicant's acceptance criteria. To prevent this, always have the supplicant validate the server certificate and select the proper trusted root CA certificate that is used for the validation. This provides an additional check that validates the name of the certificate being sent to the client.

Client Certificates

A common mistake that people make is to confuse the root CA certificate that is installed on the supplicants with client-side certificates. A client certificate is an entirely different animal within a PKI infrastructure. As shown in Figure 4.22, client certificates can also be installed on a WLAN supplicant and be used as client credentials with some types of EAP authentication. The most commonly deployed protocol that uses client certificates is EAP-TLS, which will be discussed in greater detail later in this chapter. Server certificates and a root CA certificate are still used in conjunction to validate the RADIUS server; however, the client certificate is used as the validation credentials for the supplicant. Adding client certificates into the mix with 802.1X/EAP security does provide an extra level of security but also adds an extra level of management and cost. Client certificates or smart cards with EAP-TLS security are often used in verticals such as military, government, and the finance industry.

FIGURE 4.22 Client certificate – supplicant

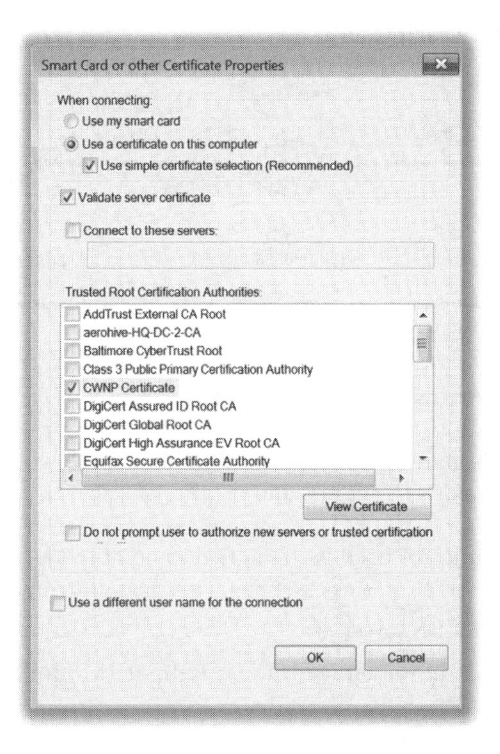

Using a public CA is usually cost-prohibitive because every client certificate costs several hundred dollars. Therefore, a private CA with an internal PKI is used to create and manage the client certificates. Management of client-side certificates requires much more time and proper skill sets, which many administrators might not have. Furthermore, the client certificates need to be provisioned onto the WLAN clients along with the root CA certificates. The provisioning of client certificates to company devices can be automated via a GPO in Windows. However, challenges also remain for distribution of client certificates for employee personal devices and devices using a non-Windows OS. As previously stated, for non-Windows OS devices, an MDM or certificate onboarding solution is usually needed for certificate distribution.

Shared Secret

A *shared secret* is used between the authenticator and the authentication server for the RADIUS protocol exchange. As shown in Figure 4.23, Layer 2 EAP protocol communications occur between the supplicant and the AS. The 802.11 frames use what is called *EAP over LAN (EAPOL)* encapsulation between the supplicant and the authenticator to carry the EAP data. On the wired side, the RADIUS protocol is used between the authenticator and the AS. The EAP data is encapsulated within a RADIUS packet. A shared secret exists between the authenticator and the AS so that they can validate each other with the RADIUS protocol.

FIGURE 4.23 Shared secret

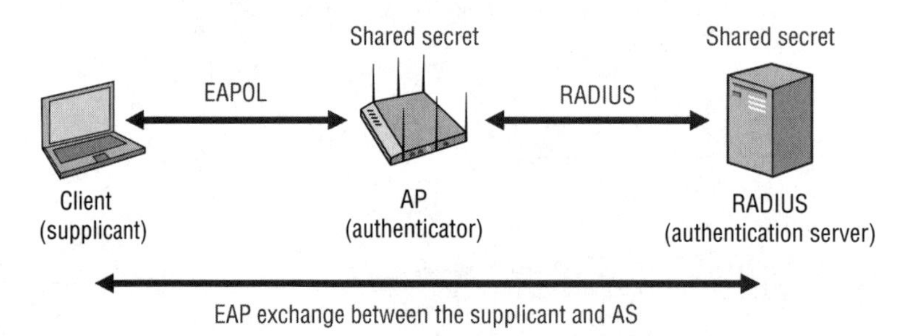

As mentioned earlier in this chapter, when configuring a RADIUS server, you need to configure each authenticator's identity within each authentication server. Typically, the authenticator is a WLAN controller or an access point. The AS would need to be configured with each authenticator's IP address and the shared secret in order to communicate with each authenticator.

Conversely, the authenticator itself is configured to point to a prioritized list of authentication servers (RADIUS server, in this example) with the following information:

- IP address of the RADIUS server
- UDP ports (1645 or 1812 for authentication; 1646 or 1813 for accounting)
- Shared secret

Usually, APs and WLAN controllers allow designation of up to three RADIUS servers for redundancy purposes.

Legacy Authentication Protocols

Before explaining EAP, it is important to discuss some legacy authentication protocols that have contributed to the history of network security. Some of these legacy protocols are still used within the stronger EAP authentication protocols. However, the legacy authentication protocols are used inside a TLS tunnel. You will see that the TLS tunnel is used to encrypt and protect the legacy authentication encryption methods. Let's take a closer look at the legacy authentication protocols relevant to WLAN security.

PAP

Password Authentication Protocol (PAP) is defined in RFC 1334 and provides no protection to the peer identity. It was originally designed for use with Point-to-Point Protocol (PPP). Although rarely used, it would be logical to use PAP inside an encrypted TLS tunnel because of its nonsecure nature.

CHAP

Challenge Handshake Authentication Protocol (CHAP) is defined in RFC 1994 and is slightly more evolved than PAP. CHAP is used with PPP as well and differs in that the password of the user identity is encrypted with an MD5 hash. Because MD5 is not considered secure by today's standards, CHAP should never be used outside of a TLS tunnel.

MS-CHAP

Microsoft developed a proprietary version of CHAP and later defined it in RFC 2433. *Microsoft Challenge Handshake Authentication Protocol (MS-CHAP)* also uses a hash of the password in a transmitted user identity. Early versions of Active Directory (AD) store an MS-CHAP compatible hash of a user's password in the AD database in lieu of the actual password. This hash can then be compared to the hash presented by the supplicant.

This initial version of MS-CHAP is considered weak, and many freely available software packages exist that can recover the password from the MS-CHAP hash. Therefore, MS-CHAP should be used only inside a TLS tunnel.

MS-CHAPv2

Of course, whenever vulnerability is discovered, a new version is ushered along. *MS-CHAPv2* is defined in RFC 2759 and was first released with the Microsoft Windows 2000 family.

MS-CHAPv2 uses a much stronger hashing algorithm and also supports mutual authentication during an MS-CHAPv2 exchange. EAP-PEAP, EAP-TTLS, and EAP-FAST all optionally use MS-CHAPv2, or in the case of EAP-LEAP, it is exclusively used. MS-CHAPv2 has also been found vulnerable and should also be used only inside a TLS tunnel.

EAP

The *Extensible Authentication Protocol (EAP)*, as defined in IETF RFC 2284, provides support for many authentication methods. EAP was originally adopted for use with PPP. EAP has since been redefined in the IETF RFC 3748 for use with 802.1X port-based access control.

As noted earlier, EAP stands for Extensible Authentication Protocol. The key word in EAP is *extensible*. EAP is a Layer 2 protocol that is very flexible, and many different flavors of EAP exist. Some, such as Cisco's Lightweight Extensible Authentication Protocol (LEAP), are proprietary, whereas others, such as EAP-TLS, are considered standard based. Some may provide for only one-way authentication, whereas others provide two-way authentication, more commonly called mutual authentication. Mutual authentication requires not only that the authentication server validate the client credentials, but also that the supplicant authenticate the validity of the authentication server. Most types of EAP that require mutual authentication use a server-side digital certificate to validate the authentication server. As stated earlier, a server certificate will also be used to create an encrypted TLS tunnel. This tunnel is used to protect the exchange of client credentials.

As you learned earlier in this chapter, 802.1X is an authorization framework with the three components of the supplicant, authenticator, and authentication server. The main purpose of an 802.1X solution is to authorize the supplicant to use network resources. The supplicant will not be allowed to communicate at the upper Layers of 3–7 until the supplicant's identity has been validated at Layer 2. EAP is the Layer 2 protocol used within an 802.1X framework.

The EAP messages are encapsulated in *EAP over LAN (EAPOL)* frames. There are five major types of EAPOL messages, as shown in Table 4.2.

TABLE 4.2 EAPOL messages

Packet type	Name	Description
0000 0000	EAP-Packet	This is an encapsulated EAP frame. The majority of EAP frames are EAP-Packet frames.
0000 0001	EAPOL-Start	This is an optional frame that the supplicant can use to start the EAP process.
0000 0010	EAPOL-Logoff	This frame terminates an EAP session and shuts down the virtual ports. Hackers sometimes use this frame for DoS attacks.

Packet type	Name	Description
0000 0011	EAPOL-Key	This frame is used to exchange dynamic keying information. For example, it is used during the 4-Way Handshake.
0000 0100	EAPOL- Encapsu-lated - ASF-Alert	This frame is used to send alerts, such as SNMP traps, to the virtual ports.

Let's review a generic EAP exchange. The two workhorses are the supplicant and the authentication server because they both use the EAP protocol to communicate with each other at Layer 2. The authenticator is the bouncer that sits between the two devices. As you have already learned, the authenticator maintains two virtual ports: an uncontrolled port and a controlled port. When open, the uncontrolled port allows EAP authentication traffic to pass through. The controlled port blocks all other traffic until the supplicant has been authenticated. When the controlled port is open, upper Layer 3–7 traffic can pass through. Dynamic IP addressing with DHCP is performed once the controlled port is opened.

As shown in Figure 4.24, 802.1X/EAP authentication works together with standard 802.11 Open System authentication and association. An 802.11 client station will actually establish a Layer 2 connection with the AP by associating and joining the basic service set (BSS). However, if 802.1X/EAP is implemented, Layer 2 is as far as the 802.11 station gets until the client also goes through the entire 802.1X/EAP process.

FIGURE 4.24 802.11 associations and 802.1X/EAP

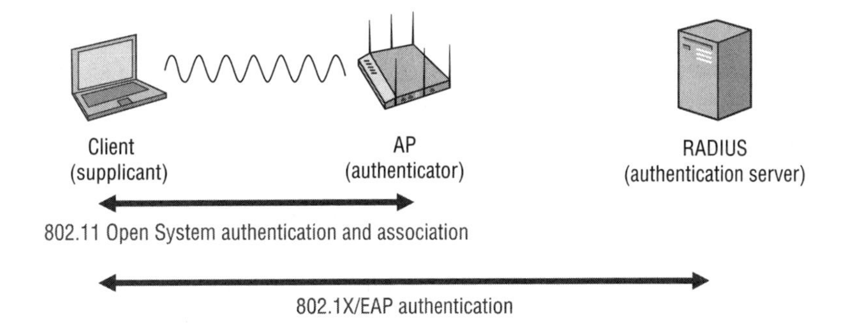

Figure 4.25 displays all the steps in a generic EAP exchange. The authenticator in this example is an access point. Please refer to the figure as we go over each of these steps.

1. The 802.11 client (supplicant) associates with the AP and joins the BSS. Both the controlled and uncontrolled ports are blocked on the authenticator.
2. The supplicant initiates the EAP authentication process by sending an 802.11 EAPOL-Start frame to the authenticator. This is an optional frame and may or may not be used by different types of EAP.

FIGURE 4.25 Generic EAP exchange

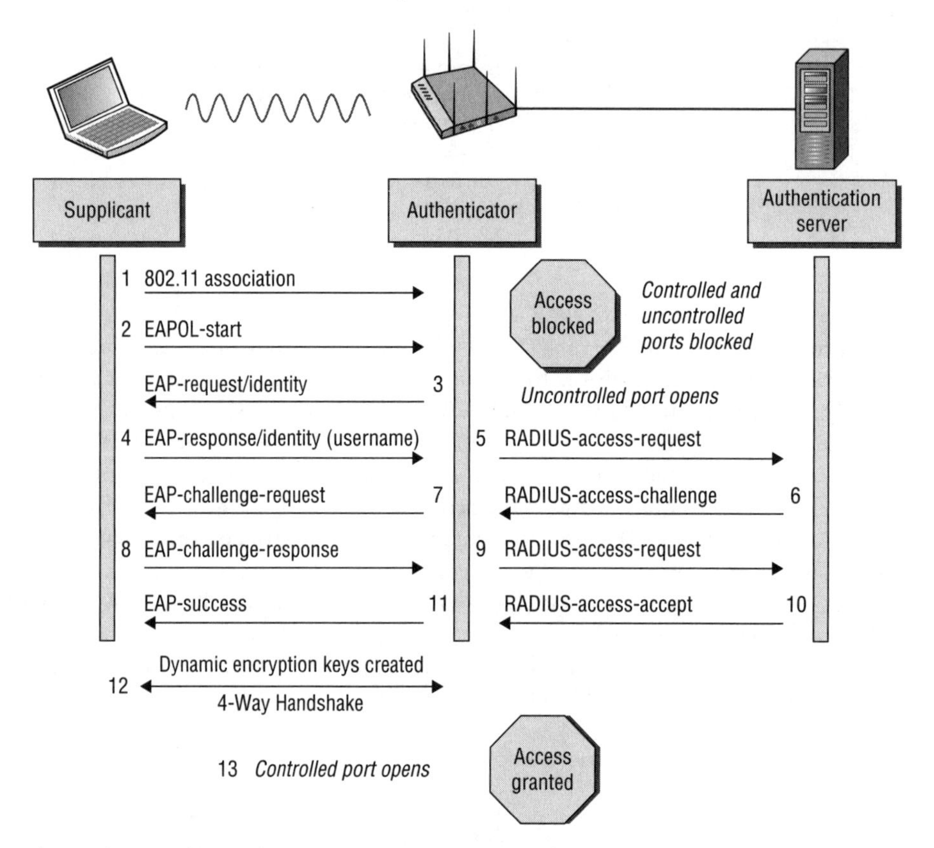

3. The authenticator sends an 802.11 EAP-Request frame requesting the identity of the supplicant. The EAP-Request Identity frame is always a required frame.
4. The supplicant sends an EAP response frame with the supplicant's identity in clear text. The username is always in clear text in the EAP-Response Identity frame. At this point, the uncontrolled port opens to allow EAP traffic through. All other traffic remains blocked by the controlled port.
5. The authenticator encapsulates the EAP response frame in a RADIUS packet and forwards it to the authentication server.
6. The AS looks at the supplicant's name and checks the database of users and passwords. The AS will then create a password challenge and send the challenge to the supplicant encapsulated in a RADIUS packet.
7. The authenticator forwards the password challenge to the supplicant in an 802.11 EAP frame.
8. The supplicant takes the password and hashes it with a hash algorithm such as MD-5 or MS-CHAPv2. The supplicant then sends the hash response in an EAP frame back to the AS.

9. The authenticator forwards the challenge response in a RADIUS packet to the AS.
10. The AS runs an identical hash and checks to see if the response is correct. The AS will then send either a success or failure message back to the supplicant.
11. The authenticator forwards the AS message to the supplicant in an EAP-Success frame. The supplicant has now been authenticated.
12. The final step is the 4-Way Handshake negotiation between the authenticator and the supplicant. This is a complex process used to generate dynamic encryption keys. This process is discussed in great detail in Chapter 5.
13. Once the supplicant has completed Layer 2 EAP authentication and created dynamic encryption keys, the controlled port is unblocked. The supplicant is then authorized to use network resources. If using IP, the next step the supplicant will take is obtain an IP address using DHCP.

You should notice that in step 4, the supplicant's username is seen in clear text. This might be considered a security risk. You should also notice that in steps 6–9, the supplicant's password credentials are validated using a weak challenge/hash response. This frame exchange can be captured using a WLAN protocol analyzer such as Wireshark. This is a security risk because the hash algorithms have been cracked and they are susceptible to offline dictionary attacks. In other words, the presentation of supplicant identity and validation of supplicant credentials is a security risk. Wouldn't it be better if steps 4–9 were protected in an encrypted tunnel? Since its original adoption, a number of weaknesses were discovered with some EAP authentication methods. The most secure EAP methods used today employ *tunneled authentication* to pass identity credentials (usernames and passwords) similar to what you find with web-based e-commerce transactions. We will now discuss the major EAP protocols, including the ones that use tunneled authentication.

Weak EAP Protocols

Older legacy EAP protocols exist that are highly susceptible to a variety of attacks, including social engineering and offline dictionary attacks. These authentication protocols had their day in the sun, but they should be viewed as absolutely unacceptable solutions in enterprise WLANs now that more secure EAP protocols are available. We will now discuss two legacy protocols called EAP-MD5 and EAP-LEAP.

EAP-MD5

EAP-Message Digest5 (EAP-MD5) is a fairly simple EAP type and is conceptually similar to the generic EAP method described in the previous section. EAP-MD5 was for a long time used for port authentication on wired networks and therefore was one of the very first EAP types used with WLANs. Many organizations were already using EAP-MD5, and it was a logical progression to leverage it for WLANs because RADIUS servers already supported it. However, EAP-MD5 has several major weaknesses:

One-Way Authentication Only the supplicant is validated; the server is not validated. Mutual authentication is needed to create dynamic encryption keys. If EAP-MD5 is the chosen authentication method, the encryption method is static WEP or no encryption at all.

Username in Clear Text The supplicant's username is always seen in clear text, as illustrated in Figure 4.25 earlier. If a hacker knows the identity of the user, the hacker can attempt to get the password using social engineering techniques. Therefore, EAP-MD5 is vulnerable to social engineering attacks.

Weak MD5 Hash The supplicant password is hashed using the MD5 hash function, which was once considered secure enough for its intended use in PPP networks. MD5 was never designed to be used over a wireless medium, and it is easily broken using a variety of hacker tools available today. Therefore, EAP-MD5 is highly susceptible to offline dictionary attacks.

Because of these three major weaknesses, EAP-MD5 should never be used in an enterprise WLAN environment now that stronger EAP protocols exist.

EAP-LEAP

EAP-Lightweight Extensible Authentication Protocol (EAP-LEAP), also known simply as LEAP, was a hugely successful EAP type used in the enterprise for many years. LEAP was easy to deploy using the same identity credentials to which end users were already intimately accustomed—that is, usernames and passwords. Furthermore, it was introduced in 2000 when security weaknesses were discovered with static WEP encryption. LEAP was also used to generate dynamic WEP keys, as described in Chapter 5. Dynamic WEP was a predecessor to TKIP/RC4 and CCMP/AES dynamic encryption. LEAP was a big step forward in security at the time.

Unlike EAP-MD5, LEAP does perform a type of pseudo-mutual authentication. Most documentation claims that LEAP supports mutual authentication. The password hash comparison algorithm used with MS-CHAP and MS-CHAPv2 validates that each side has the same password using a mutual authentication process. This is not the same mutual authentication that we've discussed where the supplicant validates the authentication server identity.

LEAP is not a TLS tunneled authentication method and is not an open standard; it must be licensed from Cisco. Figure 4.26 depicts the entire EAP-LEAP authentication process.

LEAP has several major weaknesses:

Username in Clear Text The supplicant's username is always seen in clear text in the initial EAP-Response frame sent by the supplicant to the authentication server. If a hacker knows the identity of the user, the hacker can attempt to get the password using social engineering techniques. Therefore EAP-LEAP is vulnerable to social engineering attacks.

Weak MS-CHAPv2 Hash The supplicant password is hashed using the MS-CHAPv2 hash function. MS-CHAPv2 has also been found to be vulnerable. A widely available program called ASLEAP, developed by Joshua Wright, can perform an offline dictionary attack on the hash function and easily obtain the password. Therefore, EAP-LEAP is highly susceptible to offline dictionary attacks.

FIGURE 4.26 EAP-LEAP

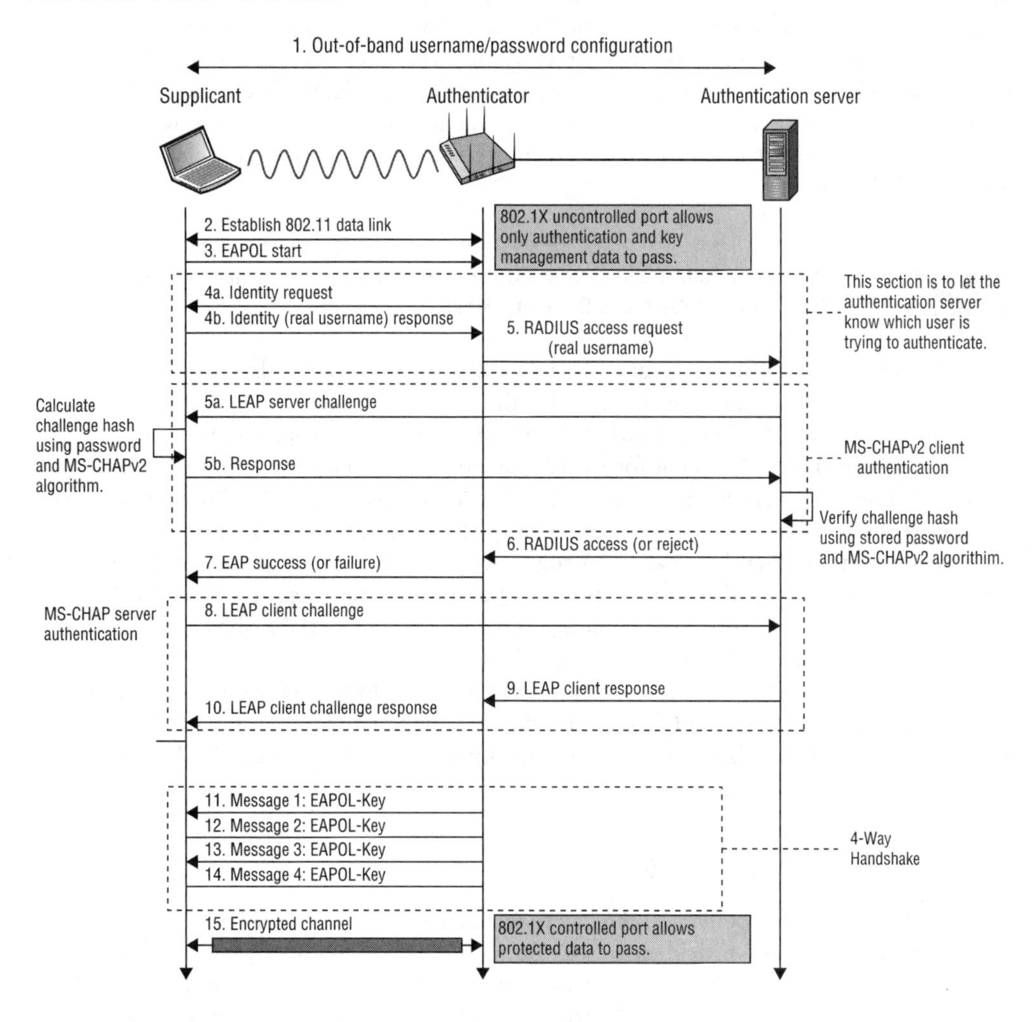

Pseudo-Mutual Authentication There has been some confusion with LEAP over the years with respect to mutual authentication. Since LEAP uses MS-CHAPv2, the MS-CHAPv2 protocol itself supports a *form* of mutual authentication. With respect to the grand scheme of WLAN security and what a WLAN security professional should look for in a security protocol, this form of mutual authentication buys very little.

The biggest problem with both EAP-MD5 and LEAP is that the prize is absolutely huge to a hacker—a username and password. The username and password is likely the same username and password used for Windows Active Directory or whatever network operating system might be installed. With a high-gain antenna sitting from quite far away, an attacker can pick up this exchange and gain full access credentials for WLAN users in mass. Combine this with a little knowledge of who the IT administrator is, and an attacker can have the keys to the kingdom in very short order. Or what about the CEO? That username and password just might give an attacker access to the CEO's email account. What's more, there is no way this attack can be detected because it can be performed completely passively and offline.

The success of an offline dictionary attack is only as strong as the size of the offline dictionary file. The use of strong and complex passwords might defeat this attack, but the bulk of the end-user population does not use passwords anywhere near strong enough to defeat the offline attack. Furthermore, enforcing strong password policies comes with significant end-user challenges.

A more detailed discussion about offline dictionary attacks can be found in Chapter 12, "Wireless Security Risks."

LEAP is a Cisco proprietary protocol, and despite claims of mutual authentication, the authentication server in reality is never authenticated. LEAP relies on the mutual authentication properties of MS-CHAPv2, which is simply the peer challenge and response.

When configuring a supplicant for LEAP, put simply, you only provide a username and password. There is no configuration for validating a server-side certificate or any information about the AS. An authentication server can easily be impersonated when a supplicant participates in a LEAP authentication. A rogue access point configured for a different authentication server—perhaps one that records usernames and passwords to a log file—would result in a supplicant merrily sending along its user credentials, unaware that the credentials are being intercepted.

As a WLAN security professional, you should never deploy LEAP within a modern-day WLAN given the options available to you today. If your organization is currently running LEAP, provide a quick migration path for your organization to a WPA/WPA2-based, TLS-tunneled EAP method.

Strong EAP Protocols

It should be clear at this point that the key to enterprise wireless security is leveraging 802.1X/EAP. The service provided by 802.1X, as you have also already learned, blocks traffic until a successful authentication transaction occurs. 802.1X is simply a vehicle that enables port blocking and provides the framework for the authentication transaction to occur. The stronger methods of EAP use TLS-based authentication and/or TLS-tunneled authentication. We have already referenced tunneled EAP authentication types earlier in this chapter. In this section, we are going to explore the most common of these types and their inner workings in detail.

Unlike EAP-MD5 and EAP-LEAP, which have only one supplicant identity, EAP methods that use tunneled authentication have two supplicant identities. These two supplicant identities are often called the *outer identity* and the *inner identity*. The outer identity is effectively a bogus username, and the inner identity is the true identity of the supplicant. The outer identity is seen in clear text outside the encrypted TLS tunnel, whereas the inner identity is protected within the TLS tunnel.

As you can see in Figure 4.27, the original EAP standard requires that there always be a clear-text value in the initial EAP-Response frame sent by the supplicant to the authentication server. This clear-text value is the outer identity that travels outside the TLS tunnel. The default value used by most supplicants is "anonymous."

FIGURE 4.27 Outer identity

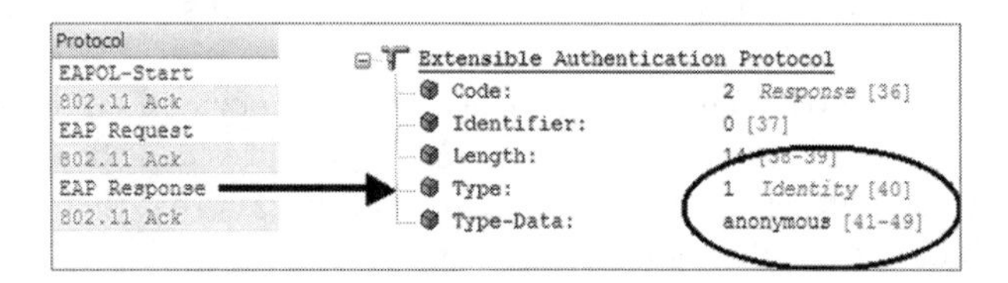

Although the default value used by many supplicants for the outer identity is "anonymous," the outer identity is usually a configurable setting. Keep in mind that this is not the real username. Some WLAN administrators use funny names such as Donald Duck or Mickey Mouse. Other WLAN administrators use a facility code identifying a group of supplicants. The facility code could be used for troubleshooting efforts of 802.1X/EAP supplicant failures and can help you quickly narrow down the facility where the problem was occurring. Other WLAN administrators use the outer identity as a social engineering honeypot. It should be noted that the default setting for the outer identity in some supplicants is the real username. In other words, the real username is used for both the inner and the outer identity. Some WLAN administrators do not necessarily view this as a security risk because company naming conventions are very easy to guess. The TLS tunnel will always encrypt the inner identity, but more importantly the TLS tunnel is used to protect the password challenge and response that is susceptible to offline dictionary attacks. Do not confuse the encryption used by the TLS tunnel with Layer 2 encryption that is used to protect the payload of 802.11 data frames. The encrypted TLS tunnel is created and exists only for a few milliseconds. The whole purpose of tunneled authentication is to provide a secure channel to protect the username/password identity credentials. The username/password credentials are encrypted inside the TLS tunnel. The TLS tunnel is *not* used to encrypt 802.11 data frames. We will now discuss versions of EAP that support tunneled authentication.

Tunneled EAP and a Social Engineering Honeypot

In computer terminology, a *honeypot* is a trap set for potential hackers to detect and possibly counteract unauthorized access of a computer network. EAP methods, such as EAP-PEAP or EAP-TTLS that use tunneled authentication, will always have an outer identity that can be seen in clear text with a WLAN protocol analyzer. A common strategy is to set a social engineering honeypot using the outer identity. The WLAN administrator configures all of the company's supplicants with the same value for the outer identity. The value could be "David Coleman" but there is no user by that name who works at the company. Employees at the company are trained to alert security if anyone ever inquires about an employee named David Coleman. If someone inquires about the imaginary David Coleman, a social engineering attack is occurring and can be further investigated.

EAP-PEAP

EAP-Protected Extensible Authentication Protocol (EAP-PEAP), also known simply as PEAP, creates an encrypted TLS tunnel within which the supplicant's inner identity is validated. Thus the term "protected" is used because the supplicant's identity and credentials are always encrypted inside the TLS tunnel that is established.

PEAP is probably the most common and most widely supported EAP method used in WLAN security. That is, it is the most popular EAP type that is considered highly secure. The confusion regarding PEAP usually revolves around the fact that there are multiple flavors of PEAP, including these three major versions:

- EAP-PEAPv0 (EAP-MSCHAPv2)
- EAP-PEAPv0 (EAP-TLS)
- EAP-PEAPv1 (EAP-GTC)

PEAP is often referred to as "EAP inside EAP" authentication because the inner authentication protocol used inside the TLS tunnel is also another type of EAP. PEAPv0 and PEAPv1 both refer to the outer authentication method and are the mechanisms that create the secure TLS tunnel to protect subsequent authentication transactions. The EAP protocol enclosed within parentheses is the inner EAP protocol used with each of these three flavors of EAP-PEAP. The main difference between these three major flavors of EAP is simply the inner EAP protocol that is used within the TLS tunnel.

We will discuss the differences between these three versions of PEAP later in this section. First, let's discuss how all flavors of PEAP operate. A key point is that in order to establish the TLS tunnel, a server-side certificate is required for all flavors of PEAP. As shown in Figure 4.28, the EAP-PEAP process involves two phases. Please refer to the figure as we discuss the two phases of EAP-PEAP. Please note that the steps are summarized and do not match the numbers in Figure 4.28. Although there are three major types of PEAP, we will use EAP-PEAPv0 (EAP-MSCHAPv2) as the example in Figure 4.28.

Phase 1

1. The authenticator sends an EAP frame requesting the identity of the supplicant.
2. The supplicant responds with an EAP response frame with the clear-text outer identity that is not the real username and is a bogus username.
3. At this point, the uncontrolled port opens on the authenticator to allow EAP traffic through. All other traffic remains blocked by the controlled port. The authenticator forwards the outer identity response to the AS.
4. The outer identity response cannot inform the AS about the actual identity of the supplicant. It simply informs the AS that a supplicant wants to be validated.
5. The AS sends the server certificate down to the supplicant. The supplicant uses the root-CA certificate to validate the server-side certificate and therefore authenticates the authentication server.
6. An encrypted point-to-point TLS tunnel is created between the supplicant and the authentication server. Once the TLS tunnel is established, Phase 2 can begin.

FIGURE 4.28 EAP-PEAP process

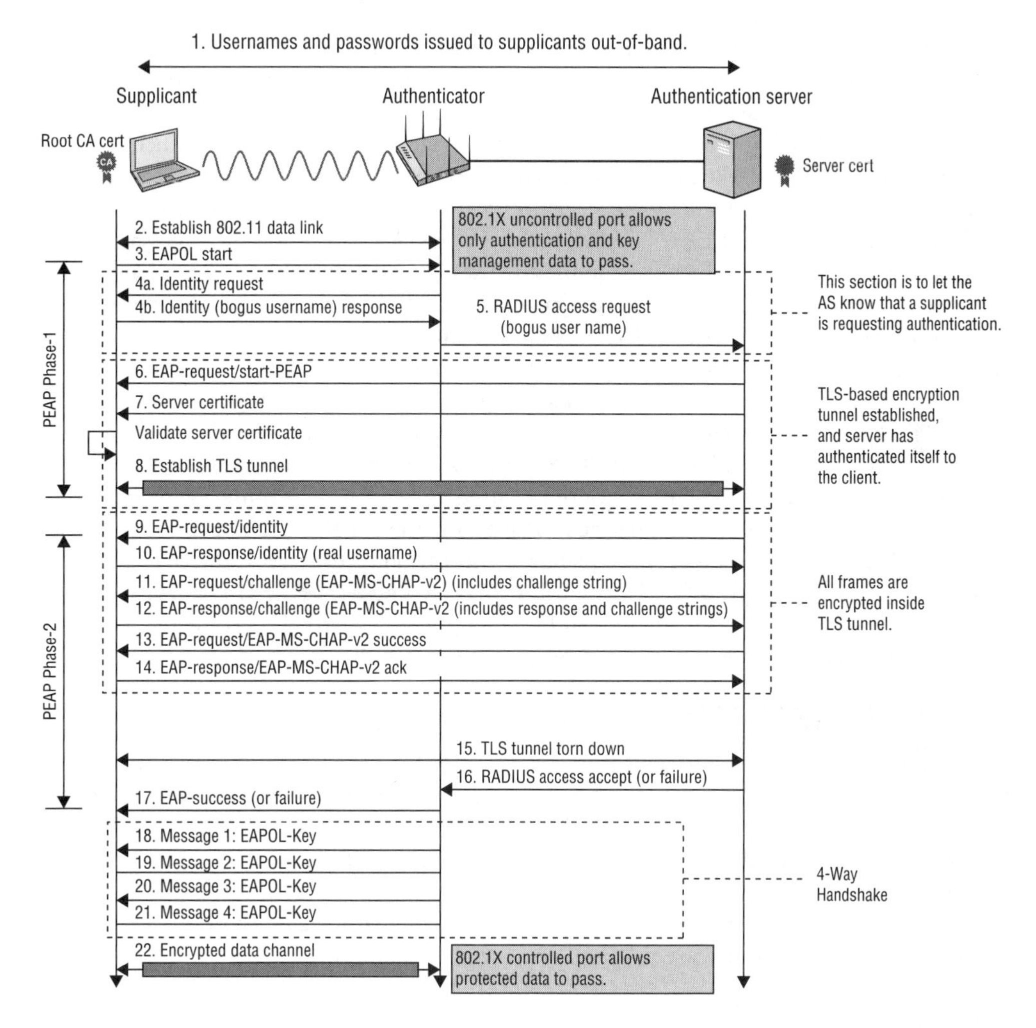

Phase 2

1. The AS requests the real identity of the supplicant.
2. The supplicant responds with the inner identity, which is the real username. The real username is now hidden because it is encrypted inside the TLS tunnel.
3. The remaining steps in Phase 2 involve a password challenge from the AS and a hashed response from the supplicant using an authentication protocol within the tunnel. The supplicant username/password credentials are validated by the authentication server. The entire exchange is encrypted within the TLS tunnel.

The whole point of Phase 2 is to validate the supplicant username/password credentials while encrypted within the TLS tunnel. The inner identity, or real username, is protected and therefore hidden. The password challenge and response exchange is also encrypted and protected. Whatever authentication method that is used inside the tunnel is also protected; therefore, any offline dictionary attacks are ineffective in obtaining the password.

PEAP has an interesting history. It began as a joint proposal by Cisco, Microsoft, and RSA Security. It is reported that Microsoft and Cisco didn't completely agree on every detail and subsequently Microsoft implemented PEAPv0 using MS-CHAPv2 as the inner authentication method. MS-CHAPv2 is Microsoft's own version of CHAP. Because Microsoft is the dominant player in both client and server operating systems providing built-in support, this has led to the success of EAP-PEAPv0 (EAP-MS-CHAPv2). Cisco split from the original specification, PEAPv1, which predominantly uses EAP-GTC as the inner authentication method. Supplicants and RADIUS servers may support all three type of PEAP; however, Microsoft's EAP-PEAPv0 (EAP-MS-CHAPv2) is the most widely supported, including supplicants that are native to other operating systems such as Mac OS X, iOS, and Android OS.

We will now discuss the differences between the various versions of PEAP. As mentioned earlier, PEAP is often referred to as "EAP inside EAP" authentication because the inner authentication protocol used inside the TLS tunnel is also another type of EAP. The only real difference is the inner EAP protocol that is used within the TLS tunnel. PEAPv0 and PEAPv1 both refer to the outer authentication method and are the mechanisms that create the secure TLS tunnel to protect subsequent authentication transactions. PEAPv0 supports inner EAP methods of EAP-MSCHAPv2 and EAP-TLS whereas PEAPv1 supports the inner EAP method of EAP-GTC. All versions of PEAP require a server-side certificate, and all versions of PEAP operate using the two phases described earlier.

EAP-PEAPv0 (EAP-MSCHAPv2)

As mentioned earlier, Microsoft's *EAP-PEAPv0 (EAP-MSCHAPv2)* is the most common form of PEAP. The protocol used for authentication inside the tunnel is EAP-MSCHAPv2. What is the difference between the normal MS-CHAPv2 protocol and the EAP-MSCHAPv2 protocol? For all practical purposes, they basically operate in the same manner and use the same hash algorithm. However, it should be noted that EAP-MSCHAPv2 is considered a separate protocol. The credentials used for this version of PEAP are usernames and passwords. Client-side certificates are not used and are not supported.

EAP-PEAPv0 (EAP-TLS)

This is another type of PEAP from Microsoft. *EAP-PEAPv0 (EAP-TLS)* uses the EAP-TLS protocol for the inner-tunnel authentication method. EAP-TLS requires the use of a client-side certificate. The client-side certificate is validated inside the TLS tunnel. No username is required for validation because the client-side certificate serves as the user credentials. This flavor of EAP is rarely used because the standards-based EAP-TLS is the usual choice when client-side certificates are also deployed.

EAP-PEAPv1 (EAP-GTC)

Cisco's version of PEAP uses yet another different type of EAP for the inner tunnel authentication. *EAP-PEAPv1 (EAP-GTC)* uses *EAP-Generic Token Card (EAP-GTC)* for the inner-tunnel authentication protocol. EAP-GTC is defined in the IETF RFC 3748 and was developed to provide interoperability with existing security token device systems that use one-time passwords (OTP) such as RSA's SecurID solution. The EAP-GTC method is intended for use with security token devices, but the credentials can also be a clear-text username and password. In other words, when EAP-GTC is used within a PEAPv1 tunnel, normally the credentials are a simple clear-text username and password.

Other versions of PEAP also exist. Other EAP protocols, such as EAP-POTP, can also be used for the inner-tunnel authentication method. EAP-POTP will be discussed later in this chapter.

Initially there were numerous compatibility issues because of the different types of PEAP. However, in practice you will be hard-pressed to find a case where you need to consider what PEAP version you are using anymore. Most RADIUS server vendors now support all versions of PEAP. Likewise, some supplicants will support multiple versions of PEAP. As stated earlier, almost all supplicants support EAP-PEAPv0 (EAP-MSCHAPv2) as the default version of PEAP.

EAP-TTLS

EAP-Tunneled Transport Layer Security (EAP-TTLS) was originally designed by Certicom and Funk Software (which is now owned by Juniper Networks) and is defined in RFC 5281. As with PEAP, it also uses a TLS tunnel to protect less-secure inner authentication methods. As shown in Figure 4.29, EAP-TTLS also uses two phases of operation very similar to EAP-PEAP.

The differences between EAP-TTLS and EAP-PEAP are fairly minor when analyzing them from a high level. The biggest difference is that EAP-TTLS supports more inner authentication methods, such as the legacy methods of PAP, CHAP, MS-CHAP, and MS-CHAPv2. EAP-TTLS also supports the use of EAP protocols as the inner authentication method. Figure 4.29 shows EAP-MSCHAPv2 being used for inner authentication; however, multiple authentication methods are supported within the TLS tunnel.

Remember that EAP-PEAP *only* supports EAP protocols for inner authentication whereas EAP-TTLS supports just about anything for inner authentication. Server-side certificates are required with EAP-TTLS to create the TLS tunnel, and client-side certificates are optional. Depending on the type of inner authentication method used, client-side certificates can be used as the protected supplicant credentials within the tunnel. Normally, however, the supplicant credentials are usernames and passwords because they are far easier to implement.

EAP-TTLS has been widely deployed, and it is likely to be encountered in many enterprise WLANs. While EAP-TTLS is almost identical to EAP-PEAP, it may not have the same native support in the operating systems of some supplicant devices. All major enterprise RADIUS servers seem to have built-in support for EAP-TTLS.

FIGURE 4.29 EAP-TTLS process

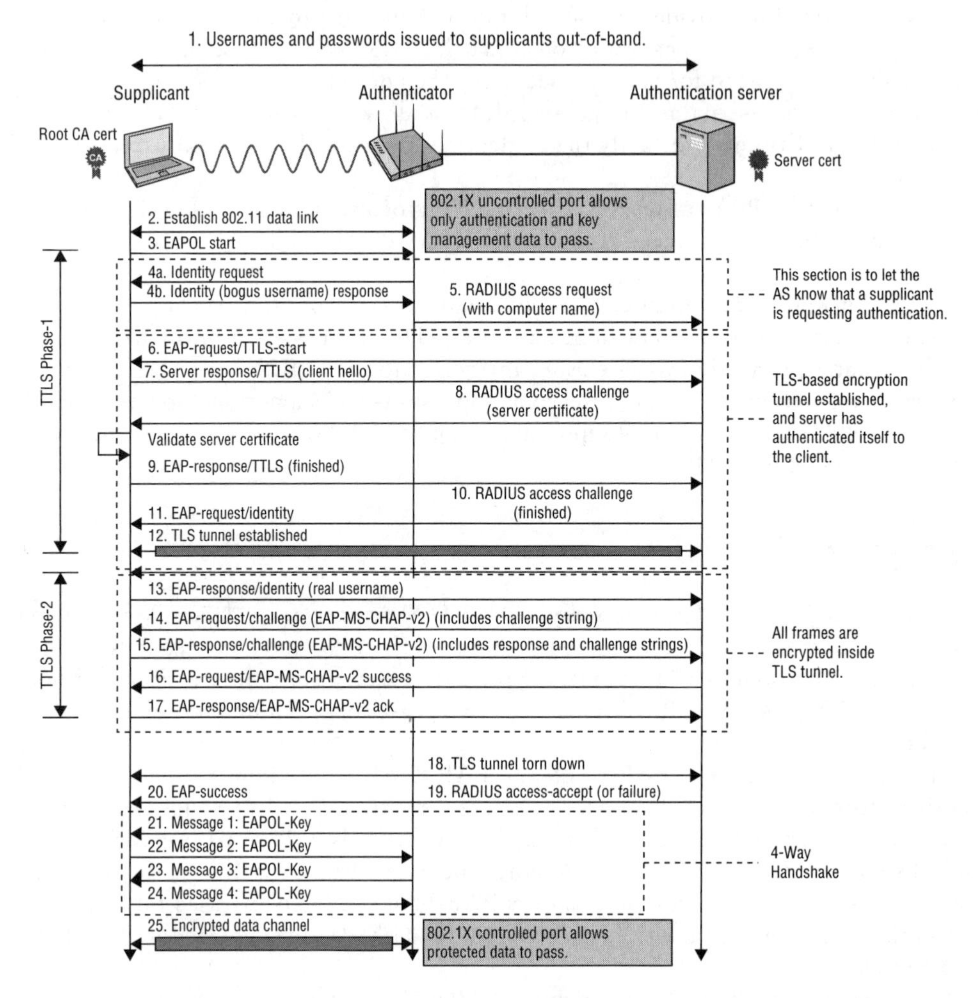

EAP-TLS

EAP Transport Layer Security (EAP-TLS) is defined in RFC 5216 and is a widely used security protocol. It is largely considered one of the most secure EAP methods used in WLANs today. This, however, comes at a cost.

EAP-TLS requires the use of client-side certificates in addition to a server certificate. Implementing a server-side certificate is not much of a burden, and EAP-TLS has the same server-side certificate requirement as EAP-PEAP and EAP-TTLS. The problem is that having a unique digital client certificate for each client requires a great deal of planning, infrastructure, and staff time relative to all the other EAP methods.

The biggest factor when deciding to implement EAP-TLS is whether an enterprise PKI infrastructure is already in place. This would usually, and optimally, include separate servers in a high-availability server cluster. Furthermore, access to these servers is critically important and must be guarded just like a master password list. Quite literally, these servers will have the certificate store, which includes the private keys for the entire PKI infrastructure. Therefore, even though EAP-TLS is widely considered secure, it can only be as secure as the certificate store.

Most enterprises consider managing, securing, and maintaining a PKI to be an unwanted burden. Therefore, unless one was already in place, EAP-TLS would not likely be a WLAN security professional's first recommendation. EAP-TLS is often used in verticals such as the banking industry where the cost and time for providing proper security is an afterthought. Figure 4.30 depicts the EAP-TLS process.

FIGURE 4.30 EAP-TLS process

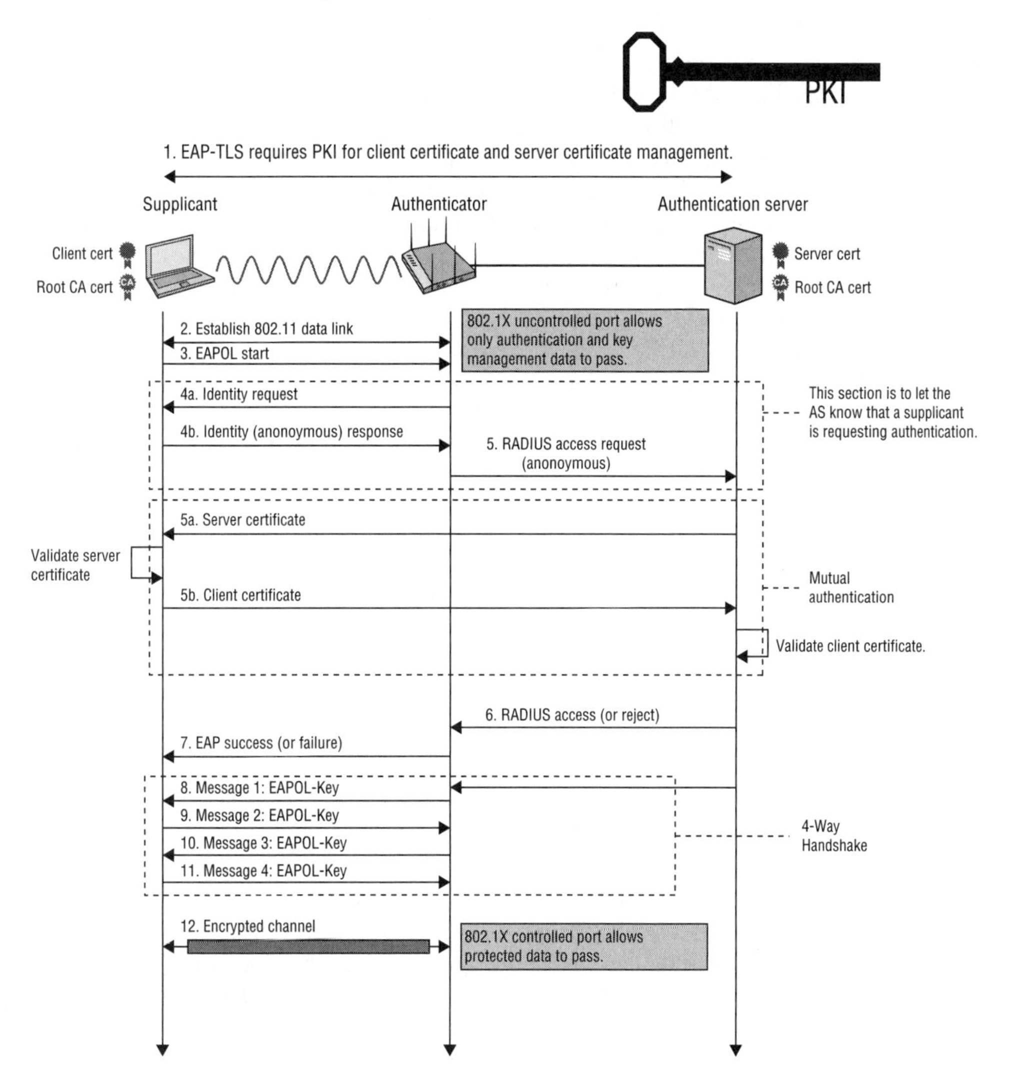

There are a few noteworthy points about the EAP-TLS protocol. From this process, you will notice that the AS presents its server-side certificate to the client first. If this certificate is not from a source that the client trusts, the supplicant can end the conversation immediately. The same goes for the AS. If the AS does not like the identity of the client-side certificate, it will reject the authentication attempt after the client has been given the first right of refusal.

An AS configured for EAP-TLS will typically allow validation of the client-side certificate by comparing one or more of the following in the user certificate:

- Certificate Subject Alternative Name (SAN)
- Certificate Subject Common Name (CN)
- Certificate Binary—a binary comparison to what's included with the user object in LDAP or Active Directory database to what the supplicant presents

Also, there isn't necessarily a tunnel where an inner authentication takes place like with the other TLS tunnel-based EAP types. However, there is an optional *privacy* mode where a TLS handshake can be established before the client identity is passed. When privacy mode is established, a typical TLS tunnel is also created as in PEAP and EAP-TTLS. However, the privacy method is generally not implemented by vendors.

Using client certificates is optional with EAP-TTLS and some flavors of EAP-PEAP that also use tunneled authentication. However, using a tunnel for a client certificate is not necessary. It is typically recommended to deploy EAP-TLS when using client-side certificates because of the wide support for the protocol.

An important point to mention is Microsoft's implementation of Certificate Autoenrollment. Microsoft built a feature into their Active Directory solution when using their Certificate Authority (CA) product (free with Windows Server products) to auto-publish client certificates. This means that the biggest problem with EAP-TLS—deploying certificates to each end-user device—has been mostly automated with Microsoft end-user devices. Keep in mind that only domain objects will be applicable for this feature and the client-side certificate will have to be installed manually with other non-Microsoft WLAN clients.

EAP-FAST

Cisco initially developed the *EAP-Flexible Authentication via Secure Tunneling (EAP-FAST)* protocol, and it has been a proprietary protocol until fairly recently. No, "FAST" doesn't mean that it is necessarily faster than other EAP types. IETF RFC 4851 was meant to define a standard for EAP-FAST, and the protocol is also recognized as part of the Wi-Fi Alliance WPA2 interoperability certification.

EAP-FAST is clearly stated in public documents by Cisco that it was designed to be a convenient, easy-to-implement replacement for LEAP. When it was discovered that LEAP can be easily cracked, something had to be done quickly. The whole point of EAP-FAST was to create a secure method of EAP authentication that would be resilient against offline dictionary attacks, yet not require the use of certificates. EAP-FAST provides for both mutual authentication and tunneled authentication just like EAP-PEAP and EAP-TTLS. However, EAP-FAST does not use standards-based X.509 digital certificates to create the TLS tunnel. Instead, EAP-FAST uses PACs.

PACs

A *Protected Access Credential (PAC)* is a cousin of a digital certificate. Actually, it is a shared secret. Since EAP-FAST is the only EAP type that uses PACs, it is prudent to discuss a bit of the EAP-FAST process with respect to how it is used in authentication server identity validation for the supplicant.

A PAC can consist of three components:

Shared Secret—The PAC-Key A preshared key between the client and the authentication server.

Opaque Element—The PAC-Opaque A variable-length data element sent during tunnel establishment where the AS can decode the required information to validate the client's identity.

Other Information—PAC-Info A variable-length data element that minimally provides the authority identity of the PAC issuer (usually the master RADIUS server, which would be the PAC Authority). May also contain the PAC-Key lifetime.

As shown in Figure 4.31, EAP-FAST operates in three phases:

Phase 0 This phase is used for automatic PAC provisioning. Each supplicant is automatically sent a unique client PAC using an anonymous Diffie-Hellman exchange. After each client station initially receives a PAC, clients will skip Phase 0 in subsequent logins. Phase 0 is an optional phase, and all PACs can also be installed on the clients manually.

Phase 1 During this phase, the supplicant sends the outer bogus identity to let the AS know that a client seeks validation. The client and the AS negotiate using symmetric key encryption from the PAC shared secret (PAC-Opaque). The result of this negotiation is the establishment of an encrypted TLS tunnel.

Phase 2 The supplicant is then validated within the encrypted tunnel. EAP-FAST supports several inner authentication methods, including client-side certificates, just as EAP-TLS does. The authentication protocol normally used inside the tunnel is EAP-GTC when username and passwords serve as the client identity information. A token-based solution is also a possibility.

Let's discuss manual and automatic PAC provisioning in a little more detail. The client PACs referenced earlier are created on the RADIUS server using a server-side master key and are unique to each client identity. After they have been created, they must be *installed* on each supplicant much like a client-side certificate is done. The client PACs can be manually installed by the WLAN administrator on each separate machine, and there will be no need for the optional Phase 0. Clients that already have a PAC file would proceed directly to Phase 1. However, if the WLAN administrator enables *automatic PAC provisioning* in Phase 0, the client PACs are installed automatically.

FIGURE 4.31 EAP-FAST process

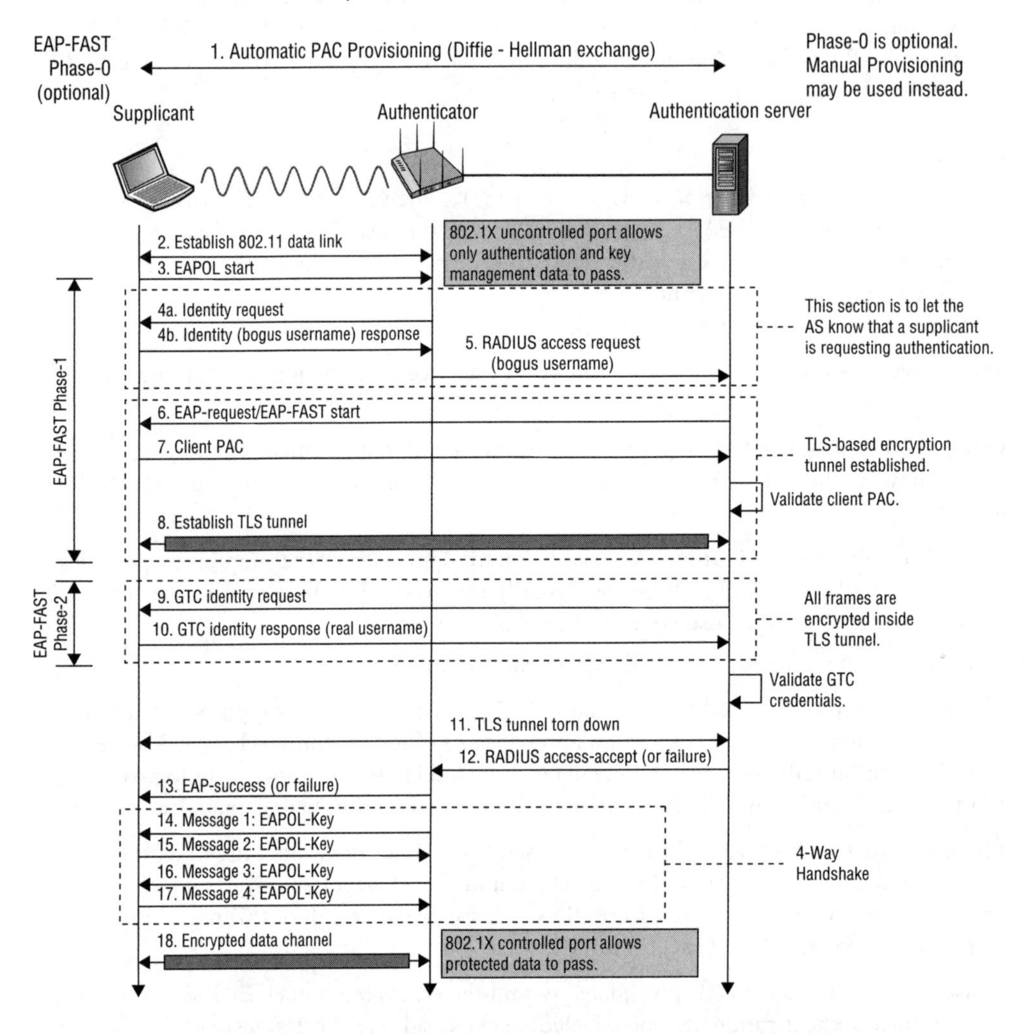

Auto-provisioning might sound great, but the problem is, how does the client know that it is talking to a valid RADIUS server? If you use EAP-FAST with auto-provisioned PAC files, you are allowing your clients to auto-provision from an unknown and perhaps untrusted server.

Automatic PAC provisioning is performed using an anonymous Diffie-Hellman exchange whereby the client simply has to trust the person providing the PAC. This subjects EAP-FAST to man-in-the-middle attacks during Phase 0. Despite the risks of Phase 0, most organizations using EAP-FAST typically deploy using automatic PAC provisioning because they are seeking the convenience provided by EAP-FAST. Auto-provisioning is a configurable option on the RADIUS server. However, the problem is that most client supplicants do not have any administrative control over enforcing this, even if the RADIUS server did not allow automatic PAC provisioning.

As mentioned earlier, if EAP-FAST is deployed without the optional phase 0, the IT administrator will have to deploy PAC files manually to each machine. Doing this sacrifices the convenience of EAP-FAST. If you were to do all of that, why would you not just use a more standardized and much more widely acceptable protocol like EAP-TLS or EAP-PEAP? The amount of effort and coordination involved to accomplish this is ridiculous. It would be far easier and logical to use EAP-PEAP, EAP-TTLS, or EAP-TLS. Furthermore, an EAP-FAST supplicant must use the appropriate PAC file from its PAC file storage for the appropriate user identity. Since PAC files are based on user identities, a different client identity cannot use the same PAC file or authentication will fail. Even though EAP-FAST is not proprietary, it is normally deployed in enterprise environments that use a Cisco infrastructure.

Some organizations will use Phase 0 and automatic PAC provisioning during the initial installation of the WLAN and then disable automatic PAC provisioning. However, it should be noted that the only way EAP-FAST can be considered relatively secure is if, and only if, each PAC was deployed manually and automatic PAC provisioning was disabled.

There is another point to mention about EAP-FAST. A key component of this protocol revolves around PACs and distributing them to clients. What is truly unique about EAP-FAST is that the PAC file contains a shared secret *instead* of a certificate. When using a PKI, asymmetric encryption must be used to decrypt the client identity response. In other words, the AS will use its private key, that only it has, to decrypt the supplicant's response that is encrypted using the AS public certificate. The computational overhead involved in this asymmetric encryption process is expensive. EAP-FAST uses a symmetric encryption algorithm based on the shared secret of each client's unique PAC. Therefore, there is technically a minor performance advantage to using EAP-FAST.

Despite the fact that EAP-FAST can now be considered a mature protocol, it has not really caught on outside of Cisco-specific deployments. There is not a lot of widespread support for the EAP-FAST protocol in supplicants that instead support the EAP protocols that use certificates. EAP-PEAPv0 (EAP-MSCHAPv2) remains the most widely used protocol for 802.1X WLAN security when server-side certificates are the only requirement. If client-side certificates are also mandated, EAP-TLS is usually the chosen Layer 2 authentication protocol used within an 802.1X authorization framework.

The CWSP exam tests heavily on the differences of all the various EAP types, both strong and weak. Make sure you understand the processes and capabilities of each Layer 2 EAP protocol. The following chart shows an in-depth comparison of most of the major EAP methods. Please use this chart when studying for the exam.

	EAP-MD5	EAP-LEAP	EAP-TLS	EAP-TTLS	PEAPv0 (EAP-MSCHAPv2)	PEAPv0 (EAP-TLS)	PEAPv1 (EAP-GTC)	EAP-FAST
Security Solution	RFC-3748	Cisco proprietary	RFC-5216	RFC 5281	IETF draft	IETF draft	IETF draft	RFC 4851
Digital Certificates—Client	No	No	Yes	Optional	No	Yes	Optional	No
Digital Certificates—Server	No	No	Yes	Yes	Yes	Yes	Yes	No
Client Password Authentication	Yes	Yes	N/A	Yes	Yes	No	Yes	Yes
PACs—Client	No	No	No	No	No	No	No	Yes
PACs—Server	No	No	No	No	No	No	No	Yes
Credential Security	Weak	Weak (depends on password strength)	Strong	Strong	Strong	Strong	Strong	Strong (if Phase 0 is secure)
Encryption Key Management	No	Yes	Yes	Yes	Yes	Yes	Yes	Yes
Mutual Authentication	No	Debatable	Yes	Yes	Yes	Yes	Yes	Yes
Tunneled Authentication	No	No	Optional	Yes	Yes	Yes	Yes	Yes
Wi-Fi Alliance Supported	No	No	Yes	Yes	Yes	No	Yes	Yes

Miscellaneous EAP Protocols

Many other flavors of EAP also exist. For example, there are proprietary protocols such as AirFortress EAP and Cogent Systems Biometrics Authentication EAP. We mentioned earlier that PEAPv1-EAP-GTC can be used with one-time password (OTP) token devices. IETF RFC 4793 defines *EAP-Protected One-Time Password Protocol (EAP-POTP),* which is another EAP method suitable for use with OTP token devices. EAP-POTP may be used as a better alternative for an internal authentication method inside the TLS tunnel of other protocols such as EAP-PEAP or EAP-TTLS.

Some EAP protocols are intended for use in the mobile phone industry. In the sections that follow, we will discuss two EAP methods that are intended for use with cellular networks.

EAP-SIM

EAP-Subscriber Identity Module (EAP-SIM) was primarily developed for the mobile phone industry and more specifically for second-generation (2G) mobile networks. Many of us who have mobile phones are familiar with the concept of a *Subscriber Identity Module (SIM)* card. A SIM card is an embedded identification and storage device very similar to a smart card. SIM cards are smaller and fit into small mobile devices like cellular or mobile phones with a 1:1 relationship to a device at any given time. The *Global System for Mobile Communications (GSM)* is a second-generation mobile network standard. EAP-SIM is outlined in the IETF RFC 4186, and it specifies an EAP mechanism that is based on 2G mobile network GSM authentication and key agreement primitives. For mobile phone carriers, this is a valuable piece of information that can be utilized for authentication. EAP-SIM does not offer mutual authentication, and key lengths are much shorter than the third-generation mechanisms used in third-generation (3G) mobile networks.

EAP-AKA

*EAP-Authentication and Key Agreement (EAP-AKA)*is an EAP type primarily developed for the mobile phone industry and more specifically for 3G mobile networks. EAP-AKA is outlined in the IETF RFC 4187, and it defines the use of the authentication and key agreement mechanisms already being used by the two types of 3G mobile networks. The 3G mobile networks include the *Universal Mobile Telecommunications System (UTMS)* and *CDMA2000*. AKA typically runs in a SIM module. The SIM module may also be referred to as a User Subscriber Identity Module (USIM) or Removable User Identity Module (R-UIM), which, as discussed earlier, is very similar to a smart card.

AKA is based on challenge-response mechanisms and symmetric cryptography and runs in the USIM or RUIM module. Key lengths can be substantially longer, and mutual authentication has now been included.

As WLANs are used more and more for voice communications, a protocol that can extend from the mobile network carriers into an enterprise WLAN can provide some advantages. *Fixed Mobile Convergence (FMC)* is a growing market segment targeted at large enterprises whereby dual-mode mobile phones (Wi-Fi-enabled mobile phones) can roam from a mobile network carrier to a WLAN and maintain a call session state. The promise is that with EAP-AKA being standardized in 802.11-based networks, a single user identifier would authenticate the device for both networks.

In mid-2009, the Wi-Fi Alliance announced the inclusion of EAP-AKA into the WPA2 interoperability suite. Both EAP-SIM and EAP-AKA are protocols that can be used with cellular devices.

A growing trend with public access networks is the use of 802.1X/EAP with *Hotspot 2.0*. Hotspot 2.0 is a Wi-Fi Alliance technical specification that is supported by the Passpoint certification program. With Hotspot 2.0, the client device is equipped by an authentication provider with one or more credentials, such as a SIM card, username/password pair, or X.509 certificate. Passpoint devices can query the network prior to connecting in order to identify potential authentication providers that are available on that network. Both EAP-SIM and EAP-AKA are protocols that can be used with cellular devices.

As WLANs are used more and more for voice communications, a protocol that can extend from the mobile network carriers into an enterprise WLAN can provide some advantages. *Fixed Mobile Convergence (FMC)* is a growing market segment targeted at large enterprises whereby dual-mode mobile phones (Wi-Fi-enabled mobile phones) can roam from a mobile network carrier to a WLAN and maintain a call session state. It is still early for enterprise WLANs to support FMC, and it will be interesting to see how EAP-AKA will play a role with these devices. The promise is that with EAP-AKA being standardized in 802.11-based networks, a single user identifier would authenticate the device for both networks.

EAP-TEAP

A more recent specification of EAP was defined in RFC-7170 for *Tunneled Extensible Authentication Protocol (TEAP)*. This protocol is an enhanced version of EAP-FAST that uses PACs for credentials. TEAP also uses tunneled authentication, and multiple inner authentication protocols are supported within the TLS tunnel. TEAP allows for multiple credentials to be authenticated within a single EAP transaction. TEAP has the ability to link the credentials of the machine and the user together in a process known as *EAP chaining*. For example, EAP-TLS could be run twice as an inner authentication method, first using machine credentials then with user credentials.

EXERCISE 4.1

802.1X/EAP Frame Exchanges

In this exercise, you will use a protocol analyzer to view the 802.11 frame exchanges used during an 802.1X/EAP authentication process.

1. To perform this exercise, you need to first download the files `EAP_MD5.PCAP`, `EAP_LEAP.PCAP`, `EAP_PEAP.PCAP`, `EAP_TTLS.PCAP`, and `EAP_TLS.PCAP` from the book's online resource area, which can be accessed at `www.sybex.com/go/cwsp2e`.

2. After the file is downloaded, you will need packet analysis software to open it. If you do not already have a packet analyzer installed on your computer, you can download Wireshark from `www.wireshark.org`.
3. Using the packet analyzer, open the `EAP_MD5.PCAP` file. Most packet analyzers display a list of capture frames in the upper section of the screen, with each frame numbered sequentially in the first column.
4. Notice in frames 7–13 that Open System authentication and association occurs prior to the EAP exchange. Observe the EAP frame exchange in packets 15–25 using EAP-MD5. Notice the lack of EAPOL-key frames. This is because EAP-MD5 uses one-way authentication and dynamic encryption keys are not created.
5. Click packet 15 to observe the frame details. Typically in the lower section of the screen is the packet details window. This section contains annotated details about the selected frame. In the lower window, locate and expand the field called 802.1X Authentication. Observe that the packet type is an EAPOL-Start frame.
6. Click packet 17 to observe the frame details. In this window, locate and expand the field called 802.1X Authentication. Observe that the packet type is an EAP-Packet frame.
7. Open the `EAP_LEAP.PCAP` file.
8. Click packet 7 to observer the frame details. In the lower window, locate and expand the field called Extensible Authentication Protocol. Observe that the identity is this EAP-Response frame, which can be seen in clear text. The identity is "airspy." The real username is always seen in clear text when LEAP is used.
9. Open the `EAP_PEAP.PCAP` file.
10. Click packet 13 to observe the frame details. In the lower window, locate and expand the field called Extensible Authentication Protocol. Observe that the identity is this EAP-Response frame, which can be seen in clear text. The identity is "administrator." This is the outer identity that is always seen in clear text when PEAP is used. This is a bogus username. The real username is hidden inside the encrypted TLS tunnel.
11. Notice the frames 17-46, which are EAPOL-packet frames used to create the TLS tunnel and encrypt the authentication exchange for the inner identity.
12. Click packet 19 to observe the frame details. In the lower window, locate and expand the field called Secure Sockets Layer. Note the TLS handshake using the server certificate. This is to create the TLS tunnel.
13. Click packet 31 to observe the frame details. In the lower window, locate and expand the field called Secure Sockets Layer. Note the application data is encrypted.
14. Observe the EAPOL-Key frames in packets 47–53. These are the frames used to create dynamic encryption keys following the authentication process. Once the supplicant gets validated and the keys are created, the controlled port becomes unblocked.

EXERCISE 4.1 *(continued)*

15. Open the EAP_TTLS.PCAP file and observe the EAP-TTLS frame exchange. Notice the similarity to PEAP.
16. Open the EAP_TLS.PCAP file and observe the EAP-TLS frame exchange.
17. Click packet 29 to observe the frame details. In the lower window, locate and expand the field called Secure Sockets Layer. Notice the exchange of the client certificate. The supplicant uses the client certificate as the authentication credential that must be validated.

Summary

802.1X/EAP enterprise methods currently exist that allow for secure WLAN communication. Hopefully, enough information was presented in this chapter to create more awareness regarding 802.1X framework and Layer 2 EAP authentication methods.

As a WLAN security professional, your ability to understand the complexities of the many protocols involved in WLAN security is paramount. Each protocol has its advantages and drawbacks, which must be mapped to the requirements of each implementation. Furthermore, a proper deployment of each security type must be performed. Simply using a strong EAP type doesn't mean security is achieved.

The key thing to keep in mind is that you need to employ the most secure method possible for your organization based on the abilities of the client devices, end users, and systems being deployed.

Exam Essentials

Explain the concept of credentials. Understand the differences between something you are, something you have, and something you know. Explain the importance of multifactor authentication.

Understand the concept of AAA. Explain in detail how authentication is used to validate identity, authorization is used to grant access, and accounting is used for a paper trail.

Describe the 802.1X framework. Explain the roles of the 802.1X components of the supplicant, authenticator, and authentication server. Understand the concept of controlled and uncontrolled virtual ports.

Define the various types of supplicant identity credentials. Describe the many different types of credentials that can be used by a supplicant. This includes username/passwords, client certificates, machine credentials, and many more.

Describe the role of server-side and root CA certificates. Understand how a server-side certificate is used to validate the authentication server. Understand that the other purpose of the server-side certificate is to create an encrypted TLS tunnel.

Describe the role of client-side and root CA certificates. Understand that client-side certificates can be used with some versions of EAP as a client credential during the authentication process. Realize that EAP-TLS is the most common protocol deployed when client certificates are mandated.

Explain all the Layer 2 EAP methods. Be able to explain all the capabilities of each EAP method as well as the differences. Understand why different EAP methods may be used in different situations.

Review Questions

1. Which of these types of EAP use tunneled authentication? (Choose all that apply.)
   A. EAP-LEAP
   B. EAP-PEAPv0 (EAP-MSCHAPv2)
   C. EAP-PEAPv1 (EAP-GTC)
   D. EAP-FAST
   E. EAP-TLS (privacy mode)

2. Which of these types of EAP require a client-side X.509 digital certificate to be used as the supplicant credentials? (Choose all that apply.)
   A. EAP-TTLS
   B. EAP-PEAPv0 (EAP-MSCHAPv2)
   C. EAP-PEAPv0 (EAP-TLS)
   D. EAP-FAST
   E. EAP-TLS (privacy mode)
   F. EAP-TLS (nonprivacy mode)

3. Which of these types of EAP use three phases of operation? (Choose all that apply.)
   A. EAP-TTLS
   B. EAP-PEAPv0 (EAP-MSCHAPv2)
   C. EAP-PEAPv0 (EAP-TLS)
   D. EAP-FAST
   E. EAP-TLS (privacy mode)
   F. EAP-TLS (nonprivacy mode)

4. Which of these types of EAP require a server-side certificate to create an encrypted TLS tunnel?
   A. EAP-TTLS
   B. EAP-PEAPv0 (EAP-MSCHAPv2)
   C. EAP-PEAPv0 (EAP-TLS)
   D. EAP-FAST
   E. EAP-PEAPv1 (EAP-GTC)
   F. EAP-LEAP

5. Which of these types of EAP are susceptible to offline dictionary attacks? (Choose all that apply.)
   A. EAP-SIM
   B. EAP-MD5
   C. EAP-PEAPv0 (EAP-TLS)
   D. EAP-FAST

E. EAP-PEAPv1 (EAP-GTC)

F. EAP-LEAP

6. What is the difference between the inner and outer identity?

   A. Only the authentication server provides its credentials in the outer identity response.

   B. The inner identity is only for authentication server credentials provided to the supplicant.

   C. The inner identity must correspond to the outer identity for realm-based authentications.

   D. The outer identity is in plain text; the inner identity is securely transmitted inside a TLS tunnel.

   E. The outer identity is only for authentication server credentials provided to the supplicant.

7. How does a RADIUS server communicate with an authenticator? (Choose all that apply.)

   A. UDP ports 1812 and 1813

   B. TCP ports 1645 and 1646

   C. Encrypted TLS tunnel

   D. Encrypted IPsec tunnel

   E. RADIUS IP packets

   F. EAPOL frames

8. In a point-to-point bridge environment where 802.1X/EAP is used for bridge authentication, what device in the network acts as the 802.1X supplicant?

   A. Nonroot bridge

   B. WLAN controller

   C. Root bridge

   D. RADIUS server

   E. Layer 3 core switch

9. Which Layer 2 protocol is used for authentication in an 802.1X framework?

   A. PAP

   B. MS-CHAPv2

   C. EAP

   D. CHAP

   E. MS-CHAP

10. Which of these types of EAP offers support for legacy authentication protocols within the inner TLS tunnel to validate supplicant credentials?

    A. EAP-TLS

    B. EAP-TTLS

    C. EAP-FAST

    D. EAP-PEAPv0

    E. EAP-PEAPv1

11. When you are using an 802.11 wireless controller solution, which device would you consider the authenticator?
    A. Access point
    B. RADIUS database
    C. LDAP
    D. WLAN controller
    E. VLAN

12. For an 802.1X/EAP solution to work properly, which two components must both support the same type of EAP? (Choose two.)
    A. Supplicant
    B. Authorizer
    C. Authenticator
    D. Authentication server

13. What does 802.1X/EAP provide when implemented for WLAN security? (Choose all that apply.)
    A. Access to network resources
    B. Verification of access point credentials
    C. Dynamic authentication
    D. Dynamic encryption-key generation
    E. Verification of user credentials

14. Chris has been hired as a consultant to secure the Harkins Corporation's WLAN infrastructure. Management has asked him to choose a WLAN authentication solution that will best protect the company's network resources from unauthorized users. The company is also looking for a strong dynamic encryption solution for data privacy reasons. Management is also looking for the cheapest solution as well as a solution that is easy to administer. Which of these WLAN security solutions does Chris decide meets all of the objectives required by management? (Choose the best answer.)
    A. EAP-TLS and TKIP/RC4 encryption
    B. EAP-TLS and CCMP/AES encryption
    C. EAP-PEAPv0 (MSCHAPv2) and CCMP/AES encryption
    D. EAP-PEAPv0 (EAP-TLS) and CCMP/AES encryption
    E. EAP-FAST/manual provisioning and CCMP/AES encryption
    F. EAP-MD5 and CCMP/AES encryption

15. What type of credential is used by the authenticator and authentication server to validate each other?
    A. Server-side X.509 digital certificate
    B. PAC

C. Client-side X.509 digital certificate

D. Username and password

E. Security token

F. Shared secret

16. Which of these types of EAP is designed for a Fixed Mobile Convergence (FMC) authentication solution over an 802.11 WLAN and a 3G cellular telephone network?

A. EAP-SIM

B. EAP-GTC

C. EAP-PEAPv0 (EAP-TLS)

D. EAP-AKA

E. EAP-Fortress

F. EAP-TTLS

17. What are some of the supplicant credentials that could be validated by an authentication server? (Choose all that apply.)

A. Server-side X.509 digital certificate

B. PAC

C. Client-side X.509 digital certificate

D. Username and password

E. Security token

F. Smart card

18. When 802.1X/EAP is properly deployed, which of these external databases can a RADIUS server query for proxy authentication?

A. Active Directory

B. E-Directory

C. Open Directory

D. All of the above

19. Which of these inner authentication EAP types is intended to be used with an 802.1X framework that uses security token devices as the supplicant credentials? (Choose all that apply.)

A. EAP-GTC

B. EAP-MSCHAPv2

C. EAP-POTP

D. EAP-LEAP

E. EAP-PEAP

F. EAP-TTLS

20. WLAN administrator Tammy O'Connell has been tasked with securing the corporate WLAN with 802.1X/EAP security. She has chosen to use the EAP-PEAPv0 (MSCHAPv2) protocol on the RADIUS servers and supplicants. Tammy created a root certificate using the company's internal private Certificate Authority (CA) solution. She also created a server certificate, which was signed by the internal private CA. The wireless clients connecting to the WLAN include a mixture of corporate Windows laptops. Employees will also be connecting with personal devices such as Android phones and tablets. What other steps must Tammy take to ensure full functionality with 802.1X/EAP security? (Choose all that apply.)

A. Install the server certificate on the RADIUS server.

B. Distribute and install the server certificate to Windows laptops with GPO.

C. Distribute and install the root CA certificate to Windows laptops with GPO.

D. Distribute and install the server certificate to employee devices with GPO.

E. Distribute and install the root CA certificate to employee devices with GPO.

F. Distribute and install the server certificate to employee devices with MDM.

G. Distribute and install the root CA certificate to employee devices with MDM.

Chapter 5

802.11 Layer 2 Dynamic Encryption Key Generation

IN THIS CHAPTER, YOU WILL LEARN ABOUT THE FOLLOWING:

✓ **Advantages of dynamic encryption**

- Dynamic WEP
- Robust security network (RSN)
- RSN information element
- Authentication and key management (AKM)
- RSNA key hierarchy
- 4-Way Handshake
- Group Handshake
- PeerKey Handshake
- TDLS Peer Key Handshake
- Passphrase-to-PSK mapping
- Roaming and dynamic keys

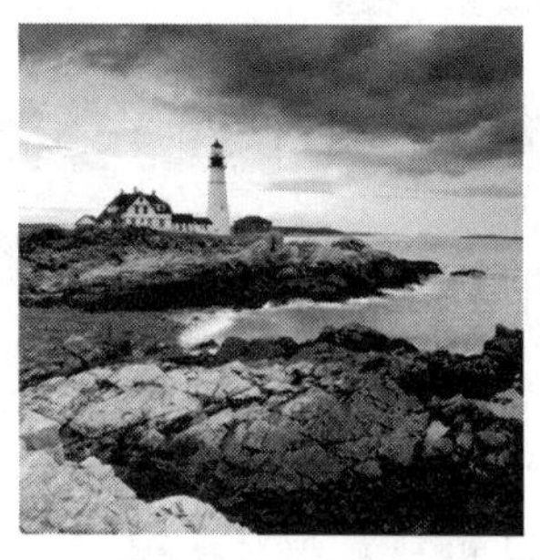

The 802.11-2012 standard defines two classes of security methods using pre-RSNA and RSNA algorithms. Pre-RSNA methods use static WEP encryption and the legacy authentication methods that were discussed in Chapter 2, "Legacy 802.11 Security." RSNA security methods use either TKIP/ARC4 or CCMP/AES encryption, dynamic key management procedures, and the PSK and 802.1X authentication methods that were discussed in Chapter 4, "802.1X/EAP Authentication." In this chapter, you will learn that there is a symbiotic relationship between PSK/802.1X authentication and the generation of dynamic encryption keys. You will also learn about dynamic WEP, which was a short-term, nonstandard method of dynamic encryption, as well as the standards-based dynamic encryption key management methods that are widely deployed in today's enterprise WLANs.

Advantages of Dynamic Encryption

Although the 802.1X/EAP framework does not require encryption, it highly suggests the use of encryption to provide data privacy. You have already learned that the purpose of 802.1X/EAP is authentication and authorization. If a supplicant is properly authenticated by an authentication server using a Layer 2 EAP protocol, the supplicant is allowed through the controlled port of the authenticator and communication at the upper layers of 3–7 can begin for the supplicant. 802.1X/EAP protects network resources so that only validated supplicants are authorized for access. However, as shown in Figure 5.1, an outstanding by-product of 802.1X/EAP can be the generation and distribution of dynamic encryption keys. Later in this chapter, you will see that dynamic encryption keys can also be generated as a by-product of PSK authentication.

EAP protocols that utilize mutual authentication provide "seeding material" that can be used to generate encryption keys dynamically. Mutual authentication is required to generate unique dynamic encryption keys. EAP-TLS, EAP-TTLS, EAP-FAST, EAP-LEAP, EAP- PEAP, and other versions of EAP utilize mutual authentication and can provide the seeding material needed for dynamic encryption key generation. EAP-MD5 cannot generate dynamic keys because EAP-MD5 uses only one-way authentication.

Legacy WLAN security involved the use of static WEP keys. The use of static keys is typically an administrative nightmare, and when the same static key is shared among multiple users, the static key is easy to compromise via social engineering. The first advantage of using dynamic keys rather than static keys is that they cannot be compromised by social engineering attacks because the users have no knowledge of the keys. The second advantage of dynamic keys is that every user has a different and unique key. If a single user's encryption key was somehow compromised, none of the other users would be at risk because every user has a unique key. The dynamically generated keys are not shared between the users.

FIGURE 5.1 802.1X/EAP and dynamic keys

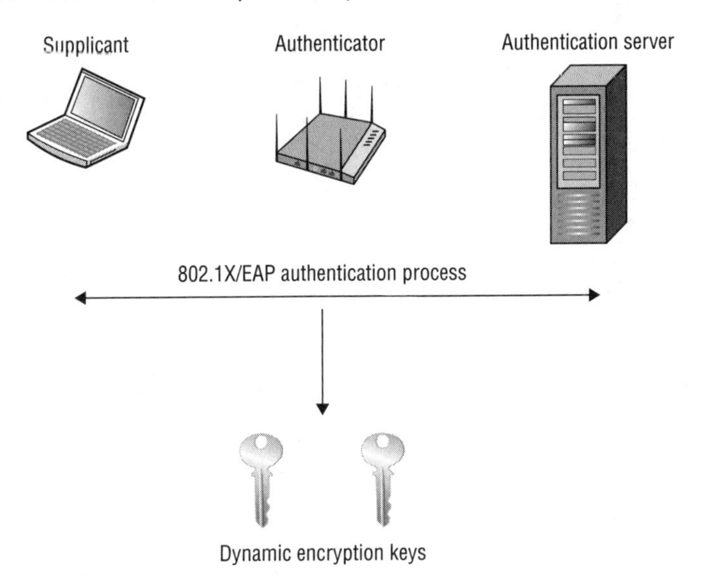

In 2004, the 802.11i security amendment was ratified, defining stronger encryption and better authentication methods. The 802.11i amendment, which is now part of the 802.11-2012 standard, fully defines robust security network association (RSNA), which is discussed later in this chapter. RSNA security methods use either TKIP/ARC4 or CCMP/AES encryption. Before TKIP/ARC4 or CCMP/AES was used, WLAN vendors offered a dynamic key generation security solution using WEP encryption. Many of these solutions were proprietary and did not offer vendor interoperability. Dynamic WEP has never been defined as any part of 802.11 security and was intended only as a stop-gap dynamic encryption solution until the stronger encryption methods of TKIP/ARC4 or CCMP/AES became available.

Dynamic WEP encryption keys were generated as a by-product of the 802.1X/EAP process. Dynamic WEP was a nonstandard and legacy encryption solution that was mostly used with autonomous access points prior to the availability of TKIP/ARC4 and CCMP/AES dynamic encryption. Within the 802.1X framework, the autonomous AP would be considered the authenticator. As shown in step 2 of Figure 5.2, after an EAP frame exchange where mutual authentication is required, both the authentication server and the supplicant now have information about each other due to the mutual authentication exchange of credentials. As shown in step 2, this newfound information is used as *seeding material* or *keying material* to generate a matching dynamic encryption key for both the supplicant and the authentication server. These dynamic keys are generated *per session per user*, meaning that every time a supplicant authenticates, a new key is generated and every user has a unique and separate key. This dynamic WEP session key is often referred to as the *unicast key* because it is the dynamically generated WEP key that is used to encrypt and decrypt all unicast 802.11 data frames.

FIGURE 5.2 Dynamic WEP process

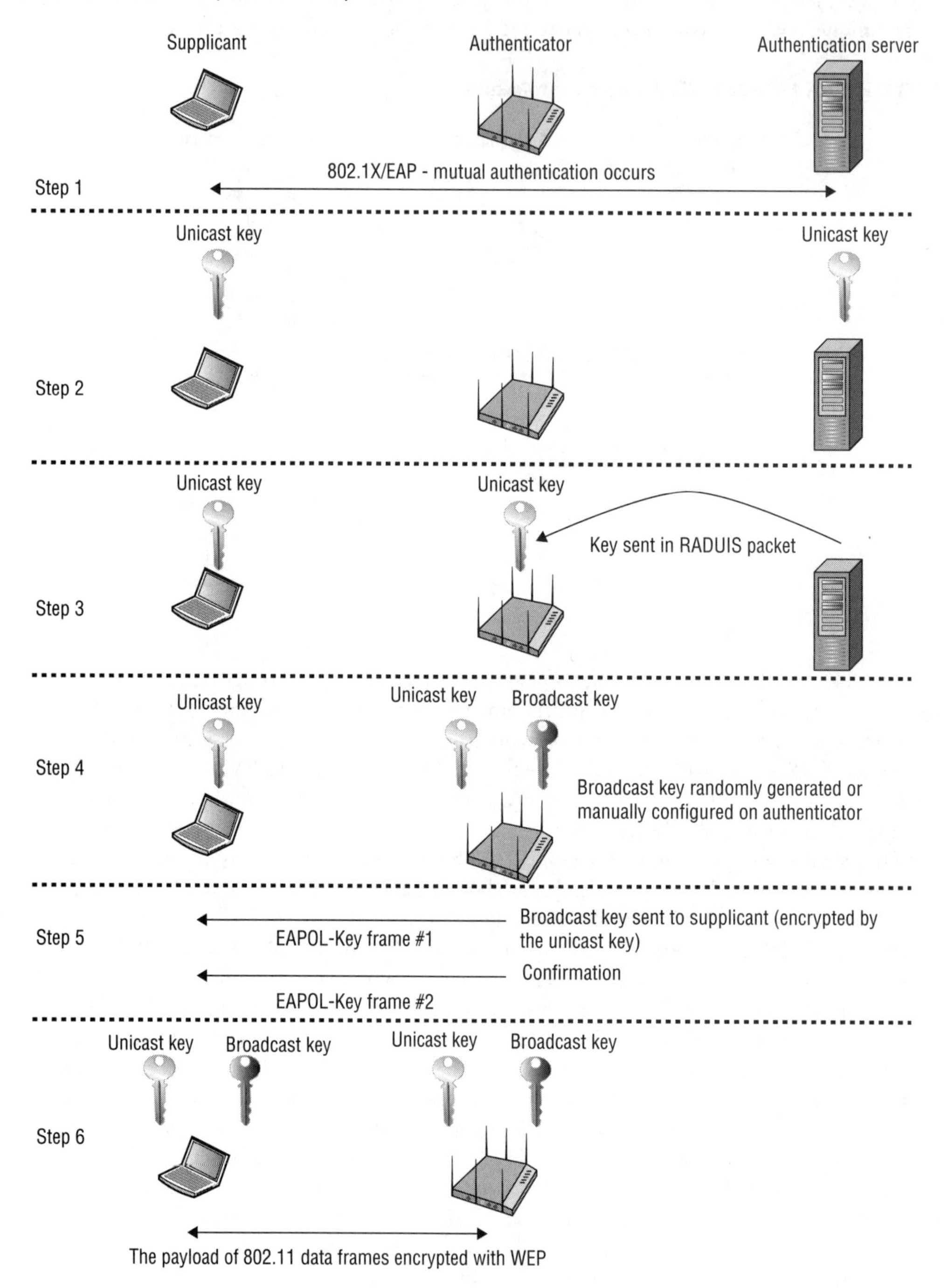

As shown in step 3, after the unicast key is created, the authentication server delivers its copy of the unicast key encapsulated inside a RADIUS packet to the authenticator. The access point and the client station now both have unique unicast keys that can be used to encrypt and decrypt unicast 802.11 data frames. A second key exists on the access point known as the *broadcast key*. As depicted in step 4, the broadcast key can be manually configured on the access point or can be randomly generated. The broadcast key is used to encrypt and decrypt all broadcast and multicast 802.11 data frames. Each client station has a unique and separate unicast key, but every station must share the same broadcast key. As step 5 shows, the broadcast key is delivered from the access point to the client station in a unicast frame encrypted with the client station's unique unicast key. When dynamic WEP is deployed, a two EAPOL-Key frame exchange always follows the EAP frame exchange. The two EAPOL-Key frames are both sent by the authenticator to the supplicant. The first EAPOL-Key frame carries the broadcast key from the access point to the client. The second EAPOL-Key frame is effectively a confirmation that the keys are installed and that the encryption process can begin. Once the client station has both the unicast and broadcast keys, the *MAC service data unit (MSDU)* payload of all 802.11 data frames will then be protected using WEP encryption.

EXERCISE 5.1

Dynamic WEP

In this exercise, you will use a protocol analyzer to view the 802.11 frame exchanges used to create dynamic WEP keys.

1. To perform this exercise, you need to first download the `PEAP_WEP.PCAP` file from the book's online resource area, which can be accessed at `www.sybex.com/go/cwsp2e`.
2. After the file is downloaded, you will need packet analysis software to open it. If you do not already have a packet analyzer installed on your computer, you can download Wireshark from `www.wireshark.org`.
3. Using the packet analyzer, open the `PEAP_WEP.PCAP` file. Most packet analyzers display a list of capture frames in the upper section of the screen, with each frame numbered sequentially in the first column.
4. Observe that the EAP frame exchange in packets 346–389 is using EAP-PEAP. The access point (authenticator) MAC address is 00:12:43:CB:0F:30. The client radio (supplicant) MAC address is 00:40:96:A3:0C:45.
5. Notice the two EAPOL-Key frames in packets 391 and 393 that are being sent from the access point to the client.
6. Notice that all 802.11 data frames are encrypted using a WEP key that was dynamically generated.
7. Click on packet 395. Typically the packet details window appears in the lower section of the screen. This section contains annotated details about the selected frame. In this window, locate and expand the IEEE QoS Data Frame header. Locate the WEP parameters field to verify the use of WEP encryption.

Is Dynamic WEP Encryption Secure?

In the past, the generation and distribution of dynamic WEP keys as a by-product of the EAP authentication process had many benefits and was preferable to the use of static WEP keys. When using dynamic WEP, static keys were no longer used and keys did not have to be entered manually. Also, every user had a separate and independent key. If a user's dynamic unicast key was compromised, only that one user's traffic could be decrypted. However, a dynamic WEP key can still be cracked and, if compromised, can indeed be used to decrypt data frames. The numerous WEP cracking tools such as Aircrack-ng can obtain a WEP key in a matter of minutes. Therefore, dynamic WEP still has severe data privacy risks. Dynamic WEP should only be used with legacy WLAN equipment that does not support the use of TKIP/ARC4 or CCMP/AES encryption. Most WLAN vendors have a key interval setting or reauthentication interval setting available on the access point. The configuration setting forces clients to reauthenticate at timed intervals and thus create new dynamic WEP keys. Because WEP can be cracked in such a short time, the key interval settings should be set at about 10 minutes or less on the legacy WLAN equipment. Please understand that dynamic WEP is not the same as RSNA dynamic key management. Later in this chapter, you will learn about RSNAs that define the creation of stronger and safer dynamic TKIP/ARC4 or CCMP/AES encryption keys that can also be generated as a by-product of the EAP authentication process and PSK authentication. The radios in modern-day WLAN equipment use dynamically generated CCMP/AES encryption keys.

Robust Security Network (RSN)

The 802.11i amendment, which was ratified and published as IEEE Std. 802.11i-2004, defined stronger encryption and better authentication methods. The 802.11i security amendment is now part of the 802.11-2012 standard. The 802.11-2012 standard defines what is known as a robust security network (RSN) and robust security network associations (RSNAs).

A security association is a set of policies and keys used to protect information. A *robust security network association (RSNA)* requires two 802.11 stations (STAs) to establish procedures to authenticate and associate with each other as well as create dynamic encryption keys through a process known as the *4-Way Handshake*. This association between two stations is referred to as an RSNA. In other words, any two radios must share dynamic encryption keys that are unique between those two radios. CCMP/AES encryption is the mandated encryption method, whereas TKIP/ARC4 is an optional encryption method. As you have already learned, only CCMP/AES can be used for 802.11n and 802.11ac data rates.

IEEE 802.11-2012 Clause 11

All aspects of robust security network mechanisms can be found in clause 11 of the 802.11-2012 standard.

Anyone who has passed the CWNA certification exam is familiar with the WLAN topologies of a basic service set (BSS) and an independent basic service set (IBSS). The *basic service set (BSS)* is the cornerstone topology of an 802.11 network. The communicating devices that make up a BSS are solely one AP with one or more client stations. Client stations join the AP's wireless domain and begin communicating through the AP. Stations that are members of a BSS have a Layer 2 connection and are called *associated*. The 48-bit (6-octet) MAC address of an access point's radio is known as the *basic service set identifier (BSSID)*. The BSSID address is the Layer 2 identifier of each individual BSS. Most often, the BSSID is the MAC address of the access point. Do not confuse the BSSID address with the SSID. The *service set identifier (SSID)* is the logical WLAN name that is user configurable, while the BSSID is a Layer 2 MAC address used to identify a basic service set (BSS).

When RSN security associations are used within a BSS, all of the client station radios have unique encryption keys that are shared with the radio of the access point. As shown in Figure 5.3, all the client stations have undergone a unique RSNA process called the 4-Way Handshake where the access point and each client radio has either a unique dynamic TKIP/ARC4 or CCMP/AES key that is shared between the client radio and the access point radio. This key is called the pairwise transient key (PTK) and is used to encrypt/decrypt unicast traffic. All the stations share a broadcast key called the group temporal key (GTK), which is used to encrypt/decrypt all broadcast and multicast traffic. You will learn more about the PTK and GTK keys later in this chapter in the section "4-Way Handshake."

FIGURE 5.3 RSNA within a BSS

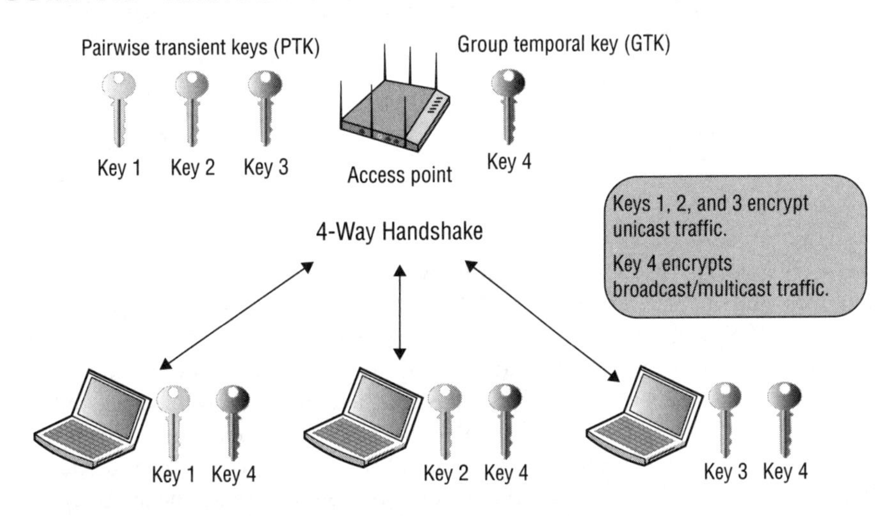

The 802.11 standard also defines a WLAN topology called an *independent basic service set (IBSS)*. The radio cards that make up an IBSS network consist solely of client stations (STAs), and no access point is deployed. An IBSS network that consists of just two STAs is analogous to a wired crossover cable. An IBSS can, however, have multiple client stations in one physical area communicating in an ad hoc fashion. As

you can see in Figure 5.4, all the stations within an IBSS have undergone a unique RSNA process (called the 4-Way Handshake) with each other, because all unicast communications are peer to peer. Each station has either a unique dynamic TKIP/ARC4 or CCMP/AES pairwise transient key (PTK) that is shared with any other station within the IBSS. In an IBSS, each STA defines its own group temporal key (GTK), which is used for its broadcast/multicast transmissions. Each IBSS station will use either the 4-Way Handshake or the Group Key Handshake to distribute its transmit GTK to its peer stations. PSK authentication is used within the IBSS to seed the 4-Way Handshake. Therefore, every time a client joins an IBSS with a peer station, the client must reauthenticate and create new keys.

FIGURE 5.4 RSNA within an IBSS

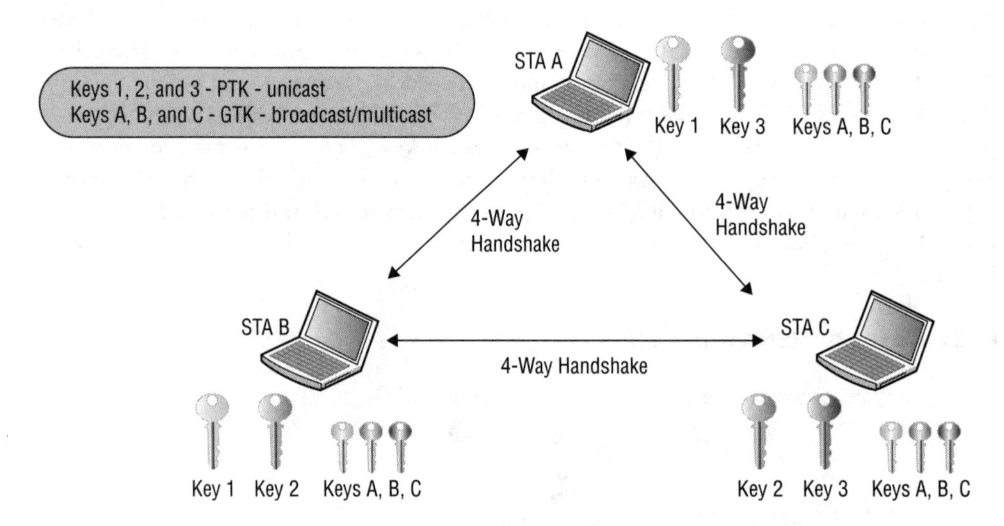

A *robust security network (RSN)* is a network that allows for the creation of only robust security network associations (RSNAs). In other words, a basic service set (BSS) where all the stations are using only TKIP/ARC4 or CCMP/AES dynamic keys for encryption would be considered an RSN. Robust security only exists when all devices in the service set use RSNAs. As shown in Figure 5.5, all the stations within the BSS have established an RSNA that resulted in either TKIP/ARC4 or CCMP/AES unique dynamic keys. Because only RSNA security is in use, the pictured BSS would be considered a robust security network. Note that the group temporal key (GTK) uses TKIP/ARC4 to encrypt/decrypt broadcast and multicast traffic within the robust security network BSS. Although the majority of modern-day WLANs radios use CCMP/AES encryption, older radios that only support TKIP/ARC4 still exist. If only one TKIP/ARC4 client station is associated within the BSS, the lowest common dominator method of TKIP encryption would be used for broadcast and multicast traffic.

FIGURE 5.5 Robust security network

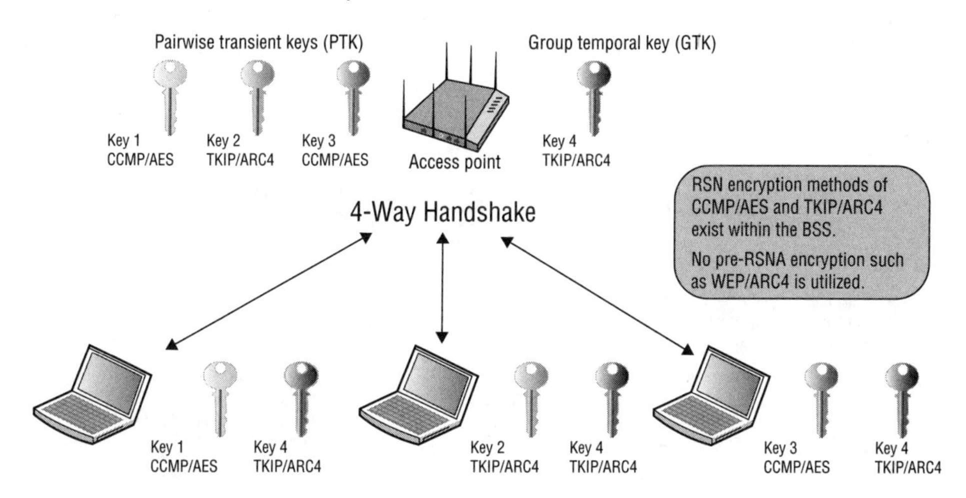

As you learned in Chapter 2, a pre-RSN security network uses static WEP encryption and legacy authentication methods. A WLAN that uses dynamic WEP encryption keys would also be considered as using pre-RSN security, but the use of dynamic WEP was never defined by either the IEEE or the Wi-Fi Alliance. The 802.11-2012 standard does allow for the creation of *pre-robust security network associations (pre-RSNAs)* as well as RSNAs. In other words, legacy security measures can be supported in the same basic service set (BSS) along with RSN-security-defined mechanisms. A *transition security network (TSN)* supports RSN-defined security as well as legacy security, such as WEP, within the same BSS. As you can see in Figure 5.6, some of the stations within the BSS have established an RSNA that resulted in either TKIP/ARC4 or CCMP/AES unique dynamic keys. However, some of the stations are using static WEP keys for encryption. Because both RSNAs and pre-RSNAs are in use, the pictured BSS would be considered a transition security network. Note that the group temporal key (GTK) uses a static WEP key to encrypt/decrypt broadcast and multicast traffic within the transition security network BSS. Although the majority of modern-day WLAN radios use CCMP/AES encryption, older radios that only support WEP still exist. If only one WEP client station is associated within the BSS, the lowest common dominator method of WEP encryption would be used for broadcast and multicast traffic.

As you learned earlier in this chapter, each WLAN has a logical name (SSID) and each WLAN BSS has a unique Layer 2 identifier, the basic service set identifier (BSSID). The BSSID is typically the MAC address of the access point's radio card if only one SSID is being transmitted. However, most WLAN vendors offer the capability to transmit multiple SSIDs from an access point radio. Client devices typically get confused when they see multiple SSIDs all tied to the same MAC address; therefore, if multiple SSIDs are transmitted from the same AP radio, multiple BSSIDs are also needed. As shown in Figure 5.7, the multiple BSSIDs are effectively virtual MAC addresses that are incremented or derived from the actual physical MAC address of the AP radio. Effectively, multiple basic service sets exist within the same coverage cell area of the access point. As shown in Figure 5.7, because each SSID may have been configured with different security, an RSN WLAN, a pre-RSNA WLAN, and a TSN WLAN can all exist within the same coverage area of an access point.

FIGURE 5.6 Transition security network

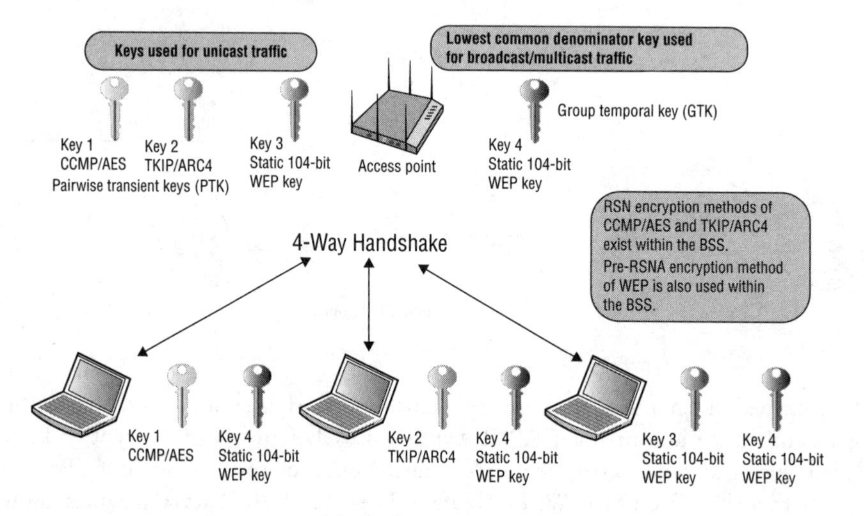

FIGURE 5.7 RSN, pre-RSN, and TSN within the same AP cell

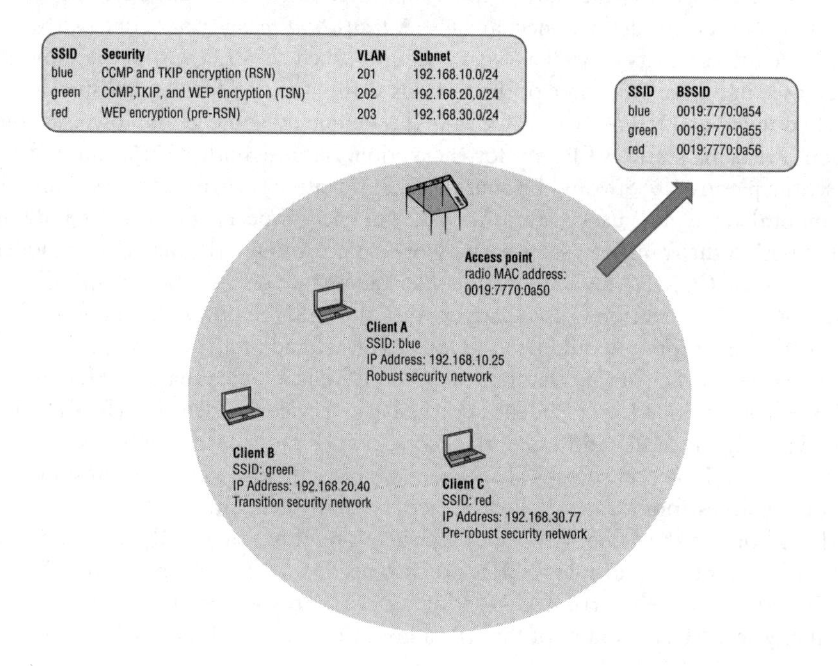

Robust Security Networks vs. Transition Security Networks

Most 802.11 radios manufactured between the years of 1997 and 2004 only supported legacy pre-RSNA security that used static WEP/ARC4 encryption. The 802.11i security amendment was ratified in 2004 and full WLAN vendor support for RSNA security became a reality in 2005. All major WLAN vendors fully support RSNA-capable equipment. Any station (STA) that is able to create RSNAs would be considered RSNA-capable. RSNA-capable devices can also use pre-RSNA security to maintain backward compatibility. Therefore, modern WLAN devices support the use of dynamic CCMP/AES encryption, dynamic TKIP/ARC4 encryption, and legacy WEP/ARC4 encryption.

As mentioned in Chapter 1, "WLAN Security Overview," the Wi-Fi Alliance designates the *Wi-Fi Protected Access 2 (WPA2)* security certification. WPA2-certified devices can support CCMP/AES, TKIP/ARC4, and WEP/ARC4 encryption. The earlier *Wi-Fi Protected Access (WPA)* certified devices only support TKIP/ARC4 and WEP/ARC4 encryption. Any devices that are not WPA- or WPA2-certified will only support WEP/ARC4 encryption.

The majority of WLAN devices that are deployed in enterprise environments are RSNA-capable devices. Therefore, the existence of transition security networks (TSNs) that support both RSN-defined security as well as legacy WEP security is not very commonplace. If client radios that only support WEP encryption are still deployed, they are long overdue for replacement. A more common scenario is that there are still devices deployed in the enterprise that only support TKIP encryption. Technically a WLAN that supports the use of TKIP encryption is still considered a robust security network. However, TKIP is now also consider to be a legacy method of encryption and will not work with 802.11n and 802.11ac data rates. If client radios that only support TKIP are deployed, they should also be replaced. Whenever possible, deploying a "pure" robust security network using strictly CCMP/AES encryption is highly recommended.

RSN Information Element

Within a BSS, how can client stations and an access point notify each other about their RSN capabilities? RSN security can be identified by a field found in certain 802.11 management frames. This field is known as the *robust security network information element (RSNIE)* and is often referred to simply as the *RSN information element*. An information element is an optional field of variable length that can be found in 802.11 management frames. The RSN information element can identify the encryption capabilities of each station. The RSN information element will also indicate whether 802.1X/EAP authentication or preshared key (PSK) authentication is being used.

The RSN information element field is always found in four different 802.11 management frames: beacon management frames, probe response frames, association request frames, and reassociation request frames. The RSN information element can also be found in reassociation response frames if 802.11r capabilities are enabled on an AP and roaming client. Within a basic service set, an access point and client stations use the RSN information element within these four management frames to communicate with each other about their security capabilities prior to establishing association. As shown in Figure 5.8, access points will use beacons and probe response frames to inform client stations of the AP security capabilities.

FIGURE 5.8 Access point RSN security capabilities

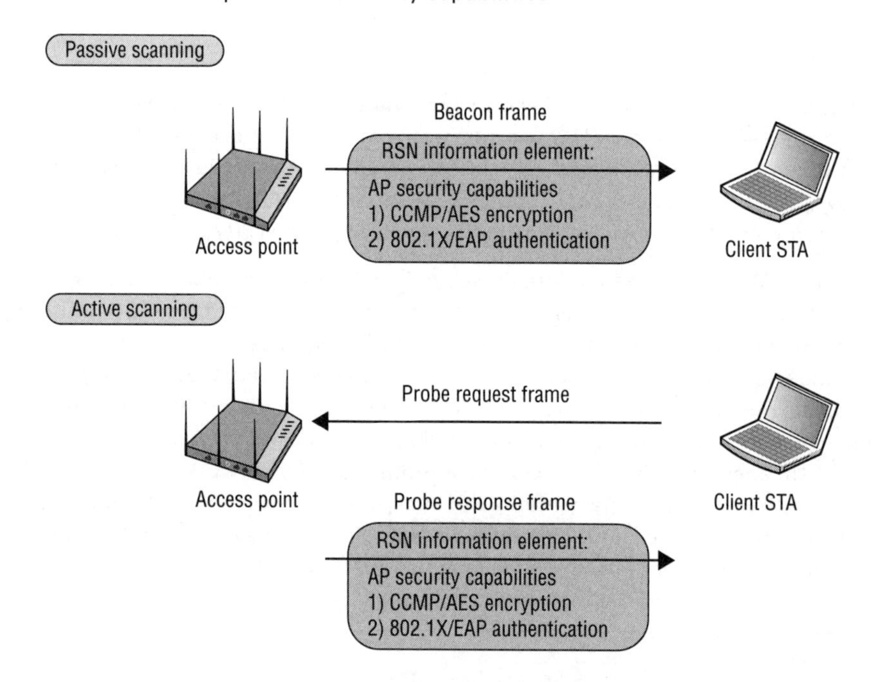

As you can see in Figure 5.9, client stations use the association request frame to inform the access point of the client station security capabilities. When stations roam from one access point to another access point, they use the reassociation request frame to inform the new access point of the roaming client station's security capabilities. The security capabilities include supported encryption cipher suites and supported authentication methods.

All 802.11 radios will use one cipher suite for unicast encryption and another cipher suite for encrypting multicast and broadcast traffic. Pairwise unicast encryption keys are created that are unique between two stations—the AP and a single client station. The pairwise cipher suite of the RSN information element contains the cipher suite information used by stations for unicast traffic. The cipher suite selector 00-0F-AC-04 (CCMP) is the default cipher suite value. The cipher suite selector 00-0F-AC-02 (TKIP) is optional. A group key is also created that is shared by all stations for broadcast and multicast traffic. The Group Cipher Suite field of the RSN information element contains the cipher suite information used by the BSS to protect broadcast/multicast traffic. The actual creation process of the pairwise and group keys is discussed later in this chapter in the section "4-Way Handshake."

FIGURE 5.9 Client station RSN security capabilities

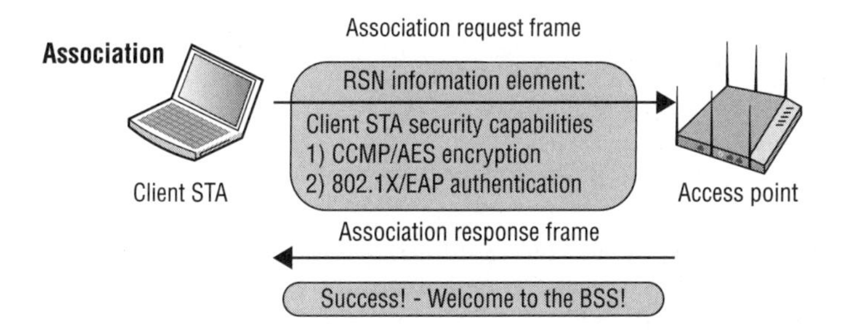

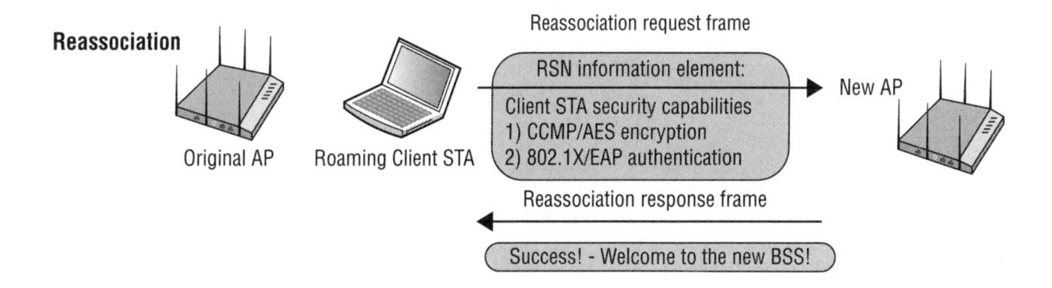

Figure 5.10 shows a capture of a beacon frame from an access point configured to use only CCMP/AES encryption. The RSN information element indicates that CCMP/AES encryption is being used for both the group cipher suite and the pairwise cipher suite. The access point is using the RSN information element to inform all client stations that the AP will be using CCMP encryption for all broadcast/multicast and will also be using CCMP/AES encryption for any unicast traffic. Any client station must support those exact ciphers for the client stations to be able to establish a robust secure network association (RSNA) with the AP and to create dynamic encryption keys. In other words, the client stations must support CCMP/AES encryption to be allowed to join the AP's basic service set.

Figure 5.11 shows a beacon frame from an access point configured to support both CCMP/AES and TKIP/ARC4 encryption. The RSN information element indicates that TKIP encryption is being used for the group cipher suite. CCMP/AES is the default cipher for the pairwise cipher suite. The access point is using the RSN information element to inform all client stations that the AP will support either CCMP/AES or TKIP/ARC4 encryption for any unicast traffic. However, only TKIP/ARC4 encryption can be used for

FIGURE 5.10 RSN Information element—CCMP pairwise and CCMP group cipher

```
RSN Information
  Element ID:            48  RSN Information [66]
  Length:                20 [67]
  Version:               1 [68-69]
  Group Cipher OUI:      00-0F-AC [70-72]
  Group Cipher Type:     4 CCMP - default in an RSN [73]
  Pairwise Cipher Count:1 [74-75]
  PairwiseKey Cipher List
    Pairwise Cipher OUI:   00-0F-AC-04  CCMP - default in an RSN [76-79]
  AuthKey Mngmnt Count: 1 [80-81]
  AuthKey Mngmnt Suite List AKMP Suite OUI=00-AC-01 802.1X Authentication
  RSN Capabilities=%0000000000101000
```

broadcast/multicast traffic. In this situation, the client stations must support either CCMP or TKIP to be allowed to join the AP's basic service set. Because all the stations share a single group encryption key for broadcast and multicast traffic, the lowest common denominator must be used for the group cipher. In this case, the group cipher is TKIP.

FIGURE 5.11 RSN information element—CCMP pairwise and TKIP group cipher

```
RSN Information
  Element ID:            48  RSN Information [66]
  Length:                20 [67]
  Version:               1 [68-69]
  Group Cipher OUI:      00-0F-AC {70-72]
  Group Cipher Type:     2 TKIP [73]
  Pairwise Cipher Count:1 [74-75]
  PairwiseKey Cipher List
    Pairwise Cipher OUI: 00-0F-AC-04  CCMP - default in an RSN [76-79]
  AuthKey Mngmnt Count: 1 [80-81]
  AuthKey Mngmnt Suite List
    AKMP Suite OUI:          00-AC-01 802.1X Authentication
  RSN Capabilities=%0000000000101000
```

The cipher suite selectors 00-0F-AC-01 (WEP-40) and 00-0F-AC-05 (WEP-104) are used as a group cipher suite in a transition security network (TSN) to allow pre-RSNA devices to join a BSS. For example, an access point might support CCMP, TKIP, and WEP encryption. WPA2-capable clients will use CCMP encryption for unicast traffic between the client STA and the AP. WPA-capable clients will use TKIP encryption for unicast traffic between the client STAs and the AP. Legacy clients will use WEP encryption for unicast traffic between the client STAs and the AP. All of the clients will use WEP encryption for the broadcast and multicast traffic. Because all the stations share a single group encryption key for broadcast and multicast traffic, the lowest common denominator must be used for the group cipher. In the case of a TSN, the group cipher is WEP.

Figure 5.12 shows a capture of a beacon frame from an access point configured to support CCMP/AES, TKIP/ARC4, and static WEP/ARC4 encryption using a 40-bit static key. The RSN information element indicates that WEP-40 encryption is being used for the group cipher suite. CCMP/AES is the default cipher for the pairwise cipher suite. The access point is using the RSN information element to inform all client stations that the AP can support CCMP/AES, TKIP/ARC4 encryption, or WEP-40 for any unicast traffic. However, only WEP-40 encryption can be used for broadcast/multicast traffic. In this situation, the client stations can support CCMP/AES, TKIP/ARC4, or WEP-40 and be allowed to join the AP's basic service set.

FIGURE 5.12 RSN information element—CCMP pairwise and WEP-40 group cipher

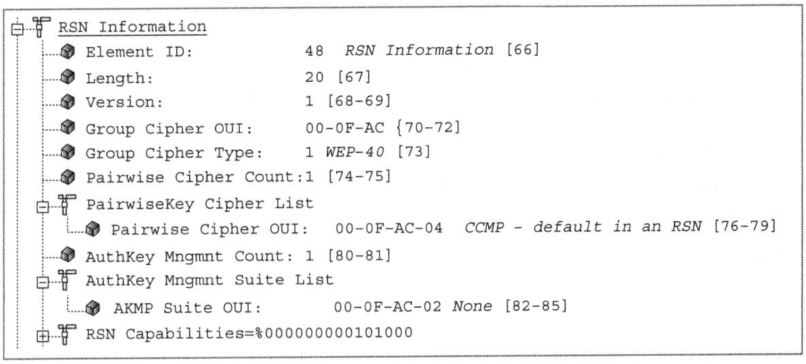

```
RSN Information
  Element ID:              48  RSN Information [66]
  Length:                  20  [67]
  Version:                 1  [68-69]
  Group Cipher OUI:        00-0F-AC {70-72]
  Group Cipher Type:       1  WEP-40  [73]
  Pairwise Cipher Count:1  [74-75]
  PairwiseKey Cipher List
    Pairwise Cipher OUI:     00-0F-AC-04   CCMP - default in an RSN [76-79]
  AuthKey Mngmnt Count: 1  [80-81]
  AuthKey Mngmnt Suite List
    AKMP Suite OUI:          00-0F-AC-02 None [82-85]
  RSN Capabilities=%0000000000101000
```

The RSN information element can also be used to indicate what authentication methods are supported. The authentication key management (AKM) suite field in the RSN information element indicates whether the station supports either 802.1X authentication or PSK authentication. If the AKM suite value is 00-0F-AC-01, authentication is negotiated over an 802.1X infrastructure using an EAP protocol. If the AKM suite value is 00-0F-AC-02 (PSK), then PSK is the authentication method that is being used.

Figure 5.13 shows a capture of an association request frame from a client station configured to 802.1X/EAP. The AKM suite field in the RSN information element indicates that 802.1X is the chosen authentication method.

FIGURE 5.13 RSN information element—AKM suite field: 802.1X

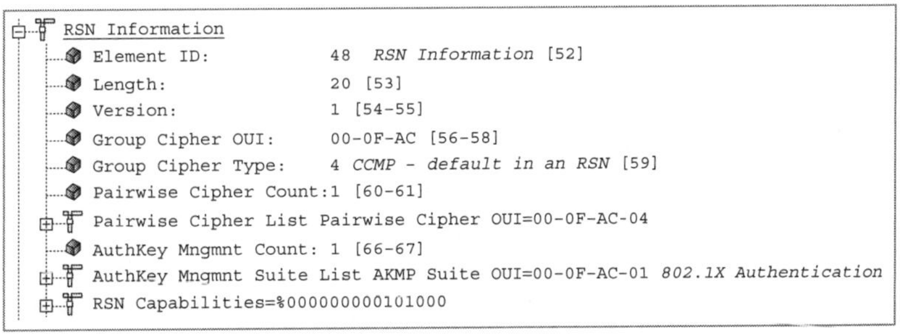

```
RSN Information
  Element ID:              48  RSN Information [52]
  Length:                  20  [53]
  Version:                 1  [54-55]
  Group Cipher OUI:        00-0F-AC [56-58]
  Group Cipher Type:       4  CCMP - default in an RSN [59]
  Pairwise Cipher Count:1  [60-61]
  Pairwise Cipher List Pairwise Cipher OUI=00-0F-AC-04
  AuthKey Mngmnt Count: 1  [66-67]
  AuthKey Mngmnt Suite List AKMP Suite OUI=00-0F-AC-01 802.1X Authentication
  RSN Capabilities=%0000000000101000
```

Client radios and access point radios use the RSN information element to inform each other about their RSN security capabilities. The access point uses beacons and probe response frames to educate clients about the authentication and encryption methods that are supported by the AP. Client stations use association request and reassociation request frames to educate the AP about the authentication and encryption methods that are supported by the client. After the exchange of RSN information, the radios can proceed with PSK or 802.1X authentication and then create dynamic encryption keys.

We further discuss the RSN information element in Chapter 7, "802.11 Fast Secure Roaming."

Authentication and Key Management (AKM)

The 802.11-2012 standard defines *authentication and key management (AKM)* services. AKM services consist of a set of one or more algorithms designed to provide authentication and key management, either individually or in combination with higher-layer authentication and key-management algorithms, which are often outside the scope of the 802.11-2012 standard. Non-IEEE-802 protocols may be used for AKM services. Many of these non-IEEE-802 protocols are defined by other standards organizations, such as the Internet Engineering Task Force (IETF). In Chapter 4, you learned about the various EAP protocols used within an 802.1X framework for authentication. EAP protocols are also used during AKM services. An *authentication and key management protocol (AKMP)* can be either a preshared key (PSK) or an EAP protocol used during 802.1X authentication. The 802.11-2012 standard also defines a third type of AKMP called *Simultaneous Authentication of Equals (SAE)*, which could be the future replacement for PSK authentication. SAE is discussed in greater detail in Chapter 6, "PSK Authentication."

As you have previously learned, the main goal of 802.1X/EAP within a WLAN is twofold:

Authentication Validate the credentials of a client station seeking access to network resources via the WLAN portal.

Authorization Grant access for the client station to network resources via the WLAN portal.

You also have learned that the goal of encryption is to provide data privacy for the MSDU payload in 802.11 data frames. AKM services require both authentication processes and the generation and management of encryption keys. Although authentication and encryption have different goals and are different processes, they are linked together in AKM services. In other words, an authentication process is necessary to generate dynamic encryption keys. The 802.1X/EAP and PSK authentication processes generate the seeding material needed to create dynamic encryption keys. Furthermore, until dynamic encryption keys are created, the controlled port of an 802.1X authenticator will not open. As shown in Figure 5.14, a symbiotic relationship exists between authentication and dynamic encryption. Authorization is not finalized until encryption keys are created and encryption keys cannot be created without authentication.

FIGURE 5.14 Authentication and key management (AKM)—overview

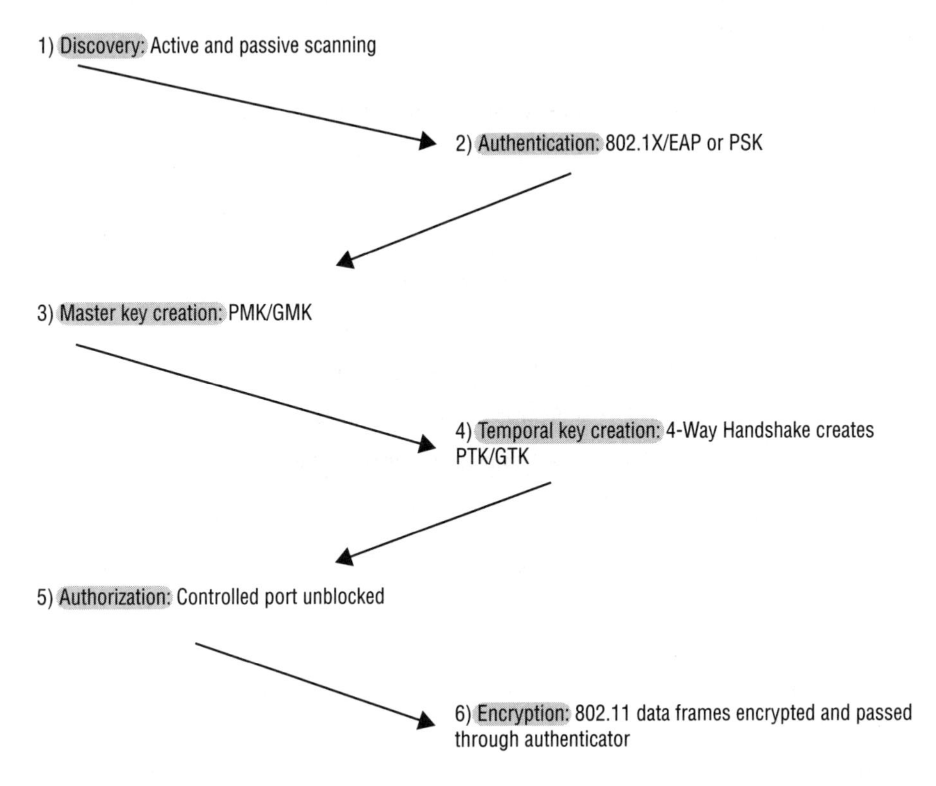

When an 802.1X/EAP authentication solution is used, AKM operations include the following:

Secure Channel The 802.11-2012 standard makes the assumption that the authenticator and authentication server (AS) have established a secure channel. The security of the channel between the authenticator and the AS is outside the scope of the 802.11-2012 standard. Authentication credentials must be distributed to the supplicant and authentication server prior to association. Many of the processes were discussed in Chapter 4.

Discovery As shown in Figure 5.15, a client station discovers the access point's security requirements by passively monitoring for beacon frames or through active probing. The access point's security information can be found in the RSN information element field inside beacon and probe response frames. The client station security requirements are delivered to the AP in association and reassociation frames.

Authentication As shown in Figure 5.16, the authentication process starts when the AP's authenticator sends an EAP-Request or the client station supplicant sends an EAPOL-Start message. As you learned in Chapter 4, EAP authentication frames are then exchanged between the supplicant and authentication server via the authenticator's uncontrolled port. The supplicant and the authentication server validate each other's credentials. The controlled port remains blocked.

FIGURE 5.15 Authentication and key management (AKM)—discovery component

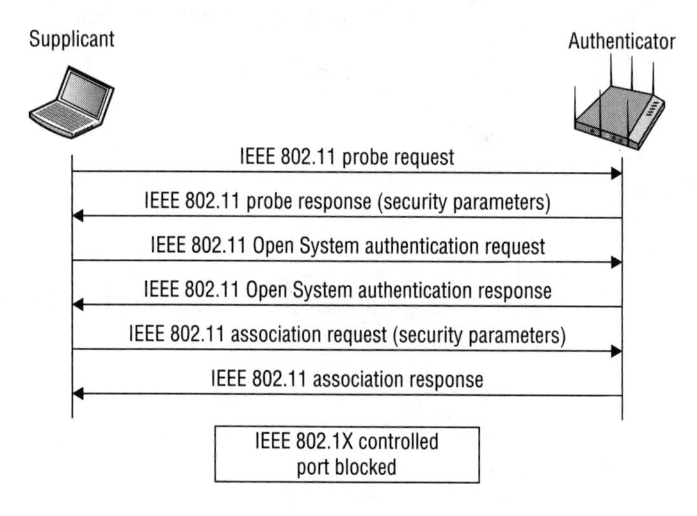

FIGURE 5.16 Authentication and key management (AKM)—authentication and master key generation component

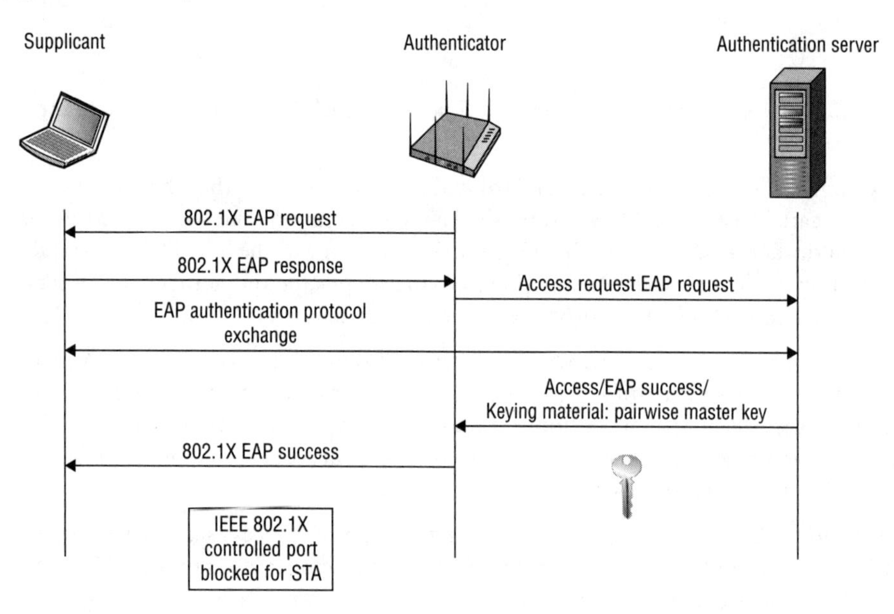

Master Key Generation As you can see in Figure 5.16, the supplicant and authentication server generate a master encryption key called the pairwise master key (PMK). The PMK is sent from the authentication server to the authenticator over the secure channel described earlier. The controlled port is still blocked.

Temporal Key Generation and Authorization As shown in Figure 5.17, a 4-Way Handshake frame exchange between the supplicant and the authenticator utilizing EAPOL-Key frames is used to generate temporary encryption keys that are used to encrypt and decrypt the MSDU payload of 802.11 data frames. The 4-Way Handshake will be discussed in detail in the next section of this chapter. Once the temporal keys are created and installed, the controlled port of the authenticator opens, and the supplicant can then send encrypted 802.11 data frames through the controlled port onward to network resources.

FIGURE 5.17 Authentication and key management (AKM)—temporal key generation and authorization

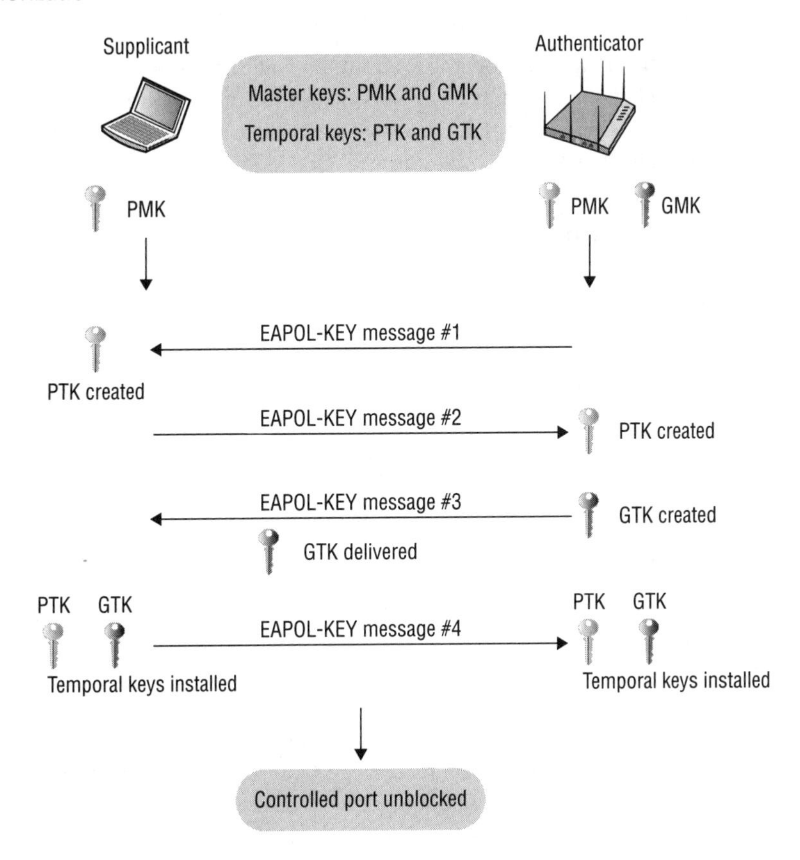

EXERCISE 5.2

Authentication and Key Management

In this exercise, you will use a protocol analyzer to view all the 802.11 frame exchanges used during AKM services.

1. To perform this exercise, you need to first download the `AKM.PCAP` file from the book's online resource area, which can be accessed at `www.sybex.com/go/cwsp2e`.
2. After the file is downloaded, you will need packet analysis software to open it. If you do not already have a packet analyzer installed on your computer, you can download Wireshark from `www.wireshark.org`.
3. Using the packet analyzer, open the `AKM.PCAP` file. Most packet analyzers display a list of capture frames in the upper section of the screen, with each frame numbered sequentially in the first column.
4. Observe the beacon and probing frames in packets 196–199. Normal open system authentication and association occurs during frames 201–207.
5. Observe packets 209–253 which show the EAP authentication frames that are exchanged between the supplicant and authentication server via the authenticator's uncontrolled port. The supplicant and the authentication server validate each other's credentials. The controlled port remains blocked.
6. Observe packets 255–261. A 4-Way Handshake exchange between the supplicant and the authenticator utilizing EAPOL-Key frames is used to generate temporary encryption keys. Notice that all 802.11 data frames are now encrypted.

As you can see, a good portion of AKM is the authentication process. In the next sections of this chapter, you will learn in greater detail about the generation of both master and temporary keys, which is the other key component of AKM. Remember that authentication and encryption key management are dependent on each other.

RSNA Key Hierarchy

AKM services also include the creation of encryption keys. Some of the encryption keys are derived from the authentication process, some of the keys are master keys, and some are the final keys that are used to encrypt/decrypt 802.11 data frames. As Figure 5.18 shows, a total of five keys make up a top-to-bottom hierarchy that is needed to establish a final robust security network association (RSNA). This key hierarchy includes a key derived from either an 802.1X/EAP authentication or derived from PSK authentication. One set of keys is considered to be *group* keys, which are keys that are used to protect multiple destinations. Another set of keys is considered to be *pairwise*. A pairwise relationship can be defined as two entities that are associated with each other: for example, an access point

(AP) and an associated station (STA), or two stations communicating in an independent basic service set (IBSS).

FIGURE 5.18 RSN key hierarchy

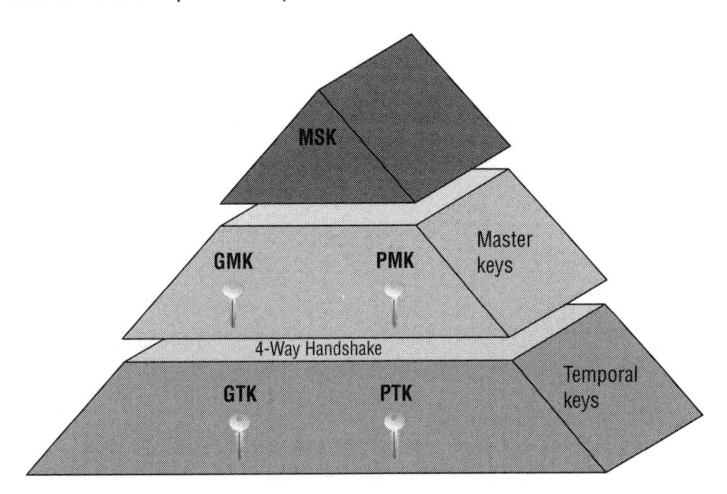

Master Session Key (MSK)

At the top of the RSNA key hierarchy is the *master session key (MSK)*, which is also sometimes referred to as the *AAA key*. The MSK is generated either from an 802.1X/EAP process or is derived from PSK authentication. You will learn how the master session key (MSK) is derived from PSK authentication later in this chapter in the section "Passphrase-to-PSK Mapping." When an 802.1X/EAP infrastructure is deployed, both the authentication server and the supplicant will know information about each other after the mutual authentication exchange of credentials. The MSK is the keying material that is derived from the EAP process and exported by the EAP method to the supplicant and authentication server. The MSK is at least 64 octets in length. How the MSK is generated from the EAP process is outside of the scope of the 802.11-2012 standard and is EAP method specific.

Master Keys

After the creation of the MSK as a result of 802.1X/EAP, two master keys are created. Try to think of the MSK as seeding material or keying material that is a result of 802.1X/EAP mutual authentication. The MSK seeding material is then used to create a master key called the *pairwise master key (PMK)*.

The PMK is derived from the MSK seeding material. The PMK is simply computed as the first 256 bits (bits 0–255) of the MSK. Because the PMK is derived from the MSK seeding material, a PMK now resides on both the supplicant and the authentication server. Effectively a portion of the MSK seeding material becomes the PMK. A new, unique PMK is generated every time a client authenticates or reauthenticates. It is very important to understand that when 802.1X/EAP is used, every client's PMK is unique to that individual client.

As shown in Figure 5.19, the PMK is then sent from the authentication server over a secure channel to the authenticator. The security of the channel between the authenticator and the AS is outside the scope of the 802.11-2012 standard. A PMK is now installed on both the client station, which is the supplicant, and the access point, which is the authenticator.

FIGURE 5.19 Master keys

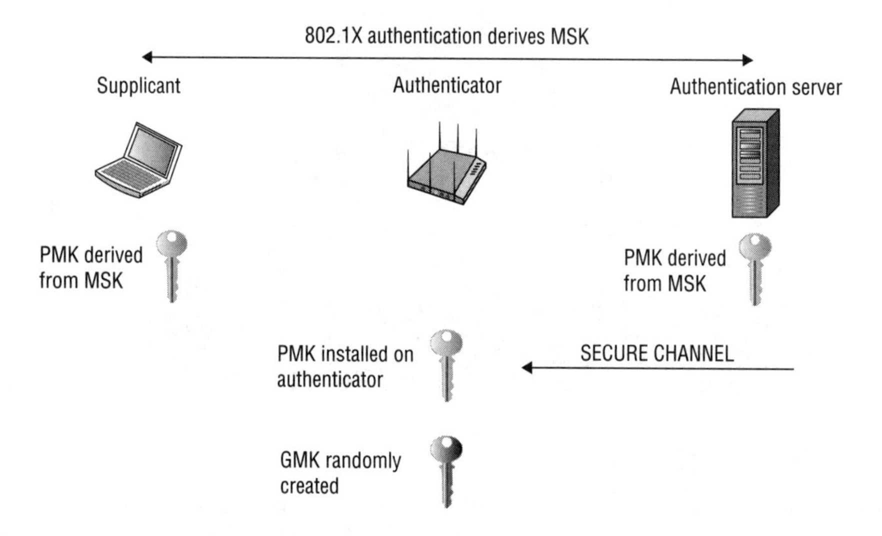

Another master key, called the *group master key (GMK)*, is randomly created on the access point/authenticator. Any GMK may be regenerated at a time interval configured on the AP to reduce the risk of the GMK being compromised.

Keep in mind that master keys are not used to encrypt or decrypt 802.11 data. The master keys are now the seeding material for the 4-Way Handshake process. The 4-Way Handshake is the final process that is used to create the keys that are used to encrypt and decrypt data. The keys generated from the 4-Way Handshake are called the pairwise transient key (PTK) and the group temporal key (GTK). The pairwise master key (PMK) is used to create the pairwise transient key (PTK), and the group master key (GMK) is used to create the group temporal key (GTK). In other words, the master keys are used to produce the temporal keys that are used to encrypt 802.11 data frames. The 4-Way Handshake process used to create the temporal encryption keys can begin when the GMK is created and installed on the authenticator, and the PMK is created and installed on both the supplicant and authenticator.

Temporal Keys

As you will learn, the 4-Way Handshake process creates temporal keys that are used by the client station and the access point to encrypt and decrypt 802.11 data frames. The *pairwise transient key (PTK)* is used to encrypt all unicast transmissions between a client station and an access point. As discussed earlier in this chapter, each PTK is unique between each individual client station and the access point. Every client station possesses a unique PTK for unicast transmissions between the client STA and the AP. PTKs are used between a single supplicant and a single authenticator.

The *group temporal key (GTK)* is used to encrypt all broadcast and multicast transmissions between the access point and multiple client stations. Although the GTK is dynamically generated, it is shared among all client STAs for broadcast and multicast frames. The GTK is used between all supplicants and a single authenticator.

As shown in Figure 5.20, the pairwise transient key (PTK) is derived from the pairwise master key (PMK). The PTK is composed of three sections:

Key Confirmation Key (KCK) The KCK is used to provide data integrity during the 4-Way Handshake and Group Key Handshake.

Key Encryption Key (KEK) The KEK is used by the EAPOL-Key frames to provide data privacy during the 4-Way Handshake and Group Key Handshake.

Temporal Key (TK) The TK is the temporal encryption key used to encrypt and decrypt the MSDU payload of 802.11 data frames between the supplicant and the authenticator.

FIGURE 5.20 Pairwise transient key (PTK)

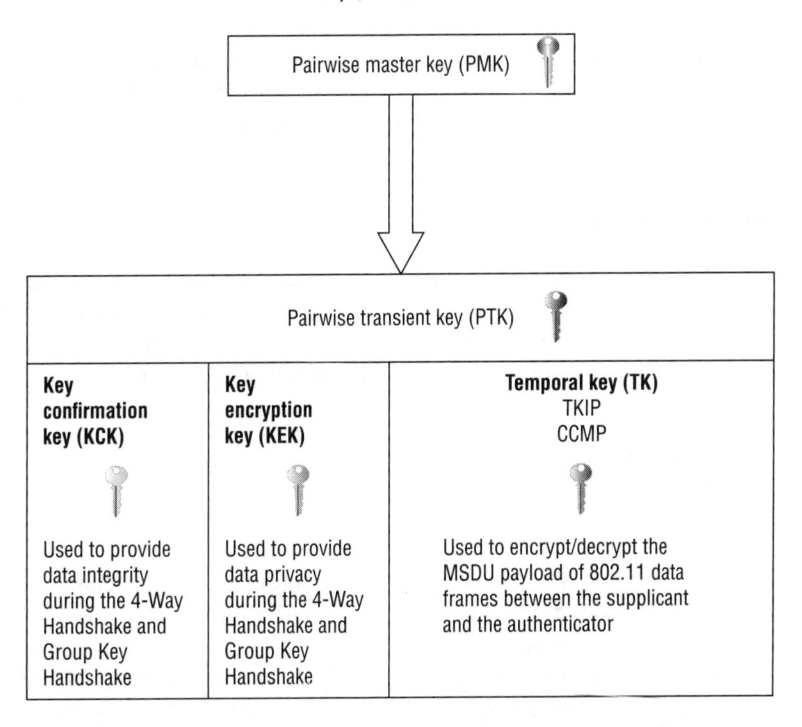

As shown in Figure 5.21, the group temporal key (GTK) is derived from the group master key (GMK). The GTK is a temporal key used to provide data privacy for broadcast/multicast communication. GTKs are used between a single authenticator and all the supplicants that are communicating with the authenticator.

FIGURE 5.21 Group temporal key (GTK)

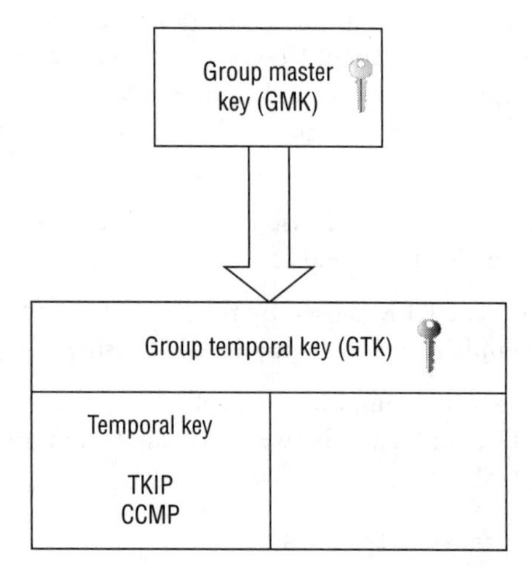

It should be understood that the PTK/GTKs used for encryption are either CCMP/AES or TKIP/ARC4 as defined by the 802.11-2012 standard. However, the 4-Way Handshake can also be used to generate keys for proprietary encryption methods.

4-Way Handshake

The 802.11-2012 standard requires that EAPOL-Key frames be used to exchange cryptographic information between radios. Most of these exchanges are between a client STA supplicant and an authenticator, which is usually an access point. EAPOL-Key frames are used during a variety of frame exchanges that are used to create dynamic encryption keys:

- 4-Way Handshake
- Group Key Handshake
- PeerKey Handshake
- TDLS PeerKey Handshake

As already mentioned, the 4-Way Handshake is a final process used to generate pairwise transient keys for encryption of unicast transmissions and a group temporal key for encryption of broadcast/multicast transmissions.

The 4-Way Handshake uses four EAPOL-Key frame messages between the authenticator and the supplicant for six major purposes:

- Confirm the existence of the PMK at the peer station.
- Ensure that the PMK is current.

- Derive a new pairwise transient key (PTK) from the PMK.
- Install the PTK on the supplicant and the authenticator.
- Transfer the GTK from the authenticator to the supplicant and install the GTK on the supplicant and, if necessary, the authenticator.
- Confirm the selection of the cipher suites.

802.1X/EAP authentication is completed when the access point sends an EAP-Success frame and the AP can now initiate the 4-Way Handshake. Keep in mind that the authentication process has already generated the master keys (PMK and GMK), which will be used by the 4-Way Handshake to derive the temporal keys.

Before we explain the 4-Way Handshake process, it is necessary to define several key terms. The 4-Way Handshake uses pseudo-random functions. A *pseudo-random function (PRF)* hashes various inputs to derive a pseudo-random value. The PMK is one of the inputs combined with other inputs to create the pairwise transient key (PTK). Some of the other inputs used by the pseudo-random function are called nonces. A *nonce* is a random numerical value that is generated one time only. A nonce is used in cryptographic operations and is associated with a given cryptographic key. In the case of the 4-Way Handshake, a nonce is associated with the PMK. A nonce is only used once and is never used again with the PMK. Two nonces are created by the 4-Way Handshake: the *authenticator nonce (ANonce)* and the *supplicant nonce (SNonce)*.

To create the pairwise transient key, the 4-Way Handshake uses a pseudo-random function that combines the pairwise master key, a numerical authenticator nonce, a supplicant nonce, the authenticator's MAC address (AA), and the supplicant's MAC address (SPA).

The following is a simplified depiction of the formula used by the pseudo-random function (PRF) to derive a pairwise transient key:

PTK = PRF (PMK + ANonce + SNonce + AA + SPA)

As Figure 5.22 shows, the 4-Way Handshake consists of the following steps:

4-Way Handshake Message 1 The authenticator and supplicant each randomly create their respective nonces. The authenticator sends an EAPOL-Key frame containing an ANonce to the supplicant. The supplicant now has all the necessary inputs for the pseudo-random function. The supplicant derives a PTK from the PMK, ANonce, SNonce, and MAC addresses. The supplicant is now in possession of a pairwise transient key that can be used to encrypt unicast traffic.

4-Way Handshake Message 2 The supplicant sends an EAPOL-Key frame containing an SNonce to the authenticator. The authenticator now has all the necessary inputs for the pseudo-random function. The supplicant also sends its RSN information element capabilities to the authenticator and a message integrity code (MIC). The authenticator derives a PTK from the PMK, ANonce, SNonce, and MAC addresses. The authenticator also validates the MIC. The authenticator is now in possession of a pairwise transient key that can be used to encrypt unicast traffic.

4-Way Handshake Message 3 If necessary, the authenticator derives a GTK from the GMK. The authenticator sends an EAPOL-Key frame to the supplicant containing the

ANonce, the authenticator's RSN information element capabilities, and a MIC. The EAPOL-Key frame may also contain a message to the supplicant to install the temporal keys. Finally, the GTK will be delivered inside this unicast EAPOL-Key frame to the supplicant. The confidentiality of the GTK is protected because it will be encrypted with the PTK.

4-Way Handshake Message 4 The supplicant sends the final EAPOL-Key frame to the authenticator to confirm that the temporal keys have been installed.

Controlled Port Unlocked The virtual controlled port opens on the authenticator, and now, encrypted 802.11 data frames from the supplicant can pass through the authenticator and on to their final destination. All unicast traffic will now be encrypted with the PTK, and all multicast and broadcast traffic will now be encrypted with the GTK.

FIGURE 5.22 The 4-Way Handshake

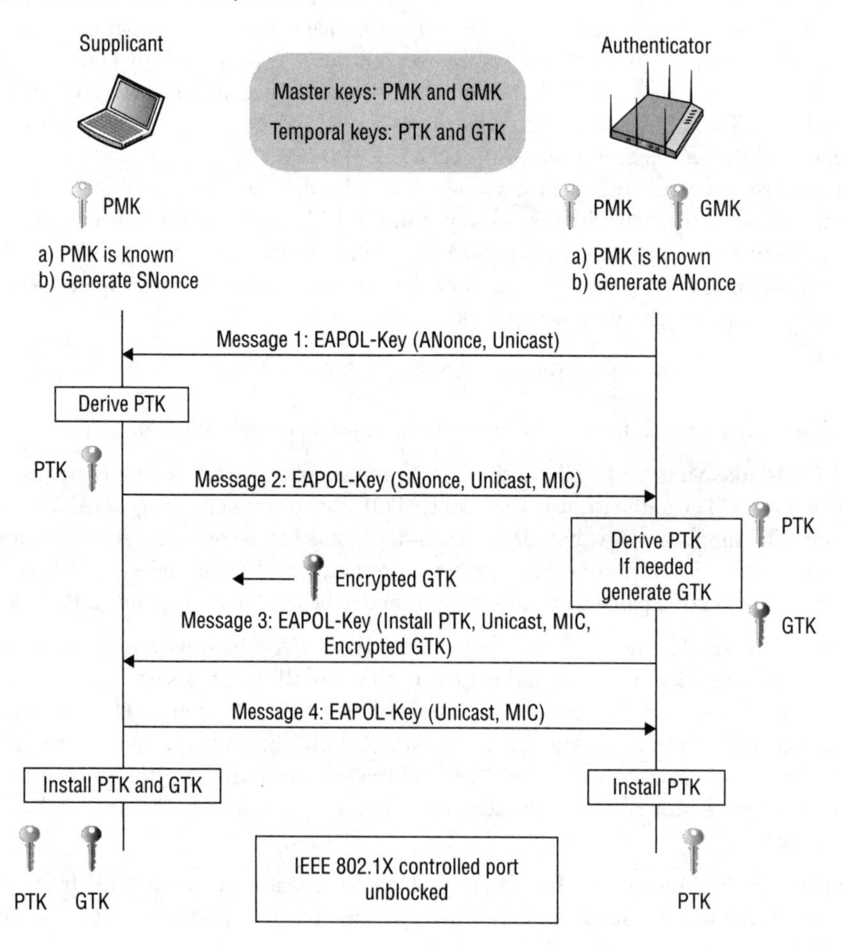

EXERCISE 5.3

The 4-Way Handshake

In this exercise, you will use a protocol analyzer to view the 4-Way Handshake EAPOL-Key frames that are used to generate the temporal keys used for encryption.

1. To perform this exercise, you need to first download the `4WAY_HANDSHAKE.PCAP` file from the book's online resource area, which can be accessed at `www.sybex.com/go/cwsp2e`.
2. After the file is downloaded, you will need packet analysis software to open it. If you do not already have a packet analyzer installed on your computer, you can download Wireshark from `www.wireshark.org`.
3. Using the packet analyzer, open the `4WAY_HANDSHAKE.PCAP` file. Most packet analyzers display a list of capture frames in the upper section of the screen, with each frame numbered sequentially in the first column.
4. Observe the EAP-Success frame at packet 66. At this point, 802.1X/EAP authentication is completed, and the AP can now initiate the 4-Way Handshake. The access point (authenticator) MAC address is 00:12:43:CB:0F:30. The client station (supplicant) MAC address is 00:40:96:A3:0C:45.
5. Observe the EAPOL-Key frames of the 4-Way Handshake in packets 68, 70, 72, and 74. Open the first EAPOL-Key frame in packet 68. Notice that the AP is sending the client station an ANonce.
6. Open the second EAPOL-Key frame in packet 70. Notice that the client station is sending the AP an SNonce, an RSN information element, and a MIC.
7. Open the third EAPOL-Key frame in packet 72. Notice the AP is sending the supplicant a MIC and instructions to install the temporal keys.
8. Open the fourth EAPOL-Key frame in packet 74. The supplicant is now sending a message to the authenticator that the temporal keys are installed.

Group Key Handshake

The 802.11-2012 standard also defines a two-frame handshake that is used to distribute a new group temporal key (GTK) to client stations that have already obtained a PTK and GTK in a previous 4-Way Handshake exchange. The *Group Key Handshake* is used only to issue a new group temporal key (GTK) to client stations that have previously formed security associations. Effectively, the Group Key Handshake is identical to the last two frames of the 4-Way Handshake. Once again, the purpose of the Group Key Handshake is to deliver a new GTK to all client stations that already have an original GTK generated by an earlier 4-Way Handshake.

The authenticator can update the GTK for a number of reasons. For example, the authenticator may change the GTK on disassociation or deauthentication of a client station. WLAN vendors may also offer a configuration setting to trigger the creation of a new GTK based on a timed interval.

As shown in Figure 5.23, the Group Key Handshake consists of the following steps:

Group Key Handshake Message 1 The authenticator derives a new GTK from the GMK. The new GTK is sent in a unicast EAPOL-Key frame to the supplicant. The confidentiality of the new GTK is protected because it will be encrypted with the original PTK from the initial 4-Way Handshake. The authenticator also sends a message integrity code (MIC). The supplicant validates the MIC when it receives the EAPOL-Key frame. The supplicant decrypts and installs the new GTK.

Group Key Handshake Message 2 The supplicant sends an EAPOL-Key frame to the authenticator to confirm that the GTK has been installed. The supplicant also sends a message integrity code (MIC). The authenticator validates the MIC when it receives the EAPOL-Key frame.

FIGURE 5.23 The Group Key Handshake

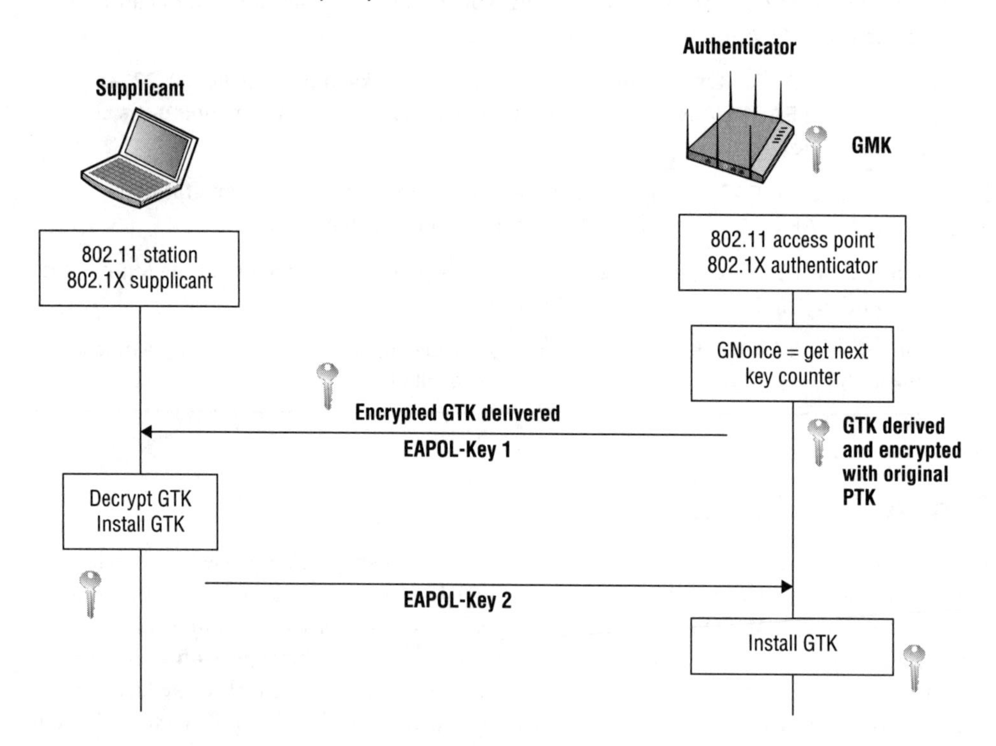

Please do not confuse Group Key Handshake with the two EAPOL-Key frame exchange that is used to distribute dynamic WEP keys. Although both handshakes use a two EAPOL-Key frame exchange, each handshake has an entirely different purpose. Also remember that dynamic WEP is proprietary and that the two EAPOL-Key frame exchange used by dynamic WEP is not an RSN security association.

PeerKey Handshake

Most WLAN communications do not involve peer-to-peer applications between clients; however, peer-to-peer connectivity within a BSS is possible. If two client stations are associated to an AP, peer-to-peer communications from one client station to another client station can occur as long as the traffic is forwarded through the AP. One client station would have to send a unicast frame to the AP, which would then be forwarded through the AP to a peer client station (STA).

Unlike a typical peer-to-peer communications within a BSS, the 802.11-2012 standard defines a method that gives client stations the option to securely communicate with each other in a BSS without sending their frames through the access point. After client STAs have already established individual security associations with an access point, a *station-to-station link (STSL)* can also be established. An STSL is a direct link established between two stations while associated to a common access point.

The client stations within the BSS use a *PeerKey Handshake* management protocol to create PeerKeys that are unique to two client STAs so that they can communicate directly and securely in a station-to-station link (STSL) within the BSS. The PeerKey Handshake is used to establish security for data frames passed directly between two STAs associated with the same AP. The AP must establish an RSNA with each STA prior to the PeerKey Handshake.

The PeerKey Handshake is actually two different handshakes:

SMK Handshake This frame exchange is used by the two peer stations to create a master key called the *STSL master key (SMK)*. One of the client stations must initiate this exchange through the AP to create the SMK with another client station that is also associated to the AP.

4-Way STK Handshake This frame exchange uses the SMK as seeding material to create an *STSL transient key (STK)*. The STK is the final key that is used to encrypt the unicast communications between the two peer client stations while they are still associated to the AP.

As shown in 5.24, the peer stations will remain associated to the AP and use the STK for secure STSL communications. The stations can also still securely communicate within the AP using their original unique pairwise transient keys (PTKs).

Although originally defined by the 802.11i-2004 security amendment, widespread use of the PeerKey Handshake never happened. The 802.11z-2010 amendment defined enhanced mechanisms for tunneled direct link setup (TDLS) between two peer stations within a BSS.

FIGURE 5.24 Station-to-station link (STSL)

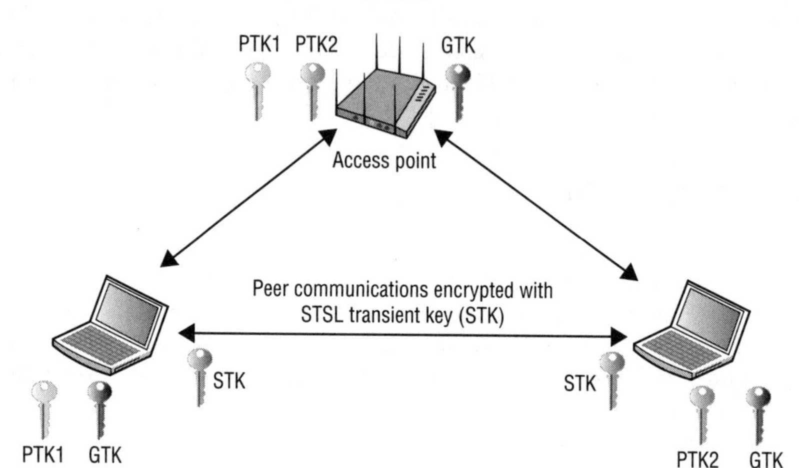

TDLS Peer Key Handshake

The 802.11z-2010 amendment that defined enhanced mechanisms for direct link communications between two client stations while connected to an AP is now part of the 802.11-2012 standard. 802.11z defines a *tunneled direct link setup (TDLS)* security protocol. The Wi-Fi Alliance also introduced Wi-Fi CERTIFIED TDLS as a certification program for devices using TDLS to connect directly to one another after they have joined a traditional Wi-Fi network (BSS). One of the intended goals of TDLS is to boost performance for applications such as multimedia streaming between WLAN clients by bypassing congested APs. For example, two TDLS-linked devices can temporarily leave the channel of the AP to which they are associated, and dynamically switch to a 40 MHz channel in the 5 GHz band for a private conversation. Even when the private direct link is switched to a different channel, the client STAs periodically switch back to the home channel to maintain connectivity with the AP, which is unaware of the TDLS setup. TDLS is currently being used for streaming high-definition video between clients and Wi-Fi-enabled televisions.

The TDLS Peer Key security protocol is executed between the two non-AP client stations that intend to establish an RSNA for direct-link communication. The *TDLS Peer Key (TPK)* Handshake occurs as part of the TDLS direct-link setup procedure. The TPK security association (TPKSA) is the result of the successful completion of the TDLS Peer Key Handshake protocol, which derives keys for providing confidentiality and data origin authentication.

The TDLS Peer Key Handshake consists of three messages between a TDLS initiator client station and a TDLS responder client station. As shown in Figure 5.25, this three-way handshake exchange is communicated through the AP to which the clients are associated. At the end of the exchange, each client station has a copy of the unicast *TDLS Peer Key (TPK)* and the clients can now have an encrypted direct link for sidebar communications. Keep in mind that both clients remain associated with the original AP and the clients can still communicate with other devices using normal BSS communications.

FIGURE 5.25 TDLS Peer Key Handshake

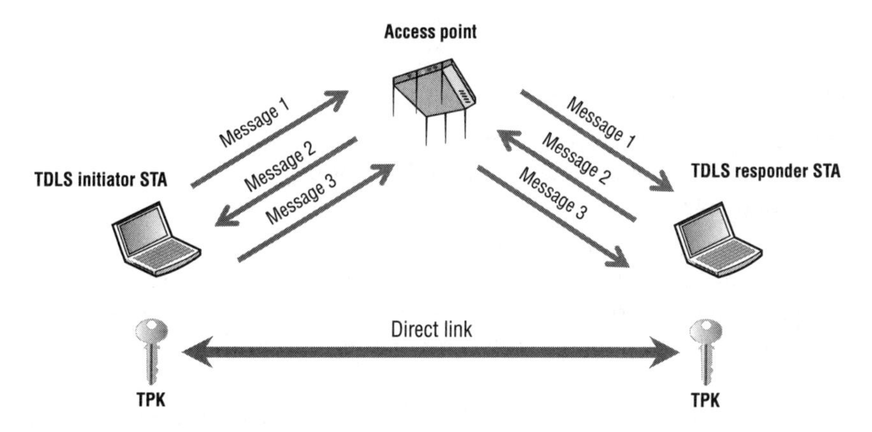

RSNA Security Associations

Security associations can be defined as group policies and keys used to protect information. A robust security network association (RSNA) requires two 802.11 stations (STAs) to establish procedures to authenticate and associate with each other as well as create dynamic encryption keys through a 4-Way Handshake. This association between two STAs is referred to as an RSNA. The 802.11-2012 standard defines multiple RSNA security associations that are established during the many procedures already discussed in this chapter:

Pairwise Master Key Security Association (PMKSA) The conditions resulting from a successful 802.1X authentication exchange between the supplicant and authentication server or from a preshared key (PSK).

Pairwise Transient Key Security Association (PTKSA) The conditions resulting from a successful 4-Way Handshake exchange between the supplicant and authenticator.

Group Temporal Key Security Association (GTKSA) The conditions resulting from a successful group temporal key (GTK) distribution exchange via either a 4-Way Handshake or a Group Key Handshake.

STSL Master Key Security Association (SMKSA) An SMKSA is the result of a successful SMK Handshake by the initiator STA. It is derived from parameters provided by the access point and the client stations.

STSL Transient Key Security Association (STKSA) The STKSA is a result of the successful completion of the 4-Way STK Handshake. This security association is bidirectional between the initiator and the peer STAs. The STKSA is used to create session keys to protect the station-to-station link.

TDLS Peer Key Security Association (TPKSA) The conditions resulting from a successful completion of the TDLS Peer Key Handshake. Two non-AP STAs establish a secure direct-link communication while remaining in a BSS. The security association is bidirectional between the TDLS initiator STA and the TDLS responder STA. The TPKSA is used to create session keys to protect this TDLS session.

Passphrase-to-PSK Mapping

As discussed earlier, an authentication and key management protocol (AKMP) can either be derived from an EAP protocol used during 802.1X or by a preshared key (PSK). When a PSK authentication solution is used, AKM operations include the following:

Discovery A client station discovers the access point's security requirements by passively monitoring for beacon frames or through active probing. The access point's security information can be found in the RSN information element field inside beacon and probe response frames. The client station security requirements are delivered to the AP in association and reassociation frames.

Negotiation The client STA associates with an AP and negotiates a security policy. The preshared key (PSK) becomes the pairwise master key (PMK).

Temporal Key Generation and Authorization The 4-Way Handshake exchange between the supplicant and the authenticator utilizing EAPOL-Key frames is used to generate temporary encryption keys that are used to encrypt and decrypt the MSDU payload of 802.11 data frames. Once the temporal keys are created and installed, the controlled port of the authenticator opens and the supplicant can now send traffic through the controlled port onward to network resources.

The PSK authentication used during RSNA is often known by the more common name of WPA-Personal or WPA2-Personal. PSK authentication was initially intended as a simple security solution for home users who do not have the resources to deploy full-blown 802.1X/EAP security with a RADIUS server. However, PSK authentication is commonplace in many enterprise deployments despite known security risks. WPA2-Personal and PSK authentication are discussed in greater detail in Chapter 6.

Vendor Terminology

Many individuals often confuse Shared Key authentication with the preshared key (PSK) authentication used in WPA/WPA2-Personal. As discussed in Chapter 2, Shared Key authentication is a legacy 802.11 authentication method that requires the use of a static WEP key. Do not confuse Shared Key authentication with PSK authentication.

Furthermore, WLAN vendors often add to the confusion with the use of their own terminology. Vendors have many names for PSK authentication, including WPA/WPA2-Passphrase, WPA/WPA2-PSK, and WPA/WPA2-Preshared Key. The correct Wi-Fi Alliance terminology is WPA/WPA2 Personal. All of these terms refer to PSK authentication.

A WPA/WPA2 preshared key is a static key that is configured on the access point and all the clients. The same static PSK is used by all members of the basic service set (BSS). The RSNA PSK is 256 bits in length or 64 characters when expressed in hex. Most end users are not comfortable configuring an AP and client stations with a long, 64-character hexadecimal key. Most end users are, however, very comfortable configuring short ASCII passwords or passphrases. Therefore, a *passphrase-PSK mapping* formula is defined by the 802.11-2012 standard to allow end users to use a simple ASCII passphrase that is then converted to the 256-bit PSK. The PSK is generated using a password-based key generation function (PBKDF).

Here is the formula to convert a passphrase to a PSK:

PSK = PBKDF2(*PassPhrase*, *ssid*, *ssidLength*, 4096, 256)

- The *PassPhrase* is a sequence of between 8 and 63 ASCII-encoded characters. The limit of 63 is mandated so as to differentiate between an ASCII passphrase and a PSK that is 64 hexadecimal characters.
- Each character in the passphrase must have an encoding in the range of 32 to 126 (decimal), inclusive.
- *ssid* is the SSID of the ESS or IBSS where this passphrase is in use, encoded as an octet string used in the beacon and probe response frames for the ESS or IBSS.
- *ssidLength* is the number of octets of the *ssid*.
- 4096 is the number of times the passphrase is hashed.
- 256 is the number of bits output by the passphrase mapping.

As you can see, a simple passphrase is combined with the SSID and hashed to produce a 256-bit PSK.

Throughout this study guide, you will be presented with various formulas. You will not need to know these formulas for the CWSP certification exam. The formulas are in this study guide to demonstrate concepts and to be used as reference material.

Roaming and Dynamic Keys

Every time a client station roams from one access point to another, the client STA will send a reassociation request frame to initiate the roaming handoff. The client station uses the RSN information element to inform the access point about the client's security capabilities, including supported encryption cipher suites and supported authentication methods.

Every time a client roams, unique encryption keys must be generated using a 4-Way Handshake process between the access point and the client STA. As you have already learned, either an 802.1X or PSK authentication process is needed to produce the pairwise master key (PMK) that seeds the 4-Way Handshake. Therefore, every time a client roams, the client must reauthenticate.

Roaming can be especially troublesome for VoWiFi and other time-sensitive applications when using a WPA-Enterprise or WPA2-Enterprise security solution, which requires the use of a RADIUS server. Due to the multiple frame exchanges between the authentication server and the supplicant, an 802.1X/EAP authentication often takes 700 milliseconds or longer for the client to authenticate. VoWiFi requires a handoff of 150 milliseconds or less to avoid a degradation of the quality of the call or, even worse, a loss of connection. An ideal roaming handoff time for VoWiFi is 50 milliseconds. One advantage of using WPA/WPA2-Personal is that PSK authentication does not have the latency issues of 802.1X/EAP. PSK authentication only requires the 4-Way Handshake exchange, and the roaming handoff can occur in typically 40 to 60 milliseconds.

The 802.11r-2008 amendment is known as *the fast basic service set transition (FT)* amendment and is now part of the IEEE 802.11-2012 standard. The technology is more often referred to as *fast secure roaming* because it defines faster handoffs when roaming occurs between cells in a WLAN using 802.1X/EAP. You can find a detailed discussion about fast BSS transition methods in Chapter 7.

Summary

It is important to remember that authentication and key management (AKM) uses the authentication and encryption processes together to provide authorized protection for the WLAN portal as well as data privacy for the 802.11 data frames. The authentication and encryption key generation processes are linked together and are dependent on each other. Many vendors initially linked authentication and encryption together using dynamic WEP. However, dynamic WEP security was proprietary and still susceptible to WEP-cracking attacks. The 802.11-2012 standard now defines robust security network associations (RSNAs), which require two 802.11 stations (STAs) to establish procedures to authenticate and associate with each other as well as create dynamic encryption keys through the 4-Way Handshake process. RSNAs use either TKIP/ARC4 or CCMP/AES encryption protocols. Dynamic keys prevent social engineering attacks, and all client STAs have unique keys.

Exam Essentials

Explain dynamic WEP. Be able to explain processes used to generate dynamic WEP keys. Understand that dynamic WEP is not the same as RSNA dynamic key management.

Describe the differences between an RSN, TSN, and pre-RSN security network. Understand that a robust security network utilizes only TKIP/ARC4 and CCMP/AES encryption. A TSN can use TKIP/ARC4, CCMP/AES, and legacy encryption such as WEP. A pre-RSN security network only uses legacy encryption.

Explain the purpose of the RSN information element field. Know what type of security capability information is in the RSNIE. Know which 802.11 management frames are used to deliver the RSNIE and how the management frames are used.

Understand the concept of AKM. Describe the symbiotic relationship between authentication and dynamic encryption. Understand that an 802.1X/EAP or PSK process provides the seeding material to generate dynamic encryption keys.

Describe the RSNA key hierarchy. Understand the relationship between all the various keys. Know the difference between master and temporal keys. Explain the difference between pairwise and group keys. Know the three keys that comprise a PTK.

Explain all phases of the 4-Way Handshake in detail. Understand that the 4-Way Handshake is used to produce the final temporal keys used for encryption. Know the purpose of each EAPOL-Key frame. Explain all of the components needed to create a PTK.

Explain the purpose of the Group Key Handshake. Understand that the Group Key Handshake is used to create a new GTK for broadcast/multicast traffic after a previous 4-Way Handshake has already occurred.

Understand the concept of RSN security associations. Explain all the different security associations, why they are needed, and when they occur.

Describe passphrase-to-PSK mapping. Explain how a PSK is derived from a passphrase and why. Understand the relationship between PSK authentication and the 4-Way Handshake.

Review Questions

1. What must occur in order for dynamic TKIP/ARC4 or CCMP/AES encryption keys to be generated? (Choose all that apply.)
   - **A.** Shared Key authentication and 4-Way Handshake
   - **B.** 802.1X/EAP authentication and 4-Way Handshake
   - **C.** Open System authentication and 4-Way Handshake
   - **D.** PSK authentication and 4-Way Handshake

2. Which encryption types can be used to encrypt and decrypt unicast traffic with the pairwise transient key (PTK) that is generated from a 4-Way Handshake? (Choose all that apply.)
   - **A.** Temporal Key Integrity Protocol
   - **B.** 3-DES
   - **C.** Dynamic WEP
   - **D.** CCMP
   - **E.** Proprietary encryption
   - **F.** Static WEP

3. View the frame capture of the 4-Way Handshake in the graphic shown here. Which EAPOL-Key message frame is displayed?

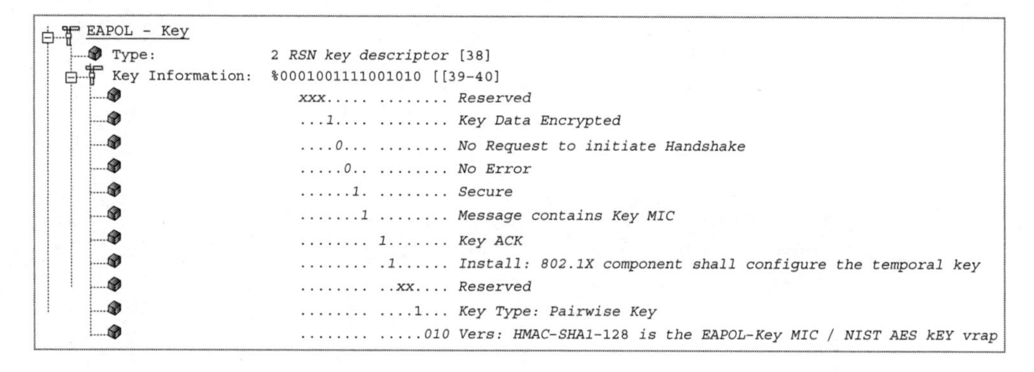

   - **A.** 4-Way Handshake message 1
   - **B.** 4-Way Handshake message 2
   - **C.** 4-Way Handshake message 3
   - **D.** 4-Way Handshake message 4

4. What are the keys that make up a pairwise transient key? (Choose all that apply.)
   - **A.** STK
   - **B.** KEK
   - **C.** SMK

D. TK

E. KCK

5. What are some of the variables that are used during the 4-Way Handshake to produce a pairwise transient key (PTK)? (Choose all that apply.)

A. Pairwise master key

B. Master session key

C. Group master key

D. Nonces

E. Authenticator MAC address

6. After viewing the frame capture in the graphic shown here, identify which type of encryption method is being used.

Source	Destination	Protocol
00:12:43:CB:0F:30	00:40:96:A3:0C:45	802.11 Ack
00:12:43:CB:0F:30	00:40:96:A3:0C:45	EAP Request
00:40:96:A3:0C:45	00:12:43:CB:0F:30	802.11 Ack
00:40:96:A3:0C:45	00:12:43:CB:0F:30	EAP Response
00:12:43:CB:0F:30	00:40:96:A3:0C:45	802.11 Ack
00:12:43:CB:0F:30	00:40:96:A3:0C:45	EAP Success
00:40:96:A3:0C:45	00:12:43:CB:0F:30	802.11 Ack
00:12:43:CB:0F:30	00:40:96:A3:0C:45	EAPOL-Key
00:40:96:A3:0C:45	00:12:43:CB:0F:30	802.11 Ack
00:12:43:CB:0F:30	00:40:96:A3:0C:45	EAPOL-Key
00:40:96:A3:0C:45	00:12:43:CB:0F:30	802.11 Ack

A. TKIP

B. CCMP

C. xSec

D. Fortress

E. WEP

F. AES

7. After viewing the frame capture in the graphic shown here, identify which type of security network is being used.

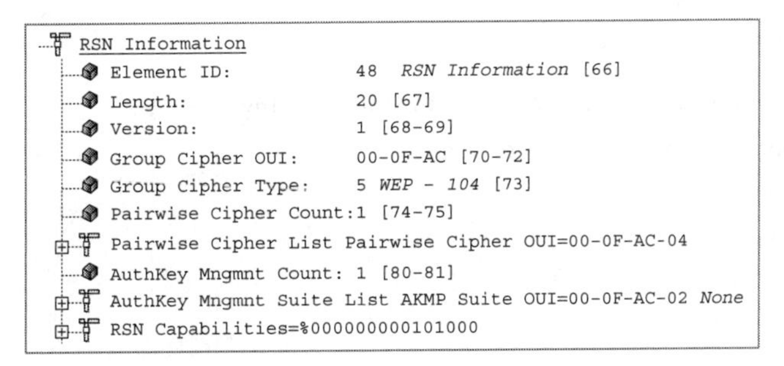

A. Robust security network
B. Rotund security network
C. Transition security network
D. WPA security network

8. In a robust security network (RSN), which 802.11 management frames are used by an access point to inform client STAs about the RSNA security capabilities of the access point and effectively the BSS? (Choose all that apply.)
A. Beacon management frame
B. Probe request frame
C. Probe response frame
D. Association request frame
E. Reassociation response frame
F. Reassociation request frame
G. Association response frame

9. The ___________ key is used to encrypt/decrypt unicast 802.11 frames, and the ___________ key is used to encrypt/decrypt broadcast and multicast 802.11 frames.
A. Group Master, Group Temporal
B. Pairwise Master, Group Temporal
C. Master Session, Pairwise Transient
D. Pairwise Transient, Group Temporal
E. Pairwise Master, Pairwise Transient

10. In a robust security network (RSN), which 802.11 management frames are used by client stations to inform an access point about the RSNA security capabilities of the client STAs? (Choose all that apply.)
A. Beacon management frame
B. Probe request frame
C. Probe response frame
D. Association request frame
E. Reassociation response frame
F. Reassociation request frame
G. Association response frame

11. Bob's access point has been configured with the following settings:

```
Management VLAN 800 - Interface:
192.168.80.5/24

User VLANS:

VLAN 201
```

```
VLAN 202
VLAN 203
SSIDs:
SSID-1: (employee) security: (802.1X/EAP/CCMP) - VLAN 201 - BSSID
(00:08:12:43:0F:30)
SSID-2 (voice) security: (PSK/TKIP and WEP) - VLAN 202 - BSSID
(00:08:12:43:0F:31)
SSID-3: (guest) security: (WEP) - VLAN 203 - BSSID (00:08:12:43:0F:32)
```

Based on the settings on Bob's access point, what type of WLAN security exits within the coverage area of the AP? (Choose all that apply.)

A. Closed security network

B. Transition security network

C. Pre-RSNA security network

D. Open security network

E. Robust security network

12. Which authentication methods provide the seeding material that is needed by the 4-Way Handshake to create temporal keys for encrypting 802.11 MSDU payloads? (Choose all that apply.)

A. Shared Key authentication

B. PSK

C. 802.1X/EAP

D. Captive Portal

E. WPA2-Personal

13. Client stations must authenticate and create new dynamic encryption keys under which conditions? (Choose all that apply.)

A. When probing a BSS

B. When joining a BSS

C. When joining an IBSS

D. When roaming to a new BSS

E. When leaving a BSS

14. When RSN security is in place, which security handshake is used to distribute keys to encrypt broadcast/multicast traffic even though a previous security association has already occurred?

A. PeerKey Handshake

B. Group Key Handshake

C. 4-Way Handshake

D. 2-Way Handshake

15. When two client stations are already associated to an AP, which handshake is used to create a different unicast key that the two client stations can use for a private conversation while they remain associated to the AP?

A. Mesh Group Key Handshake

B. Group Key Handshake

C. 4-Way Handshake

D. 2-Way Handshake

E. TDLS Peer Key Handshake

16. After viewing the frame capture shown here, identify the type of authentication method being used.

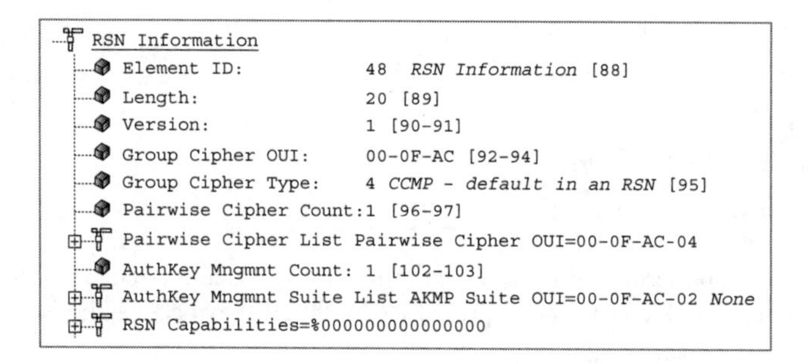

A. EAP-TTLS

B. Open System

C. PSK

D. EAP-TLS

E. PEAP

17. What type of RSNA security association is established as the result of a successful 802.1X authentication exchange between the supplicant and authentication server, or from a preshared key (PSK)?

A. PMKSA

B. GMKSA

C. SMKSA

D. PTKSA

18. What are the advantages of using dynamic encryption keys instead of static keys? (Choose all that apply.)

A. Every client STA has a unique unicast key.

B. Dynamic encryption uses compression.

C. All client STAs share a broadcast/multicast key.

D. Authentication is optional with dynamic encryption.

19. What operations must occur before the virtual controlled port of the authenticator becomes unblocked? (Choose all that apply.)

    A. 802.1X/EAP authentication

    B. 4-Way Handshake

    C. 2-Way Handshake

    D. RADIUS proxy

20. What are some of the purposes of the 4-Way Handshake? (Choose all that apply.)

    A. Transfer the GTK to the supplicant.

    B. Derive a PTK from the PMK.

    C. Transfer the GMK to the supplicant.

    D. Confirm cipher suites.

PSK Authentication

IN THIS CHAPTER, YOU WILL LEARN ABOUT THE FOLLOWING:

✓ **WPA/WPA2-Personal**

- Preshared keys (PSK) and passphrases
- WPA/WPA2-Personal risks
- Entropy
- Proprietary PSK

✓ **Simultaneous Authentication of Equals (SAE)**

This chapter discusses preshared key (PSK) authentication, which was originally intended as a simple security solution to be used in small office/home office (SOHO) environments. PSK authentication is also known by the more common name of WPA-Personal or WPA2-Personal. Although 802.1X/EAP is the preferred security method for the enterprise, PSK authentication is used in many enterprises despite its weaknesses. We will explore the PSK security mechanisms, risks, and the importance of entropy. We will also discuss per-user/per-device implementations of PSK that some WLAN vendors offer as an enhancement to static PSK authentication solutions. Finally, we will discuss a proposed replacement for PSK authentication called Simultaneous Authentication of Equals (SAE).

WPA/WPA2-Personal

As you know, the IEEE 802.11i security amendment, which was ratified and published as IEEE Standard 802.11i-2004, defined a *robust security network (RSN)* using stronger encryption and better authentication methods. The 802.11i amendment is now part of the IEEE 802.11-2012 standard. As you learned in Chapter 5, "802.11 Layer 2 Dynamic Encryption Key Generation," the 802.11-2012 standard defines *authentication and key management (AKM)* services. AKM services require both authentication processes and the generation and management of encryption keys. An *authentication and key management protocol (AKMP)* can be either a *preshared (PSK)* or an EAP protocol used during 802.1X/EAP authentication. 802.1X/EAP requires a RADIUS server and advanced skills. The average home Wi-Fi user has no knowledge of 802.1X/EAP and does not have a RADIUS server in their home. PSK authentication is meant to be used in small office/home office (SOHO) environments because the stronger enterprise 802.1X/EAP authentication solutions are not available. Therefore, the security used in SOHO environments is PSK authentication. WPA/WPA2-Personal is the same thing as PSK authentication.

Most SOHO wireless networks are secured with WPA/WPA2-Personal mechanisms. Prior to the IEEE ratification of the 802.11i amendment, the Wi-Fi Alliance introduced the *Wi-Fi Protected Access (WPA)* certification as a snapshot of the not-yet-released 802.11i amendment, supporting only TKIP/ARC4 dynamic encryption-key generation. 802.1X/EAP authentication was required in the enterprise, and a passphrase authentication method called *WPA-Personal* was required in a SOHO environment.

WPA-Personal allows an end user to enter a simple ASCII character string, dubbed a passphrase, anywhere from 8 to 63 characters in size. Behind the scenes, a "passphrase to PSK mapping" function takes care of the rest. Therefore, all the user has to know is a single, secret passphrase to allow access to the WLAN.

In June 2004, the IEEE 802.11 TGi working group formally ratified 802.11i, which added support for CCMP/AES encryption. The Wi-Fi Alliance therefore revised the previous WPA specification to WPA2, incorporating the CCMP/AES cipher. Therefore, the only practical difference between WPA and WPA2 has to do with the encryption cipher. WPA-Personal and WPA2-Personal both use PSK authentication methods; however, WPA-Personal specifies TKIP/ARC4 encryption and *WPA2-Personal* specifies CCMP/AES. WLAN vendors have many names for PSK authentication, including WPA/WPA2-Passphrase, WPA/WPA2-PSK, and WPA/WPA2-Preshared Key. For example, Figure 6.1 shows a Linksys Wi-Fi router that uses the term WPA Pre-shared Key. The correct Wi-Fi Alliance terminology is WPA/WPA2-Personal. All of these terms refer to 802.11 PSK authentication.

FIGURE 6.1 Vendor terminology example

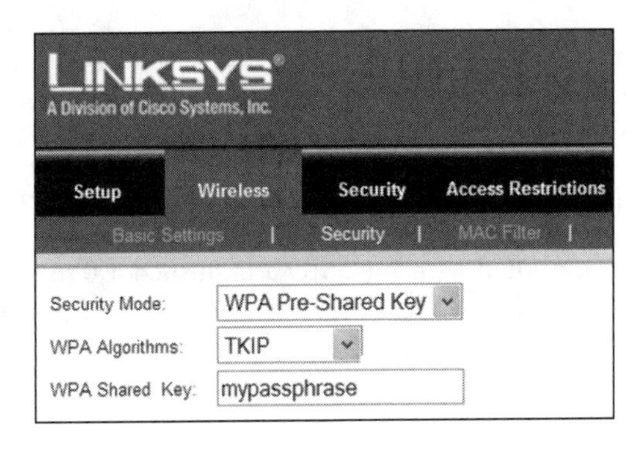

Preshared Keys (PSK) and Passphrases

A preshared key (PSK) used in a robust security network is 256 bits in length, or 64 characters when expressed in hex. A PSK is a static key that is configured on the access point and on all the clients. The same static PSK is used by all members of the basic service set (BSS). The problem is that the average home user is not comfortable with entering a 64-character hexadecimal PSK on both a SOHO Wi-Fi router and a laptop client utility. Even if home users did enter a 64-character PSK on both ends, they probably would not be able to remember the PSK and would have to write it down. Most home users are, however, very comfortable configuring short ASCII passwords or passphrases. As shown in Figure 6.2, the home user enters a *passphrase,* which is an 8- to 63-character string entered into the client software utility on the end-user device and also at the access point. The passphrase must match on both ends.

FIGURE 6.2 Client configured with passphrase

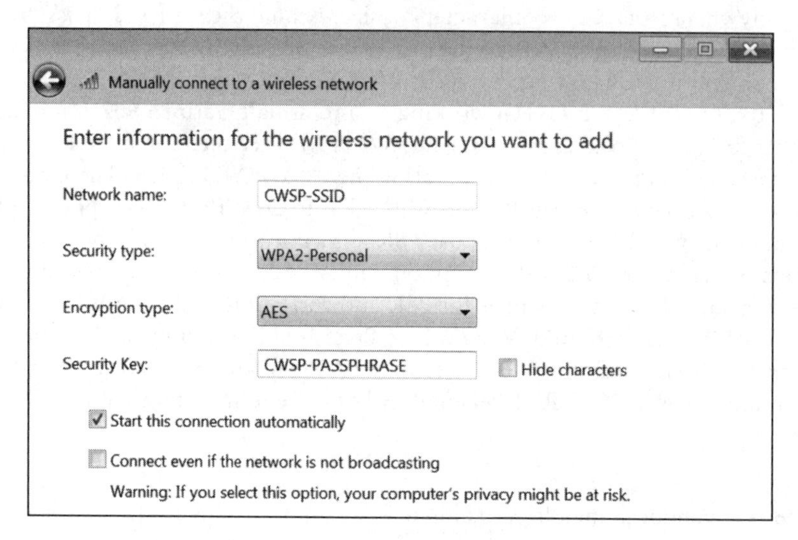

As you learned in Chapter 5, a *passphrase-PSK mapping* formula is defined by the 802.11-2012 standard to allow end users to use a simple ASCII passphrase that is then converted to the 256-bit PSK. Here is a quick review of the formula to convert the passphrase to a PSK:

PSK = PBKDF2(PassPhrase, ssid, ssidLength, 4096, 256)

A simple passphrase is combined with the SSID and hashed 4,096 times to produce a 256-bit (64-character) PSK. Table 6.1 illustrates some examples of how the formula uses both the passphrase and SSID inputs to generate the PSK.

TABLE 6.1 Passphrase-PSK mapping

Passphrase (8–63 characters)	SSID	256-bit /64-character PSK
carolina	cwsp	db119f5a0f86166dc2c12f6f6dcc2ac2121f1decf83c83d 54468bd35d71e477f
certification	cwsp	e7b3a9521a6c71236ff47cfa310f942b1a9cbb04307191ed 98ef11b84bf29be6
Victoria	cwsp	538e225e1d04cce3a274d2c1ca748f3e37b5cbdfe31c04e b925ec51657b3e86d

The whole point of the passphrase-PSK mapping formula is to simplify configuration for the average home end user. Most people can remember an 8-character passphrase as opposed to a 256-bit PSK.

The 256-bit PSK is also used as the *pairwise master key (PMK)*. The PMK is the seeding material for the 4-Way Handshake that is used to generate dynamic encryption keys. Therefore, the PSK in WPA/WPA2-Personal mode is quite literally the same as the PMK.

Previously, you learned that the 802.1X/EAP process generates a PMK and that every supplicant has a unique PMK. The same cannot be said when PSK authentication is used. The 256-bit PSK is the PMK. Because every client station uses the same PSK, or passphrase that is converted to a PSK, every client station has the same pairwise master key (PMK). This presents a serious security risk.

If a hacker is able to obtain the passphrase maliciously, the hacker could then use the passphrase-PSK mapping formula to re-create the 256-bit PSK. As you can see in Figure 6.3, most protocol analyzers are aware of the passphrase-PSK mapping formula and use the SSID and passphrase inputs to re-create the 256-bit PSK. Remember that the PSK is also used as the pairwise master key (PMK). An online 256-bit PSK generator can be found at `www.wireshark.org/tools/wpa-psk.html`.

FIGURE 6.3 Passphrase-PSK mapping

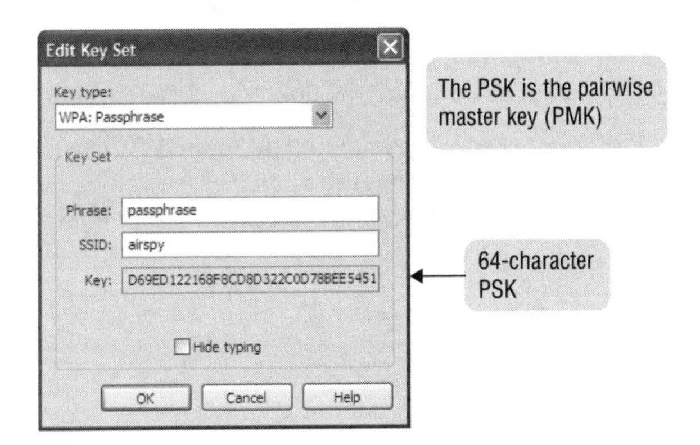

The *4-Way Handshake* uses a pseudo-random function to combine the PMK with the Authenticator Nonce (ANonce), Supplicant Nonce (SNonce), Authenticator Address (AA), and Supplicant Address (SPA) to create a pairwise transient key (PTK). As you learned in previous chapters, the PTK is used by the client station and the AP to encrypt/decrypt unicast 802.11 data frames.

The ANonce, SNonce, and the MAC addresses of the client station and the access point are all seen in clear text during the 4-Way Handshake frame exchange. Once the hacker has the PMK derived from the passphrase, the final step is to capture a 4-Way Handshake exchange between a client station and an AP with a protocol analyzer. The attacker now has all the variables needed to duplicate the pairwise transient key (PTK). If the hacker can duplicate the PTK, the hacker can then decrypt any unicast traffic between the AP and the individual client station that performed the 4-Way Handshake.

EXERCISE 6.1

Passphrase-PSK Mapping

In this exercise, you will use a protocol analyzer and passphrase-PSK mapping to regenerate the pairwise master key (PMK). You will then use a capture of a client station's 4-Way Handshake EAPOL-Key frames together with the PMK to re-create the temporal keys that are used for encryption. You will then be able to decrypt the data frames using the temporal key.

1. To perform this exercise, you need to first download the `PASSPHRASE_PSK.PCAP` file from the book's online resource area, which can be accessed at `www.sybex.com/go/cwsp2e`.
2. After the file is downloaded, you will need packet analysis software to open it. If you do not already have a packet analyzer installed on your computer, you can download Wireshark from `www.wireshark.org`.
3. Using the Wireshark packet analyzer, open the `PASSPHRASE_PSK.PCAP` file. Most packet analyzers display a list of capture frames in the upper section of the screen, with each frame numbered sequentially in the first column.
4. Observe the 4-Way Handshake frame exchange in packets 18–24. Click on frame 34 and observe that CCMP is used to encrypt the 802.11 QoS Data frame.
5. To create a 256-bit PSK, open a web browser and go to the URL `www.wireshark.org/tools/wpa-psk.html`. As shown here, enter the passphrase value of **certification** and the SSID value of **cwsp** and click the Generate PSK button. Highlight and copy the 256-bit PSK.

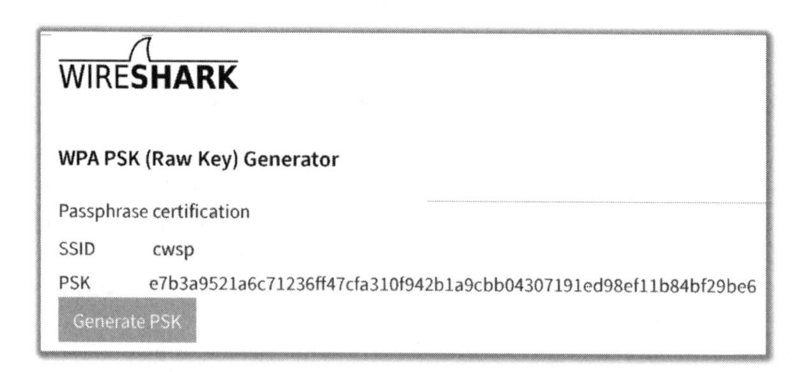

6. In Wireshark, click Edit ➤ Preferences and expand Protocols. Click and highlight the IEEE 802.11 protocol and then choose the Enable Decryption check box as shown here.

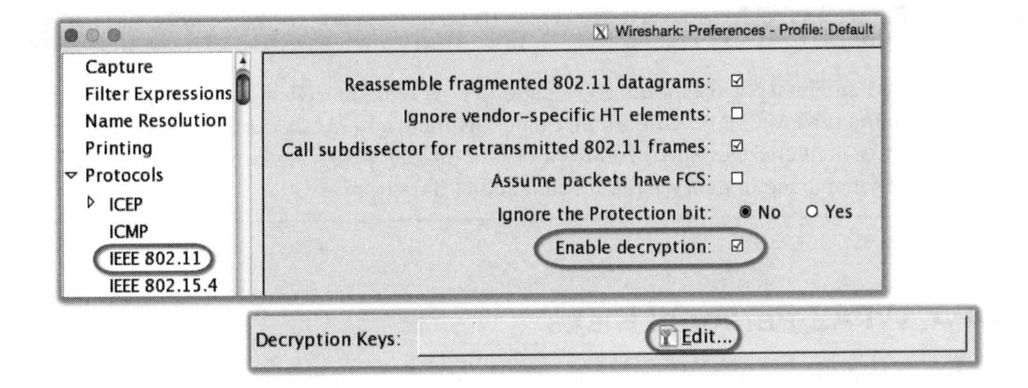

7. Next to Decryption Keys click the Edit button. Click the New button. Choose wpa-psk from the Key Type drop-down. For the Key value, paste the 256-bit PSK that was created earlier. Click OK. Click Apply and OK to close the encryption key window. Click Apply and OK to close the protocol window.

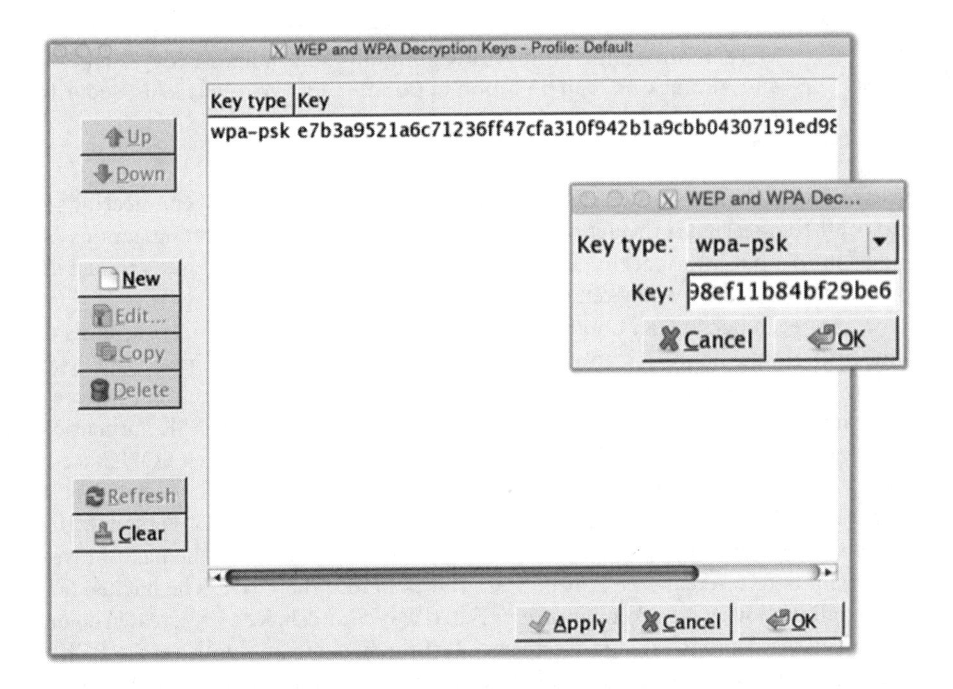

8. Notice that all of the previously encrypted 802.11 data frames are now decrypted. Scroll to packet 34 to view a decrypted DHCP discovery request packet.

EXERCISE 6.1 ***(continued)***

9. The SSID and passphrase inputs were used to create the pairwise master key (PMK). Combining the PMK along with the ANonce, SNonce, and MAC addresses found in the 4-Way Handshake enabled you to re-create the pairwise transient key (PTK). The PTK was then used to decrypt the unicast 802.11 data frames.

WPA/WPA2-Personal Risks

The risks involved with WPA/WPA2-Personal are basically twofold: network resources can be placed at risk and the encryption keys can be compromised. First, let us discuss how network resources can be placed at risk. WPA/WPA2-Personal uses the weak PSK authentication method that is vulnerable to an offline brute-force dictionary attack. WLAN auditing software such as coWPAtty and Aircrack-ng can be used for malicious purposes to obtain weak passphrases using an offline brute-force dictionary attack. We will discuss how to protect against these types of attacks later in this chapter.

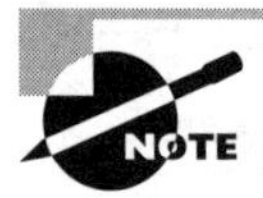

More information about WLAN auditing tools, such as coWPAtty and Aircrack-ng, can be found in Chapter 13, "Wireless LAN Security Auditing."

The easier way to obtain a WPA/WPA2 passphrase is through social engineering techniques. Social engineering is the act of manipulating people into performing actions or divulging confidential information. The simplest way to get the passphrase is to just ask someone what it is and they may tell you.

As you can see, a hacker can obtain the passphrase using social engineering skills or an offline brute-force dictionary attack. Once the passphrase is compromised, the hacker can use any client station to associate with the WPA/WPA2-Personal configured access point. The hacker is then free to access network resources. This is why PSK authentication was never intended for use in the enterprise and was intended for use as a SOHO security method.

We have already discussed the second risk involved with WPA/WPA2-Personal. If a hacker is able to obtain the passphrase maliciously, the hacker can use the passphrase-PSK mapping formula to re-create the 256-bit PSK and therefore the PMK. The hacker can use the PMK combined with a packet capture of the 4-Way Handshake to re-create the unicast encryption key. As demonstrated in Exercise 6.1, if the hacker can duplicate the PTK, the hacker can then decrypt any unicast traffic between the AP and the individual client station that performed the 4-Way Handshake. Also keep in mind that WPA-Personal uses legacy TKIP encryption, which is considered to be a weak form of dynamic encryption. As mentioned many times in this book, TKIP is not supported for 802.11n and 802.11ac data rates. If PSK authentication is chosen, only WPA2-Personal devices that use CCMP encryption should be deployed.

Entropy

In digital communications, *entropy* is a measure of uncertainty associated with a random variable. The complete explanation of the information theory on which entropy is based is well beyond the scope of this book, but we will attempt to explain entropy as it relates to the use of passphrases.

The random act of flipping a coin can be used to visualize entropy. The final result of a coin flip will either be "heads" or "tails," but there is no way to tell what the result will be each time the coin is flipped. The measure of uncertainty associated with the randomness of a coin flip is equal to one bit of entropy. Another example of entropy is when you roll a single die, it will land on one of six sides. As shown in Figure 6.4, the measure of uncertainty associated with the random roll of a single die is equal to 2.58 bits of entropy. A random variable has a certain number (N) of outcomes. A coin flip has two possible outcomes whereas the roll of a die has six possible results. The roll of the die has more uncertainty than a coin flip; therefore, it has more bits of entropy. The more randomness there is to a situation, the more measureable bits of uncertainty exist. The formula to calculate entropy is log2(N), where N is the number of possible outcomes. Thus, the formula for the role of the die is log2(6), which is equal to 2.58 bits. To confirm this, you can calculate the inverse formula, which is 2^2.58, which is approximately 6 (we rounded down to 2.58).

FIGURE 6.4 Entropy bits

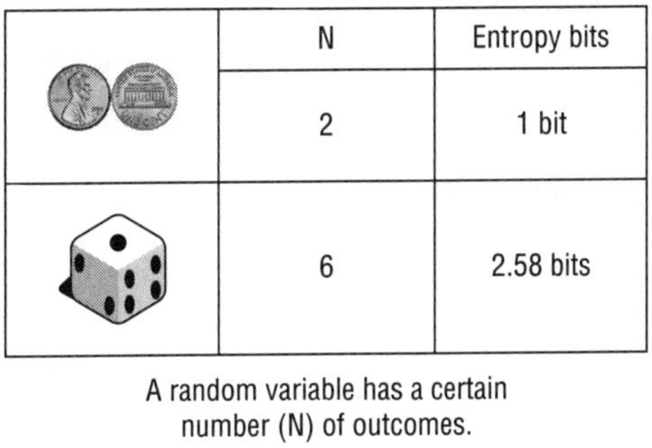

	N	Entropy bits
(coins)	2	1 bit
(die)	6	2.58 bits

A random variable has a certain number (N) of outcomes.

As shown in Table 6.2, entropy bits can also be applied to individual characters in a password or passphrase. For example, if a personal identification number (PIN) was created from the numerical digits, each character of the PIN has 10 possibilities (0–9). Based on complex informational theory, each character created strictly from numerical digits would have 3.3 bits of entropy. If the single-case letters were used in a PIN, each character of the PIN has 26 possibilities (a–z). Based on complex informational theory, each character created strictly from numerical digits would have 4.7 bits of entropy.

TABLE 6.2 Entropy bit

Symbol set	N of symbols	Entropy bits per character
Numerical digits (0–9)	10	3.32 bits
Single-case letters (a–z)	26	4.7 bits
Single-case letters and digits (a–z, 0–9)	36	5.17 bits
Mixed-case letters and digits (a–z, A–Z, 0–9)	62	5.95 bits
Mixed-case letters, digits, and symbols (Standard U.S. keyboard)	94	6.55 bits

Passwords or passphrases by themselves have an entropy value of zero. It is the method you use to select the passphrase that contains the entropy. When strong passwords or passphrases are created, they should include a combination of uppercase and lowercase letters, numbers, and special symbols. A strong passphrase will have more entropy bits per character than a password created from a name or phone number. The more entropy (measured in bits) that is contained within the method you use to create your passphrase, the more difficult it will be to guess the passphrase.

Now let's discuss how many entropy bits are contained in a passphrase used in WPA/WPA2-Personal. The following is a quote from the original 802.11i-2004 amendment:

> A passphrase typically has about 2.5 bits of security per character, so the passphrase mapping converts an *n* octet password into a key with about 2.5*n* + 12 bits of security. Hence, it provides a relatively low level of security, with keys generated from short passwords subject to dictionary attack. Use of the key hash is recommended only where it is impractical to make use of a stronger form of user authentication. A key generated from a passphrase of less than about 20 characters is unlikely to deter attacks.
>
> — *802.11i-2004 amendment*

What this quotation means is that each character in a passphrase is worth only 2.5 bits of entropy because most users choose easy-to-remember words as opposed to a mix of alphanumeric and punctuation characters. Therefore, an 8-character passphrase would only contain 20 bits of entropy before the passphrase-to-PSK mapping hash algorithm is calculated. Because the passphrase-to-PSK mapping mixes the SSID with the passphrase, approximately 12 extra bits of entropy are added. Therefore, a typical 8-character passphrase used in a WPA/WPA2-Personal configuration would have a total of 32 bits of entropy.

Please do not confuse entropy security bits with the number of digital bits in a PSK. The passphrase-to-PSK mapping formula will always convert any passphrase of 8 to

63 characters into a 256-bit PSK. Remember that entropy bits are a measure of uncertainty associated with a random variable. The random variable is basically the strength of the passphrase.

The biggest question is the number of bits of entropy needed to protect current data communications. Current information technology security best practices state that 96 bits of entropy should be safe; however, 128 bits might be needed in the future. This all sounds pretty scary if you are using a simple passphrase for your home Wi-Fi router—and you should be scared. As mentioned earlier, the simple passphrases used by most users are susceptible to brute-force offline dictionary attacks. Both the IEEE and the Wi-Fi Alliance recommend a passphrase of 20 characters or more, which would give you at least 62 bits of entropy. The entropy of the passphrase increases significantly if a combination of uppercase and lowercase letters, numbers, and special symbols are used in the passphrase. The entropy of a passphrase also increases with each extra character that is added to the passphrase.

In summary, always use, at a minimum, a 20-character passphrase and mix up the characters for your home Wi-Fi solution. Sadly, most people will use their dog's name or their telephone number for their passphrase. In that case, the bigger risk is a social engineering attack because the passphrase can be easily guessed. Passphrases were never intended for the enterprise. Enterprise APs usually give you the choice of entering either a passphrase or a 64-character PSK. If, for some reason, static PSK authentication must be used in the enterprise, at the minimum, use a strong 20-character or longer passphrase.

Is There an Easy Way to Create a Passphrase with Strong Entropy?

The Gibson Research Corporation maintains a free website that will randomly generate unique cryptographic-strength password strings: `https://www.grc.com/passwords.htm`.

Numerous freeware software tools exist that can be used to assist end users with creating strong passphrases with increased entropy. These tools can be used to increase the strength of passwords, passphrases, and even the shared secrets that are used between RADIUS servers and an authenticator.

Proprietary PSK

Although the use of passphrases and PSK authentication is intended for use in a home environment, in reality WPA/WPA2-Personal is often still used in the enterprise. For example, even though fast secure roaming (FSR) mechanisms are now possible, many older VoWiFi phones and other handheld devices still do not yet support 802.1X/EAP. As a result, the strongest level of security used with these devices is PSK authentication. Cost issues may also drive a small business to use the simpler WPA/WPA2-Personal solution as opposed to installing and configuring a RADIUS server for 802.1X/EAP.

The biggest problem with using PSK authentication in the enterprise is social engineering. The PSK is the same on all WLAN devices. If end users accidentally give the PSK to a hacker, WLAN security is compromised. If an employee leaves the company, all the devices have to be reconfigured with a new PSK. Because the passphrase or PSK is shared by everyone, a very strict policy should be mandated stating that only the WLAN security administrator knows the passphrase or PSK. That, of course, creates another administrative problem because of the work involved in manually configuring each device.

Several enterprise WLAN vendors have come up with a creative solution to using WPA/WPA2-Personal that solves some of the biggest problems using a single passphrase for WLAN access. Each computing device will have its own unique PSK for the WLAN. Therefore, the MAC address of each STA will be mapped to a unique WPA/WPA2-Personal passphrase. A database of unique PSKs mapped to usernames or client stations must be stored on all access points or on a centralized management server. The individual client stations are then assigned individual PSKs that are created either dynamically or manually. As shown in Figure 6.5, the multiple per-user/per-device PSKs can be tied to a single SSID. The PSKs that are generated can also have an expiration date. Unique time-based PSKs can also be used in a WLAN environment as a replacement for more traditional username/password credentials.

FIGURE 6.5 Per-user/per-device PSK

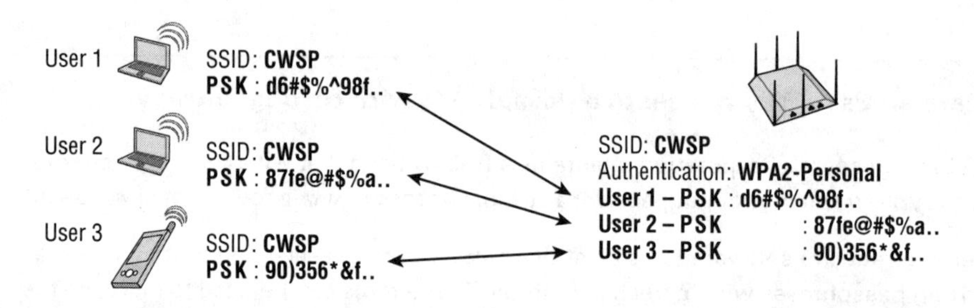

Three WLAN vendors offering *proprietary PSK* solutions, which provide the capability of unique PSKs for each user or each device, are Aerohive Networks, Ruckus Wireless, and Xirrus. Proprietary PSK solutions provide a way to implement unique per-user/per-device credentials without the burden of deploying a more complex 802.1X/EAP solution. Social engineering and brute-force dictionary attacks are still possible, but they are harder to accomplish if strong and lengthy passphrase credentials are implemented. If a unique PSK is compromised, an administrator only has to revoke the single PSK credential and no longer has to reconfigure all access points and end-user devices.

Some WLAN client devices have limited support for 802.1X/EAP. In situations such as these, proprietary PSK solutions may be of benefit for those classes of devices and a vast improvement over standard static PSK authentication. A proprietary PSK solution offers unique user or device credentials that standard PSK cannot provide. Additionally, proprietary PSK solutions with unique credentials do not require anywhere near the complex configuration needed for 802.1X/EAP. Another advantage of a per-user/per-device implementation of PSK authentication is that every user or device has a unique PMK that is derived from the unique PSK. Keep in mind that these proprietary implementations of per-user/per-device PSK authentication are still susceptible to offline brute-force dictionary attacks. Although the passphrases are unique, they should still be 20 characters or more.

Proprietary implementations of PSK authentication are not meant to be a replacement for 802.1X/EAP. However, multiple use cases for per-user and per-device PSK credentials have gained popularity in the enterprise. Some of these use cases include:

Legacy Devices Supplement to 802.1X/EAP security to provide legacy devices with unique PSK credentials.

Personal Devices Bring your own device (BYOD) security to provide unique PSK credentials for employee personal devices.

Guest Access Provide unique PSK credentials for access to the guest WLAN.

IoT Devices Machines and sensors with 802.11 radios often do not support 802.1X/EAP. Provide Internet of Thing (IoT) devices with unique PSK credentials.

Simultaneous Authentication of Equals (SAE)

The 802.11s amendment was ratified in 2011 with the intention of standardizing mesh networking of 802.11 WLANs. The 802.11s amendment is now part of the 802.11-2012 standard. The amendment defined a *Hybrid Wireless Mesh Protocol (HWMP)* that 802.11 mesh portals and mesh points could use to dynamically determine the best path selection for traffic flow through a meshed WLAN. HWMP and other mechanisms defined by the 802.11s-2011 amendment have not been embraced by WLAN vendors because of competitive reasons. The majority of WLAN vendors do not want their APs to accept mesh communications from a competitor's APs. As a result, the major WLAN vendors offer proprietary mesh solutions using their own mesh protocols and metrics.

The 802.11s-2011 amendment also defined RSN security methods that could be used by mesh portals and mesh points. An *Authenticated Mesh Peering Exchange (AMPE)* is used to securely create and exchange pairwise master keys (PMKs). 802.1X/EAP could be one method used to derive a PMK in a mesh environment. This method is not ideal for mesh points because the RADIUS server resides on the wired network. Therefore, the 802.11s-2011 amendment proposed a new peer-to-peer authentication method called *Simultaneous Authentication of Equals (SAE)*. SAE is based on a *Dragonfly key exchange*. Dragonfly is a patent-free and royalty-free technology that uses a *zero-knowledge proof* key exchange, which means a user or device must prove knowledge of a password without having to reveal a password.

Although SAE has yet to be implemented for 802.11 mesh networks, the Wi-Fi Alliance views SAE as a more secure replacement for PSK authentication. As you have already learned, PSK authentication is susceptible to brute-force dictionary attacks and is very insecure when weak passphrases are used. With the current implementation of PSK authentication, brute-force dictionary attacks can be circumvented by using very strong passphrases between 20 and 63 characters. Dictionary attacks against strong passphrases could take years before they are successful. However, dictionary attacks are feasible and more easily achieved via the combined resources of distributed cloud computing. The bigger worry is that most users usually only create an 8-character passphrase, which can usually be compromised in several hours or even minutes. The ultimate goal of SAE is to prevent dictionary attacks altogether. As of this writing, the Wi-Fi Alliance has a proposed program of interoperability certification for SAE.

Some of the proposals from the Wi-Fi Alliance regarding an SAE certification program include the following:

- WEP and TKIP must not be used for SAE.
- For transition purposes, WPA-2 Personal and SAE must be supported simultaneously within the same BSS.
- For transition purposes, WPA2-Personal devices and SAE devices should use the same passphrase.

Think of SAE as a more secure PSK authentication method. The goal is to provide the same user experience by still using a passphrase. However, the SAE protocol exchange protects the passphrase from a brute-force dictionary attacks. The passphrase is never sent between 802.11 stations during the SAE exchange.

As shown in Figure 6.6, an SAE process consists of a commitment message exchange and a confirmation message exchange. The commitment exchange is used to force each radio to commit to a single guess of the passphrase. The confirmation exchange is used to prove that the password guess was correct. SAE authentication frames are used to perform these exchanges. The passphrase is used in SAE to deterministically compute a secret element in the negotiated group, called a password element, which is then used in the authentication and key exchange protocol.

FIGURE 6.6 SAE authentication exchange

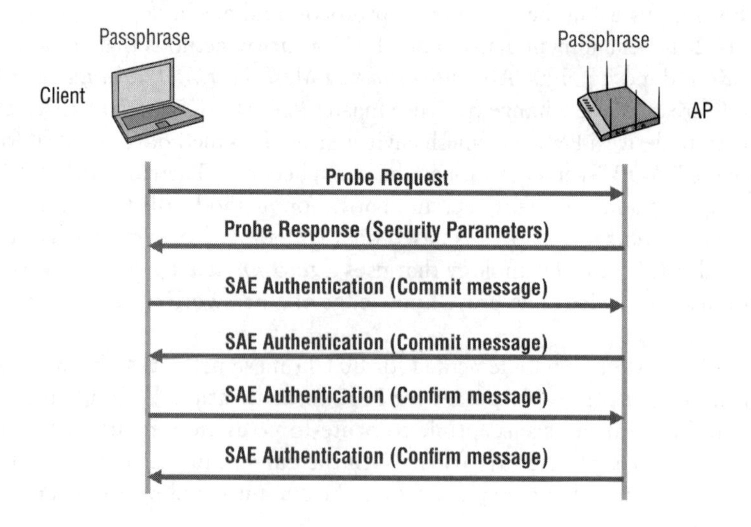

The original intent of the 802.11s amendment was to generate a pairwise master key (PMK) from the SAE exchange, which would then be used to derive a *mesh temporal key (MTK)* to be used for encrypting unicast traffic between mesh APs. The Wi-Fi Alliance wants to also use an SAE exchange between an AP and client station. As depicted in Figure 6.7, once the SAE exchanges are complete, a unique pairwise master key (PMK) is derived and installed on both the 802.11 AP and the client station. SAE authentication is performed prior to association. Once the PMK is created and the association process completes, the AP and the client can then commence a 4-Way Handshake to create a pairwise transient key (PTK).

FIGURE 6.7 SAE authentication, association, and 4-Way Handshake

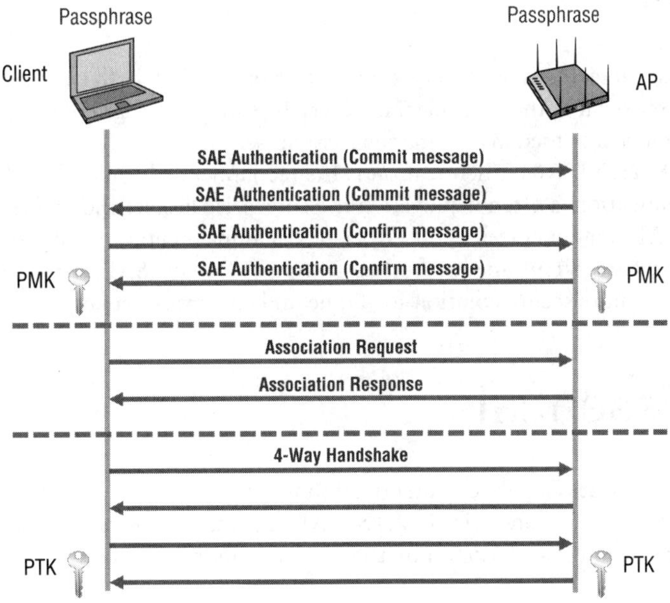

What about roaming when using SAE authentication? There are two potential methods:

Option 1 A client station could roam to a new AP with the following sequence of frame exchanges: Probe Request/Response frame exchange, SAE authentication frame exchange, reassociation frame exchange, and then a 4-Way Handshake.

Option 2 A client station can be SAE authenticated to many APs simultaneously by completing the SAE protocol with any number of APs while still being associated to another AP. In other words, a client could perform an SAE commit and confirm exchange with a potential roaming target prior to roaming to the target AP. This creates a PMK on neighboring APs. When the client roams, the PMK is already on the target AP and all the client has to do is a reassociation frame exchange and a 4-Way Handshake when it actually roams. This option would be slightly faster when roaming if the client station was SAE pre-authenticated to the AP that is the roaming target.

As previously mentioned, the goal of SAE is to address the previous weaknesses with PSK authentication. Because the SAE exchange allows for only one guess of the passphrase, brute-force offline dictionary attacks are no longer viable. An attacker with a protocol analyzer will not be able to determine the passphrase or the PMK from listening to the SAE exchange. Additionally, the SAE exchange is also resistant to forging and replay attacks. Even if the passphrase is compromised, it could not be used to re-create any previously generated PMKs.

Summary

WPA/WPA2-Personal methods have come a long way from WEP in both end-user configuration convenience and the security they provide. However, the entropy strength of a passphrase still depends on the complexity used in creating the passphrase.

Even though 802.1X/EAP authentication is the recommended solution in the enterprise, static PSK authentication is often deployed in the enterprise despite the security risks. As a result, some WLAN vendors offer per-user/per-device implementations of PSK authentication. In the near future, Simultaneous Authentication of Equals (SAE) may replace PSK authentication as a more secure solution for home and enterprise settings.

Exam Essentials

Explain PSK authentication. PSK authentication is meant to be used in SOHO environments because the stronger enterprise 802.1X/EAP authentication solutions are not available. PSK authentication is also known by the more common name of WPA-Personal or WPA2-Personal.

Define entropy and how it relates to PSK authentication. Passwords or passphrases by themselves have an entropy value of zero. Entropy is a measure of uncertainty associated with a random variable. The method you use to select a passphrase can be measured in terms of randomness.

Understand passphrase-to-PSK mapping. Explain how the passphrase-to-PSK formula uses a hash algorithm to mix the SSID with the passphrase. Understand the security weaknesses associated with this method.

Explain the advantages of proprietary PSK solutions. Proprietary PSK solutions prevent social engineering and employee sharing of a passphrase. Proprietary PSK solutions greatly limit the ability of malicious users to decrypt user traffic and simplify administration.

Review Questions

1. What can happen when a hacker compromises the preshared key used during PSK authentication? (Choose all that apply.)
   - A. Decryption
   - B. Spoofing
   - C. Encryption cracking
   - D. Access to network resources
   - E. Denial of service
2. When configuring an 802.11 client station for WPA2-Personal, what security credentials can be used? (Choose all that apply.)
   - A. Server-side certificate
   - B. 64 hex-character PSK
   - C. 8-63 character PSK
   - D. Token card
   - E. Client-side certificate
   - F. 64-character passphrase
   - G. 8-to-63 character passphrase
3. What inputs are used by passphrase-PSK mapping to create a final 256-bit PSK during 802.11 PSK authentication? (Choose all that apply.)
   - A. BSSID
   - B. SNonce
   - C. SSID
   - D. Client MAC address
   - E. AP MAC address
   - F. Passphrase
   - G. ANonce
4. Tammy, the WLAN security engineer, has recommended to management that WPA-Personal security not be deployed within the ACME Company's WLAN. What are some of the reasons for Tammy's recommendation? (Choose all that apply.)
   - A. Static passphrases and PSKs are susceptible to social engineering attacks.
   - B. WPA-Personal is susceptible to brute-force dictionary attacks, but WPA-Personal is not at risk.
   - C. WPA-Personal uses static encryption keys.
   - D. WPA-Personal uses weaker TKIP encryption.
   - E. 802.11 data frames can be decrypted if the passphrase is compromised.

5. When considering the 4-Way Handshake that is used to create dynamic encryption keys, what is the main difference between 802.1X/EAP and PSK authentication? (Choose all that apply.)
   A. 802.1X/EAP supplicants all use the same PMK.
   B. Clients that use PSK authentication all use the same PTK.
   C. 802.1X/EAP supplicants all use a different PMK.
   D. Clients that use PSK authentication all use a different PTK.
   E. 802.1X/EAP supplicants all use a different PTK.
   F. Clients that use PSK authentication all use the same PMK.

6. What is the Wi-Fi Alliance recommendation for the number of characters used in a passphrase for WPA/WPA2-Personal security?
   A. 6 characters
   B. 8 characters
   C. 10 characters
   D. 12 characters
   E. 20 characters

7. Don has been hired as a consultant to secure the Maxwell Corporation's WLAN infrastructure. Management has asked him to choose a WLAN authentication solution that will best protect the company's branch offices from unauthorized users. The branch offices will use older VoWiFi phones that do not support opportunistic key caching (OKC) or fast BSS transition (FT). The company is also looking for the strongest dynamic encryption solution for data privacy reasons. Management is also looking for the cheapest solution as well as a solution that is easy to administer. Which of these WLAN security solutions meets all the objectives required by management? (Choose the best two answers.)
   A. WPA-Personal
   B. WPA2-Personal
   C. WPA-Enterprise
   D. WPA2-Enterprise
   E. Proprietary PSK

8. When considering the 4-Way Handshake that is used to create dynamic encryption keys, what can be said about both 802.1X/EAP authentication and proprietary per-user/per-device implementation of PSK authentication? (Choose all that apply.)
   A. 802.1X/EAP supplicants all use the same PMK.
   B. Clients that use per-user/per-device PSK authentication all use the same PMK.
   C. 802.1X/EAP supplicants all use a different PMK.
   D. Clients that use per-user/per-device PSK authentication all use a different PMK.
   E. Clients that use per-user/per-device PSK authentication all use the same PTK.

9. The IEEE 802.11-2012 standard requires an authentication and key management protocol (AKMP) that can be either a preshared (PSK) or an EAP protocol used during 802.1X/EAP authentication. What is another name for PSK authentication? (Choose all that apply.)
   - A. Wi-Fi Protected Setup
   - B. WPA/WPA2-Personal
   - C. WPA/WPA2-PSK
   - D. WPA/WPA2-Preshared Key
   - E. WPA/WPA2-Passphrase

10. Which of these terms best describes a measure of uncertainty associated with a random variable?
    - A. Entropy
    - B. Encryption
    - C. Encapsulation
    - D. Encoding

11. After viewing the image shown here, identify the type of WLAN security method being used.

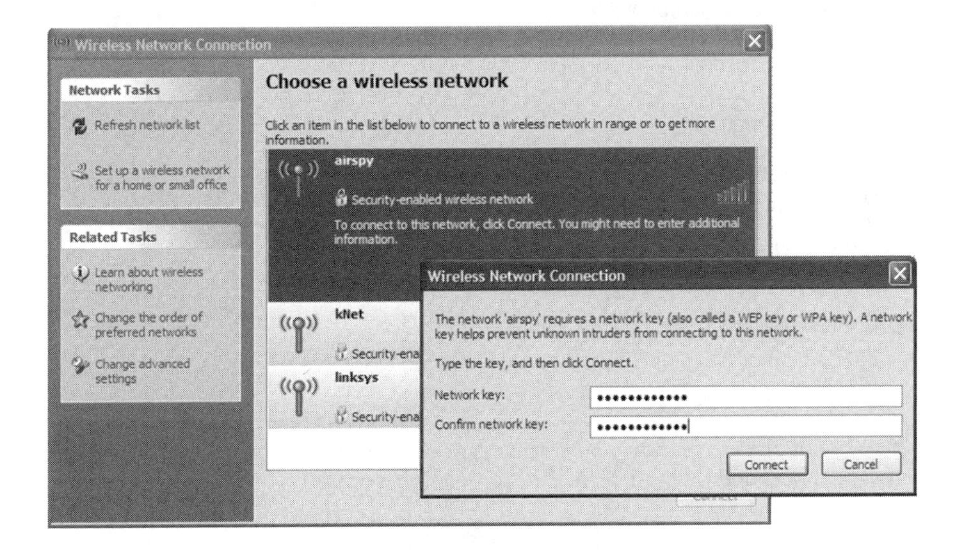

   - A. EAP-TLS
   - B. CCMP/AES
   - C. Dynamic PSK
   - D. EAP-TTLS
   - E. None of the above

12. Which of these passphrases is the least susceptible to a brute-force offline dictionary attack?
    - A. 20-character passphrase using numerical digits (0–9)
    - B. 20-character passphrase using single-case letters (a–z)
    - C. 20-character passphrase using single-case letters and digits (a–z, 0–9)
    - D. 20-character passphrase using mixed-case letters and digits (a–z, A–Z, 0–9)
13. Based on the IEEE's estimation of entropy strength for each character used in the average WPA2 passphrase, how many entropy bits of security would be in an 8-character passphrase?
    - A. 8 bits
    - B. 12 bits
    - C. 20 bits
    - D. 32 bits
    - E. 64 bits
14. Which of these security methods is being considered by the Wi-Fi Alliance as a replacement for PSK authentication?
    - A. Per-user/per-device PSK
    - B. Wi-Fi Protected Setup (WPS)
    - C. Simultaneous Authentication of Equals (SAE)
    - D. EAP-PSK
    - E. WPA2 Personal
15. What are some of the advantages of using SAE authentication over PSK authentication? (Choose all that apply.)
    - A. Protects against brute-force dictionary attacks.
    - B. Protects against forgery and replay attacks.
    - C. Protects against rogue APs and clients.
    - D. PMKs cannot be compromised or regenerated.
    - E. PMKs are no longer needed.
16. After viewing the image shown here, what type of credentials can be strengthened with the software application that is depicted? (Choose all that apply.)

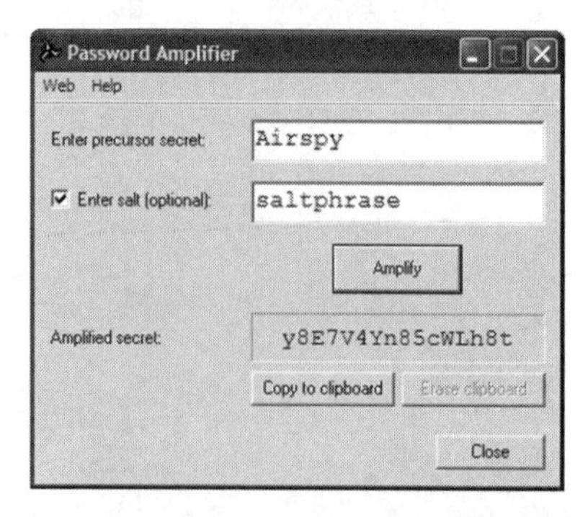

A. Passwords
B. Shared secrets
C. Passphrases
D. Pairwise master keys
E. Pairwise transient keys
F. Group temporal keys

17. The ACME Company is using WPA2-Personal to secure handheld barcode scanners that are not capable of 802.1X/EAP authentication. Because an employee was recently fired, all the barcode scanners and APs had to be reconfigured with a new static 64-bit PSK. What type of WLAN security solution may have avoided this administrative headache?
A. MAC filter
B. Hidden SSID
C. Change the default settings
D. Proprietary PSK

18. If an 802.1X/EAP solution is not available in the enterprise, which of these security credentials should be used instead?
A. MAC filter
B. WPA passphrase with at least 62 bits of entropy
C. WPA2 passphrase with at least 62 bits of entropy
D. Static WEP key

19. Which of these Wi-Fi Alliance security certifications specify mechanisms not defined by the IEEE-2012 standard?
A. WPA-Personal
B. WPA-Enterprise
C. WPA2-Personal
D. WPA2-Enterprise
E. Wi-Fi Protected Setup

20. Which one of these use cases for a per-user/per-device implementation of PSK authentication is not recommended?
A. Unique credentials for BYOD devices
B. Unique credentials for IoT devices
C. Unique credentials for guest WLAN access
D. Unique credentials for legacy enterprise devices without 802.1X/EAP support
E. Unique credentials for enterprise devices with 802.1X/EAP support

Chapter 7

802.11 Fast Secure Roaming

IN THIS CHAPTER, YOU WILL LEARN ABOUT THE FOLLOWING:

✓ **Roaming basics**

- Client roaming thresholds
- AP-to-AP handoff

✓ **RSNA**

- PMKSA
- PMK caching
- Preauthentication

✓ **Opportunistic key caching (OKC)**

✓ **Proprietary FSR**

✓ **Fast BSS transition (FT)**

- Information elements
- FT initial mobility domain association
- Over-the-air fast BSS transition
- Over-the-DS fast BSS transition

✓ **802.11k**

✓ **802.11v**

✓ **Voice Enterprise**

✓ **Layer 3 roaming**

✓ **Troubleshooting**

One of the main reasons that Wi-Fi networks have spread like wildfire is the fact that 802.11 technology provides mobility. In today's world, end users demand the freedom provided by WLAN mobility. Corporations also realize productivity increases if end users can access network resources wirelessly. Mobility requires that client stations have the ability to transition from one access point to another while maintaining network connectivity for the upper-layer applications. This ability is known as *roaming*. A perfect analogy is the roaming that occurs when using a cell phone. When you are talking on a cell phone to your best friend while riding in a car, your phone will roam between cellular towers to allow for seamless communications and hopefully an uninterrupted conversation. Similarly, seamless roaming between access points allows for WLAN mobility, which is the heart and soul of true wireless networking and connectivity. This chapter will cover the basic as well as advanced handoff mechanisms needed for seamless roaming.

This chapter will also explore the relationship between roaming and security. Although mobility is paramount to any WLAN, maintaining security is just as important. Ideally, all client stations should use WPA2-Enterprise level security when roaming between access points. However, when using WPA2-Enterprise security, which involves the use of a RADIUS server, roaming can be especially troublesome for VoWiFi and other time-sensitive applications. Due to the multiple frame exchanges between the authentication server and the supplicant, an 802.1X/EAP authentication can take 700 milliseconds (ms) or longer for the client to authenticate. VoWiFi requires a handoff of 150 ms or less to avoid a degradation of the quality of the call or, even worse, a loss of connection. Therefore, faster, secure roaming handoffs are required.

This chapter will discuss a variety of methods that allow for *fast secure roaming (FSR)*. The 802.11i security amendment, which is now part of the 802.11-2012 standard, defines two fast secure roaming mechanisms called preauthentication and PMK caching. Most WLAN vendors have used an enhanced method of FSR called opportunistic PMK caching. Since the ratification of the 802.11r-2008 amendment, even more complex FSR methods have begun to find their way into the enterprise. As FSR gains broader acceptance, time-sensitive applications, such as voice and video, will be assured of having stronger authentication security without a disruption in communications.

History of 802.11 Roaming

Before we can delve into the relationship between roaming and security, we must first discuss the basics of roaming as well as a little history. The original legacy 802.11 standard, for the most part, only defined roaming as a Layer 2 process known as the reassociation service. The *reassociation service* enables an established association between an access point (AP) and a client station (STA) to be transferred from one AP to another AP. Reassociation allows a client station to move from one basic service set (BSS) to another; therefore, a more technical term used for roaming is *BSS transition*. However, the 802.11-2012 standard does not define two very important processes for BSS transition: client roaming thresholds and AP-to-AP handoff communications.

Client Roaming Thresholds

In standard WLAN environments, client stations always initiate the roaming process known as reassociation. In simpler words, clients make the roaming decision and access points do not tell the client when to roam. What causes the client station to roam is a set of proprietary rules determined by the manufacturer of the wireless radio, usually defined by *received signal strength indicator (RSSI)* thresholds. RSSI thresholds usually involve signal strength, signal-to-noise ratio (SNR), and bit-error rate. As the client station communicates on the network, it continues to look for other access points via probing and will hear received signals from other APs. As shown in Figure 7.1, as the client station moves away from the original access point with which it is associated, the signal drops to a received signal level of –75 dBm. In the meantime, the client hears a nearby AP with a stronger signal of –70 dBm. The 5 dB difference might be the roaming threshold that triggers the client station to attempt to connect to the new target access point. The client sends a frame, called the *reassociation request frame*, to start the roaming procedure.

FIGURE 7.1 Client roaming decision

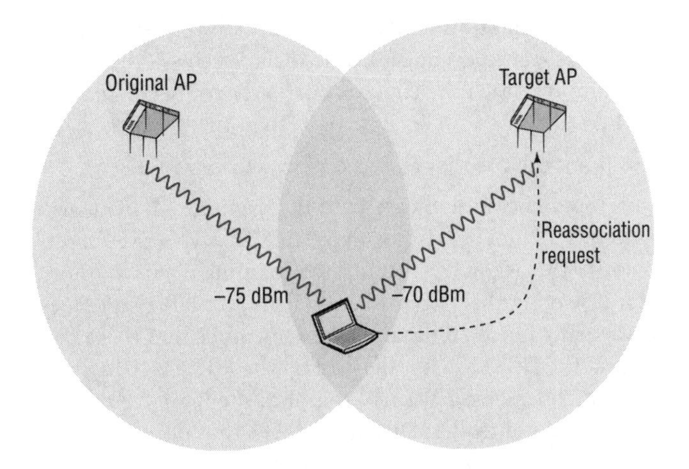

Vendors have different thresholds that kick off the client reassociation process. As mentioned in the earlier example, a 5 dB difference might trigger a client to attempt a forward roam. However, the client might need to hear a 10 dB stronger signal to roam back to the client's original AP after an initial roam. A higher roaming threshold for a backward roam prevents clients from ping-ponging back and forth between APs.

There is no standard on what triggers a client to initiate the reassociation process. In other words, two client stations could be associated with the same AP and could be receiving the same strength signal, yet one of them may roam before the other because they have different vendor RSSI roaming thresholds. The roaming thresholds in handheld devices tend to encourage roaming more often than the roaming thresholds of WLAN radios found in laptops. Some WLAN vendors attempt to encourage or discourage roaming by manipulating the client station with the use of management frames. However, it should be understood that ultimately the roaming decision is made by the client station. The bottom line is that clients make the roaming decision, and all client roaming thresholds are proprietary.

AP-to-AP Handoff

The 802.11-2012 standard also does not define AP-to-AP handoff communications. As a station roams, the original access point and the target access point should communicate with each other across the *distribution system medium (DSM)* and help provide a clean transition between the two APs. The DSM is also simply referred to as the *distribution system (DS)*. The DS is typically an 802.3 wired network. The AP-to-AP handoff communications involves two primary tasks:

- The target AP informs the original AP that the client station is roaming.
- The target AP requests the client's buffered packets from the original AP.

Let's discuss how basic roaming mechanisms work. As shown in Figure 7.2, roaming occurs after the client and the access point have exchanged reassociation frames, as described in the following steps:

1. In the first step, the client station sends a reassociation request frame to the target access point. The reassociation request frame includes the BSSID (MAC address) of the access point radio to which the client is currently connected (we will refer to this as the original AP).
2. The target access point then replies to the station with an ACK.
3. The target AP attempts to communicate with the original AP by using the distribution system medium (DSM). The DSM is normally an 802.3 Ethernet network. The target AP attempts to notify the original AP about the roaming client to inform the original AP that the client is leaving the original BSS. The target AP also requests that the original AP forward any buffered data. Please remember that these communications between the APs via the DSM are not defined by the 802.11-2012 standard and are proprietary. In a controller-based WLAN solution, the inter-AP communications might instead occur within the controller. Distributed APs that do not require a controller will communicate with each other at the edge of the network.
4. If this communication is successful, the original access point will use the DSM to forward any buffered data to the target access point.
5. The target access point then sends a reassociation response frame to the client via the wireless network.

6. The client sends an ACK to the target access point. The client has now joined the BSS of the target AP.
7. If the reassociation is not successful, the client will retain its connection to the original AP and either will continue to communicate with the original AP or attempt to roam to another access point.

FIGURE 7.2 Reassociation

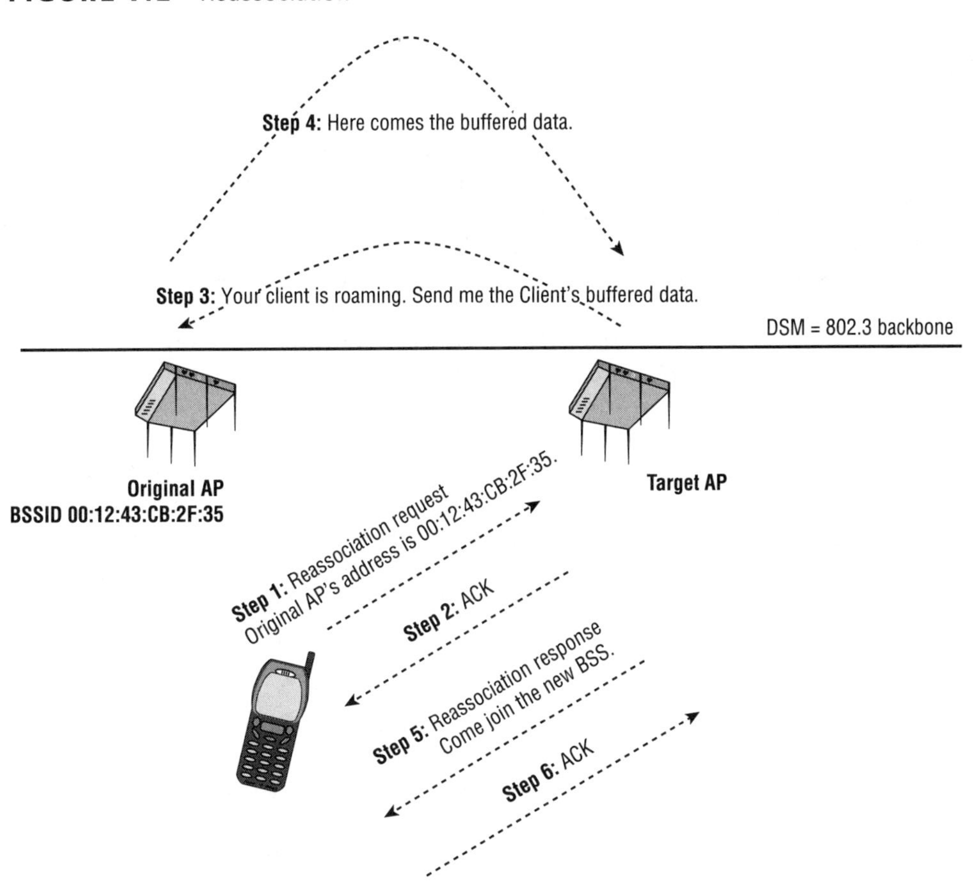

As the client station roams, the original AP and the new AP should communicate with each other across the DSM and help provide a clean transition between the two. Many manufacturers provide this handoff, but it is not officially part of the 802.11 standard, so each vendor does it using its own proprietary method. In WLAN controller–based solutions, the roaming handoff mechanisms usually occur within the WLAN controller. In controller-less environments, the roaming handoff mechanisms occur at the edge of the network where the conversation is directly between the cooperative APs that do not require a WLAN controller.

Now that you have learned the basics of roaming, we will now begin to discuss the relationship between roaming and security. In the remaining sections of this chapter, we will examine the creation and forwarding of dynamic encryption keys between multiple access points and roaming clients.

RSNA

Let's review a few things you already learned in Chapter 5, "802.11 Dynamic Encryption Key Generation." The 802.11i amendment, which was ratified and published as IEEE Std. 802.11i-2004, defined stronger encryption and better authentication methods. The 802.11i security amendment is now part of the 802.11-2012 standard. The 802.11-2012 standard defines what is known as a robust security network (RSN) and robust security network associations (RSNAs).

Keep in mind that a *robust security network association (RSNA)* requires two 802.11 stations (STAs) to establish procedures to authenticate and associate with each other as well as create dynamic encryption keys through the *4-Way Handshake* process. This security association between two stations is referred to as an RSNA. In other words, any two radios must share dynamic encryption keys that are unique between those two radios. CCMP/AES encryption is the mandated encryption method, whereas TKIP/ARC4 is an optional encryption method. Note that the term STA is usually used to refer to a wireless client, but officially it refers to any addressable wireless device. Since an AP is also an addressable wireless device, it is also a station. However, an AP is a special type of station that provides access to the distribution system. Therefore, it is simply referred to as an AP. When referring to an RSNA between two 802.11 stations, realize that this connection is between a client STA and an AP STA.

Every time a client roams, unique encryption keys must be generated using a 4-Way Handshake process between the access point and the client STA. As you have already learned, either an 802.1X/EAP or PSK authentication process is needed to produce the *pairwise master key (PMK)* that seeds the 4-Way Handshake. Therefore, every time a client roams, the client must reauthenticate. An 802.1X/EAP framework requires multiple EAPOL frames to be exchanged between the supplicant and the authentication server (AS). If a client station has to reauthenticate using 802.1X/EAP every time, the roaming process will be secure but the delay will be significant. This delay will disrupt the communications of time-sensitive applications such as voice and video. A typical 802.1X/EAP roaming handoff can take 700 ms, which far exceeds the needed handoff time of 150 ms or less for VoWiFi. Secure roaming handoff times will be even worse if the RADIUS server resides across a WAN link. The 802.11i security amendment is now part of the 802.11-2012 standard and defines two fast secure roaming mechanisms called *preauthentication* and *PMK caching*, which will be discussed later in this chapter.

PMKSA

Robust security network associations (RSNAs) can be broken down into several subtypes. *A pairwise master key security association (PMKSA)* is the result of a successful IEEE 802.1X/EAP authentication exchange between a supplicant and authentication server (AS) or from a preshared key (PSK) authentication. In other words, once the PMK has been created, a bidirectional security association exists between the authenticator and the supplicant. As you learned in Chapter 5, the PMK is the seeding material for the 4-Way Handshake that creates the *pairwise transient key (PTK)*, which is used for encryption and decryption of unicast traffic. Once the 4-Way Handshake creates the final encryption keys, a *pairwise transient key security association (PTKSA)* exists between the authenticator and the supplicant. In Chapter 5, we discussed in great detail how the 4-Way Handshake process results in a PTKSA. In this chapter, we will focus more on pairwise master key security associations (PMKSAs).

RSN security can be identified by a field found in certain 802.11 management frames. This field is known as the *robust security network information element (RSNIE)*, which is often referred to simply as the *RSN information element*. An information element is an optional field of variable length that can be found in 802.11 management frames. The RSN information element field is always found in four different 802.11 management frames: beacon management frames, probe response frames, association request frames, and reassociation request frames. The RSN information element can also be found in reassociation response frames if 802.11r capabilities are enabled on an AP and roaming client. In Chapter 5, you learned about the authentication key management (AKM) suites and pairwise cipher suites found in the RSN information element. Figure 7.3 shows the entire format of the RSN information element. Notice the PMKID fields within the RSN information element.

FIGURE 7.3 RSN information element format

Element ID	Length	Version	Group Cipher Suite	Pairwise Cipher Suite Count	Pairwise Cipher Suite List	AKM Suite Count	AKM Suite List	RSN Capabilities	PMKID Count	PMKID List

A unique identifier is created for each PMKSA that has been established between the authenticator and the supplicant. The *pairwise master key identifier (PMKID)* is a unique identifier that refers to a PMKSA. The PMKID can reference the following types of pairwise master key security associations:

- A PMKSA derived from a PSK for the target AP
- A cached PMKSA from an 802.1X/EAP or SAE authentication
- A cached PMKSA that has been obtained through preauthentication with the target AP
- A PMK-R0 security association derived as part of an FT initial mobility domain association
- A PMK-R1 security association derived as part of an FT initial mobility domain association or as part of a fast BSS transition

We will discuss PMK caching and preauthentication in the next sections of this chapter. PMK-R0 and PMK-R1 security associations will be discussed in the fast BSS transition (FT) section of this chapter. The pairwise master key identifier (PMKID) is found in the RSN information element in association request frames and reassociation request frames that are sent from a client station to an AP. The PMKID is also found in FT Action frames. Figure 7.4 shows a protocol analyzer capture of a reassociation request frame's RSN information element and PMKID information. Remember that the PMKID is a unique identifier of an individual PMKSA; however, you will learn that a client station may have established multiple PMKSAs. Therefore, the PMKID Count field specifies the number of PMKIDs in the PMKID List field. The PMKID list contains 0 or more PMKIDs that the STA believes to be valid for a destination AP. The example in Figure 7.4 shows only a single PMKID.

FIGURE 7.4 Pairwise master key identifier (PMKID)

```
RSN Information
  Element ID:                48 RSN Information [46]
  Length:                    22 [47]
  Version:                   1 [48-49]
  Group Cipher OUI:          00-0F-AC [50-52]
  Group Cipher Type:         4 CCMP - default in an RSN [53]
  Pairwise Cipher Count:     1 [54-55]
  PairwiseKey Cipher List
    Pairwise Cipher OUI:     00-0F-AC-04 CCMP - default in an RSN [56-59]
  AuthKey Mngmnt Count:      1 [60-61]
  AuthKey Mngmnt Suite List
    AKMP Suite OUI:          00-0F-AC-02 None [62-65]
  RSN Capabilities:          %0000000000111100 [66-67]
                               xxxxxxxx x....... Reserved
                               ........ ..11.... GTKSA Replay Ctr: 3 - 16 replay counters
                               ........ ......0. Does not Support No Pairwise
                               ........ .......0 Does Not Support Pre-Authentication
  PMKID Count:               1
  PMKID:                     0x75C2764687C3C2826800E6B76C27545F
```

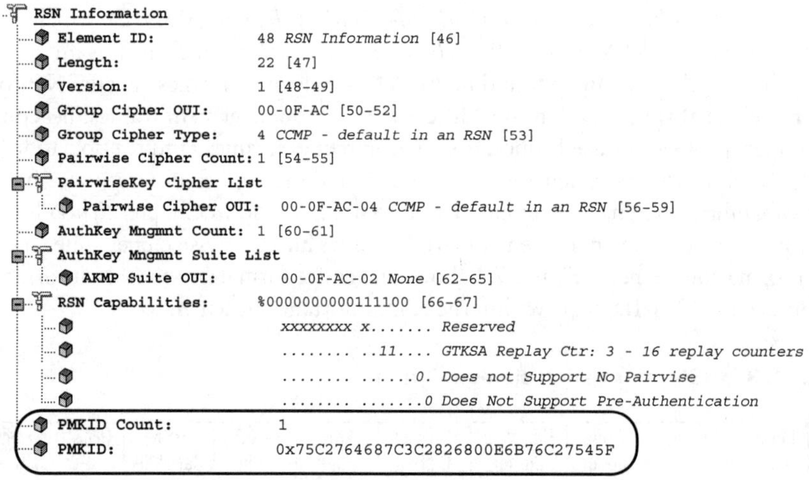

So what exactly comprises a PMKSA? To review, a pairwise master key security association is the result of a successful IEEE 802.1X/EAP authentication exchange between a supplicant and authentication server (AS) or from a preshared key (PSK) authentication. A bidirectional security association exists between the authenticator and the supplicant. The components of a PMKSA include the following:

PMK The pairwise master key that was created.

PMKID The unique identifier of the security association.

Authenticator MAC The MAC address of the authenticator.

Lifetime If the key lifetime is not otherwise specified, then the PMK lifetime is infinite.

AKMP The authentication and key management protocol.

Authorization Parameters Any parameters specified by the authentication server or local configuration. This can include parameters such as the STA's authorized SSID.

Although WLAN vendors sometimes use the terms PMK and PMKSA interchangeably, they are two separate entities. The PMK is the key that seeds the 4-Way Handshake, whereas the PMKSA consists of all the components just listed, including the PMK. You will see that the most important components of the PMKSA are the PMK, the PMKID, and the authenticator's MAC address.

The 802.11-2012 standard states that a client station can establish a PMKSA as a result of the following:

- Complete 802.1X/EAP authentication
- PSK authentication
- SAE authentication
- PMK cached via some other mechanism

Every time a client station roams to a new access point, a new PMKSA is established. Figure 7.5 depicts the creation of a new PMKSA during reassociation using 802.1X/EAP. PMK #1 is installed on the original AP and the client station. PMK #1 is then used to seed the 4-Way Handshake that is used to create the final keys used for encryption. The client roams to the target AP and reauthenticates with the RADIUS server. The new EAP exchange creates PMK #2, which is installed on the target AP and the client station. PMK #2 is then used to seed the 4-Way Handshake that is used to create the final keys used for encryption in the new BSS.

FIGURE 7.5 802.1X/EAP and PMKSA

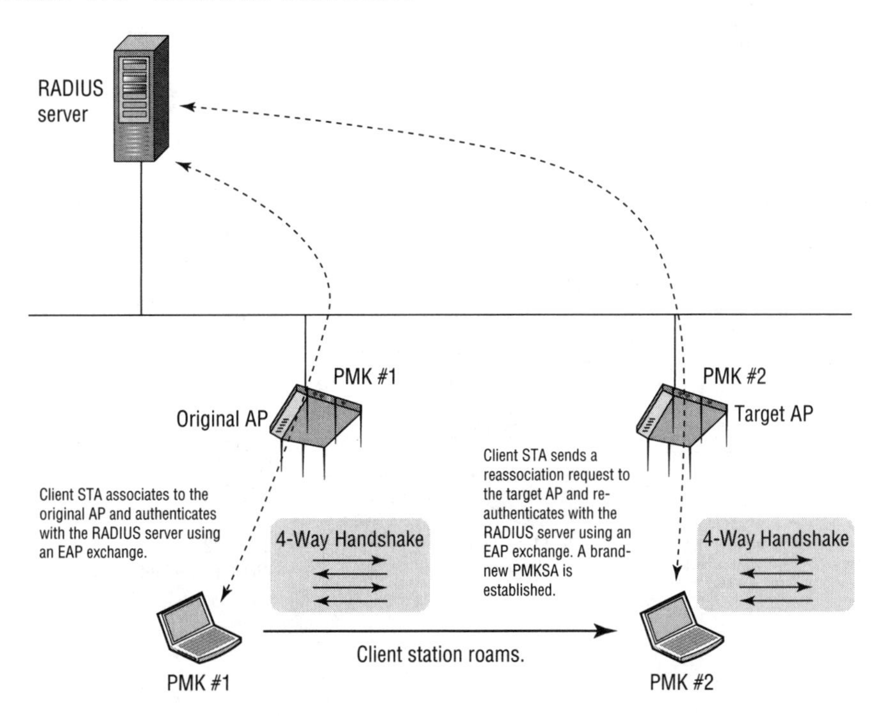

As you can see, each time the client roams to a new target AP, the client station must reauthenticate with the RADIUS server. Although a new PMKSA is established, the time needed to reauthenticate with the RADIUS server is significant. Many applications will not be affected by the roaming handoff time, but the delay will disrupt the communications of time-sensitive applications such as voice and video. Therefore, the 802.11-2012 standard defines three fast secure roaming mechanisms, called PMK caching, preauthentication, and fast BSS transition.

PMK Caching

PMK caching is a method used by APs and client stations to maintain PMKSAs for a period of time while a client station roams to a target AP and establishes a new PMKSA. An authenticator and a client station can cache multiple PMKs. For example, as shown in Figure 7.6, a client station will associate with an original AP and create an original PMK #1. The client will roam to a target AP and create a new PMK #2; however, the original AP and the client station will both cache PMK #1.

FIGURE 7.6 PMK caching

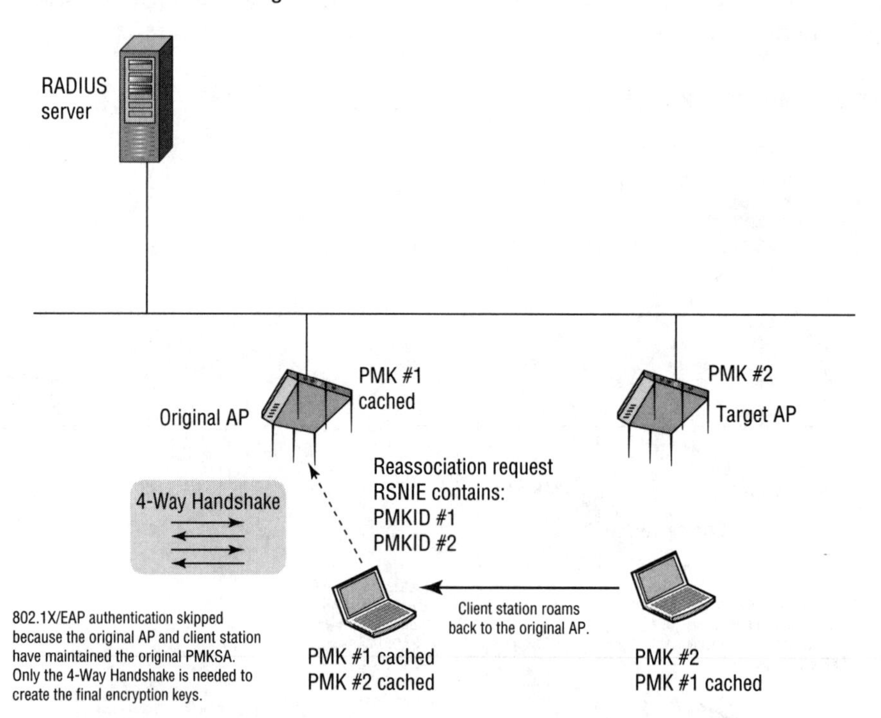

Whenever a client station roams back to the original AP, the client station will send a reassociation request frame that lists multiple PMKIDs in the RSN information element (RSNIE). In other words, the client will be informing the AP about all of the client's cached PMKs. The 802.11-2012 standard states that "An AP whose authenticator has retained the PMK for one or more of the PMKIDs can skip the IEEE 802.1X/EAP authentication and proceed with the 4-Way Handshake." In simpler words, when the client roams back to the original AP, both devices still have the original cached PMK #1 and they can skip the 802.1X/EAP exchange. The client does not need to reauthenticate and create a new PMK because the original PMK still exists. The cached original PMK is then used to seed the 4-Way Handshake.

When a client roams to an AP and the 4-Way Handshake is all that is needed, the roaming handoff is usually between 40 and 60 ms, which is fast enough for VoWiFi. Skipping the 802.1X/EAP exchange saves precious time for time-sensitive applications. PMK caching is sometimes called *fast secure roam-back* because the client station is able to roam back to the original AP and skip the 802.1X/EAP exchange. This is great if a client station roams back to an original AP where it shares a PMKSA, but how does this speed things up when the client station roams forward to a new AP? The short answer is there will not be a cached PMK on the target AP unless there is a method to create or distribute cached PMKs to target APs. Preauthentication is one such method.

Preauthentication

A client station can use *preauthentication* to establish a new PMKSA with an AP prior to roaming to a new target AP. Preauthentication allows a client station to initiate a new 802.1X/EAP exchange with a RADIUS server while associated with the original AP. The purpose of the new 802.1X/EAP authentication is to create a new PMKSA relationship with a new target AP where the client might roam. As Figure 7.7 shows, the client station sends an EAPOL-Start frame through the original AP over the distribution system (DS). An entire 802.1X/EAP exchange occurs between the client station and the RADIUS server; however, the authenticator is the new target AP. Once the client has preauthenticated, a new PMK #2 is created and cached on both the client station and the target AP. If the client station decides to roam to the target AP, the client does not need to reauthenticate and create a new PMK because a precreated cached PMK already exists. The client station roams to the target AP, and the 4-Way Handshake is all that is needed.

How do client stations know if they can even learn about APs with which they can preauthenticate? Client stations will discover new APs with both active and passive scanning. As shown in Figure 7.8, an AP can indicate to the client station that the AP is capable of preauthentication in the RSN information element sent in the AP's probe response or beacon frames.

FIGURE 7.7 Preauthentication

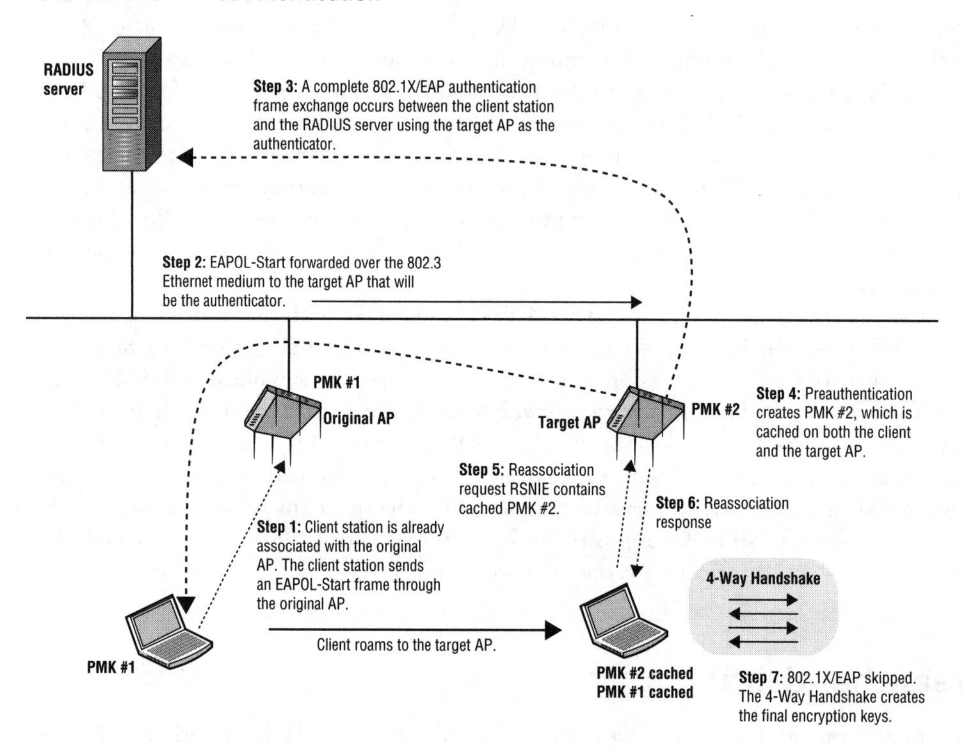

FIGURE 7.8 Preauthentication-enabled AP

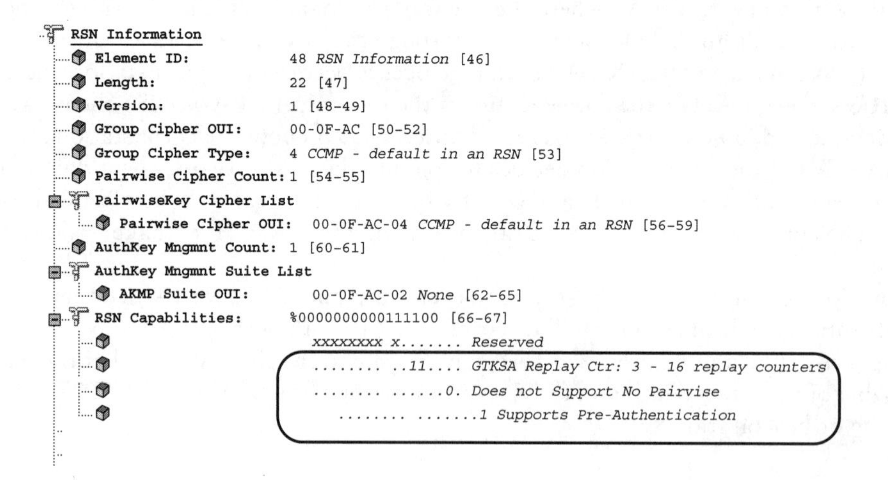

Preauthentication was originally intended for use with autonomous APs, although it can be used with WLAN controller systems as well. It should be noted that preauthentication does not scale very well because it requires all APs to create PMKSAs with all clients that might potentially roam to each AP. Every single client station would need to preauthenticate with every single AP in advance. Preauthentication would therefore place a tremendous load on the backend RADIUS server. PMK caching and preauthentication are simply not very scalable solutions, and therefore the 802.11r task group was formed to define more scalable fast secure transition methods between basic service sets. However, most WLAN vendors did not want to wait for the ratification of the 802.11r-2008 amendment, so they implemented a preview of 802.11r mechanisms called *opportunistic key caching*.

Opportunistic Key Caching (OKC)

Because PMK caching and preauthentication do not scale well, the majority of the WLAN vendors offer a fast secure roaming solution that is an enhancement of PMK caching called *opportunistic key caching (OKC)*. Although it's widely adopted, please understand that opportunistic key caching is a fast secure roaming (FSR) technique not defined by the 802.11-2012 standard. OKC allows for PMK caching between multiple APs that are under some sort of administrative control.

Unlike preauthentication, OKC does not mandate how a PMK arrives at the target AP. OKC instead allows a client station the opportunity to take advantage of a single cached PMK shared among multiple access points. OKC forwards a PMK from the original AP and then distributes it to other APs. The PMK distribution between APs is dependent on the WLAN architecture and is usually proprietary. In a WLAN controller environment, the PMKs are usually forwarded by the controller to the APs. In a non-controller environment, the PMKs are forwarded by the APs to each other using a proprietary protocol.

To understand OKC, let's first discuss the formula for a pairwise master key identifier (PMKID). The 802.11-2012 standard defines a PMK identifier as follows:

PMKID = HMAC-SHA1-128(PMK, "PMK Name" || AA || SPA)

The AA is the authenticator's MAC address, and the SPA is the supplicant's MAC address. This formula shows that a hash function combines the PMK with the access point and client station MAC addresses to create the PMKID.

What Is HMAC?

Keyed-Hash Message Authentication Code (HMAC) is a mechanism for message authentication using cryptographic hash functions. HMAC can be used with any iterative cryptographic hash function, such as SHA-1, in combination with a secret shared key. HMAC is defined by IETF RFC 2104, "HMAC: Keyed-Hashing for Message Authentication."

As you can see in Figure 7.9, opportunistic key caching takes advantage of the PMKID formula. In this example, the original AP forwards the original PMK and PMKID to multiple access points at the edge of the network. In a WLAN controller environment, the PMK distribution to APs is usually centralized. Refer to Figure 7.9 when reviewing these steps.

1. The client station uses full 802.1X/EAP authentication and an original PMK #1 and PMKID #1 are created for use by the original AP and the client station. The original AP and the client station perform a 4-Way Handshake.
2. The original AP caches PMK #1 and then forwards PMK #1 to the target access point.
3. The client station calculates a new PMKID #2 using the original PMK #1 + the target AP MAC address + the client MAC. The client sends a reassociation request frame to the target AP with PMKID #2 in the RSN information element.
4. The target AP looks at the MAC address of the client station that just sent the reassociation request and calculates PMKID #2 using the same formula. The target AP sends a reassociation response frame.
5. Because the PMKID #2 found in the reassociation request frame matches the PMKID #2 calculated by the target AP, reauthentication is not needed. The AP and the client station are still using the original PMK to seed the 4-Way Handshake, but they are both in possession of the newly calculated PMKID #2, which will allow for a unique security association between the two devices. After the AP sends a reassociation response frame, the 4-Way Handshake is then used to create the final encryption keys shared between the target AP and the client station. 802.1X/EAP authentication was skipped and therefore a fast roaming handoff time has occurred.

FIGURE 7.9 Opportunistic PMK caching

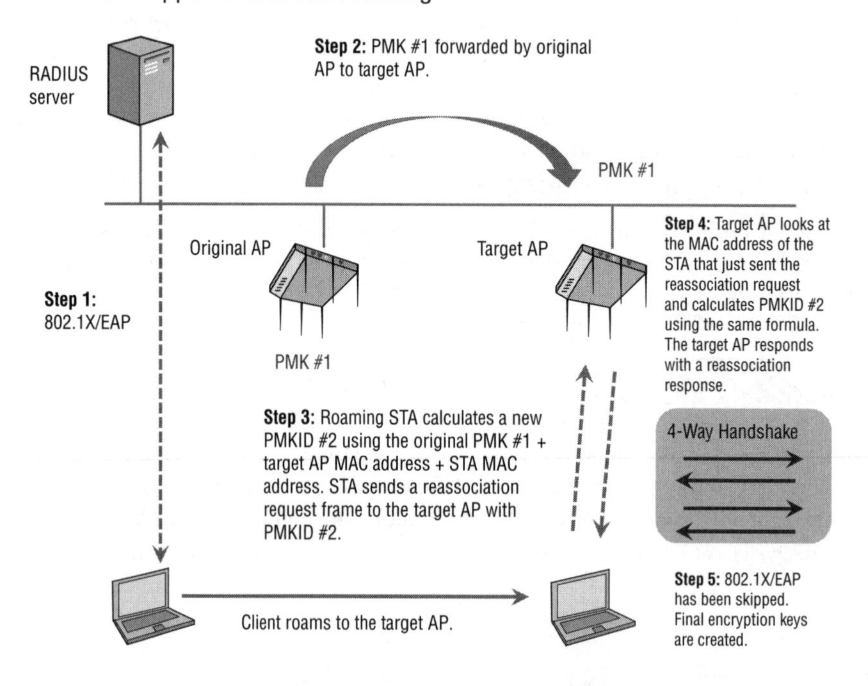

OKC effectively eliminates the need for client preauthentication, and therefore a more scalable solution is provided. OKC offers several advantages over preauthentication. OKC only requires one initial 802.1X/EAP exchange between the client and the authentication server. Therefore, OKC reduces the load that is placed on the RADIUS server. Because only a single 802.1X/EAP exchange occurs, only one original PMK is created. The OKC process uses the client station's original PMK as seeding material for all the APs. How the cached PMK is distributed between the APs is up to the WLAN vendor. PMK key distribution can occur at the edge between cooperative APs or can be centrally distributed via a WLAN controller. This entire key management process becomes more complex if a client station was to roam to a controller-based AP that tunnels traffic back to a different WLAN controller. Therefore, intercontroller handoff protocols are also entirely proprietary. Some WLAN controller vendors may not support OKC between controllers. In that case, the client station would initiate an 802.1X/EAP exchange and the whole process would start over. OKC will fail if the target AP does not have a newly calculated PMKID that matches the client station's newly calculated PMKID that is sent in a reassociation request frame. In that case, the client station would initiate an 802.1X/EAP exchange and the whole process would start over.

Figure 7.10 displays the roaming cache of a single AP with six different PMKs. Supplicant #4 is a client that is currently associated to the AP. The other supplicants are client stations that are one hop away and not associated to the AP. However, the PMKs of the other stations have already been forwarded to this AP and are cached. Any client that also supports OKC can use its original PMK when roaming to a new AP, as depicted in Figure 7.9.

FIGURE 7.10 AP roaming cache

No.	Supplicant	Authenticator	Size	UID	PMK	PMKID	Life	Age	TLC	Hop
0	000e:3b33:23ea	08ea:4476:4f14	864	2	249f*	4fd3*	-1	21033	3568	1
1	000e:3b33:30e5	08ea:4476:5117	864	2	a0af*	7928*	-1	240630	3571	1
2	000e:3b33:3a6c	08ea:4476:3e15	864	10	a40a*	2c62*	-1	21301	3600	1
3	000e:3b33:3365	08ea:446b:f717	864	10	05f6*	72e7*	-1	21667	3594	1
4	000e:3b33:30b8	08ea:4476:4fd4	864	10	c84f*	bdae*	-1	24508	3573	0
5	000e:3b33:3a66	08ea:4476:5057	864	10	8579*	5838*	-1	24337	3564	1

Real World Scenario

How Widely Is Opportunistic Key Caching (OKC) Supported?

There is no defined standard for opportunistic key caching. OKC was originally developed as a short-term fast secure roaming solution by one WLAN vendor until the 802.11r-2008 amendment was ratified and widely supported. The majority of WLAN infrastructure vendors that manufacture APs also quickly adopted OKC. WLAN clients must also support OKC for the process to work. OKC quickly became a de facto standard for fast secure roaming because adoption of 802.11r mechanisms has been extremely slow. Although OKC is supported by all the enterprise-grade AP manufacturers, many WLAN clients do not support OKC because it is not an official roaming standard. For example, OKC is supported for Apple devices that use Mac OS but not supported for iOS devices. Most of the major VoWiFi phone manufacturers also support OKC. Although OKC is natively supported in some client device operating systems, a vast number of WLAN clients simply do not support the technology.

The 802.11r-2008 amendment clearly defined fast secure roaming mechanisms and is now part of the 802.11-2012 standard. The Wi-Fi Alliance has adopted the Voice Enterprise certification that mandates 802.11r. The majority of WLAN infrastructure vendors have finally begun to support 802.11r-2008 mechanisms; however, client-side support for 802.11r is scarce.

Proprietary FSR

WLAN vendor Cisco Systems has long offered a proprietary version of fast secure roaming called Cisco Centralized Key Management (CCKM). Cisco demonstrated market leadership with CCKM; however, CCKM only works within a Cisco WLAN infrastructure. Now that FSR solutions are becoming standardized, CCKM will likely take a backseat much like the ISL Ethernet switch protocol did with the introduction of IEEE Std. 802.1Q.

CCKM falls within Cisco's licensed CCX program that other vendors may license. The current implementation of CCKM requires Cisco-compatible hardware and works with EAP-LEAP, EAP-FAST, PEAPv1 (EAP-GTC), PEAPv0 (EAP-MSCHAPv2), and EAP-TLS. CCKM uses a Cisco access point or controller to cache security credentials and effectively takes the place of the RADIUS server when the client stations authenticate. Like all FSR solutions, CCKM shortens the roaming handoff delay.

You will not be tested on CCKM or any other vendor proprietary technologies mentioned in this book. The CWSP exam is a vendor-neutral exam.

Fast BSS Transition (FT)

The 802.11r-2008 amendment is known as the *fast basic service set transition (FT)* amendment. Think of the term *fast BSS transition* as the technical name for standardized fast secure roaming. The main difference between OKC and FT is that the 802.11r-2008 amendment fully defined the key hierarchy used when creating cached keys. The fast BSS transition mechanisms originally defined in the 802.11r-2008 amendment are now found in clause 12 of the 802.11-2012 standard.

As we have previously stated, OKC key management is a preview of FT, which is also designed for a method of pairwise master key (PMK) distribution. FT mechanisms operate within a mobility domain. A *mobility domain* is a set of basic service sets (BSSs), within the same extended service set (ESS), that supports fast BSS transitions. In simpler words, a mobility domain is a group of APs that belong to the same ESS where client stations can roam in a fast and secure manner. The first time a client station enters a mobility domain, the client will associate with an AP and perform an initial 802.1X/EAP authentication. From that point forward, as the client station roams between APs, the client will be using fast BSS transitions. You will learn later in this chapter that a fast BSS transition can be over-the-air or over-the-DS.

As you learned in Chapter 4, "802.1X/EAP Authentication," an 802.1X/EAP exchange creates a *master session key (MSK)* that is used to create a pairwise master key (PMK) for non-FT roaming. FT also uses the 802.1X/EAP exchange to create the master session key, which seeds a multi-tiered key management solution. As shown in Figure 7.11, after the supplicant and the RADIUS server exchange credentials, a first-level pairwise master key called the PMK-R0 is created from the master session key and sent to the authenticator and the WLAN client. Depending on the WLAN architecture, the 802.1X/EAP authenticator can either be an AP or a WLAN controller.

FIGURE 7.11 802.1X/EAP and first-level pairwise master key (PMK-RO)

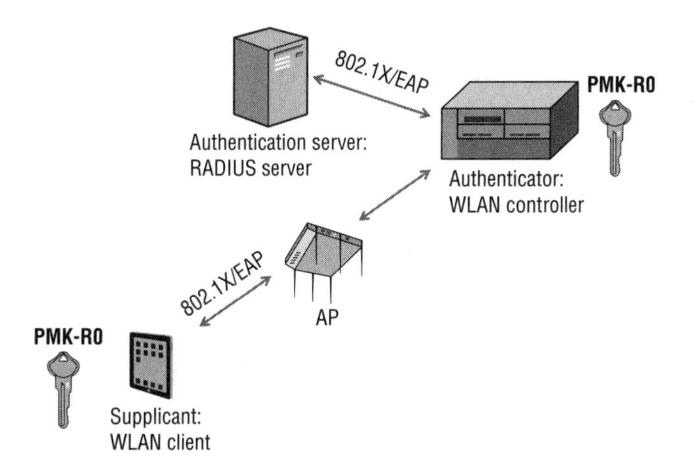

FT mechanisms introduce multiple layers of PMKs that are cached in different devices. Fast BSS transition uses a three-level key hierarchy:

Pairwise Master Key R0 (PMK-R0) The first-level key of the FT key hierarchy. This key is derived from the master session key (MSK).

Pairwise Master Key R1 (PMK-R1) The second-level key of the FT key hierarchy.

Pairwise Transient Key (PTK) The third-level key of the FT key hierarchy. The PTK is the final key used to encrypt 802.11 data frames.

Fast BSS transition also assigns different roles to different devices. As shown in Table 7.1, each device is assigned a *key holder* role to manage one or more of the multiple keys used in the FT key hierarchy.

TABLE 7.1 Key holders

Device	Key holder role
Original AP or WLAN controller	Pairwise master key (PMK) R0 key holder (R0KH)
Access point	Pairwise master key (PMK) R1 key holder (R1KH)
Client station	Pairwise master key (PMK) S0 key holder (S0KH)
Client station	Pairwise master key (PMK) S1 key holder (S1KH)

The various levels of FT keys are derived and stored in different WLAN devices depending on the WLAN architecture that has been deployed. For example, in a controller-less environment, the first level PMK-R0 key is created and cached on an access point. In an environment where WLAN controllers are deployed, the first level PMK-R0 key is created and cached on a WLAN controller.

802.1X/EAP creates the master session key (MSK). The MSK is used to create the first-level master key, called a PMK-R0. In the example in Figure 7.12, the PMK-R0 is created and cached on the WLAN controller. The WLAN controller is the key holder for the first-level key. The second-level PMK-R1 keys are derived from the PMK-R0 and sent from the WLAN controller to the controller-based APs. The PMK-R1 keys are cached on the APs. The access points are the key holders for the PMK-R1 keys. The PMK-R1 keys are used to derive the PTKs, which are used to encrypt 802.11 data frames.

FIGURE 7.12 FT Key hierarchy—WLAN controller infrastructure

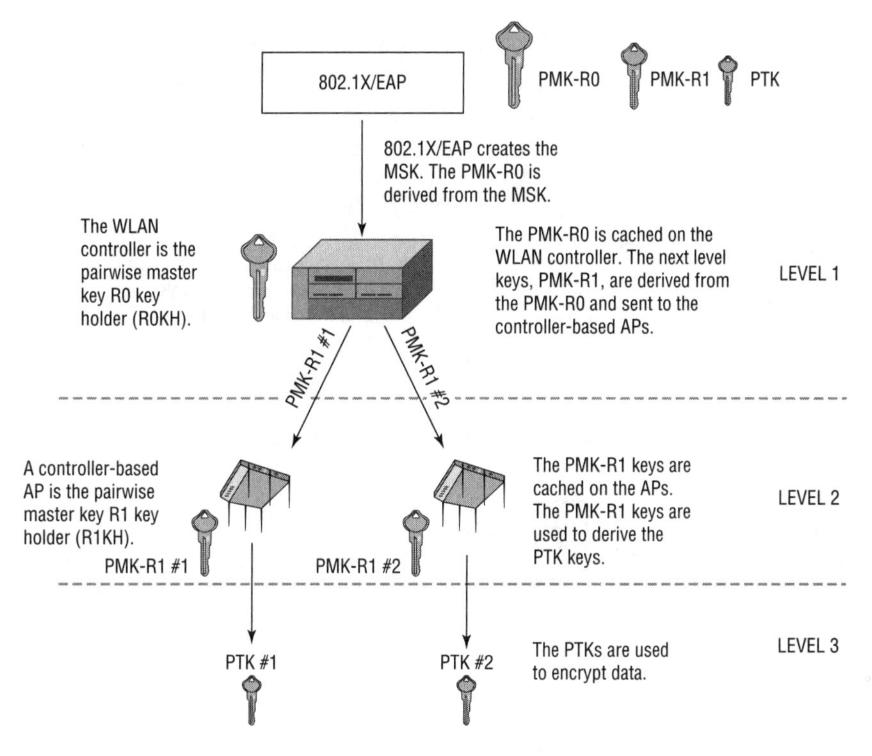

As shown in Figure 7.13, the various levels of FT keys are also derived and stored on the client station. 802.1X/EAP creates the master session key (MSK). The MSK is used to create the first-level master key, called a PMK-R0. The PMK-R0 is cached on the supplicant, which is the client station. The client station is the key holder for the first-level key. The PMK-R0 is cached on the client station. The client station derives the second-level key, PMK-R1, from the PMK-R0. The PMK-R1 key is cached on the client station. The supplicants are the key holders for the PMK-R1 keys. The PMK-R1 keys are used to derive the PTKs, which are used to encrypt unicast 802.11 data frames.

To summarize, the order of the keys is as follows:

MSK ➢ PMK-R0 ➢ PMK-R1 ➢ PTK

Fast BSS transition requires that keys sent between WLAN devices be distributed over a secure channel but does not define the actual secure method of distribution. For example, the 802.11r-2008 amendment did not specify how the PMK-R1 keys are sent from the WLAN controller to the controller-based APs. If multiple WLAN controllers are used, this becomes even more complex. There is no real definition of how keys should be exchanged between controllers; therefore, intercontroller handoff protocols are entirely proprietary. This also means that it remains highly doubtful that distribution of these keys between WLAN infrastructure devices from different WLAN vendors will be effective.

FIGURE 7.13 FT key hierarchy—supplicant

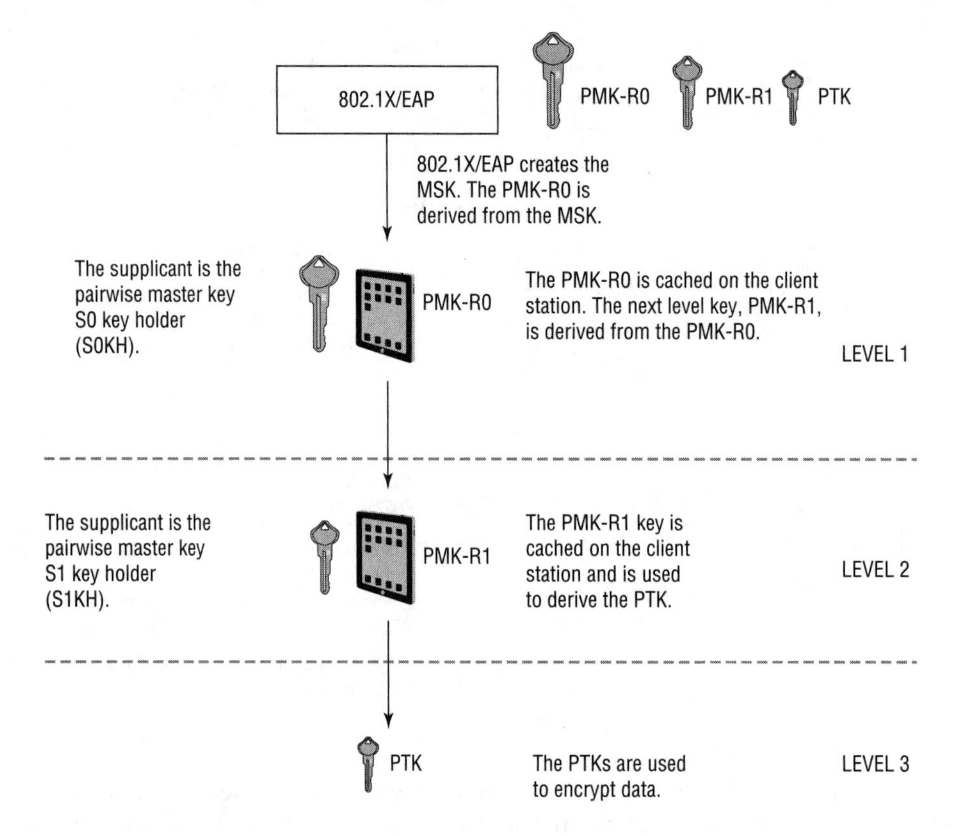

Most WLAN controller vendors encrypt/decrypt client traffic at the edge of the network using the access points. However, some WLAN vendors perform encryption at the controller level instead of at the AP level. End-to-end encryption provides data privacy between the client at the access layer and the WLAN controller that is typically deployed at the core. In that scenario, the WLAN controller functions as both the pairwise master key R0 holder (R0KH) and the pairwise master key R1 holder (R1KH).

Now let us take a look at FT key hierarchy in a distributed AP architecture that does not require a WLAN controller. The 802.1X/EAP exchange creates the master session key (MSK). The MSK is used to create the first-level master key, called a PMK-R0. In the example in Figure 7.14, the PMK-R0 is created and cached on an AP where the client first associates. The original AP is the key holder for the first-level key. The second-level PMK-R1 keys are derived from the PMK-R0 and sent from the original AP to other target APs over a secure channel. As previously stated, how the PMK-R1 keys are securely distributed is outside of the scope of the 802.11-2012 standard. The PMK-R1 keys are cached on the target APs, which are the key holders for the PMK-R1 keys. The PMK-R1 keys are used to derive the PTKs, which are used to encrypt unicast 802.11 data frames.

In Figure 7.14, the PMK-R0 is cached on the supplicant, which is the client station. The client station is the key holder for the first-level key. The PMK-R0 remains cached on the client station as the client roams between access points. From the PMK-R0, the client station derives the second-level PMK-R1 keys for each target AP. The PMK-R1 keys are also cached on the client station, which functions as the key holder for all the PMK-R1 keys. The unique PMK-R1 keys are used to derive unique PTKs, which are used to encrypt unicast 802.11 data frames.

FIGURE 7.14 FT Key hierarchy—distributed AP infrastructure

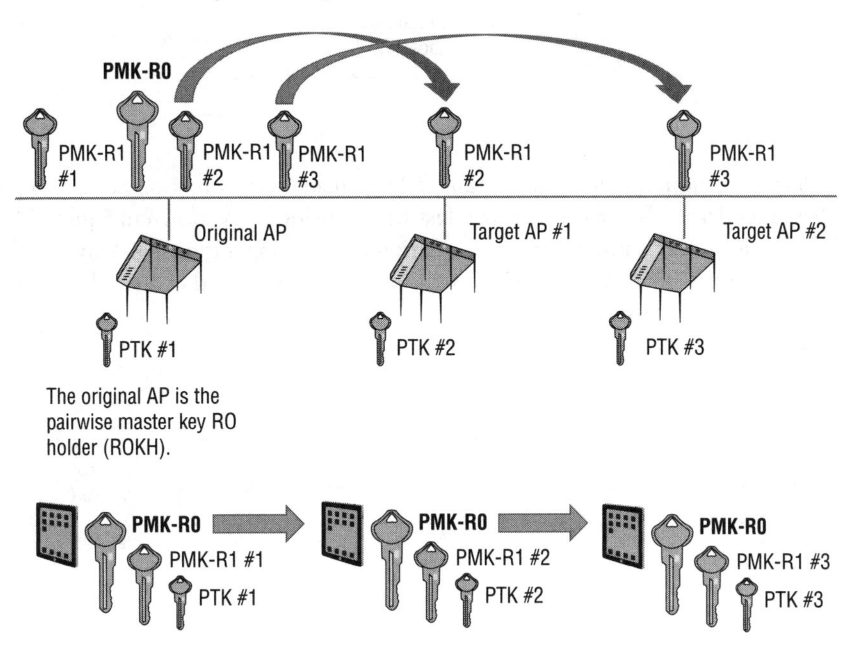

Information Elements

To achieve successful fast secure roaming, FT mechanisms still require the use of the RSN information element to indicate the specific authentication key management (AKM) suites and pairwise cipher suites that are being used between the AP and the client station. The 802.11-r-2008 amendment defined four new information elements. However, we are going to focus on just two of them.

The *mobility domain information element (MDIE)* is used to indicate the existence of a mobility domain as well as the method of fast BSS transition. As shown in Figure 7.15, the *mobility domain identifier (MDID)* field is the unique identifier of the group of APs that constitute a mobility domain. The FT capability and policy field is used to indicate whether over-the-air or over-the-DS fast BSS transition is to be performed. We will discuss the difference between over-the-air and over-the-DS fast BSS transition later in this chapter.

FIGURE 7.15 Mobility domain information element

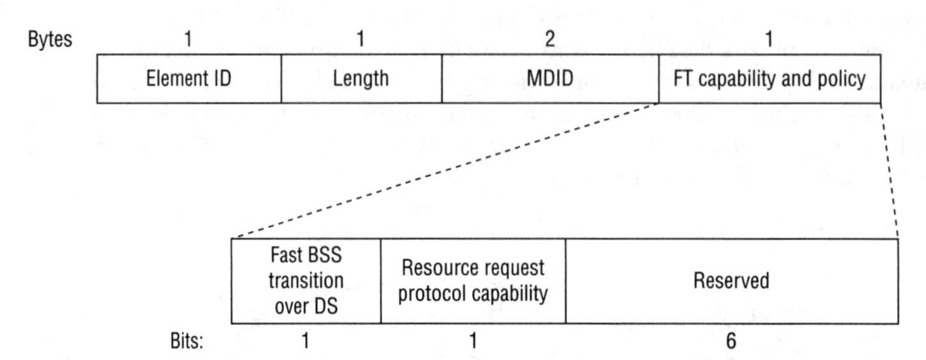

The *fast BSS transition information element (FTIE)* includes information needed to perform the FT authentication sequence during a fast BSS transition. As shown in Figure 7.16, some of the fields look very similar to the information used during a typical 4-Way Handshake exchange. In the next section, you will see how this information is used in a similar manner during the various FT processes.

FIGURE 7.16 Fast BSS transition information element

	Element ID	Length	MIC control	MIC	ANonce	SNonce	Optional parameter(s)
Octets:	1	1	2	16	32	32	Variable

FT Initial Mobility Domain Association

The *FT initial mobility domain association* is the first association in the mobility domain. As shown in Figure 7.17, the client station first exchanges the standard 802.11 Open System authentication request/response frames with the first AP. The client station and AP then use the MDIE and FTIE information in the association request/response frames to indicate future use of the FT procedures. An original 802.1X/EAP exchange between the client station and the RADIUS server must then occur so seeding material is established for a *FT 4-Way Handshake* that occurs only during the first association. The PTK and GTK encryption keys are created during the FT 4-Way Handshake and the 802.1X/EAP controlled port is unblocked. The original 802.1X/EAP exchange also creates the master session key (MSK) that is used for the FT key hierarchy. As you can see, the FT initial mobility domain association is not much different than any initial association used by clients that do not support fast BSS transition. The main difference is that extra information, such as the MDIE and FTIE, is communicated during an FT initial mobility domain association.

FIGURE 7.17 FT initial mobility domain association

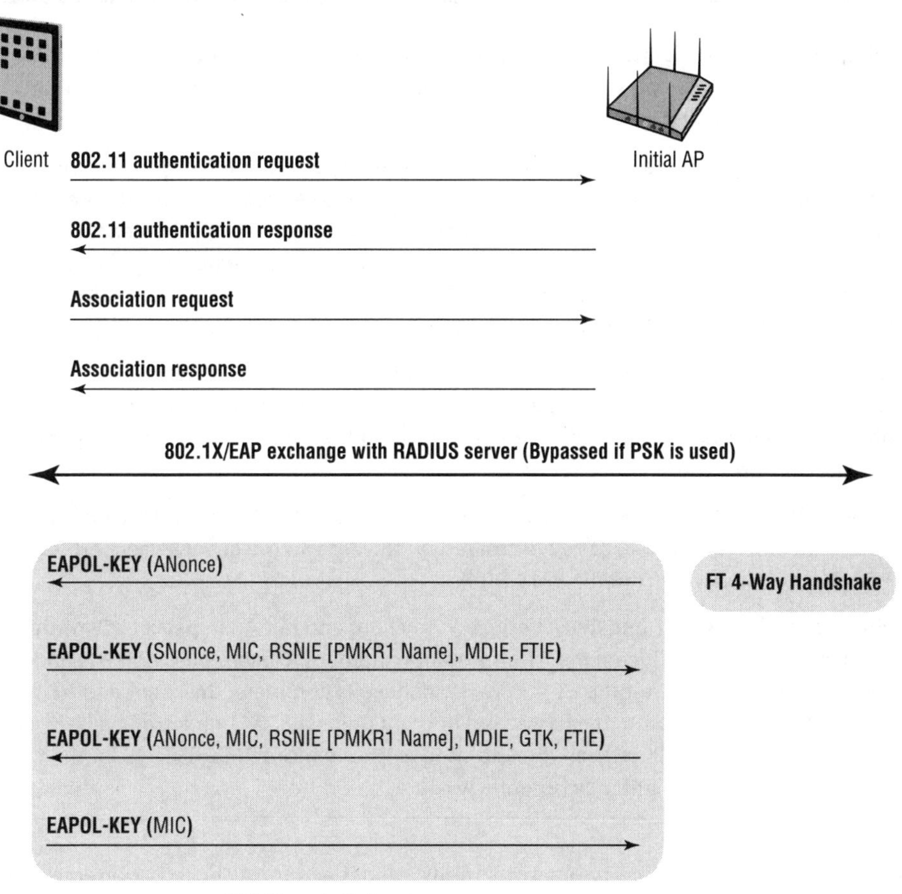

EXERCISE 7.1

FT Initial Mobility Domain Association

In this exercise, you will use a protocol analyzer to view the 802.11 frame exchanges used by a client and AP for an FT initial mobility domain association.

1. To perform this exercise, you need to first download the following file from the book's online resource area, which can be accessed at `www.sybex.com/go/cwsp2e`.

   `FT.PCAP`

2. After the capture file has been downloaded, you will need packet analysis software to open the file. If you do not already have a packet analyzer installed on your computer, you can download Wireshark from `www.wireshark.org`.

EXERCISE 7.1 *(continued)*

3. Using the packet analyzer, open the FT.PCAP file. Most packet analyzers display a list of capture frames in the upper section of the screen, with each frame numbered sequentially in the first column.

4. Click packet 4, which is a beacon frame transmitted by an AP. Typically in the lower section of the screen is the packet details window. This section contains annotated details about the selected frame. In this window, locate and expand the beacon frame. In the body of the beacon frame, expand the tagged parameters and locate the mobility domain information element (MDIE). The access point uses the AP to inform FT-capable clients of the mobility domain identifier (MDID).

5. Click packet 11, which is an association request from an Apple iOS device. In the body of the frame, expand the tagged parameters and locate the mobility domain information element (MDIE). The client is informing the AP that the client is capable of fast BSS transition.

6. Observe the 802.1X/EAP exchange in packets 15 through 52. After 802.1X/EAP is completed, the first-level PMK-RO key is created on the AP and the client. A second-level PMK-R1 key is derived from the PMK-R0-key.

7. Observe the FT 4-Way Handshake between the client and the AP in packets 53 through 60. Click packet 55 and locate the FT information element. Observe the PMK-R0 and PMK-R1 information. After the FT 4-Way Handshake is complete, the client and AP will use the newly created PTK to encrypt and decrypt unicast 802.11 data frames between the two radios. The PMK-RO can then be used to create more PMK-R1 keys for the target APs to which the client may potentially roam.

After the initial association, two new methods are defined for a client station to roam from the original AP to a target AP. We will now discuss these two methods of fast BSS transition.

Over-the-Air Fast BSS Transition

Let's consider all the frames that need to be exchanged between a client station and an AP. First, a client has to exchange Open System authentication frames and association frames, as was shown in Figure 7.17. This is a total of four frames, not including the ACKs. Next, a successful 802.1X/EAP exchange between the supplicant and the RADIUS server is needed. The 802.1X/EAP exchange requires multiple frames. Finally, a 4-Way Handshake exchange is needed between the AP and the client station to create the final dynamic encryption keys. We already know that the purpose of FT and other fast secure roaming mechanisms is to eliminate the need for a new 802.1X/EAP frame exchange every time a client roams. However, the initial four frames of Open System authentication and reassociation are still needed as well as the four frames used during the 4-Way Handshake.

The FT process defines a more efficient method that effectively combines the initial Open System authentication and reassociation frames with the 4-Way Handshake frames. In other words, four fewer frames are needed when a client roams, thus speeding up the roaming process. As shown in Figure 7.18, an FT protocol frame exchange is used to initiate the roaming exchange and create dynamic encryption keys. Note that the authentication request/response frames and reassociation request/response frames carry an FT authentication algorithm (FTAA) along with nonces and other information needed to create the final dynamic keys. The process shown in Figure 7.18 is known as *over-the-air fast BSS transition.* The client station communicates directly with the target AP using standard 802.11 authentication with the FT authentication algorithm. The PMK-R1 key is the seeding material for the over-the-air fast BSS transition process that creates the final pairwise transient key (PTK).

FIGURE 7.18 Over-the-air fast BSS transition

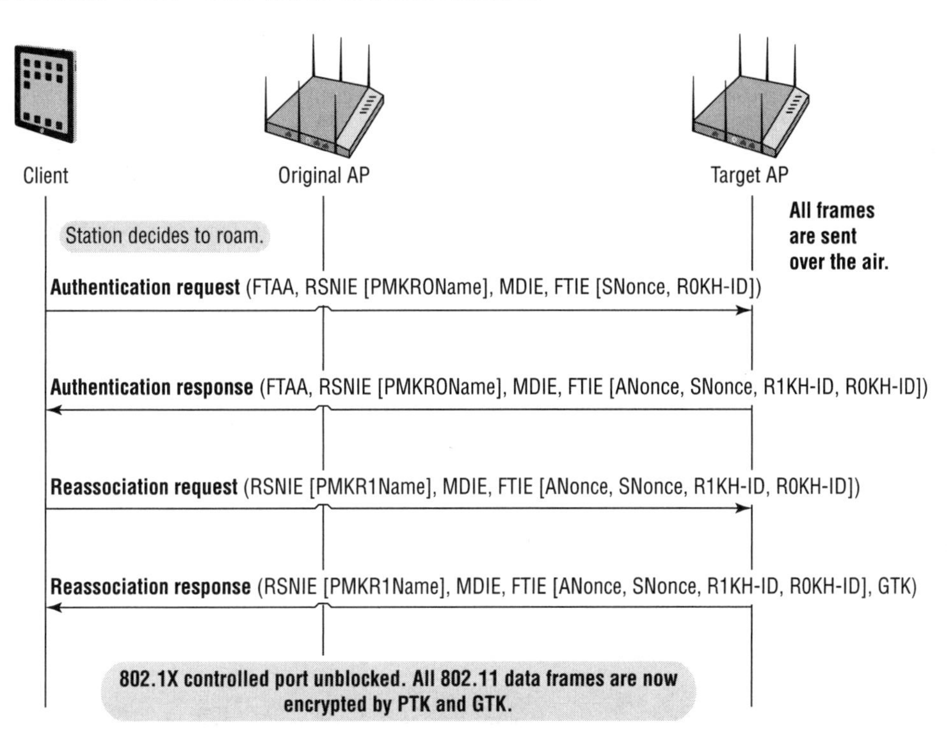

Over-the-DS Fast BSS Transition

An alternative to the FT method is *over-the-DS fast BSS transition,* which requires the use of *FT Action frames* to complete the PTK creation process. The over-the-DS process uses the FT Action frames over the wired 802.3 infrastructure. As shown in Figure 7.19, the client station sends an FT Action request frame to the original AP. The FT Action request frame is forwarded over the distribution system (DS), which is the wired infrastructure. The target AP responds back to the client station over the DS with an FT Action response frame. The reassociation request and response frames are then sent from the client station to the target AP over the air. The PMK-R1 key is the seeding material for the over-the-DS fast BSS transition exchange that creates the final pairwise transient key (PTK). Over-the-DS fast BSS transition is considered to be an optional method that may be supported by a few WLAN vendors.

FIGURE 7.19 Over-the-DS fast BSS transition

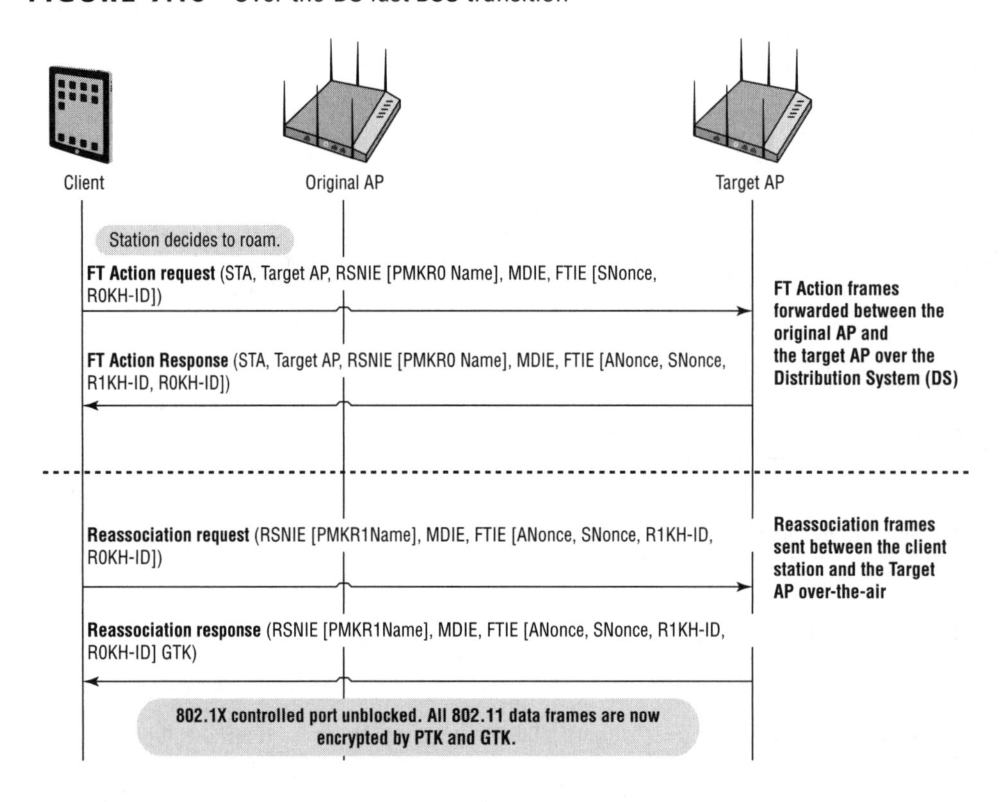

EXERCISE 7.2

Over-the-Air Fast BSS Transition

In this exercise, you will use a protocol analyzer to view the 802.11 frame exchanges used by a client to roam to a target AP using over-the-air fast BSS transition.

1. To perform this exercise, you need to first download the following file from the book's online resource area, which can be accessed at `www.sybex.com/go/cwsp2e`.

   `FT.PCAP`

2. After the capture file has been downloaded, you will need packet analysis software to open the file. If you do not already have a packet analyzer installed on your computer, you can download Wireshark from `www.wireshark.org`.
3. Using the packet analyzer, open the `FT.PCAP` file. Most packet analyzers display a list of capture frames in the upper section of the screen, with each frame numbered sequentially in the first column.
4. The over-the-air fast BSS transition in frame exchange occurs between packets 95 and 102. Click packet 95, which is an authentication request frame from the client to a target AP. In the body of the frame, expand the tagged parameters and locate the FT information element. Notice that the FT key exchange has begun.
5. To observe the second frame used for the FT key exchange, click packet 97, which is an authentication response frame from the target AP to the client. In the body of the frame, expand the tagged parameters and locate the FT information element.
6. Click packet 99, which is the reassociation request frame sent from the client to the target AP. In the body of the frame, expand the fixed parameters. Notice that the client is currently associated to the original AP with the BSSID MAC address of 08:ea:44:76:b5:68. Expand the tagged parameters and locate the FT information element. Notice that the FT key exchange continues.
7. Click packet 101, which is the reassociation response frame sent from the target AP to the roaming client. In the body of the frame, expand the fixed parameters. Notice the successful status code and the association ID. Expand the tagged parameters and locate the FT information element. Notice that the FT key exchange completes. The client has completed over-the-air fast BSS transition and has roamed to the target AP with the BSSID address of 08:ea:44:78:14:28. When the client roamed, an 802.1X/EAP exchange was not necessary. The 4-Way Handshake was also not necessary because the dynamic keys were created using the authentication and reassociation frames.
8. Observe the encrypted data frame exchange between the client and the new AP in packets 113 through 116. These unicast data frames are encrypted with the unique PTK that was generated as a result over the over-the-air fast BSS transition.

As previously mentioned, due to the multiple 802.1X/EAP frame exchanges between a RADIUS server and WLAN supplicant, roaming handoffs take way too much time. As you have learned, fast secure roaming mechanisms such as OKC or FT address this problem. So far we have discussed fast BSS transition as a solution when 802.1X/EAP is deployed. If an AP and a client both support FT, a client does not have to reauthenticate with the RADIUS server when the client roams. Furthermore, the 4-Way Handshake frames are not needed because the dynamic encryption keys are constructed in the frames used in over-the-air fast BSS transition. FT roaming is a much faster roaming handoff that should typically be less than 50 ms.

When PSK authentication is deployed, roaming handoffs are always fast because an 802.1X/EAP frame exchange is never needed. When a PSK client roams to a PSK access point, the frames exchanges consist of reassociation request/response and the 4-Way Handshake to create new keys. Typically roaming handoffs with standard PSK authentication take 40–60 ms. You should also understand that when PSK authentication is deployed, FT roaming mechanisms also occur if the clients and APs all support FT. Roaming handoff times can be shortened even more when FT is deployed. When using PSK authentication, the FT initial mobility domain association to the first AP consists of the same frames used in Figure 7.17 minus the 802.1X/EAP exchange. When an FT client roams using over-the-air fast BSS transition, the 4-Way Handshake is not needed because the encryption keys are created in the FT Action frames and reassociation request/response frames. By eliminating the 4-Way Handshake, an FT roam, when using PSK authentication, is slightly faster.

In Chapter 6, "PSK Authentication," you learned about a future replacement for PSK authentication called Simultaneous Authentication of Equals (SAE). The 802.11-2012 standard also supports fast BSS transition for SAE. Therefore, the RSN information element found in WLAN management frames supports three authentication and key management (AKM) suites:

- FT authentication using IEEE 802.1X, with FT key management
- FT authentication using PSK, with FT key management
- FT authentication over SAE with SHA-256, with FT key management

Is Fast BSS Transition (FT) Backward Compatible with Older Client Devices?

In most cases, new 802.11 technology allows for backward compatibility. When enterprise APs are deployed with the new 802.11 technology, legacy client devices will still require connectivity. For example, client devices that do not support fast BSS transition should still be able to associate to an SSID that has been configured for FT. As a matter of fact, AP configured for FT should be able to support clients that support FT and clients that support OKC on the same SSID. Clients that support neither fast roaming mechanism should also be able to associate. However, in the real world you might run into problems. When FT is configured on an access point, the AP will broadcast management frames with new information elements. For example, the mobility domain information element (MDIE) will be in all beacon and probe response frames. Unfortunately, the drivers of some older legacy client radios may not be able to process the new information in these management frames. The result is that legacy clients may have connectivity problems when an AP is configured for FT. Whenever any new 802.11 technology requires new information elements in the 802.11 management frames, there may be a possible connectivity problem with older legacy radios. Always test the legacy client population when configuring APs for fast BSS transition. If connectivity problems arise, consider using a separate SSID solely for fast BSS transition devices. However, please remember that every SSID consumes more airtime due to the Layer 2 management overhead. Additionally, as more devices begin to support FT, upgrade your client devices.

802.11k

The 802.11k-2008 amendment in conjunction with the recently ratified 802.11r-2008 amendment have the potential to improve roaming performance within secure 802.11 WLANs. 802.11k defines *radio resource measurement (RRM)* mechanisms that enable 802.11k-compliant radios to better understand the RF environment in which they exist. If an 802.11 radio is not transmitting, it can evaluate the RF environment and radio link performance while the radio is listening. Radio resource measurements can be made by an AP or a client station. The measurements can be taken locally and can be requested and obtained from another station. RRM data can be made available to upper protocol layers, where it can be used for a range of applications, such as Voice over Internet Protocol (VoIP).

Measurements such as the channel load request/report and the neighbor request/report may be used to collect pre-handoff information, which can drastically speed up handoffs between access points. The whole purpose behind RRM is that client stations and/or APs can make intelligent decisions about the most effective way to utilize the available spectrum, power, and bandwidth for desired WLAN communications.

One key component of RRM is the *neighbor report*, which is used by client stations to gain information from the associated AP about potential roaming neighbors. As defined by the 802.11k-2008 amendment, the neighbor report information assists the fast roaming process by providing a method for the client to request the associated AP to measure and report about neighboring APs available within the same mobility domain. This can speed up the client scanning process by informing the client device of nearby APs to which it may roam. The neighbor report information is typically delivered through a request/report frame exchange inside 802.11 Action frames. In the neighbor report, the client learns about multiple fields of operational information about each potential AP roaming target, including

- BSSID of the neighbor AP
- Mobility domain
- Whether the neighbor AP supports the same security as the current AP
- Quality of service (QoS)
- Automatic Power Save Delivery (APSD)
- Radio measurement
- BlockAck method

- Spectrum management
- Regulatory class
- Channel number
- PHY type (802.11a/b/g/n/ac)

As you learned earlier in this chapter, clients make the roaming decision as opposed to the APs or WLAN controllers. Clients will still look to roam to APs using active and passive scanning, and they will still evaluate locally the RF environment based on RSSI metrics. Client stations will still make the roaming decision; however, the neighbor reports will provide the client stations with additional input so the client stations can make better roaming decisions.

In addition to the neighbor report, a multitude of radio measurements from client devices can provide information to APs to generate and update the neighbor report. They include the following:

- Transmit Power Control (TPC) information enables the client device to calculate the link budget prior to association.
- Power capability elements allow the client device to compute range data as part of the monitoring of neighboring APs during an active call.
- QoS metrics provide troubleshooting data.
- Client device statistic measurements include voice diagnostics and management for QoS.
- Power constraint element (2.4 GHz and 5 GHz) provides information that reduces co-channel interference.
- Quiet-time announcement (2.4 GHz and 5 GHz) permits the collection of measurements for diagnostics and troubleshooting.

The 802.11k amendment is now part of the 802.11-2012 standard. It was originally ratified in June 2008 and was published as IEEE 802.11k-2008. It should be noted that 802.11k mechanisms will only work if supported by both AP and the clients. Most enterprise WLAN vendor APs support 802.11k and RRM mechanisms. Client-side support for 802.11k is still sparse but growing. For example, newer Apple iOS mobile devices support 802.11k mechanisms.

Real World Scenario

Should I Enable 802.11k Mechanisms on Enterprise Access Points?

Client devices that do not support 802.11k RRM mechanisms should still be able to associate to an SSID that has been configured for 802.11k. However, in the real world you might run into problems. When 802.11k is configured on an access point, the AP will broadcast management frames with multiple 802.11k information elements. For example, the RRM

Information Capabilities information element will be in all beacon and probe response frames. Unfortunately, the drivers of some older legacy client radios may not be able to process the new information in these management frames. The result is that legacy clients may have connectivity problems when an AP is configured for 802.11k. Whenever any new 802.11 technology requires new information elements in the 802.11 management frames, there may be a possible connectivity problem with older legacy radios. Always test the legacy client population when configuring APs for 802.11k. Client devices that support 802.11k will usually also support 802.11r mechanisms. If connectivity problems arise, consider using a separate SSID solely for 802.11k and 802.11r clients. However, please remember that every SSID consumes more airtime due to the Layer 2 management overhead. Additionally, as more devices begin to support 802.11k, upgrade your client devices.

EXERCISE 7.3

Radio Resource Management and Neighbor Reports

In this exercise, you will use a protocol analyzer to view the 802.11 frame exchanges used by a client and AP for an FT initial mobility domain association.

1. To perform this exercise, you need to first download the following file from the book's online resource area, which can be accessed at `www.sybex.com/go/cwsp2e`.

   `FT.PCAP`

2. After the capture file has been downloaded, you will need packet analysis software to open the file. If you do not already have a packet analyzer installed on your computer, you can download Wireshark from `www.wireshark.org`.
3. Using the packet analyzer, open the `FT.PCAP` file. Most packet analyzers display a list of capture frames in the upper section of the screen, with each frame numbered sequentially in the first column.
4. Click packet 103, which is an 802.11 Action frame transmitted by an Apple iOS client device. Typically in the lower section of the screen is the packet details window. This section contains annotated details about the selected frame. In this window, locate and expand the Action frame. In the body of the Action frame, expand the tagged parameters and notice that this Action frame is being used as a neighbour report request. The client is asking the AP if the AP has information about any neighboring APs.
5. Click packet 105 which is an 802.11 Action frame transmitted by the AP. In the body of the Action frame, expand the fixed parameters and notice that this Action frame is being used as a neighbor report response. In the body of the Action frame, expand the tagged parameters and view the neighbor report about the AP with a BSSID of 08:ea:44:76:b5:68, which is transmitting on channel 48.

802.11v

The IEEE 802.11v-2011 amendment defined *wireless network management (WNM)* as information about network resources that is exchanged between the client devices and an AP. The intended goal is to enhance overall performance of the wireless network. Whereas 802.11k provides exchange of information about the RF environment, 802.11v exchanges WNM information about surrounding existing network conditions. There are many categories of WNM information that can be potentially exchanged between clients and APs, including

- BSS Max Idle Period Management
- BSS Transition Management
- Channel Usage
- Collocated Interference Reporting
- Diagnostic Reporting
- Directed Multicast Service (DMS)
- Flexible Multicast Service (FMS)
- Multicast Diagnostic Reporting
- Event Reporting
- Location Services
- Multiple BSSID Capability
- Proxy ARP
- QoS Traffic Capability
- SSID List
- Triggered STA Statistics
- TIM Broadcast
- Timing Measurement
- Traffic Filtering Service
- U-APSD Coexistence
- WNM-Notification
- WNM-Sleep Mode

802.11v is now part of the 802.11-2012 standard. It was originally ratified in 2011 and was published as IEEE 802.11v-2011. Keep in mind that 802.11v mechanisms will work only if supported by both AP and the clients. Support for WNM mechanisms defined by 802.11v-2011 is currently not widespread, although some of the WNM capabilities are supported in WLAN vendor APs that are Voice Enterprise certified.

Voice Enterprise

In 2012, the Wi-Fi Alliance debuted a vendor-interoperability certification called *Voice Enterprise* that defines enhanced support for voice applications in the enterprise environment. Many aspects of the 802.11r, 802.11k, and 802.11v amendments are tested for Voice Enterprise certification. Effective enterprise-grade voice over Wi-Fi solutions must address two major requirements:

Voice Quality Voice quality has to be consistently good throughout the call, in all load conditions. To ensure that client devices maintain good voice quality, latency, jitter, and packet loss have to be consistently low.

Data Traffic Coexistence Wi-Fi networks are used for both voice and data applications. Therefore, voice calls must coexist and share network resources with data traffic, which often accounts for the largest portion of the network load.

In order to achieve voice quality and data traffic coexistence, the following general aspects of the Voice Enterprise certification are required:

Prioritization Support for Wi-Fi Multimedia (WMM) QoS mechanisms is mandatory. WMM enables the AP to recognize and prioritize voice traffic over other application traffic, such as Internet browsing, email reading, or large file downloads.

Bandwidth Management APs must support WMM-Admission Control, which optimizes traffic management by admitting only those traffic streams that an AP can support at a given time. WMM-AC also enables load balancing.

Seamless Transitions Across the Wi-Fi Network Support for IEEE 802.11r fast BSS transition, which allows fast AP handoffs (known as BSS transitions) even while using advanced security methods, is also mandatory. 802.11r aspects are mandatory for AP and client devices.

Network Measurement and Management Support for IEEE standards for radio resource measurement (802.11k) and wireless network management (802.11v) is also mandatory. 802.11k aspects are mandatory for AP and client devices. 802.11v mechanisms are optional.

Security WPA2-Enterprise security is mandatory for both APs and clients.

Battery Life APs must support the power save mechanism WMM-Power Save. Support for WMM-PS is optional for clients.

Performance is measured using four (802.11b) or ten (802.11a/g/n/ac) concurrent simulated voice calls, a high-speed video stream, and background data traffic, designed to represent a fully loaded enterprise wireless network environment. Testing is done while fast roaming transitions between APs are executed. Performance is measured under simulated, but realistic, network conditions. Performance of equipment submitted for Wi-Fi Voice-Enterprise certification has to meet the following thresholds to ensure that the Wi-Fi network preserves good voice call quality:

- Latency (One way delay < 50 ms)
- Jitter (< 50 ms)

- Packet Loss (< 1 percent)
- Consecutive lost packets (no more than 3)

Real World Scenario

How Widely Is Voice Enterprise Supported?

Most major WLAN enterprise vendors offer support for Voice Enterprise in their access points. Usually a WLAN administrator will be able to configure all three 802.11k/r/v mechanisms together or independently on an access point that is Voice Enterprise certified. However, client-side support for Voice Enterprise is not widespread. Any client devices that were manufactured before 2012 simply will not support 802.11r/k/v operations. The bulk of newer devices also do not support Voice Enterprise capabilities. However client-side support is growing. For example, Apple iOS devices support 802.11k/r/v: `https://support.apple.com/en-us/HT202628`. Many Android OS clients also support some Voice Enterprise capabilities. Roaming performance will gradually improve as more client devices support 802.11k/r/v mechanisms.

Layer 3 Roaming

One major consideration when designing a WLAN is what happens when client stations roam across Layer 3 boundaries. Wi-Fi operates at Layer 2 and roaming is essentially a Layer 2 process. As shown in Figure 7.20, the client station is roaming between two access points. The roam is seamless at Layer 2, but user VLANs are tied to different subnets on either side of the router. As a result, the client station will lose Layer 3 connectivity and must acquire a new IP address. Any connection-oriented applications that are running when the client reestablishes Layer 3 connectivity will have to be restarted. For example, a VoIP phone conversation would disconnect in this scenario, and the call would have to be reestablished.

Because 802.11 wireless networks are usually integrated into preexisting wired topologies, crossing Layer 3 boundaries is often a necessity, especially in large deployments. The only way to maintain upper-layer communications when crossing Layer 3 subnets is to provide a *Layer 3 roaming* solution that is based on the Mobile IP standard. *Mobile IP* is an Internet Engineering Task Force (IETF) standard protocol that allows mobile device users to move from one Layer 3 network to another while maintaining their original IP address. Mobile IP is defined in IETF Request for Comment (RFC) 5944. Layer 3 roaming solutions based on Mobile IP use some type of tunneling method and IP header encapsulation to allow packets to traverse between separate Layer 3 domains with the goal of maintaining upper-layer communications. Most WLAN vendors now support some form of Layer 3 roaming solution, as shown in Figure 7.21.

FIGURE 7.20 Layer 3 roaming boundaries

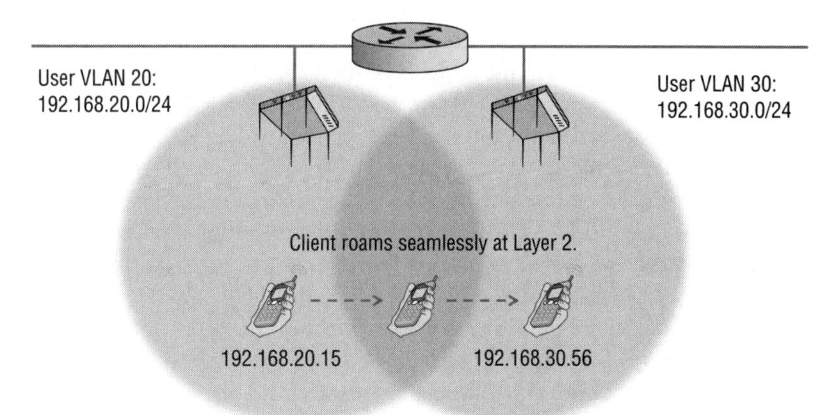

FIGURE 7.21 Mobile IP

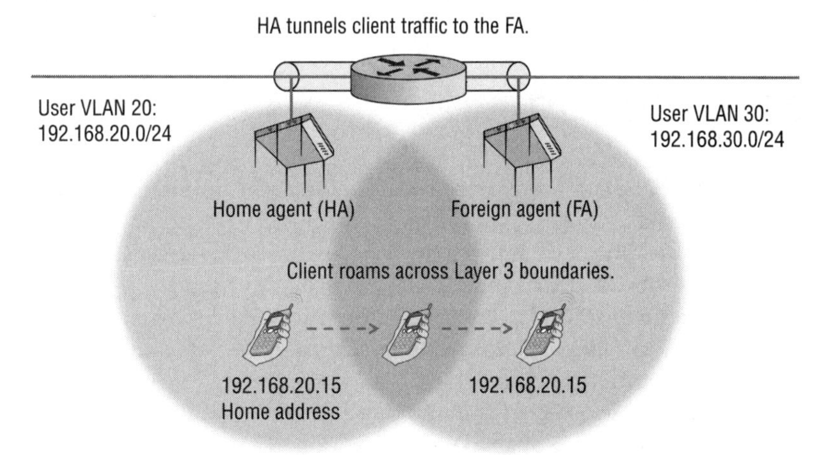

A mobile client receives an IP address also known as a *home address* on a home network. The mobile client must register its home address with a device called a *home agent (HA)*. As depicted in Figure 7.21, the client's original associated access point serves as the home agent. The home agent is a single point of contact for a client when it roams across Layer 3 boundaries. The HA shares client MAC/IP database information in a table, called a *home agent table (HAT)* with another device called the *foreign agent (FA)*.

In this example, the foreign agent is another access point that handles all Mobile IP communications with the home agent on behalf of the client. The foreign agent's IP address is known as the *care-of address*. When the client roams across Layer 3 boundaries, the client is roaming to a foreign network where the FA resides. The FA uses the HAT tables to locate the HA of the mobile client station. The FA contacts the HA and sets up a Mobile IP tunnel. Any traffic that is sent to the client's home address is intercepted by the HA and sent through the Mobile IP tunnel to the FA. The FA then delivers the tunneled traffic to the client and the client is able to maintain connectivity using the original home address. In our example, the Mobile IP tunnel is between two APs on opposite sides of a router. If the user VLANs exist at the edge of the network, tunneling of user traffic occurs between access points that assume the roles of HA and FA. The tunneling is often distributed between multiple APs. However, user VLANs may reside back in a DMZ or at the core layer of the network along with a WLAN controller. In a single WLAN controller environment, the Layer 3 roaming handoffs exist as control plane mechanisms within the single controller. In a multiple WLAN controller environment, an IP tunnel is created between controllers that are deployed in different routed boundaries with different user VLANs. One controller functions as the home agent and another functions as the foreign agent.

Although maintaining upper-layer connectivity is possible with these Layer 3 roaming solutions, increased latency is sometimes an issue. Additionally, Layer 3 roaming may not be a requirement for your network. Less complex infrastructure often uses a simpler flat Layer 2 design. Larger enterprise networks often have multiple user and management VLANs linked to multiple subnets; therefore, a Layer 3 roaming solution will be required.

Troubleshooting

The best way to ensure that seamless roaming will commence is proper design and a thorough site survey. Proper WLAN design normally requires –65 dBm primary coverage and secondary coverage of usually –5 dB lower than the primary coverage cells. The only way to determine whether proper primary and secondary coverage is in place is by conducting a coverage analysis site survey. Proper site survey procedures are discussed in detail in the *CWNA: Certified Wireless Network Administrator Official Study Guide, Fourth Edition* by David Coleman and David Westcott (Sybex, 2014).

Changes in the WLAN environment can also cause roaming headaches. RF interference will always affect the performance of a wireless network and can make roaming problematic as well. Very often new construction in a building will affect the coverage of a WLAN and create new dead zones. If the physical environment where the WLAN is deployed changes, the coverage design may have to change as well. It is always a good idea to conduct a coverage survey periodically to monitor changes in coverage patterns.

Troubleshooting roaming by using a protocol analyzer is tricky because the reassociation roaming exchanges occur on multiple channels. To troubleshoot a client roaming between channels 1, 6, and 11, you would need three separate protocol analyzers on three separate laptops that would produce three separate frame captures. Riverbed (formally CACE Technologies) offers a product called AirPcap that is a USB 802.11 radio. As you can see in Figure 7.22, three AirPcap USB radios can be configured to capture frames on channels 1, 6, and 11 simultaneously. All three radios are connected to a USB hub and save the frame captures of all three channels into a single time-stamped capture file. The AirPcap solution allows for multichannel monitoring with a single protocol analyzer.

FIGURE 7.22 AirPcap provides multichannel monitoring and roaming analysis.

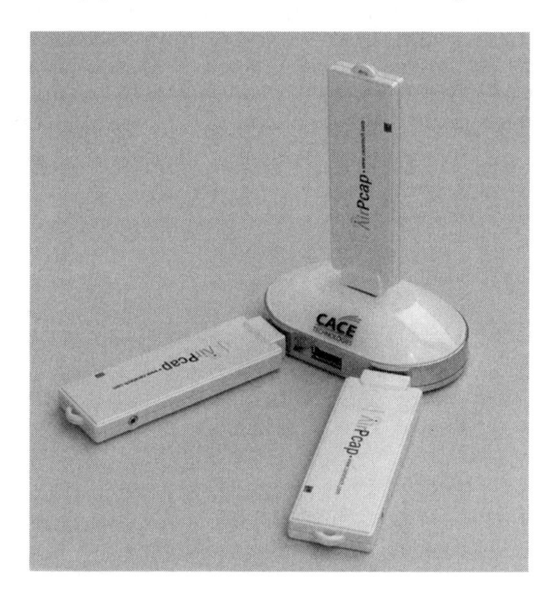

Summary

The handheld WLAN devices, such as smart phones and tablets that connect to enterprise WLANs, require mobility. Layer 2 roaming mechanisms provide mobility for WLAN client devices. Over the years, seamless roaming for WLAN client devices has improved as better Layer 2 roaming mechanisms have been defined. In this chapter, you learned about the nonstandard and standard fast secure roaming operations that provide for an enhanced roaming experience. As client-side support for Voice Enterprise and 802.11k/r/v technologies grows, roaming performance will improve.

Exam Essentials

Understand legacy roaming handoff mechanisms. Reassociation is the process clients use to move from one BSS to another. Client roaming thresholds and AP-to-AP handoff communications are proprietary.

Define the operations and limitations of legacy FSR mechanisms. Explain both the PMK caching and preauthentication FSR methods defined by the 802.11-2012 standard.

Explain opportunistic key caching (OKC). Understand how the PMKID is manipulated by OKC to accomplish fast secure roaming.

Explain fast BSS transition (FT). Define all the key hierarchy components and operations defined by the 802.11r-2008 amendment.

Understand the benefits of 802.11k-2008 radio resource measurement (RRM). Describe how RRM enables client stations to make intelligent decisions to improve roaming performance.

Explain the purpose and basics of Mobile IP. Define Mobile IP components and why a Mobile IP solution is needed when clients roam across Layer 3 boundaries.

Review Questions

1. What type of solution must be deployed to provide continuous connectivity when a client station roams across Layer 3 boundaries? (Choose all that apply.)

   A. Nomadic roaming solution

   B. Seamless roaming solution

   C. Mobile IP solution

   D. Fast secure roaming solution

2. Which pairwise master key security associations (PMKSAs) can be uniquely identified by a pairwise master key identifier (PMKID)? (Choose all that apply.)

   A. PMKSA derived from a PSK authentication

   B. PMKSA from Open System authentication

   C. Cached PMKSA from an 802.1X/EAP authentication

   D. Cached PMKSA for Mobile IP authentication

   E. Cached PMKSA from preauthentication

   F. PMK-RO or PMK-R1 derived from fast BSS transition (FT)

3. As defined by the 802.11-2012 standard, which of these authentication methods can be used by a client station to establish a pairwise master key security association (PMKSA)? (Choose all that apply.)

   A. PSK authentication

   B. WEP authentication

   C. 802.1X/EAP authentication

   D. Open authentication

   E. SAE authentication

4. Which of these methods allows an authenticator and supplicant to skip an entire 802.1X/EAP authentication and proceed with the traditional 4-Way Handshake? (Choose all that apply.)

   A. PMK caching

   B. PTK caching

   C. Opportunistic key caching

   D. Fast BSS transition

5. What is some of the operation information that an 802.11k-2008–compliant client station may receive in the neighbor report from an 802.11k-2008–compliant access point (AP)? (Choose all that apply.)

   A. BSSID of neighbor AP

   B. PHY types supported by neighbor AP

   C. APSD support of neighbor AP

   D. Channel number of neighbor AP

   E. All of the above

6. Which authentication and key management (AKM) methods can also support fast BSS transition (FT)? (Choose all that apply.)
   - **A.** MAC authentication
   - **B.** 802.1X/EAP authentication
   - **C.** Open system authentication
   - **D.** SAE authentication
   - **E.** PSK authentication

7. What are some of the components that comprise a PMKID? (Choose all that apply.)
   - **A.** Authenticator MAC address
   - **B.** Authentication server MAC address
   - **C.** PMK
   - **D.** MSK
   - **E.** Supplicant MAC address

8. Which of these 802.11 management frames contain a PMKID in the RSN information element? (Choose all that apply.)
   - **A.** Reassociation request
   - **B.** Probe response
   - **C.** FT Action
   - **D.** Beacon
   - **E.** Association request

9. Within the three-tier FT key hierarchy defined by fast BSS transition, which of these keys is cached on a WLAN controller or original associated AP?
   - **A.** PMK-R1
   - **B.** MSK
   - **C.** PTK
   - **D.** PMK-R0
   - **E.** PMK

10. What are the main goals of the Wi-Fi Alliance Voice Enterprise certification? (Choose all that apply.)
    - **A.** Data traffic coexistence
    - **B.** Higher throughput
    - **C.** Voice quality
    - **D.** Video quality
    - **E.** Quality of service

11. Which of these roaming methods requires the use of FT Action frames?
    A. Over-the-air fast BSS transition
    B. Over-the-WDS fast BSS transition
    C. Over-the-DS fast BSS transition
    D. Over-the-WLS fast BSS transition

12. Within the three-tier FT key hierarchy defined by fast BSS transition (FT), which of these keys is used to encrypt the MSDU payload of an 802.11 data frame?
    A. PMK-R1
    B. MSK
    C. PTK
    D. PMK-R0
    E. PMK

13. Which of these protocol adherence and performance metrics will most likely be defined by the Wi-Fi Alliance Voice Enterprise certification? (Choose all that apply.)
    A. Latency
    B. Jitter
    C. Retry percentage
    D. Consecutive lost packets
    E. TCP throughput
    F. Packet loss

14. Although not defined by the 802.11-2012 standard, which of these methods of fast secure roaming is supported by the majority of WLAN vendors?
    A. PMK caching
    B. Preauthentication
    C. Opportunistic key caching
    D. Over-the-air fast BSS transition

15. What aspects of roaming were not defined by the original standard? (Choose all that apply.)
    A. Security
    B. AP-to-AP handoff
    C. PTK
    D. RSSI thresholds
    E. PMK

16. What is used initially to seed the FT process that is used to create three levels of key hierarchy?
    A. PMK-R0
    B. PMK-R1
    C. PTK
    D. MSK
    E. PMK

17. WLAN administrator Alexandra Gunther recently upgraded her company's WLAN infrastructure with APs that are Voice Enterprise certified. She configured the APs for fast BSS transition (FT) and deployed new iPads that also support FT. The iPads are roaming in a fast and secure manner; however, she noticed that some older laptops can no longer connect to the WLAN. What is the cause of the problem?
    A. FT is not backward compatible with legacy client devices.
    B. FT clients and OKC clients cannot connect to the same SSID.
    C. FT and PMK caching clients cannot connect to the same SSID.
    D. The laptop radio drivers cannot decipher the FT information element in the AP beacon.

18. Which of these devices serve as a key holder for the PMK-R1 key created during a fast BSS transition? (Choose all that apply.)
    A. WLAN controller
    B. Client stations
    C. Access points
    D. RADIUS server
    E. Access layer switch

19. The ACME Company manufactures two models of WLAN controllers. The standard model uses the controller-based APs to encrypt and decrypt client traffic at the edge of the network. The deluxe model uses end-to-end encryption and the WLAN controller performs encryption/decryption of the client traffic at the core of the network. Which of these statements properly identify the key holder roles? (Choose all that apply.)
    A. Standard model: AP is the R0KH
    B. Standard model: WLAN controller is the R1KH
    C. Deluxe model: WLAN controller is the R0KH
    D. Deluxe model: WLAN controller is the R1KH
    E. Deluxe model: AP is the R1KH

20. Which key is used to seed both over-the-air and over-the-DS fast BSS transition?
    A. MSK
    B. PMK-R0
    C. PMK-R1
    D. PTK
    E. GTK

WLAN Security Infrastructure

IN THIS CHAPTER, YOU WILL LEARN ABOUT THE FOLLOWING:

✓ **802.11 services**

- Integration service (IS)
- Distribution system (DS)

✓ **Management, Control, and Data planes**

- Mangement plane
- Control plane
- Data plane
- WLAN architecture
- Autonomous WLAN architecture
- Centralized network management systems
- Cloud networking
- Centralized WLAN architecture
- Distributed WLAN architecture
- Unified WLAN architecture
- Hybrid WLAN architecture
- Enterprise WLAN routers
- WLAN mesh access points
- WLAN bridging

✓ **VPN Wireless Security**

- VPN 101
- Layer 3 VPNs
- SSL VPNs
- VPN deployment

✓ **Infrastucture management**

- Protocols for management

This chapter provides an overview of the many different WLAN architectures that are available today. In previous chapters, we discussed the authentication and encryption technologies used to provide 802.11 security. You need a good understanding of WLAN architecture so that 802.11 authentication and encryption can be properly implemented within your network design. This chapter will also discuss the use of VPNs with wireless security. Finally, we will cover WLAN network infrastructure management.

802.11 Services

Many services are defined in the 802.11-2012 standard. Six of these services are used to support MAC Service Data Unit (MSDU) delivery between access points and client stations. In this chapter, we are going to focus on two of these services and where these services are used in different types of WLAN architecture.

Integration Service (IS)

The 802.11-2012 standard defines an *integration service (IS)* that enables delivery of MSDUs between the distribution system (DS) and a non-IEEE-802.11 LAN via a portal. A simpler way of defining the integration service is to characterize it as a frame format transfer method. The portal is usually either an access point or a WLAN controller. As mentioned earlier, the payload of a wireless 802.11 data frame is the Layer 3–7 information known as the MSDU. The eventual destination of this payload is usually to a wired network infrastructure. Because the wired infrastructure is a different physical medium, an 802.11 data frame payload must be effectively transferred into an 802.3 Ethernet frame. For example, a VoWiFi phone sends an 802.11 data frame to a standalone access point. The MSDU payload of the frame is a VoIP packet with a final destination of an IP PBX that resides at the 802.3 network core. The job of the integration service is to remove the 802.11 header and trailer and then encase the MSDU VoIP payload inside an 802.3 frame. The 802.3 frame is then sent on to the Ethernet network. The integration service performs the same actions in reverse when an 802.3 frame payload must be transferred into an 802.11 frame that is eventually transmitted by the access point radio.

It is beyond the scope of the 802.11-2012 standard to define how the integration service operates. Normally, the integration service transfers data frame payloads between an 802.11 and 802.3 medium. However, the integration service could transfer an MSDU between the 802.11 medium and some sort of other medium. If 802.11 user traffic is forwarded from the edge of a network, the integration service exists in an access point. The integration service mechanism normally takes place inside a WLAN controller when 802.11 user traffic is tunneled back to a WLAN controller.

Distribution System (DS)

The 802.11-2012 standard also defines a *distribution system (DS)* that is used to interconnect a set of basic service sets (BSSs) via integrated LANs to create an extended service set (ESS). Access points by their very nature are portal devices. The DS is used to forward WLAN client traffic to the integration service or back to the wireless medium. The DS consists of two main components:

Distribution System Medium (DSM) A logical physical medium used to connect access points is known as a *distribution system medium (DSM)*. The most common example is an 802.3 medium.

Distribution System Services (DSS) System services built inside an access point are usually in the form of software. The *distribution system services (DSS)* provide switch-like intelligence. These software services are used to manage client station associations, reassociations, and disassociations. Distribution system services also use the Layer 2 addressing of the 802.11 MAC header to eventually forward the Layer 3–7 information (MSDU) either to the integration service or to another wireless client station. A full understanding of DSS is beyond the scope of the CWSP exam but is necessary at the Certified Wireless Analysis Professional (CWAP) certification level.

Management, Control, and Data Planes

Telecommunication networks are often defined as three logical planes of operation:

Management Plane The *management plane* is defined by administrative network management, administration, and monitoring. An example of the management plane would be any network management solution that can be used to monitor routers and switches and other wired network infrastructure. A centralized network management server can be used to push both configuration settings and firmware upgrades to network devices.

Control Plane The *control plane* consists of control or signaling information and is often defined as network intelligence or protocols. Dynamic Layer 3 routing protocols, such as OSPF or BGP, used to forward data would be an example of control plane intelligence found in routers. Content addressable memory (CAM) tables and Spanning Tree Protocol (STP) are control plane mechanisms used by Layer 2 switches for data forwarding.

Data Plane The *data plane*, also known as the user plane, is the location in a network where user traffic is actually forwarded. An individual router where IP packets are forwarded is an example of the data plane. An individual switch forwarding an 802.3 Ethernet frame is an example of the data plane.

In an 802.11 environment, these three logical planes of operation function differently depending on the type of WLAN architecture and the WLAN vendor. For example, in a legacy autonomous AP environment all three planes of operation existed in each standalone access point (although the control plane mechanisms were minimal). When WLAN controller solutions were first introduced in 2002, all three planes of operation were shifted into a centralized device. In modern-day deployments, the planes of operation may be divided between access points, WLAN controllers, and/or a wireless network management server (WNMS).

Do not confuse the management, control, and data planes with 802.11 MAC frame types. In this chapter, the discussion of management, control, and data planes is related to WLAN network architectural operations.

Management Plane

The functions of the *management plane* within an 802.11 WLAN are as follows:

WLAN Configuration Examples include the configuration of SSIDS, security, WMM, channel, and power settings.

WLAN Monitoring and Reporting Monitoring of Layer 2 statistics like ACKs, client associations, resassociations, and data rates occurs in the management plane. Examples of upper-layer monitoring and reporting include application visibility, IP connectivity, TCP throughput, latency statistics, and stateful firewall sessions.

WLAN Firmware Management The ability to upgrade access points and other WLAN devices with the latest vendor operational code is included here.

Control Plane

The *control plane* is often defined by protocols that provide the intelligence and interaction between equipment in a network. Here are a few examples of control plane intelligence:

Dynamic RF Coordinated channel and power settings for multiple access points are provided by the control plane. The majority of WLAN vendors implement some type of *dynamic RF* capability. Dynamic RF is also referred to by the more technical term *radio resource management (RRM).*

Roaming Mechanisms The control plane also provides support for roaming handoffs between access points. Capabilities may include L3 roaming, maintaining stateful firewall sessions of clients, and forwarding of buffered packets. Fast secure roaming mechanisms, such as opportunistic key caching (OKC) and fast BSS transition (FT), may also be used to forward master encryption keys between access points.

Client Load Balancing Collecting and sharing client load and performance metrics between access points to improve overall WLAN operations happens in the control plane.

Mesh Protocols Routing user data between multiple access points requires some sort of mesh routing protocol. Most WLAN vendors use Layer 2 routing methods to move user data between mesh access points. However, some vendors are using Layer 3 mesh routing. The 802.11s amendment has defined standardized mesh routing mechanisms, but most WLAN vendors are currently using proprietary methods and metrics.

Data Plane

The *data plane* is where user data is forwarded. The two devices that usually participate in the data plane are the AP and a WLAN controller. A standalone AP handles all data forwarding operations locally. In a WLAN controller solution, data is normally forwarded from the centralized controller, but data can also be forwarded at the edge of the network by an AP. As with the management and control planes, each vendor has a unique method and recommendations for handling data forwarding. Data forwarding models will be discussed in greater detail later in this chapter.

WLAN Architecture

While the acceptance of 802.11 technologies in the enterprise continues to grow, the evolution of WLAN architecture has kept pace. In most cases, the main purpose of 802.11 technologies is to provide a wireless portal into a wired infrastructure network. How an 802.11 wireless portal is integrated into a typical 802.3 Ethernet infrastructure continues to change drastically. WLAN vendors generally offer one of three primary WLAN architectures:

- Autonomous WLAN architecture
- Centralized WLAN architecture
- Distributed WLAN architecture

The following sections describe these three architectures in greater detail.

Autonomous WLAN Architecture

For many years, the conventional access point was a standalone WLAN portal device where all three planes of operation existed and operated on the edge of the network architecture. These APs are often referred to as *fat APs*, or *standalone APs*. However, the most common industry term for the traditional access point is *autonomous AP*.

All configuration settings exist in the autonomous access point itself, and therefore, the management plane resides individually in each autonomous AP. All encryption and decryption mechanisms and MAC layer mechanisms also operate within the autonomous

AP. The distribution system service (DSS) and integration service (IS) both function within an autonomous AP. The data plane also resides in each autonomous AP because all user traffic is forwarded locally by each individual access point. As shown in Figure 8.1, legacy autonomous APs have little shared control plane mechanisms.

FIGURE 8.1 Autonomous WLAN architecture

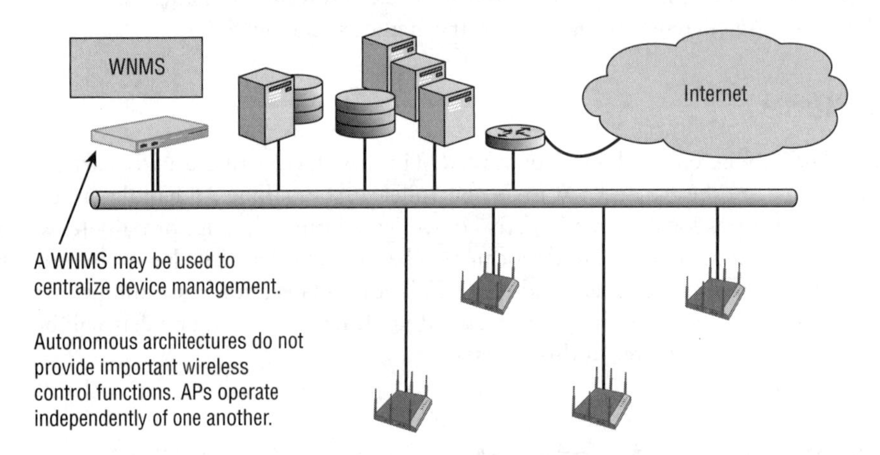

An autonomous access point contains at least two physical interfaces: usually a radio frequency (RF) radio and a 10/100/1000 Ethernet port. The majority of the time, these physical interfaces are bridged together by a virtual interface known as a *bridged virtual interface (BVI)*. The BVI is assigned an IP address that is shared by two or more physical interfaces. Access points operate as Layer 2 devices; however, they still need a Layer 3 address for connectivity to an IP network. The BVI is the management interface of an AP.

An autonomous access point typically encompasses both the 802.11 protocol stack and the 802.3 protocol stack. These APs might support the following security features:

- Multiple management interfaces, such as command line, web GUI, and SNMP
- WEP, WPA, and WPA2 security capabilities
- Filtering options, such as MAC and protocol

Autonomous APs might have some of the following advanced security features:

- Built-in RADIUS and user databases
- VPN client and/or server support
- DHCP server
- Captive web portals

Autonomous APs are deployed at the access layer and typically are powered by a Power-over-Ethernet (PoE)-capable access layer switch. The integration service within an autonomous AP translates the 802.11 traffic into 802.3 traffic. The autonomous AP was the foundation that WLAN architects deployed for many years. However, most enterprise deployments of autonomous APs were replaced by a centralized architecture utilizing a WLAN controller, which is discussed later in this chapter.

Centralized Network Management Systems

One of the challenges for a WLAN administrator using a large WLAN autonomous architecture is management. As an administrator, would you want to configure 300 autonomous APs individually? One major disadvantage of using the traditional autonomous access point is that there is no central point of management. Any intelligent edge WLAN architecture with 25 or more autonomous access points is going to require some sort of *wireless network management system (WNMS)*.

A WNMS moves the management plane out of the autonomous access points. A WNMS provides a central point of management to configure and maintain thousands of autonomous access points. A WNMS can be a hardware appliance or a software solution. WNMS solutions can be vender specific or vender neutral.

As shown in Figure 8.1, the whole point of a WNMS server was to provide a central point of management for autonomous access points, which are now considered legacy devices. That definition has changed considerably over the years. Later in this chapter, you will learn about WLAN controllers, which are used as a central point of management for controller-based APs. WLAN controllers can effectively replace a WNMS server as a central point of management for access points in small-scale WLAN deployments. However, multiple WLAN controllers are needed in large-scale WLAN enterprise deployments. Currently, most WMNS servers are now used as a central point of management for multiple WLAN controllers in large-scale WLAN enterprises. WNMS servers that are used to manage multiple WLAN controllers from a single vendor may in some cases also be used to manage other vendors' WLAN infrastructure, including standalone access points.

The term WNMS is actually outdated because many of these centralized management solutions can also be used to manage other types of network devices, including switches, routers, firewalls, and VPN gateways. Therefore, *network management server (NMS)* is now used more often. NMS solutions are usually vendor specific; however, a few exist that can manage devices from a variety of networking vendors.

The main purpose of an NMS is to provide a central point of management and monitoring for network devices. Configuration settings and firmware upgrades can be pushed down to all the network devices. Although centralized management is the main goal, an NMS can have other capabilities as well, such as RF spectrum planning and management of a WLAN. An NMS can also be used to monitor network architecture with alarms and notifications centralized and integrated into a management console. An NMS provides robust monitoring of network infrastructure as well as monitoring of wired and wireless clients connected to the network. As shown in Figure 8.2, NMS solutions usually have extensive diagnostic utilities that can be used for remote troubleshooting.

FIGURE 8.2 NMS diagnostic utilities

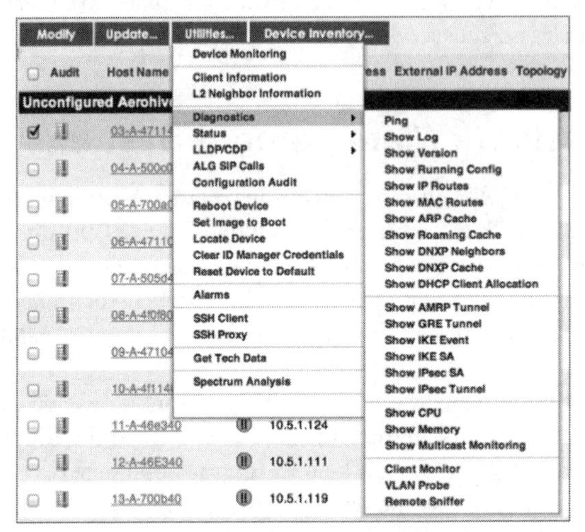

An NMS is a management plane solution; therefore, no control plane or data plane mechanisms exist within an NMS. For example, the only communications between an NMS and an access point are management protocols. Most NMS solutions use the *Simple Network Management Protocol (SNMP)* to manage and monitor the WLAN. Other NMS solutions also use the *Control and Provisioning of Wireless Access Points (CAPWAP)* as strictly a monitoring and management protocol. CAPWAP incorporates *Datagram Transport Layer Security (DTLS)* to provide encryption and data privacy of the monitored management traffic. User traffic is never forwarded by an access point to an NMS; the 802.11 client associations and traffic can be still be monitored. Figure 8.3 shows an NMS display of multiple client associations across multiple APs.

FIGURE 8.3 NMS client monitoring

10 Clients from Friday (July 1, 2016) 08:00 to Friday (July 1, 2016) 10:00

Status Health	Hostname	IP	MAC	User Name	OS Type	Channel	Usage	VLAN
	Coleman-PC	172.16.255.95	74DA3835EAB4		Windows 8/10	11	0 B	1
	aerohives-M...	172.16.255.59	A45E60CB133B		Mac OS X	11	0 B	1
	Blue-Cutter	172.16.255.63	7C5CF8A5DC0F		Windows 8/10	11	537.15 KB	1
	GreenLeaf	172.16.255.62	7C5CF8732E90		Windows 8/10	11	304.66 MB	1
	android-efdf...	172.16.255.88	301966CA471D		Android	149	0 B	1
	Davids-iPhone	172.16.255.90	B844D90E006E		Apple iOS	149	430.96 KB	1
		172.16.255.91	6C29951A2A66		CrOS	149	3.12 MB	1
	iPad	172.16.255.85	98B8E392715D		Apple iOS	149	42.97 KB	1
	charkins-mac	172.16.255.84	A4D18CCB9A80		Mac OS X El Capitan	149	0 B	1
	android-bf66...	172.16.255.87	E09971B3E389		Android	149	359.79 KB	1

10 | 20 | 50 | 100

NMS solutions can be deployed at a company datacenter in the form of a hardware appliance or as a virtual appliance that runs on VMware or some other virtualization platform. A network management server that resides in a company's own datacenter is often referred to as an on-premises NMS. NMS solutions are also available in the cloud as a software subscription service.

Cloud Networking

Cloud computing and *cloud networking* are catchphrases used to describe the advantages of computer networking functionality when provided under a *Software as a Service (SaaS)* model. The term *the cloud* essentially means a scalable private enterprise network that resides on the Internet. The idea behind cloud networking is that applications and network management, monitoring, functionality, and control are provided as a software service. Amazon is the best example of a company that provides an elastic cloud-based IT infrastructure so other companies can offer pay-as-you-go subscription pricing for enterprise applications and network services.

The most common cloud networking model is *cloud-enabled networking (CEN)*. With CEN, the management plane resides in the cloud, but data plane mechanisms such as switching and routing remain on the local network and usually in hardware. Several WLAN vendors offer cloud-enabled NMS solutions as a subscription service that manages and monitors WLAN infrastructure and clients. Some control plane mechanisms can also be provided with a CEN model. For example, WLAN vendors have begun to also offer subscription-based application services along with their cloud-enabled management solutions. Some examples of these subscriptions services include cloud-enabled guest management, NAC, and MDM solutions.

Centralized WLAN Architecture

The next progression in the development of WLAN integration is the centralized WLAN architecture. This model uses a central WLAN controller that resides in the core of the network. In the centralized WLAN architecture, autonomous APs have been replaced with *controller-based access points*, also known as lightweight APs or thin APs. Beginning in 2002, many WLAN vendors decided to move to a WLAN controller model where all three logical planes of operation would reside inside the controller. Effectively, all planes were moved out of access points and into a WLAN controller.

Management Plane Access points are configured and managed from the WLAN controller.

Control Plane Dynamic RF, load balancing, roaming handoffs, and other mechanisms exist in the WLAN controller.

Data Plane The WLAN controller exists as a data distribution point for user traffic. Access points tunnel all user traffic to a central controller.

The encryption and decryption capabilities might reside in the centralized WLAN controller or may still be handled by the controller-based APs, depending on the vendor. The distribution system services (DSS) and integration service (IS) both typically function within the WLAN controller. Some time-sensitive operations are still handled by the AP.

WLAN Controller

At the heart of the centralized WLAN architecture model is the *WLAN controller* (see Figure 8.4). WLAN controllers are often referred to as *wireless switches* because they are indeed an Ethernet-managed switch that can process and route data at the Data-Link layer (Layer 2) of the OSI model. Many of the WLAN controllers are multilayer switches that can also route traffic at the Network layer (Layer 3). However, *wireless switch* has become an outdated term and does not adequately describe the many capabilities of a WLAN controller.

FIGURE 8.4 Centralized WLAN architecture: WLAN controller

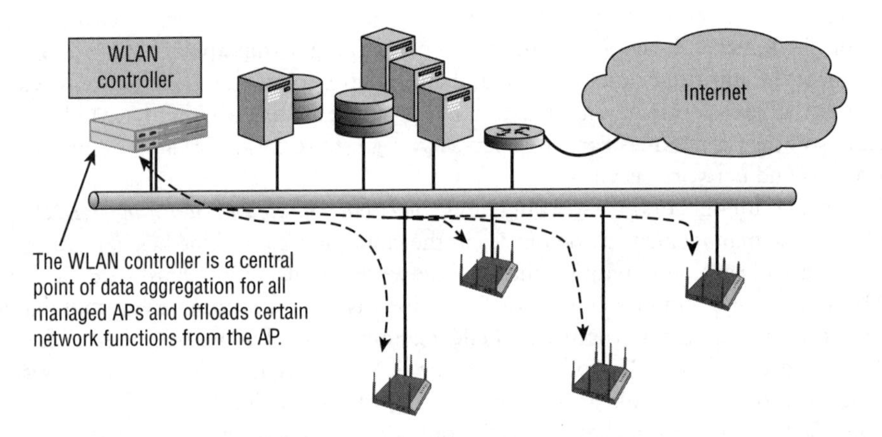

A WLAN controller may have some of these many security features:

VLANs WLAN controllers fully support the creation of VLANs and 802.1Q VLAN tagging. Multiple wireless user VLANs can be created on the WLAN controller so that user traffic can be segmented. VLANs may be assigned statically to WLAN profiles or may be assigned using a RADIUS attribute. User VLANs are usually encapsulated in an IP tunnel.

User Management WLAN controllers usually provide the ability to control the who, when, and where in terms of using role-based access control (RBAC) mechanisms.

Layer 2 Security Support WLAN controllers fully support Layer 2 WEP, WPA, and WPA2 encryption. Authentication capabilities include internal databases, as well as full integration with RADIUS and LDAP servers.

Layer 3 and 7 VPN Concentrators Some WLAN controller vendors also offer VPN server capabilities within the controller. The controller can act as a VPN concentrator or endpoint for IPsec or SSL VPN tunnels.

Captive Portal WLAN controllers have captive portal features that can be used with guest WLANs.

Internal Wireless Intrusion Detection Systems Some WLAN controllers have integrated WIPS capabilities for security monitoring and rogue AP mitigation.

Firewall Capabilities Stateful packet inspection is available with an internal firewall in some WLAN controllers.

Management Interfaces Many WLAN controllers offer full support for common management interfaces such as GUI, CLI, SSH, and so forth.

Controller Data Forwarding Models

A key feature of most WLAN controllers is that the integration service (IS) and distribution system services (DSS) operate within the WLAN controller. In other words, all 802.11 user traffic that is destined for wired-side network resources must first pass through the controller and be translated into 802.3 traffic by the integration service before being sent to the final wired destination. Therefore, controller-based access points send their 802.11 frames to the WLAN controller over an 802.3 wired connection.

The 802.11 frame format is complex and is designed for a wireless medium and not a wired medium. An 802.11 frame cannot travel through an Ethernet 802.3 network by itself. So, how can an 802.11 frame traverse between a controller-based AP and a WLAN controller? The answer is inside an IP-encapsulated tunnel. Each 802.11 frame is encapsulated entirely within the body of an IP packet. Many WLAN vendors use *Generic Routing Encapsulation (GRE)*, which is a commonly used network tunneling protocol. Although GRE is often used to encapsulate IP packets, GRE can also be used to encapsulate an 802.11 frame inside an IP tunnel. The GRE tunnel creates a virtual point-to-point link between the controller-based AP and the WLAN controller. WLAN vendors that do not use GRE use other proprietary protocols for the IP tunneling. The CAPWAP management protocol can also be used to tunnel user traffic.

As shown in Figure 8.5, the controller-based APs tunnel their 802.11 frames all the way back to the WLAN controller, from the access layer all the way back to the core layer. The distribution system service inside the controller directs the traffic, whereas the integration service translates an 802.11 data MSDU into an 802.3 frame. After 802.11 data frames have been translated into 802.3 frames, they are then sent to their final wired destination.

Most WLAN controllers are deployed at the core layer; however, they may also be deployed at either the distribution layer or even the access layer. Exactly where a WLAN controller is deployed depends on the WLAN vendor's solution, and the intended wireless integration into the preexisting wired topology. Multiple WLAN controllers that communicate with each other may be deployed at different network layers, providing they can communicate with each other.

There are two types of data forwarding methods when using WLAN controllers:

Centralized Data Forwarding Where all data is forwarded from the AP to the WLAN controller for processing, it may be used in many cases, especially when the WLAN controller manages encryption and decryption or applies security and QoS policies.

Distributed Data Forwarding Where the AP performs data forwarding locally, it may be used in situations where it is advantageous to perform forwarding at the edge and to avoid a central location in the network for all data, which may require significant processor and memory capacity at the controller.

As shown in Figure 8.5, centralized data forwarding relies on the WLAN controller to forward data. The AP and WLAN controller form an IP encapsulation tunnel, and all user data traffic is passed to the controller for forwarding (or comes from the controller). In essence, the AP plays a passive role in user data handling.

FIGURE 8.5 Centralized data forwarding

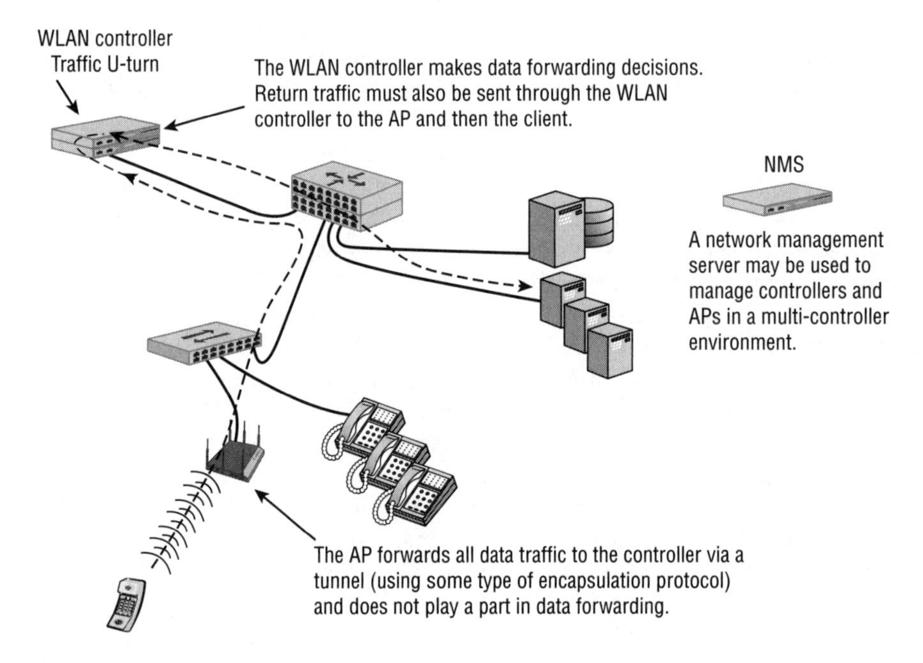

As shown in Figure 8.6, with distributed forwarding scenarios, the AP is solely responsible for determining how and where to forward user data traffic. The controller is not an active participant in these processes. This includes the application of QoS or security policies to data. Generally speaking, the device that handles the majority of MAC functions is also likely to handle data forwarding. The decision to use distributed or centralized forwarding is based on a number of factors, such as security, VLANs, and throughput. One major disadvantage of distributed data forwarding is that some control plane mechanisms may be unavailable because they exist only in the WLAN controller. Control plane mechanisms that may be lost include dynamic RF, Layer 3 roaming, firewall policy enforcement, and fast secure roaming. However, as the controller architecture has matured, some WLAN vendors have also pushed some of the control plane mechanisms back to into the APs at the edge of the network.

FIGURE 8.6 Distributed data forwarding

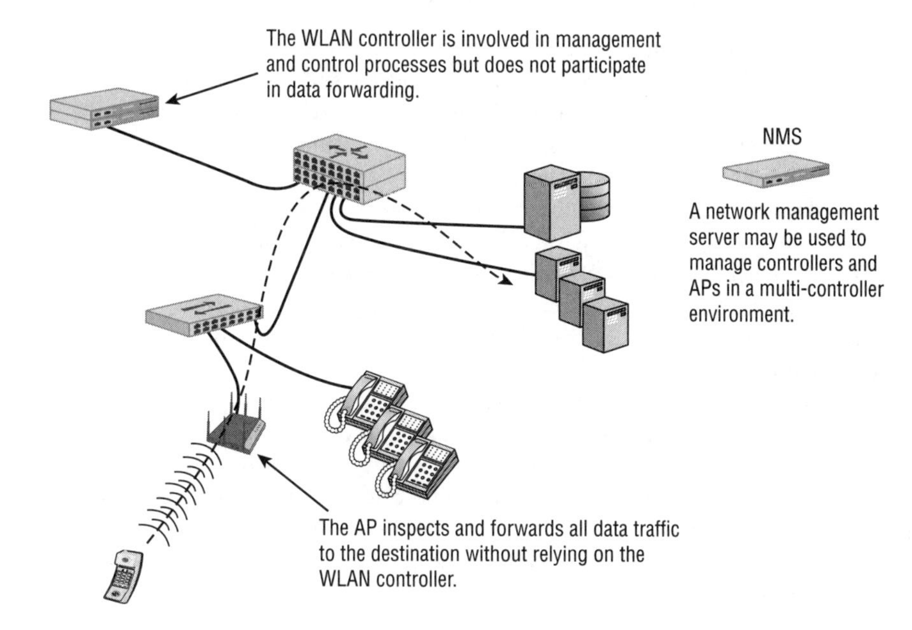

As 802.11ac technology and bandwidth become increasingly prevalent in large, enterprise networks, *centralized data forwarding* may become more difficult and expensive due to the traffic loads that can now be generated on the WLAN. Larger controllers with 10 Gbps links will become more commonplace. Additionally, WLAN controller manufacturers are now beginning to embrace *distributed data forwarding* in different ways.

Remote Office WLAN Controller

Although WLAN controllers typically reside on the core of the network, they can also be deployed at the access layer, usually in the form of a remote office WLAN controller. A remote office WLAN controller typically has much less processing power than a core WLAN controller and is also less expensive. The purpose of a remote office WLAN controller is to allow remote and branch offices to be managed from a single location. Remote WLAN controllers typically communicate with a central WLAN controller across a WAN link. Secure VPN tunneling capabilities are usually available between controllers across the WAN connection. Through the VPN tunnel, the central controller will download the network configuration settings to the remote WLAN controller, which will then control and manage the local APs. These remote controllers will allow for only a limited number of controller-based APs. Features typically include Power over Ethernet, internal firewalling, and an integrated router using NAT and DHCP for segmentation.

Distributed WLAN Architecture

A recent trend has been to move away from the centralized WLAN controller architecture toward a distributed architecture. Some WLAN vendors, such as Aerohive Networks, have designed their entire WLAN system around a distributed architecture. Some of the WLAN controller vendors now also offer a distributed WLAN architecture solution, in addition to their controller-based solution. In these systems, cooperative access points are used, and control plane mechanisms are enabled in the system with inter-AP communication via cooperative protocols. A distributed WLAN architecture combines multiple access points with a suite of cooperative protocols, without requiring a WLAN controller. Distributed WLAN architectures are modeled after traditional routing and switching design models, in that the network nodes provide independent distributed intelligence but work together as a system to cooperatively provide control mechanisms.

As shown in Figure 8.7, the protocols enable multiple APs to be organized into groups that share control plane information between the APs to provide functions such as Layer 2 roaming, Layer 3 roaming, firewall policy enforcement, cooperative RF management, security, and mesh networking. The best way to describe a distributed architecture is to think of it as a group of access points with most of the WLAN controller intelligence and capabilities mentioned earlier in this chapter. The control plane information is shared between the APs using proprietary protocols.

FIGURE 8.7 Distributed WLAN architecture

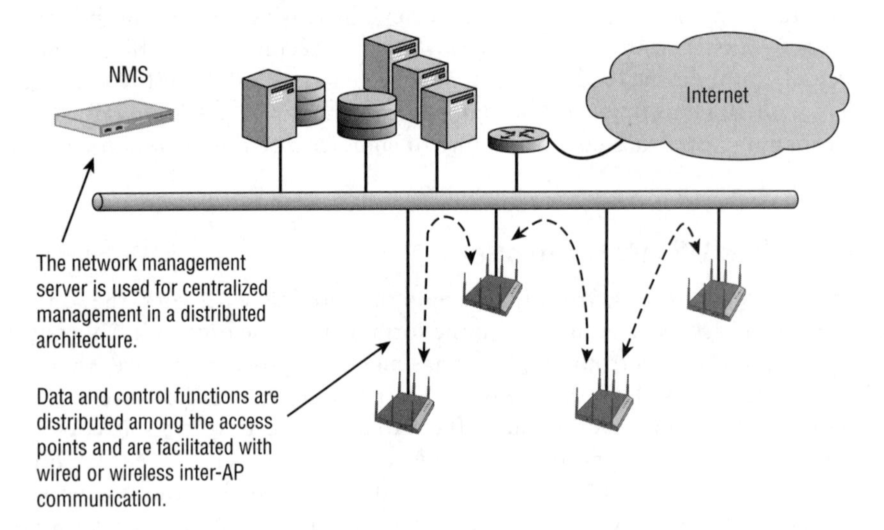

In a distributed architecture, each individual access point is responsible for local forwarding of user traffic. As mentioned earlier, since the advent of 802.11n, WLAN controller vendors have begun to offer distributed data forwarding solutions to handle traffic load. Because a distributed WLAN architecture entirely eliminates a centralized WLAN controller, all user traffic is forwarded locally by each independent AP. In a distributed architecture, the data plane resides in the access points at the edge of the network. No WLAN controller exists; therefore, the data does not need to be tunneled to the core of the network.

Although the control plane and data planes have moved back to the APs in a distributed WLAN architecture, the management plane remains centralized. Configuration and monitoring of all access points in the distributed model is still handled by an NMS server. The NMS server might be an on premise server or might be offered as a cloud-based service.

Most of the features mentioned in the earlier section about WLAN controllers can also be found in a distributed WLAN architecture even though there is no WLAN controller. For example, a captive web portal that normally resides in a WLAN controller instead resides inside the individual APs. The stateful firewall and RBAC capabilities found in a centralized WLAN controller now exist cooperatively in the APs. Back-end roaming mechanisms and dynamic RF are also cooperative. APs might also function as a RADIUS server with full LDAP integration capabilities. As mentioned earlier, all control plane mechanisms reside in the access points at the edge of the network in a distributed WLAN architecture. The APs implement control plane mechanisms cooperatively using proprietary protocols.

How VLANs are deployed in a WLAN environment depends on the design of the network, as well as the type of WLAN architecture that is in place. One very big difference between using a controller-based model versus a noncontroller model is how VLANs are implemented in the network design. In the WLAN controller model, most user traffic is centrally forwarded to the controller from the APs. Because all the user traffic is encapsulated, a controller-based AP typically is connected to an access port on an Ethernet switch that is tied to a single VLAN.

With a WLAN controller architecture, the user VLANs usually reside in the core of the network. The user VLANs are not available at the access layer switch. The controller-based APs are connected to an access port of the edge switch. The user VLANs are still available to the wireless users because all of the user VLANs are encapsulated in an IP tunnel between the controller-based APs at the edge and the WLAN controller in the core.

The noncontroller model, however, requires support for multiple user VLANs at the edge. Each access point is therefore connected to an 802.1Q trunk port on an edge switch that supports VLAN tagging. All of the user VLANs are configured in the access layer switch. The access points are connected to an 802.1Q trunk port of the edge switch. The user VLANS are tagged in the 802.1Q trunk and all wireless user traffic is forwarded at the edge of the network.

Although the whole point of a cooperative and distributed WLAN model is to avoid centrally forwarding user traffic to the core, the access points may also have IP-tunneling capabilities. Some WLAN customers require that guest VLAN traffic not cross internal networks. In that scenario, a standalone AP might forward only the guest user VLAN traffic in an IP tunnel that terminates at another standalone access point that is deployed in a DMZ. Individual APs can also function as a VPN client or VPN server using IPsec encrypted tunnels across a WAN link.

Another advantage of the distributed WLAN architecture is scalability. As a company grows at one location or multiple locations, more APs will obviously have to be deployed. When a WLAN controller solution is in place, more controllers might also have to be purchased and deployed as the AP count grows. With the controller-less distributed WLAN architecture, only new APs are deployed as the company grows. Many vertical markets such as K–12 education and retail have schools or stores at numerous locations. A distributed WLAN architecture can be the better choice as opposed to deploying a WLAN controller at each location.

Unified WLAN Architecture

WLAN architecture could very well take another direction by fully integrating WLAN controller capabilities into wired network infrastructure devices. WLAN control plane features are being integrated into wired switches, routers, firewalls, and security appliances at both the core and the edge of the network. Solutions that combine management of the wireless and wired networks are gaining in popularity. Customers are often looking for one-box solutions for distributed branch environments. This unified architecture has already begun to be deployed by some vendors and will likely grow in acceptance as WLAN deployments become more commonplace and the need for fuller seamless integration continues to rise.

Hybrid Architectures

It is important to understand that none of the WLAN architectures described in this chapter are written in stone. Many hybrids of these WLAN architectures exist among the WLAN vendors. As was already mentioned, some of the WLAN controller vendors are pushing some of the control plane intelligence back into the access points. One WLAN controller vendor has a cloud-based controller where much of the control plane intelligence exists in the cloud.

Typically, the data plane is centralized when using WLAN controllers, but distributed data forwarding is also available. With a controller-less distributed WLAN architecture, all data is forwarded locally, but the ability to centralize the data plane is a capability of a distributed WLAN architecture.

In a distributed WLAN architecture, the management plane resides in an on-premise or cloud-based network management server. With the WLAN controller model, the management plane normally exists in the WLAN controller. However, the management plane might also be pushed into an NMS that not only manages the controller-based APs but also manages the WLAN controllers.

Enterprise WLAN Routers

In addition to the main corporate office, companies often have branch offices in remote locations. A company might have branch offices across a region or an entire country, or they may even be spread globally. The challenge for IT personnel is how to provide a seamless enterprise wired and wireless solution across all locations. A distributed solution using enterprise-grade WLAN routers at each branch office is a common choice.

Keep in mind that WLAN routers are very different from access points. Unlike access points, which use a bridged virtual interface, wireless routers have separate routed interfaces. The radio card exists on one subnet whereas the WAN Ethernet port exists on a different subnet.

Branch WLAN routers have the ability to connect back to corporate headquarters with VPN tunnels. Employees at the branch offices can access corporate resources across the WAN through the VPN tunnel. Even more important is the fact that the corporate VLANs, SSIDs, and WLAN security can all be extended to the remote branch offices. An employee at a branch office connects to the same SSID that they would connect to at corporate headquarters. The wired and wireless network access policies are therefore seamless across the entire organization. These seamless policies can be extended to the WLAN routers at each branch location.

The enterprise-grade WLAN routers are very similar to the consumer-grade Wi-Fi routers that most of us use at home. However, enterprise WLAN routers are manufactured with better-quality hardware and offer a wider array of features.

The following security features are often supported by enterprise WLAN routers:

- 802.11 Layer 2 security for wireless clients
- 802.1X/EAP port security for wired clients
- Network address translation (NAT)
- Port address translation (PAT)
- Port forwarding
- Firewall
- Integrated VPN client
- 3G/4G cellular backhaul

WLAN Mesh Access Points

Almost all WLAN vendors now offer *WLAN mesh access point* capabilities. Wireless mesh APs communicate with each other by using proprietary Layer 2 routing protocols and create a self-forming and self-healing wireless infrastructure (a mesh) over which edge devices can communicate, as shown in Figure 8.8. The main purpose of a mesh WLAN is to provide wireless client access in physical areas where an Ethernet cable cannot be connected to an AP. WLAN client traffic can be sent over wireless backhaul links with an eventual destination to mesh portals that are connected to the wired network.

FIGURE 8.8 WLAN mesh network

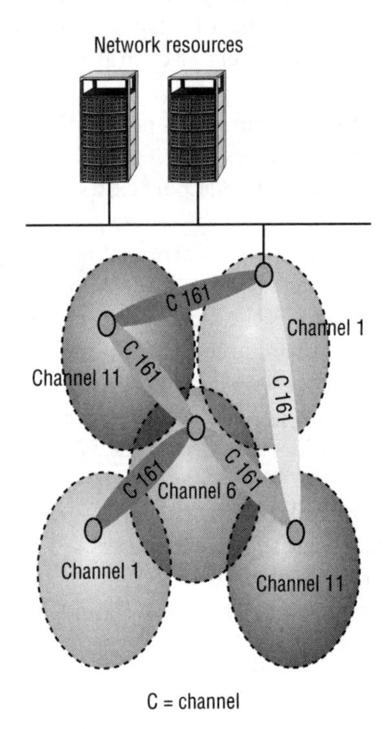

A WLAN mesh network automatically connects access points upon installation and dynamically updates traffic routes as more clients are added. Proprietary Layer 2 intelligent routing protocols determine the dynamic routes based on measurement of traffic, signal strength, data rates, hops, and other parameters.

With dual-band WLAN mesh APs, typically the 5 GHz radios are used for the mesh backhaul communications, as shown in Figure 8.8. The mesh backhaul traffic must also be encrypted. In most cases, 802.11 PSK security is used between the mesh radios to provide encryption. The PSK is usually created automatically in most mesh WLAN solutions. A very strong passphrase of 20 characters or more should be used if the WLAN vendor does offer the option to manually define mesh backhaul security. As you learned in Chapter 6, "PSK Authentication," the 802.11s-2011 amendment proposed a peer-to-peer authentication method called *Simultaneous Authentication of Equals (SAE)*. Although SAE has yet to be implemented for 802.11 mesh networks, the Wi-Fi Alliance views SAE as an eventual replacement for PSK authentication.

WLAN Bridging

When facilities are separated from each other and no physical network-capable wiring exists between them, wireless bridges are often employed. Monthly-based fees for Telco circuit costs can be mitigated with the one-time cost of a wireless point-to-point (PtP) bridge. Wireless bridges are also used between communication towers and can typically span many miles.

Bridge and backhaul links tend to have very different requirements than do typical AP functions that serve WLAN clients. The first difference is that APs typically serve in the access layer of a network. WLAN bridges operate in the distribution layer and are normally used to connect two or more wired networks together over a wireless link.

One side of the link is usually the root bridge and the other side is the nonroot bridge. The root bridge establishes the channel and beacons for the nonroot bridge to join. The nonroot bridge will then associate with the root bridge in a station-like fashion to establish the link. A point-to-multipoint (PtMP) bridge link connects multiple wired networks. The root bridge is the central bridge, and multiple nonroot bridges connect back to the root bridge.

Encryption is needed to protect the data privacy of the backhaul communications across bridge links. IPSec VPNs are often used for bridge security, which will be discussed later in this chapter. An 802.1X/EAP solution can also be used for bridge security, with the root bridge assuming the authenticator role and the nonroot bridges assuming the supplicant role. Additionally, PSK authentication is often used for WLAN bridge security and therefore a strong WPA2 passphrase of 20 characters or more is recommended.

VPN Wireless Security

Although the 802.11-2012 standard clearly defines Layer 2 security solutions, the use of upper-layer *virtual private network (VPN)* solutions can also be deployed with WLANs. VPNs are typically not recommended to provide wireless security in the enterprise due to the overhead and because faster, more secure Layer 2 solutions are now available. Although not usually a recommended practice, VPNs were often used for WLAN security because the VPN solution was already in place inside the wired infrastructure. VPNs do have their place in Wi-Fi security and should definitely be used for remote access. They are also sometimes used in wireless bridging environments. The two major types of VPN topologies are router-to-router or client-server based.

Use of VPN technology is mandatory for remote access. Your end users will take their laptops off site and will most likely use public access Wi-Fi hotspots. Because there is no security at most hotspots, a VPN solution is needed. The VPN user will need to bring the security to the hotspot in order to provide a secure, encrypted connection. It is imperative that users implement a VPN solution coupled with a personal firewall whenever accessing any public access Wi-Fi networks.

VPN 101

Before discussing the ways that VPNs are used in a WLAN, it is important to make sure that we review what a VPN is, what it does, how it works, and the components that are configured to construct one. By now you know that a VPN is a virtual private network. But what does that really mean? As shown in Figure 8.9, a VPN is essentially a private network that is created or extended across a public network. In order for a VPN to work, two computers or devices communicate to establish what is known as the VPN tunnel. Typically, a VPN client initiates the connection by trying to communicate with the VPN server.

FIGURE 8.9 VPN components

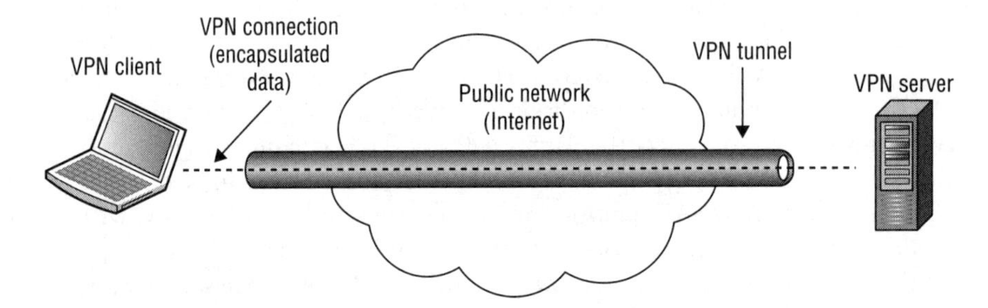

The VPN client can be a computer, router, WLAN controller, or even an AP, which you will learn about later in this chapter. When the client and the server are able to communicate with each other, the client will attempt to authenticate with the server by sending its credentials. The server will take the client's credentials and validate them. If the client's credentials are valid, the server and client will create the VPN tunnel between them. Any data that is sent from the VPN client to the VPN server is encapsulated in the VPN tunnel. The client and the server also agree on if and how the data will be encrypted. Prior to the data being encapsulated in the tunnel, the data is encrypted to make sure that, as it travels through the tunnel, it cannot be compromised. Since the underlying premise of a VPN is that the data is traveling across an insecure public network, security is one of the primary reasons for implementing a VPN.

When the client and server build the tunnel, it is their responsibility to route the data across the public network between the two devices. They take the data from the local network, encrypt and encapsulate it, and then send it to the other device, where it is decapsulated and decrypted and then placed on the local network of the other device.

Real World Scenario

VPN Analogy

To understand the concepts of VPNs and encapsulation, imagine a company that has two buildings in a city where the buildings are separated by a distance of two kilometers. Since these buildings are very large, and many people travel between the buildings, the company decides to purchase buses to transport people back and forth between the buildings. The driver of the bus is responsible for driving from the pickup point of the first building to the drop-off point of the second building. The route the driver takes may vary depending on traffic. The employees in the first building know that if they go to the pickup point and board the bus, when they exit the bus they will be at the drop-off location of the second building; just as when the data arrives at the VPN client it will be tunneled and arrive at the VPN server.

The employees are given directions instructing them to get on the bus at the pickup point and to get off the bus at the drop-off point. The employees do not have to understand the route that they are taking through the city. The employees are essentially encapsulated and secured in the climate-controlled bus as they travel between the offices.

In a VPN network, the client and server provide the transportation just as the bus does. The city streets make up the many routes that the bus can take, just as the Internet has many routes that the packets can take. The passengers travel inside the bus just as the data is transmitted inside the packets. The packet is directed through the network just as the bus is directed through the city.

Layer 3 VPNs

VPNs have several major characteristics. They provide encryption, encapsulation, authentication, and data integrity. VPNs use secure tunneling, which is the process of encapsulating one IP packet within another IP packet. The first packet is encapsulated inside the second or outer packet. The original destination and source IP address of the first packet is encrypted along with the data payload of the first packet. VPN tunneling, therefore, protects your original private Layer 3 addresses and also protects the data payload of the original packet. Layer 3 VPNs use Layer 3 encryption; therefore, the payload that is being encrypted is the Layer 4–7 information. The IP addresses of the second or outer packet are seen in clear text and are used for communications between the tunnel endpoints. The destination and source IP addresses of the second or outer packet will point to the public IP address of the VPN server and VPN client software.

The most commonly used Layer 3 VPN technology is *Internet Protocol Security (IPsec)*. IPsec VPNs use stronger encryption methods and more secure methods of authentication and are the most commonly deployed VPN solution. IPsec supports multiple ciphers, including DES, 3DES, and AES. Device authentication is achieved by using either a server-side certificate or a preshared key. IPsec VPNs require client software to be installed on the remote devices that connect to a VPN server. Most IPsec VPNs are NAT-transversal, but any firewalls at a remote site require (at a minimum) that UDP ports 4500 and 500 be open. A full explanation of IPsec technology is beyond the scope of this book, but IPsec is usually the choice for VPN technology in the enterprise.

SSL VPN

VPN technologies do exist that operate at other layers of the OSI model, including SSL tunneling. Unlike an IPsec VPN, an SSL VPN does not require the installation and configuration of client software on the end user's computer. A user connects to a *Secure Sockets Layer (SSL)* VPN server via a web browser. The traffic between the web browser and the SSL VPN server is encrypted with the SSL protocol or Transport Layer Security (TLS). TLS and SSL encrypt data connections above the Transport layer, using asymmetric cryptography for privacy and a keyed message authentication code for message reliability.

Although most IPsec VPN solutions are NAT-transversal, SSL VPNs are often chosen because of issues with NAT or restrictive firewall policies at remote locations.

VPN Deployment

VPNs are most often used for client-based security when connected to public access WLANs and hotspots that do not provide security. Because most hotspots do not provide Layer 2 security, it is imperative that end users provide their own security. VPN technology can provide the necessary level of security for remote access when end users connect to public access WLANs. Since no encryption is used at public access WLANs, a VPN solution is usually needed to provide for data privacy, as shown in Figure 8.10.

FIGURE 8.10 VPN established from a public hotspot

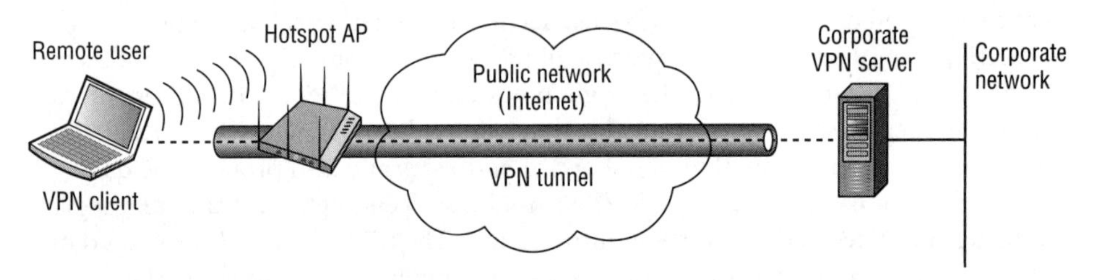

Another common use of VPN technology is to provide site-to-site connectively between a remote office and a corporate office across a WAN link. Most WLAN vendors now offer VPN client-server capabilities in either their APs or WLAN controllers. As shown in Figure 8.11, a branch office WLAN controller with VPN capabilities can tunnel WLAN client traffic and bridged wired-side traffic back to the corporate network. Other WLAN vendors can also tunnel user traffic from a remote AP or WLAN branch router to a VPN server gateway. Third-party VPN overlay solutions are often also used.

FIGURE 8.11 Site-to-site VPN

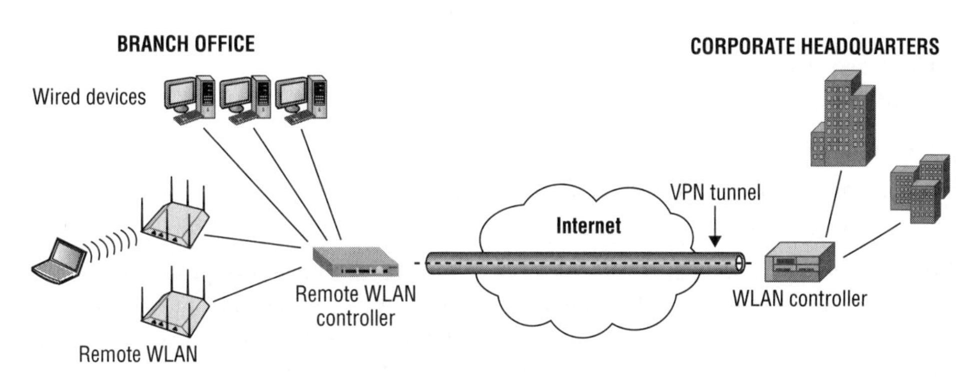

Another use of VPNs is to provide security for 802.11 WLAN bridges links. In addition to using 802.11 wireless technology to provide client access, 802.11 technology is used to create bridged networks between two or more locations. When WLAN bridges are deployed for wireless backhaul communications, VPN technology can be used to provide the necessary level of data privacy. Depending on the bridging equipment used, VPN capabilities may be integrated into the bridges, or you may need to use other devices or software to provide the VPN. Figure 8.12 shows an example of a point-to-point wireless bridge network using dedicated VPN devices. A site-to-site VPN tunnel is used to provide encryption of the 802.11 communications between the two WLAN bridges.

FIGURE 8.12 WLAN bridging and VPN security

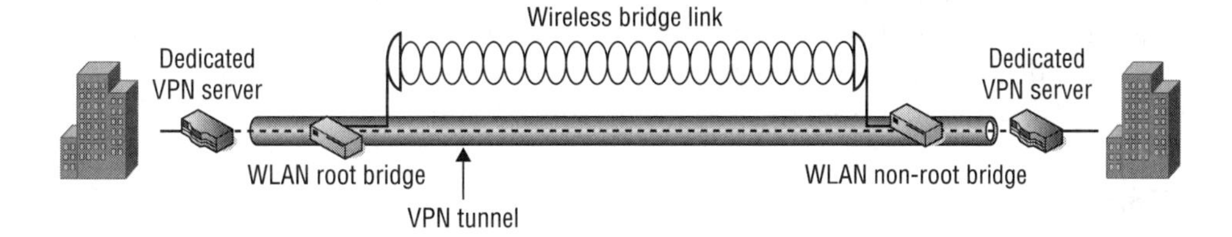

Infrastructure Management

Sound network design dictates placing network devices onto dedicated management VLANs or another out-of-band interface from regular network traffic. Enterprise network gear should allow for this functionality in order to keep infrastructure devices inaccessible to hackers or even to employees who are able to gain access to the network.

As the size of WLANs has grown over the years, so have the challenges of managing them. This includes managing the following:

- Firmware revisions
- Configurations and changes
- Monitoring and incident response
- Managing and filtering of device alerts and alarms
- Performance monitoring

To make the job of managing these devices easier, standard network protocols for device management are typically included in most WLAN hardware and can be integrated with software-based management systems that can span even the largest of networks. The more a network management system (NMS) is incorporated in network designs, the less time is spent performing mundane tasks to support the network. Without exception, every time an NMS is properly implemented, user satisfaction with the network is higher, while the cost of operating the network is lower. Equally important is incorporating the deployment of the NMS into an operational environment with the support staff. Developing processes and procedures around these systems and making them a part of the support staff's daily work are also critical.

Protocols for Management

There are many different types of protocols used for managing network devices. Simple Network Management Protocol (SNMP) has been around for quite some time and has undergone several revisions. In addition to SNMP, most devices can be configured using a command-line interface (CLI) or a graphical user interface (GUI).

The following protocols are common for managing WLANs. Some of these protocols are based on de jure standards and some are based on de facto standards. Either way, they provide the basis for WLAN management and administration.

SNMP

Simple Network Management Protocol (SNMP) is an Application layer protocol (OSI Layer 7) used to communicate directly with network devices. SNMP allows for pulling information from devices as well as pushing information to a central SNMP server based on certain, often user-configurable thresholds on network devices. A push from a device might include a message pertaining to an interface reset, a high number of errors, high network or CPU utilization, security alarms, and many other critical factors related to the healthy operations and status of devices.

SNMP is an IETF specification. You can find more information at `www.ietf.org/wg/concluded/snmpv3.html`. Additional information can also be found in RFC 3411 through 3418 for SNMP version 3.

Components

An SNMP management system contains

- Several (potentially many) nodes, each with an SNMP entity (a.k.a. agent) containing *command responder* and *notification originator* applications, which have access to management instrumentation
- At least one SNMP entity containing command generator and/or notification receiver applications (traditionally called a manager)
- A management protocol used to convey management information between the SNMP entities

Structure of Management Information

Management information is structured as a collection of managed objects contained in a database called a *management information base (MIB).*

The MIB consists of the following definitions: modules, objects, and traps. Module definitions are used when describing information modules. Object definitions are used when describing the managed objects. Trap definitions are notifications used for unsolicited transmissions of MIB information typically to an NMS.

All SNMP-capable devices have a MIB, and in that MIB should reside the configuration and status of the device. However, vendors aren't usually totally complete with their MIBs and SNMP implementations. Often you will find that certain pieces of critical information are not accessible via SNMP and therefore traps cannot be implemented using that information.

Versions and Differences

SNMP has undergone numerous revisions over the years. This section is not intended to be a complete history of SNMP, but rather an overview to guide you in knowing the differences between the various versions. Additionally, this section will help you properly implement different versions into your network designs by understanding the strengths and how to address the weaknesses.

SNMPV1

Version 1 of SNMP hit the scene in 1988. Like many other initial protocol introductions, SNMPv1 did not get it perfect the first time. SNMPv1 was designed to work over a wide range of protocols in use at the time—including IP, UDP, CLNS, AppleTalk, and IPX—but it is most commonly used with UDP.

SNMPv1 used a *community string,* which had to be known by a remote agent. Since SNMPv1 did not implement any encryption, it was subject to packet sniffing to discover the clear-text community string. Therefore, SNMPv1 was heavily criticized as being insecure. Protocol efficiency was also lacking for this initial protocol introduction. Each MIB object had to be retrieved one by one in an iterative style, which was very inefficient.

SNMPV2

When SNMPv2 was released, several areas of SNMPv1 were addressed, including performance, security, and manager-to-manager communications. Protocol performance became more efficient with the introduction of new functions such as GETBULK, which solved the iterative method of extracting larger amounts of data from MIBs.

Security was improved by the specification of a new party-based security system. Critics accused the party-based system of being too complex, and the system was not widely accepted.

SNMPv2c was later defined in RFCs 1901 through 1908 and is referred to as the *community-based* version. The community string from SNMPv1 was adopted in SNMPv2c, which essentially dropped any security improvements to the protocol. SNMPv2c does not implement encryption and is subject to packet sniffing of the clear-text community string.

SNMPV3

A great deal of security benefits were added to SNMPv3, including

- Authentication performed using SHA or MD5.
- Privacy—SNMPv3 uses DES 56-bit encryption based on the CBC-DES (DES-56) standard.
- Access control—Users and groups are used, each with different levels of privileges. Usernames and passwords replace community strings.

Although these features are optional, usually the main driver behind adopting SNMPv3 is to gain these security benefits. It is also optional to have secure authentication but disable encryption.

Even with these features, most network designers still feel it is best to implement the SNMP agents only on secure management interfaces. Specifically, VLAN segmentation and firewall filtering is usually performed on all SNMP traffic to network devices. No sufficiently complex protocol is considered completely secure, and additional safeguards are always highly recommended. If you intend to implement an NMS using SNMP, we highly recommended that you implement SNMPv3. One of the most important security concerns is most vendor equipment defaults to SNMP being enabled, with default *read* and *write* community strings. This is an *enormous* security threat to the configuration and operation of your network, and should be one of the very first lockdown steps to securing network devices.

CLI-Based Management

Command-line interfaces (CLIs) are one of the most common methods used to configure and manage network devices. It seems the age-old debate of GUI versus CLI is still present to this day and is not likely to change any time soon. GUIs do a wonderful job of presenting information, but due to browser incompatibility, JavaScript errors, GUI software bugs, time delays, and more, GUIs still drive many people back to the command line.

CLIs tend to be the raw, unedited configuration of devices and provide the ability to make specific changes quickly to device configurations. Commands issued via a CLI can even be scripted, allowing initial device configuration and even reconfigurations to be performed with a simple copy and paste into a CLI session.

CLIs can be accessed using several methods, which are dependent on the device being used. These commonly include the following:

- Serial and console ports
- Telnet
- SSH1/SSH2

Serial and Console Ports

Serial or console port interfaces can vary from manufacturer to manufacturer and even from model to model. This is extremely frustrating to network engineers. Some of them use a standard DB-9 serial connector interface, whereas others use an RJ-11 or RJ-45 interface. Furthermore, the actual cable might be a proprietary pin-out (namely for the RJ-11 and RJ-45 connectors), a NULL-modem cable, a rollover cable, or a straight-through cable. Baud rates, number of bits, flow control, parity, and other parameters also vary from device to device.

No matter what type of connector or cable you are using to manage your network device, serial or console ports should be locked down and require a user authentication mechanism. Although typically these can be thwarted by a password-recovery routine using instructions that can be readily found on the Internet, a user authentication mechanism will help deter hackers. Often, a device requires downtime in order to recover the password, and the impact of a service outage may be enough to alert staff of a physical break-in attempt to network devices. It is important to note that most password recovery routines require direct access to the device via the serial or console port. Securing network equipment in a locked data closet or computer room will help to prevent this type of attack from occurring.

Government regulations such as FIPS 140-2 may require that serial and console ports be secured with a tamper-evident label (TEL) to prevent unauthorized physical access to a WLAN infrastructure device such as a WLAN controller. TELs cannot be surreptitiously broken, removed, or reapplied without an obvious change in appearance. As shown in Figure 8.13, each TEL has a unique serial number to prevent replacement with similar labels.

FIGURE 8.13 Tamper-evident label

Telnet

Telnet is another protocol that is commonly used, but often it can only be used after a serial port configuration or initial configuration from a factory default state is performed. The IP of the device typically needs to be enabled for it to be accessed and managed via the network interface.

Telnet is heavily criticized and usually prohibited from use by enterprise security policies due to its lack of encryption. Telnet is a completely unencrypted protocol and the payload of each packet can be inspected by packet sniffing. This includes the username and password during the login sequence. We recommend that you disable Telnet after the initial device configuration. Most companies have written policies mandating that Telnet be disabled.

Secure Shell

Secure Shell (SSH) is typically used as the secure alternative to Telnet. SSH implements authentication and encryption using public-key cryptography of all network traffic traversing between the host and user device. The features of Telnet for CLI-based management apply to SSH but include added security benefits. The standard TCP port 22 has been assigned for the SSH protocol. Most WLAN infrastructure devices now support the second version of the SSH protocol, called SSH2. As a matter of policy, when WLAN devices are managed via the CLI, an SSH2-capable terminal emulation program should be used. Figure 8.14 shows the configuration screen of the popular freeware program PuTTY, which supports SSH2.

FIGURE 8.14 PuTTY freeware SSH2 client

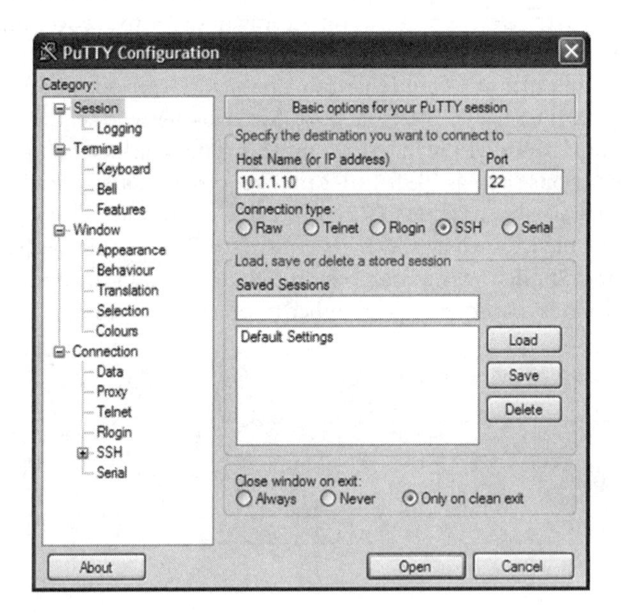

HTTPS

Hypertext Transfer Protocol Secure (HTTPS) is a combination of the Hypertext Transfer Protocol with the SSL/TLS protocol to provide encryption and secure identification. HTTPS is essentially an SSL session that uses HTTP and is implemented on network devices for management via a graphical user interface (GUI). Not all users prefer CLI-based management methods, and GUIs are commonly used where an NMS is used to manage WLAN infrastructure.

Because HTTP is transmitted in plaintext, it is susceptible to eavesdropping and man-in-the-middle attacks from modifications in transit. Some devices offer both HTTP and HTTPS, but it is important that minimal authentication be performed via HTTPS. If users of devices will be entered into the GUI, not using HTTPS is purely negligent if the device supports it.

Summary

WLAN security plays an integral part of client device performance. With wired LANs, firewalls are placed between networks where network activity needs to be policed. The security a wired-side firewall provides does not affect the installation or configuration of a wired client device.

IEEE 802.11 WLANs operate very differently. Security choices greatly affect the design, architecture, and type of clients that can run over the WLAN. In fact, the opposite is true as well. The devices themselves and their criticality to the business might dictate changes or exceptions to security policy. Because a WLAN is a network-based technology, we typically relate WLAN security and methodologies to the model of how wired LANs work. As you have gathered from your reading of this book, that couldn't be further from the truth. WLAN performance is tightly interwoven with WLAN security, RF propagation and design, as well as infrastructure features and capabilities. Hardly any technology architecture we commonly use today has this level of integration and co-dependency.

By reading this chapter, you have learned about the important components involved in enterprise security solutions from the infrastructure perspective and also understand the basic security design issues involved.

Exam Essentials

Define the three logical network planes of operation. Understand the differences between the management, control, and data plane. Be able to explain where they are used within different WLAN architectures.

Understand types of WLAN architectures. Know the different types of architectures, including autonomous, centralized, distributed, unified, and hybrid.

Understand common infrastructure device features and features that pertain to security. Know design details and security components of infrastructure and device features.

Be familiar with device management features. Know the various device management methods, features, and protocols available in WLAN devices.

Understand VPNs and WLANs. Explain VPN basics and the various ways IPsec VPNs are used for WLAN security.

Review Questions

1. Which terms best describe components of a centralized WLAN architecture where the management, control, and data planes all reside in a centralized device? (Choose all that apply.)
   - A. WLAN controller
   - B. Wireless network management system
   - C. Network management system
   - D. Distributed AP
   - E. Controller-based AP
2. Which logical plane of network operation is typically defined by protocols and intelligence?
   - A. User plane
   - B. Data plane
   - C. Network plane
   - D. Control plane
   - E. Management plane
3. Which WLAN architectural models typically require support for 802.1Q tagging at the edge on the network when multiple user VLANs are required? (Choose all that apply.)
   - A. Autonomous WLAN architecture
   - B. Centralized WLAN architecture
   - C. Distributed WLAN architecture
   - D. None of the above
4. What type of WLAN security is normally used to encrypt and provide data privacy for 802.11 traffic that traverses across mesh backhaul links?
   - A. 802.1X/EAP
   - B. SAE
   - C. PSK
   - D. IPsec
   - E. VRRP
5. Which protocols can be used to tunnel 802.11 user traffic from access points to WLAN controllers or other centralized network servers? (Choose all that apply.)
   - A. IPsec
   - B. GRE
   - C. CAPWAP
   - D. DTLS
   - E. VRRP

6. Which of these WLAN architectures may require the use of an NMS server to manage and monitor the WLAN?

   A. Autonomous WLAN architecture

   B. Centralized WLAN architecture

   C. Distributed WLAN architecture

   D. All of the above

7. How are IPsec VPNs used to provide security in combination with 802.11 WLANs?

   A. Client-based security on public access WLANs

   B. Point-to-point wireless bridge links

   C. Connectivity across WAN links

   D. All of the above

8. What are of some of the common security capabilities often found in a WLAN controller?

   A. VPN server

   B. Firewall

   C. RADIUS server

   D. WIPS

   E. All of the above

9. What is the traditional data forwarding model for 802.11 user traffic when WLAN controllers are deployed?

   A. Distributed data forwarding

   B. Autonomous forwarding

   C. Proxy data forwarding

   D. Centralized data forwarding

   E. All of the above

10. What are some of the security capabilities found in an enterprise WLAN router that is typically deployed in remote branch locations? (Choose all that apply.)

    A. Integrated WIPS server

    B. Integrated VPN server

    C. Integrated NAC server

    D. Integrated firewall

    E. Integrated VPN client

11. What are the best and most often used security solutions used to provide data privacy across 802.11 WLAN bridge links? (Choose all that apply.)

A. 802.1X/EAP

B. Captive web portal

C. Firewall

D. IPsec VPN

E. PSK

F. Mobile device management

12. Which of these protocols can be used to manage WLAN infrastructure devices?

A. HTTP

B. SSH

C. SNMP

D. Telnet

E. HTTPS

F. SNMP

G. All of the above

13. What components make up a distribution system? (Choose all that apply.)

A. HR-DSSS

B. DSS

C. DSM

D. DSSS

E. WIDS

14. Which management protocols are often used between a network management server (NMS) and remote access points for the purpose of monitoring a WLAN? (Choose all that apply.)

A. IPsec

B. GRE

C. CAPWAP

D. DTLS

E. SNMP

15. What are of some of the common security capabilities often integrated within in access points deployed in a distributed WLAN architecture?

A. Captive web portal

B. Firewall

C. Integrated RADIUS

D. WIPS

E. All of the above

16. What are some of the major differences between SNMPv3 and SNMPv2? (Choose all that apply.)
    - A. SNMPv3 requires username/passwords.
    - B. SNMPv3 requires community strings.
    - C. SNMPv3 uses 56-bit DES encryption to encrypt packets.
    - D. SNMPv3 uses 128-bit AES encryption to encrypt packets.
17. What is the biggest security risk associated with Simple Network Management Protocol (SNMP)?
    - A. DES encryption can be cracked.
    - B. Weak data integrity.
    - C. Lack of vendor support.
    - D. Default community strings.
18. What layers of the OSI model are protected by encryption with an IPsec VPN security solution?
    - A. Layers 2–7
    - B. Layers 3–7
    - C. Layers 4–7
    - D. Layers 5–7
    - E. Layers 6–7
19. What are the available form factors for network management server (NMS) solutions? (Choose all that apply.)
    - A. Hardware appliance
    - B. Virtual appliance
    - C. Software subscription service
    - D. Integrated access point
20. What planes of operation reside in the access points of a distributed WLAN architecture? (Choose all that apply.)
    - A. Radio plane
    - B. Data plane
    - C. Network plane
    - D. Control plane
    - E. Management plane

Chapter 9

RADIUS and LDAP

IN THIS CHAPTER, YOU WILL LEARN ABOUT THE FOLLOWING:

✓ **LDAP**

✓ **RADIUS**

- Authentication and authorization
- Accounting
- RADIUS configuration
- LDAP proxy
- RADIUS deployment models
- RADIUS proxy
- RADUS proxy and realms
- RADIUS failover
- WLAN devices as RADIUS servers
- Captive web portal and MAC authentication
- RadSec

✓ **Attribute value pairs**

- Vendor specific attributes
- VLAN assignment
- Role based access control
- LDAP attributes

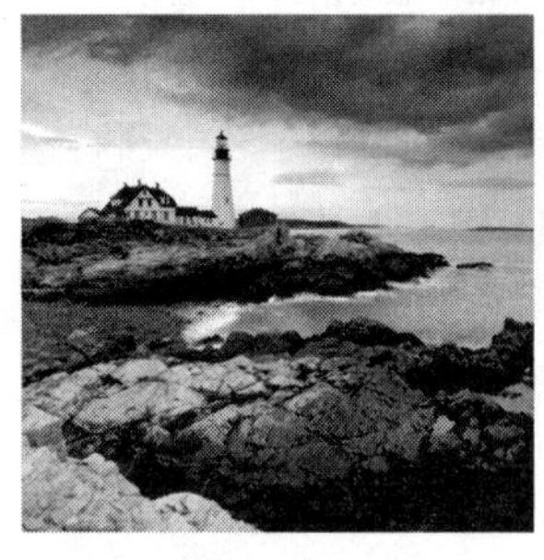

In earlier chapters, you learned about the 802.1X authorization framework and the Extensible Authentication Protocol (EAP) that is used for enterprise WLAN security. 802.1X/EAP uses an authentication server to validate the credentials of a WLAN client and then to authorize access for the WLAN to network resources. The authentication server is normally a RADIUS server that directly communicates with an existing Lightweight Directory Access Protocol (LDAP) database. This chapter will provide an in-depth review of RADIUS and LDAP deployment scenarios for WLANs. You will also learn about how RADIUS attribute value pairs can be leveraged to provide different access policies for different groups of WLAN users and devices.

LDAP

Lightweight Directory Access Protocol (LDAP) is an application protocol for providing directory services over an IP network. The current version, LDAPv3, is defined in IETF RFC 4511. A *directory service* is an infrastructure used to share information about network resources such as files, folders, computers, users, groups, and so on. A directory service could be considered a database or data store, although a directory service is different when compared to relational databases. Distributed directory information services use a hierarchical structure that can be accessed and managed using LDAP. In most IP networks, LDAP is used to provide access to a data store of usernames and passwords. Applications such as RADIUS can be used to query an LDAP server to validate user or device credentials. LDAP sessions normally use TCP or UDP port 369; however, LDAP over SSL uses port 636.

Within an 802.1X authorization framework, the *authentication server (AS)* validates the credentials of a supplicant that is requesting access and notifies the authenticator that the supplicant has been authorized. The authentication server maintains a user database or may proxy with an external user database to authenticate user or device credentials. In almost all cases, a RADIUS server functions as the authentication server. The RADIUS server may hold a master native user database but usually will instead query to a preexisting external database.

Any LDAP-compliant database can be queried by the RADIUS authentication server. Active Directory is the most commonly used external LDAP database, but a RADIUS server can also query LDAP-compliant databases such as eDirectory or OpenLDAP. As shown in Figure 9.1, typically a RADIUS server performs authentication server duties and the RADIUS server initiates a proxy query to an LDAP-compliant database, such as Active Directory. This is referred to as *proxy authentication.*

FIGURE 9.1 Proxy authentication

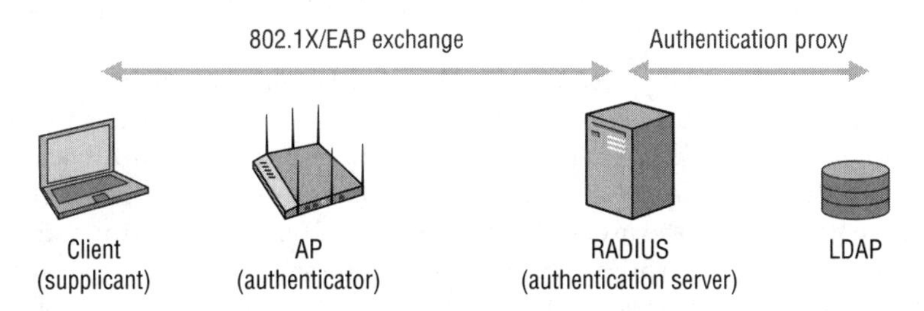

LDAP can be used with numerous forms of directory service implementations. Some of the most widely deployed are as follows:

Active Directory Microsoft introduced *Active Directory* with Windows Server 2000 and has continued to enhance it with subsequent releases of its server OS platforms. Active Directory is a hierarchical directory service that is based on LDAP. Active Directory also incorporates a DNS-naming component based on the Internet DNS structure and uses Kerberos. Windows Active Directory is the most popular implementation of directory services in the enterprise.

eDirectory NetQ's *eDirectory* was originally developed by Novell in 1993 and was known as Novell Directory Services (NDS). eDirectory supports multiple architectures, including Novell NetWare, Microsoft Windows, Red Hat Linux, and multiple types of Unix. eDirectory is a hierarchical, object-oriented database used to represent assets in an organization in a logical tree.

OpenLDAP *OpenLDAP* is an open source implementation of LDAP maintained by the OpenLDAP Foundation at `www.OpenLDAP.org`. OpenLDAP supports most modern-day computer architectures, including Windows and various forms of Linux and Unix.

Other flavors of LDAPv3-compliant directory services include Apple's Open Directory and Apache Directory Server. Open Directory is the directory services framework used by Mac OS X and Mac OS X Server. Apache Directory Server is open source.

RADIUS

Remote Authentication Dial-in User Service (RADIUS) is a networking protocol that provides authentication, authorization, and accounting (AAA) capabilities for computers to connect to and use network services. RADIUS authentication and authorization is defined in IETF RFC 2865. Accounting is defined in IETF RFC 2866. RADIUS servers are sometimes referred to as AAA servers. RADIUS was developed back in 1991 as a client/server authentication and accounting protocol. It grew to be widely used by ISPs for dial-up users and later logically extended to VPN dial-up users. RADIUS developed critical mass and had the ability to extend, or rather *broker,* authentication to many different user databases. This includes LDAP, Active Directory, SQL databases, flat files, and native RADIUS users.

Prior to RADIUS servers being used for WLAN security, many organizations already were maintaining a RADIUS infrastructure for dial-up users with a modem bank. Using RADIUS servers for WLAN security later became a logical choice. The IEEE 802.11-2012 standard does not dictate the use of a RADIUS server. However, the IEEE 802.11-2012 WLAN standard does dictate the use of the IEEE 802.1X-2004 standard for authentication and port control within an enterprise *robust security network (RSN).* As you learned in Chapter 4, "802.1X/EAP Authentication," 802.1X is a port-based access control standard that defines the mechanisms necessary to authenticate and authorize devices to network resources. RADIUS servers are usually one of the main components of an 802.1X authorization framework.

Authentication and Authorization

RADIUS clients often get confused with Wi-Fi clients (supplicants). Instead, RADIUS clients are the devices that communicate directly with a RADIUS server using the RADIUS protocol. RADIUS clients are the authenticators within an 802.1X/EAP framework. To make things even more confusing, RADIUS clients are also sometimes referred to as the *network access server (NAS).* When discussing 802.1X/EAP, the terms *authenticator*, *NAS*, and *RADIUS client* are all synonymous.

As you learned in Chapter 4, when 802.1X/EAP security is used with WLANs, either an AP or a WLAN controller functions as the authenticator. As shown in Figure 9.2, all the RADIUS communications are between the RADIUS server and the AP, which is functioning as the RADIUS client. RADIUS packets are used to encapsulate EAP frames during the authentication exchange. Think of the RADIUS protocol as a transport mechanism for EAP authentication conversations between the supplicant and the authentication server. The RADIUS communications occur strictly between the RADIUS client (AP) and the RADIUS server.

FIGURE 9.2 RADIUS and 802.1X/EAP

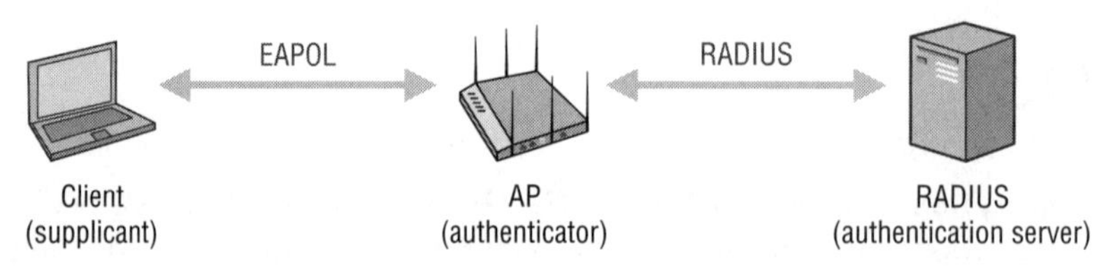

The WLAN supplicant sends an EAP request to the AP to gain access to network resources. The AP then forwards the EAP request encapsulated in a RADIUS Access-Request packet to the RADIUS server. As shown in Figure 9.3, the RADIUS server can respond in three different ways:

RADIUS Access-Challenge The RADIUS server requests additional information from the user or device such as a secondary password, PIN, token, or hash response. Many flavors of EAP protect this information in an SSL/TLS tunnel between the supplicant and the authentication server, but remember that RADIUS communications are between the AP and the RADIUS server.

RADIUS Access-Accept The user or device is granted access. The Access-Accept packet can also contain RADIUS attributes, which can define exactly what type of access is granted to the user or device. RADIUS attributes will be discussed in greater detail later in this chapter.

RADIUS Access-Reject The user or device is denied access to all requested network resources. Access could be denied because of incorrect user credentials such as the password. The user account may also have expired or not exist.

FIGURE 9.3 RADIUS protocol communications

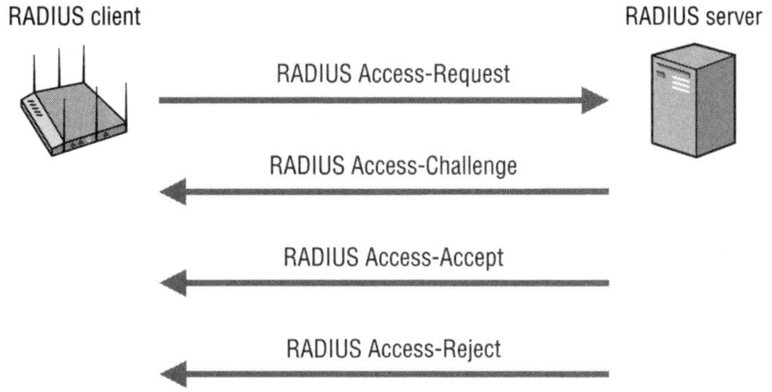

Accounting

RFC 2866 defines how the RADIUS protocol can be used to deliver accounting information from the RADIUS client to a RADIUS accounting server. Once again, an access point usually functions as the RADIUS client (also called NAS) when 802.1X/EAP is deployed for WLAN security. Accounting is used to track network access of users and devices. For example, an accounting trail can record information such as who authenticated, when they authenticated, session time, session activities, and much more.

As shown in Figure 9.4, after network access is granted to a user or device, a RADIUS Accounting-Request packet is sent by the RADIUS client to the RADIUS server to signal the start of the user network access. The packet contains an `Acct-Status-Type` attribute with the value of `start`. Some of the accounting information in the start packet includes the username, IP address, NAS information, and a unique session identifier. The AP that is functioning as the NAS will periodically send RADIUS Accounting-Request packets to update the RADIUS server about the active session. This packet contains an `Acct-Status-Type` attribute with the value `interim-update`. Information about the current session and data usage can then be tracked. When network access is concluded, the RADIUS client sends a final RADIUS Accounting-Request packet that contains an `Acct-Status-Type` attribute with the value `stop` to the RADIUS server. Information such as why the session was terminated and time of termination can be beneficial when troubleshooting 802.1X/EAP problems.

FIGURE 9.4 RADIUS accounting

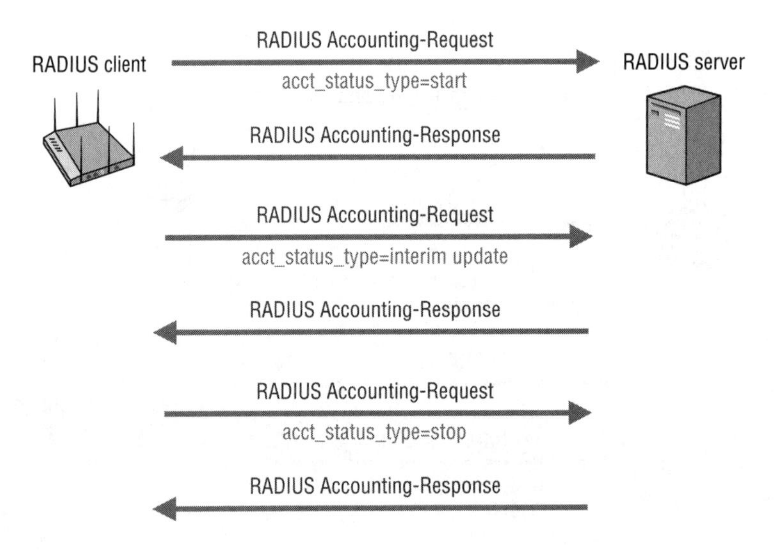

RADIUS Configuration

When used within an 802.1X framework, a RADIUS server must be configured to communicate with the RADIUS client. The RADIUS server needs to be configured with the IP addresses of any APs functioning as authenticators along with a shared secret in order to communicate with the server. As shown in Figure 9.5, most enterprise RADIUS servers will allow you to designate the subnet on which the APs reside. The shared secret configured on both the authenticator and authentication server is not used for any past part of supplicant validation. The shared secret is only for the RADIUS client (AP)–to–RADIUS server communication link. Both the RADIUS client and the RADIUS authentication server should be configured to use UDP port 1812 for authentication communications and UDP port 1813 for accounting communications.

FIGURE 9.5 RADIUS server configuration

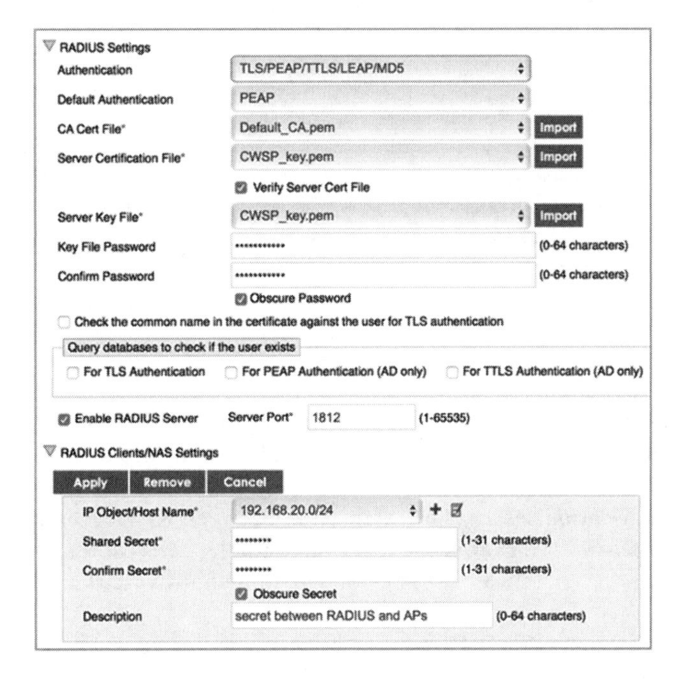

As also shown in Figure 9.5, the RADIUS server must also be configured for the flavors of the EAP protocol that will be used. A RADIUS server can support multiple types of EAP simultaneously; however, the selected EAP type must match on the WLAN supplicant if 802.1X/EAP authentication is to be successful. Most flavors of EAP use tunneled authentication and the proper certificates must also be selected. A server certificate must be installed on the RADIUS server and the Root CA certificate must also be designated. Remember that the Root CA certificate will also need to be provisioned on all the supplicants. As seen in Figure 9.6, the RADIUS server must also be configured for integration with an LDAP database such as Active Directory.

FIGURE 9.6 RADIUS and LDAP integration

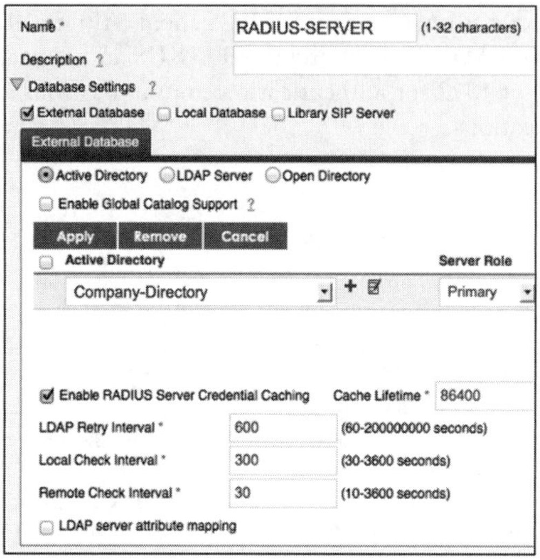

LDAP Proxy

As previously mentioned, when 802.1X/EAP is the chosen WLAN security solution, the RADIUS server may use an internal native user database, but usually will instead query to a preexisting LDAP database. Any LDAP-compliant database can be queried by the RADIUS authentication server. The interaction between RADIUS and LDAP very often depends on the solution that is chosen. For example, Microsoft's Network Policy Server (NPS) operates RADIUS as a service on the same hardware that houses Active Directory (AD).

When using a third-party RADIUS server, the server will need to be joined to the AD domain as a computer object. This will allow the RADIUS server to access the Active Directory user store in order to authenticate users. The RADIUS server can use a standard domain account to perform the LDAP queries between the RADIUS server and Active Directory. When RADIUS servers are integrated with other flavors of LDAPv3 directory services, an LDAP account will be needed by the RADIUS server to perform the queries.

There may be one or more user LDAP databases, and you should confirm support for the exact database technology. For example, if you are an ISP or hotspot provider, the user database may reside in a SQL server, which may limit the available choices of RADIUS server products that are applicable to your design requirements. Usually one database is specified to be the primary over the others and configured in a priority search order.

Real World Scenario

Integrating Different User Databases

Entering the same user information into more than one database is possible. When this occurs, the search order can be paramount. For example, a company has just merged with another. Each division has its own Windows Active Directory domain. In the design of the new network, a single RADIUS server services both sets of users that reside in the two different Active Directory domains.

In Domain 1, a user named John Smith (username: jsmith) exists along with Jane Smith (username: jsmith) in Domain 2. In this case, the first user database in the priority search order would be used for all authentications for the username jsmith. If Jane was in the second domain in the priority search order, her authentication would fail every time because it would not match John Smith's password.

RADIUS Deployment Models

In almost all enterprise environments, RADIUS queries an existing LDAP database. Exactly how RADIUS and LDAP are deployed together depends on the how many locations and numbers of users exist within an organization. As with any type of network design, scaling and redundancy are important considerations when deploying RADIUS servers.

Single-Site Deployment

A single-site deployment is the simplest of all models. If your entire organization resides at a single location and is the hub of all communications, only a single RADIUS server or cluster of redundant servers is needed. The LDAP database also resides at the single-site location. Figure 9.7 depicts a single-site RADIUS deployment. Single-site deployments should be considered when

- All WLAN users are located at a single site.
- A central user/device database is located at the site.
- One or more RADIUS servers query to the onsite LDAP database.
- All EAP communications between the supplicant and the RADIUS server are local.
- All the LDAP queries between the RADIUS server and LDAP database are also local.

FIGURE 9.7 Single-site deployment

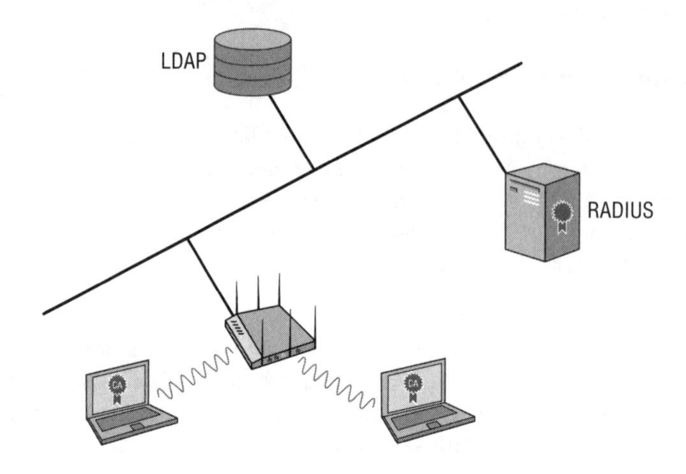

The benefits of a single-site deployment are that the RADIUS server(s) can locally proxy authentications against any type of backend authentication database, including Active Directory, LDAP, and others. Often in single-site deployments, the internal database of the RADIUS server is used instead of an LDAP database. However, even with a single-site deployment, because of performance and future scalability, we recommend that you use an external LDAP database on a separate server.

Distributed Autonomous Sites

If an organization has WLANs at multiple locations or different cities, RADIUS can be scaled in a number of ways to support the multiple sites. As shown in Figure 9.8, a distributed autonomous sites scenario can replicate the LDAP database from a central site to each autonomous site.

A distributed autonomous scenario is defined by the following:

- Multiple WLANs exist at different locations.
- The central authentication database is replicated to LDAP servers at each autonomous site.
- Each remote site has one or more RADIUS servers that proxy to the on-site LDAP database.
- EAP communications between the supplicant and the RADIUS server remain local.
- LDAP queries between the RADIUS server and the replicated LDAP databases are also local.

The main benefit to this scenario is that all of the wireless users authenticate against the local replicated LDAP database and do not have to authenticate across a WAN link. If the WAN link goes down, the users can still successfully authenticate and access the wireless network. However, if the WAN link goes down, replication between the authentication databases cannot occur. The downside to this deployment model is that it is very expensive.

FIGURE 9.8 Distributed autonomous sites

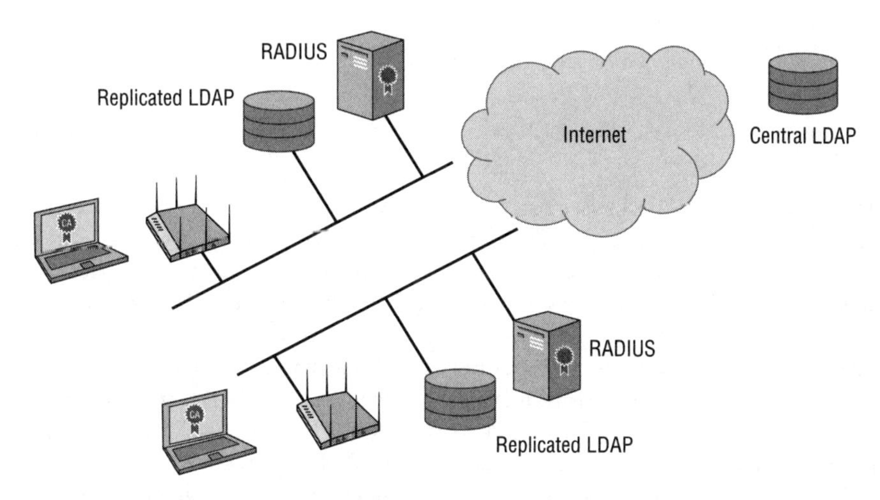

Distributed Sites, Centralized RADIUS and LDAP

Another approach to scaling RADIUS with multiple sites is to use a centralized authentication architecture. As shown in Figure 9.9, all RADIUS servers and the LDAP database are located at a central site, such as corporate headquarters.

FIGURE 9.9 Distributed sites; centralized RADIUS and LDAP

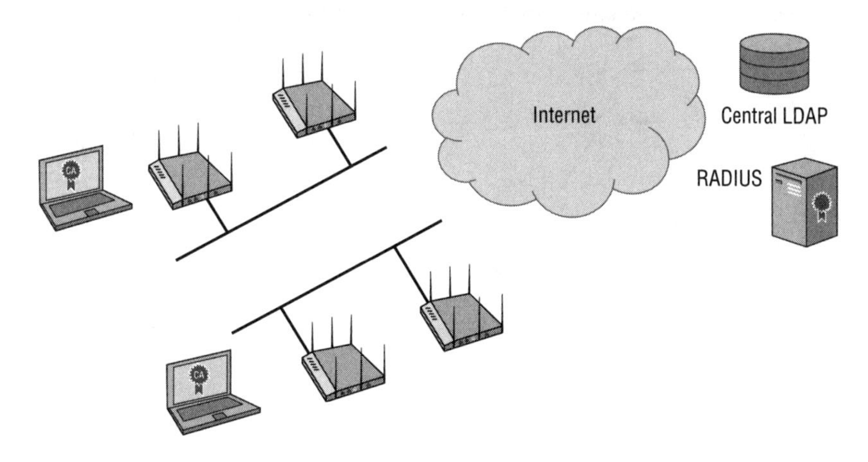

A distributed site with a centralized RADIUS and LDAP architecture is defined by the following:

- Multiple WLANs exist at different locations.
- The LDAP database is located at a central site.

- One or more RADIUS servers are also located at the central site, and they query the central authentication database.
- EAP communications between the supplicant and the RADIUS server occur across the WAN link.
- LDAP queries between the RADIUS server and LDAP database occur at the central location.

The main benefit to this design is cost, because RADIUS servers and database servers are not deployed at the remote locations. The biggest concern with this type of design model is if the WAN link goes down, no new wireless users can authenticate and the users would not be able to access the local wireless network. Performance bottlenecks can also occur that might affect latency. This design model can also negatively impact roaming for any WLAN clients that do not support fast secure roaming mechanisms such as opportunistic key caching (OKC) or fast BSS transition (FT). A typical 802.1X/EAP frame exchange between a supplicant and RADIUS server can take 700 ms but will very often take multiple seconds across a WAN link. If a WLAN client does not have fast secure roaming capabilities, it will have to reauthenticate across the WAN link every time the client roams.

Distributed Sites with RADIUS, Centralized LDAP

Probably a better design than the previous model is a mixture of a distributed and centralized authentication architecture. As shown in Figure 9.10, this scenario also uses an LDAP database at a central location. However, one or more RADIUS servers are deployed at each remote location.

FIGURE 9.10 Distributed sites with RADIUS, centralized LDAP

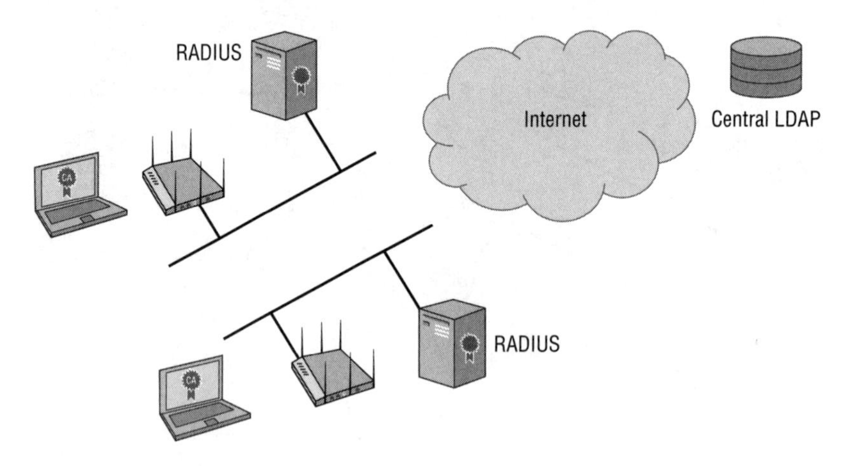

A distributed site with RADIUS, centralized LDAP architecture is defined by the following:

- Multiple WLANS exist at different locations.
- The LDAP database is located at a central site.

- One or more RADIUS servers are located at each remote site and proxy the LDAP queries to the central LDAP database.
- EAP communications between the supplicant and the RADIUS server are local.
- LDAP queries between the RADIUS servers and centralized LDAP database occur across the WAN link.
- LDAP credential caching on the RADIUS server might be an option.

The design can be more expensive than the previous model, but it offers more flexibility. Latency problems that might affect roaming are not as prevalent because the EAP authentication exchange is local and only the LDAP queries occur across the WAN link. The biggest concern with this type of design is if the WAN link goes down, no new wireless user can authenticate and they will not be able to access the WLAN.

However, some RADIUS servers also offer the capability to locally cache LDAP credentials after an initial authentication by a user. If the WAN link goes down, the RADIUS server will first attempt to query the centralized LDAP server but will eventually fall back to the locally cached LDAP credentials on the RADIUS server. When the WAN link goes back up, the RADIUS server will once again query the centralized database. Any new user that has not authenticated prior to the WAN link failure will still not be able to access the WLAN because they would not have a cached LDAP credential.

RADIUS Proxy

RADIUS servers can also be set up to proxy to authentication requests to other RADIUS servers. Do not confuse a RADIUS-to-RADIUS proxy with the RADIUS-to-LDAP proxy authentication discussed earlier. Consider a single-site deployment with a RADIUS server and LDAP database. Perhaps the vendor RADIUS solution requires that the IP addresses of the APs that are functioning as authenticators all be configured individually. Most APs receive their management IP address via DHCP. Most administrators do not want to configure static IP addresses on access points, nor do they want to enter multiple IP addresses of the authenticators in the RADIUS configuration. Perhaps the RADIUS server licensing restricts the number of allowed authenticators to 50 APs yet you have 200 APs that need to perform RADIUS communications.

Depending on the vendor, an AP, switch, or some other type of networking device might have the dual-function capability to operate as a RADIUS proxy. As shown in Figure 9.11, the majority of APs would send their RADIUS packets to a networking device that is functioning as a RADIUS proxy. The devices that are functioning as the proxies would then forward all RADIUS packets to the real RADIUS server. Only a single IP address of the RADIUS proxy device and a shared secret needs to be designated on the RADIUS server. Notice in Figure 9.11 that the RADIUS server will still query an LDAP database. Although a deployed RADIUS-to RADIUS proxy is possible, keep in mind that the LDAP proxy authentication from a RADIUS server to an LDAP database is much more common. Another scenario for a RADIUS-to-RADIUS proxy might be that the main RADIUS server was using a native user database and RADIUS servers at other locations were proxies to the central RADIUS server with the native database. The good news is that RADIUS can be deployed and scaled in the enterprise many different ways—including globally, which will be discussed in the next section.

FIGURE 9.11 RADIUS proxy

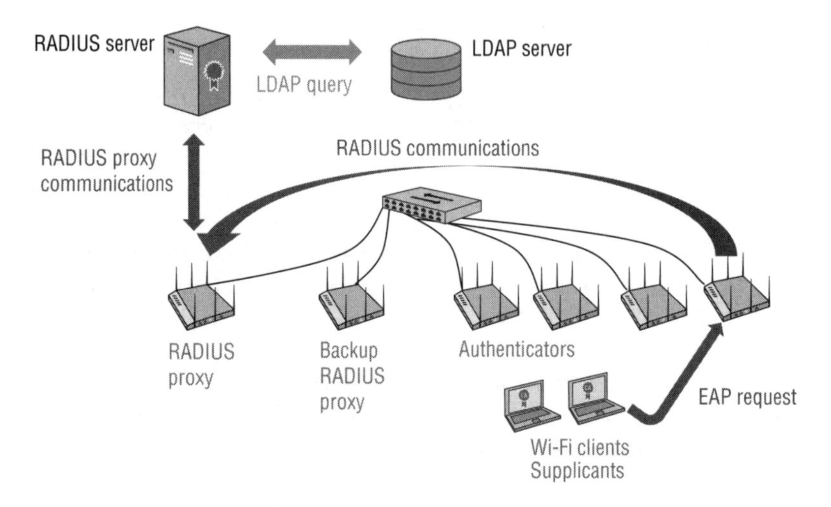

RADIUS Proxy and Realms

The RADIUS protocol can make use of networking *realms*, which identify where the RADIUS server should forward the RADIUS requests across the Internet and between ISPs. Realm naming formats are defined in RFC 4282.

As you just learned, a RADIUS server can be a proxy to one or more centralized RADIUS servers. A username can be in the form `DOMAIN/username` or `username@domain`. This is valuable information that can tell the first authentication server which final destination authentication server to request authentication from. In this case, *domain* is synonymous with the term realm as referred to previously.

For example, if an AP was configured in a regional office or subsidiary of a large enterprise, the AP might point to an authentication server located at that facility or perhaps in a nearby datacenter. If an employee with a common name, let's say Charles, from the parent company was traveling on a business trip and visiting the subsidiary's office, when Charles logged in with `CWNP/charles`, the authentication would know that it needed to contact the remote CWNP user account RADIUS server. In this situation, the first authentication server that the AP pointed to would have just performed a proxy authentication to the final authentication server where Charles's user account resided.

When a domain is used, we commonly refer to this as a *realm-based authentication*. This method of authentication allows a user to authenticate to a realm or sub-realm. When a RADIUS server receives an AAA request for a username containing a realm, the server will reference a table of configured realms. If the realm is known, the server will then proxy the request to the home RADIUS server for that domain. As shown in Figure 9.12, the realm in this case is the domain being sent in the supplicant identity. Authentication of the user is directed to a RADIUS server for CompanyX based on the domain value supplied.

FIGURE 9.12 Realm-based authentication

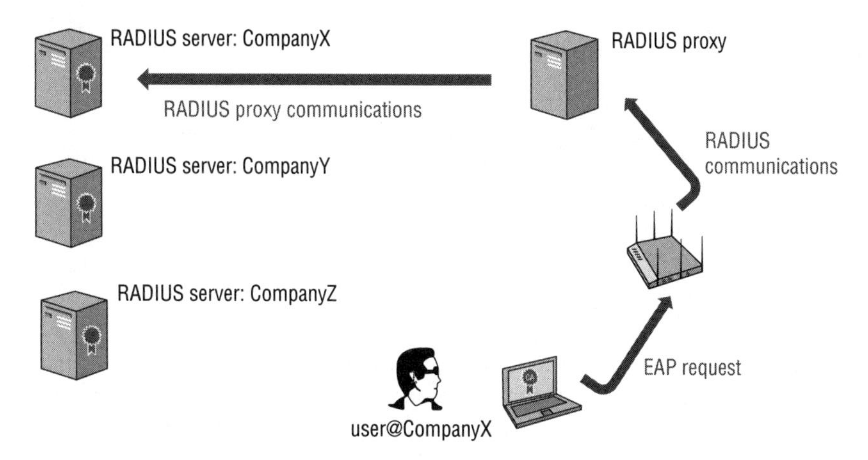

Real World Scenario

eduroam

A global example of realm-based authentication is *eduroam* (education roaming). The technology behind eduroam is based on the IEEE 802.1X standard and a hierarchy of RADIUS proxy servers. eduroam is a secure, worldwide roaming access service developed for the international research and higher education community. eduroam allows students, researchers, and staff from participating institutions to obtain Internet connectivity across campus and when visiting other participating institutions. Authentication of users is performed by their home institution, using the same credentials as when they access the network locally. Depending on local policies at the visited institutions, eduroam participants may also have additional resources at their disposal. The eduroam Architecture for Network Roaming is defined in RFC 7593. More information about the eduroam service can be found at `www.eduroam.org`.

RADIUS Failover

Networking Design 101 dictates redundancy for network services. In the enterprise, multiple RADIUS servers should be deployed for redundancy purposes. As shown in Figure 9.13, when deploying an AP or WLAN controller as a RADIUS client, you can configure multiple

RADIUS servers. The first server in the list is usually considered the primary server. When a RADIUS server is determined to be unavailable based on the method used by the WLAN infrastructure, it will move to the next server in the priority list.

FIGURE 9.13 RADIUS failover

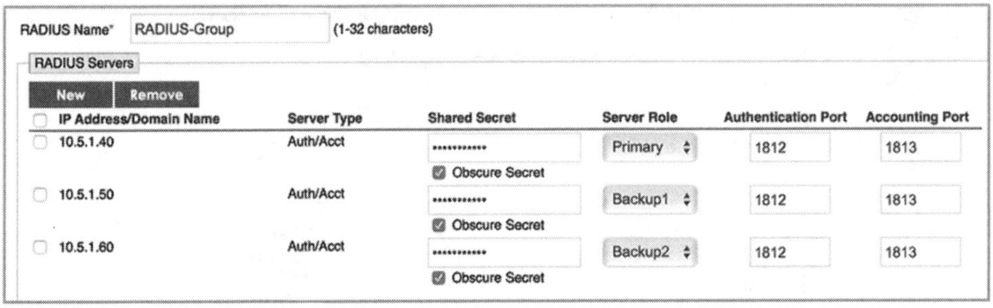

RADIUS Name* RADIUS-Group (1-32 characters)

RADIUS Servers

New Remove

IP Address/Domain Name	Server Type	Shared Secret	Server Role	Authentication Port	Accounting Port
10.5.1.40	Auth/Acct	••••••••••• Obscure Secret	Primary	1812	1813
10.5.1.50	Auth/Acct	••••••••••• Obscure Secret	Backup1	1812	1813
10.5.1.60	Auth/Acct	••••••••••• Obscure Secret	Backup2	1812	1813

Typically, a hold-down timer will be enacted that monitors the availability of that RADIUS server until it is considered stable in order to be sent authentication or accounting transactions again. Depending on the settings or logic of the authenticator functionality in the AP or WLAN controller, the original RADIUS server may not revert to the primary role automatically. Even if the other servers are geographically far away, resulting in added latency, this may not cause any other problem other than slowness of all RADIUS transactions as long as the network connection is still reliable. We highly recommend that you monitor for RADIUS failover events on your APs and WLAN controllers.

WLAN Devices as RADIUS Servers

Enterprise WLAN vendors offer solutions where either a standalone AP or a WLAN controller can dual-function as a RADIUS server and perform direct LDAP queries, thus eliminating the need for an external RADIUS server. Additionally some switch vendors also offer dual-functionality where an access layer switch might also operate as a RADIUS server. These solutions are often popular when there are multiple locations. Cost savings can be realized in distributed sites with RADIUS and centralized architecture. Keep in mind that the number of simultaneous RADIUS authentications that a dual-function device can handle will be much smaller than a dedicated RADIUS server such as Microsoft NPS or Cisco ACS. Additionally, since the main function of either a standalone AP or a WLAN controller is not RADIUS, these devices most often support advanced RADIUS server features.

Captive Web Portal and MAC Authentication

Most of the discussion in this chapter has revolved around using RADIUS with 802.1X/EAP authentication for strong WLAN security. However, a RADIUS server can also be used to authorize WLAN users and devices with weak authentication methods such as *captive web portal* authentication and MAC address authentication.

RADIUS servers are often used with a captive portal web page to authenticate guest users for guest WLANs. A captive web portal solution will query a RADIUS server with a username and password using a weak authentication protocol such as MS-CHAPv2. The native database of the RADIUS server is normally used to validate the guest users. Captive web portal authentication is also often used together with *bring your own device (BYOD)* solutions for validating employee credentials. The employee database would most likely be Active Directory, which would in turn be queried by the RADIUS server. More detailed information about captive web portal authentication can be found in Chapter 10, "Bring Your Own Device (BYOD) and Guest Access."

A RADIUS server can also be used for simple authentication of 802.11 devices based on an allowed list of MAC addresses. When an 802.11 client device attempts to associate, an AP queries the RADIUS server with a RADIUS Access-Request message. The RADIUS server can admit or deny the device based on the WLAN client MAC address, responding with either a RADIUS Access-Accept message or a RADIUS Access-Reject message.

MAC authentication does not require client-side configuration and is typically used with legacy WLAN devices that might not support PSK or 802.1X/EAP. Because MAC addresses can be very easily spoofed, MAC authentication is rarely used by itself for WLAN security. However, MAC authentication is sometimes bound together with either captive web portal authentication or 802.1X/EAP. Combining MAC authentication along with 802.1X/EAP could be considered a simple method of multifactor authentication.

RadSec

The RADIUS protocol depends on the unreliable transport protocol UDP using port 1812 and has some potential security weaknesses. RADIUS is based on the MD5 algorithm, which has been proven to be insecure. Therefore, the IETF RFC 6614 defines Transport Layer Security (TLS) for RADIUS. The defined protocol is often called *RadSec* or *RADIUS over TLS*. RadSec is the next-generation RADIUS transport that relies on TCP and TLS for reliable and secure transport with integrity verification.

The main focus of RadSec is to provide a means of securing the communication between RADIUS/TCP peers using TLS encryption. The default destination port number for RadSec is TCP/2083. Authentication, accounting, and dynamic authorization changes do not require separate ports. RadSec can be used for RADIUS packets that traverse through different administrative domains and networks. The academia roaming access service, eduroam, has already begun to use RadSec globally.

Attribute-Value Pairs

An attribute is a portion of information that determines the properties of a field or tag in a database. Attributes usually come in name/value pairs like `name="value"`. An *attribute-value pair (AVP)* is a representation of data in computer systems and applications. An attribute-value pair can be used to store and provide data in a database—for example, an attribute called `last name` followed by its value pair, which is the actual last name of a person. The IETF designates an original set of 255 standard RADIUS attributes that can be used to communicate AAA information between a RADIUS client and a RADIUS server. The attribute-value pairs (AVPs) carry data in the RADIUS request and response packets. Figure 9.14 shows a packet capture of a RADIUS Access-Accept packet with a series of AVPs.

FIGURE 9.14 RADIUS attribute-value pairs

```
▿ Radius Protocol
    Code: Access-Accept (2)
    Packet identifier: 0x6 (6)
    Length: 97
    Authenticator: fbba6a784c7decb314caf0f27944a37b
    [Time from request: 0.000114000 seconds]
  ▿ Attribute Value Pairs
    ▹ AVP: l=6 t=Framed-IP-Address(8): Assigned
    ▹ AVP: l=6 t=Framed-MTU(12): 576
    ▹ AVP: l=6 t=Service-Type(6): Framed(2)
    ▹ AVP: l=21 t=Reply-Message(18): Hello
    ▿ AVP: l=6 t=EAP-Message(79) Last Segment[1]
        EAP fragment
      ▿ Extensible Authentication Protocol
          Code: Success (3)
          Id: 1
          Length: 4
    ▿ AVP: l=18 t=Message-Authenticator(80): b9c4ae6213a71d32125ef7ca4e
        Message-Authenticator: b9c4ae6213a71d32125ef7ca4e4c6360
```

RFC 3579 defines how the EAP protocol is supported with RADIUS using attributes. To support EAP within RADIUS, two standard attributes are used:

(79) EAP-Message This attribute is used to encapsulate EAP frames with RADIUS packets. Only one EAP frame can be encapsulated within a single RADIUS packet. A RADIUS client such as an access point or WLAN controller forwards the `EAP-Message` attributes within a RADIUS Access-Request packet to the RADIUS server. The RADIUS server can return `EAP-Message` attributes in Access-Challenge, Access-Accept, and Access-Reject packets.

(80) Message-Authenticator An attacker could potentially spoof the EAP-Success, EAP-Failure, and other EAP frames. This attribute prevents attackers from modifying EAP within a RADIUS packet using per-packet data integrity checks. Therefore, the `Message-Authenticator` attribute must be used to protect all Access-Request, Access-Challenge, Access-Accept, and Access-Reject packets containing an `EAP-Message` attribute.

While most of the RADIUS attributes are used for authentication and authorization purposes, some are specifically dedicated for use with RADIUS accounting. As mentioned earlier in this chapter, RADIUS accounting packets use the `(40) Acct-Status-Type` attribute with values of `start`, `stop`, or `interim update`.

Vendor-Specific Attributes

RADIUS *vendor-specific attributes (VSAs)* are derived from the IETF attribute `(26) Vendor-Specific`. This attribute allows a vendor to create any additional 255 attributes however they wish. Data that is not defined in standard IETF RADIUS attributes can be encapsulated in the (26) Vendor-Specific attribute. This allows vendors to support their own extended attributes otherwise not suitable for general use. VSAs allow RADIUS client vendors, such as the manufacturers of access points and switches, to support their own proprietary RADIUS attributes. The vendors typically offer free VSA dictionary files, which can be easily imported into most popular commercial RADIUS servers as well as the open source FreeRADIUS.

VLAN Assignment

A common WLAN design strategy is to link a single user VLAN to a unique SSID. Most WLAN vendors allow a radio to broadcast as many as 16 SSIDs. However, broadcasting 16 SSIDs is a bad practice because of the Layer 2 overhead created by the 802.11 management and control frames for each SSID. The broadcast of 16 SSIDs will result in degraded performance. The best practice is to never broadcast more than 3 or 4 SSIDs.

What if you want your employees segmented into multiple VLANs? Can a single employee SSID be mapped to multiple VLANs? RADIUS attributes can be leveraged for VLAN assignment when using 802.1X/EAP authentication on the employee SSID. As you have already learned, when a RADIUS server provides a successful response to an authentication request, the Access-Accept response can contain a series of attribute-value pairs (AVPs). One of the most popular uses of RADIUS AVPs is assigning users to VLANs, based on the identity of the user authenticating. Instead of segmenting users to different SSIDs that are each mapped to a unique user VLAN, all the users can be associated to a single SSID and assigned to different VLANs.

RADIUS servers can be configured with different access policies for different groups of users. The RADIUS access policies are usually mapped to different LDAP groups. Figure 9.15 shows a RADIUS server access policy defined for a specific LDAP group. The access policy uses three standard IETF RADIUS attributes to assign a specific VLAN to an authenticated user. Notice that the (81) Tunnel-Private-Group-ID attribute has a value of 10. Upon authentication, any user that belongs to the LDAP group mapped to this policy will join VLAN 10.

FIGURE 9.15 RADIUS AVPs for VLAN assignment

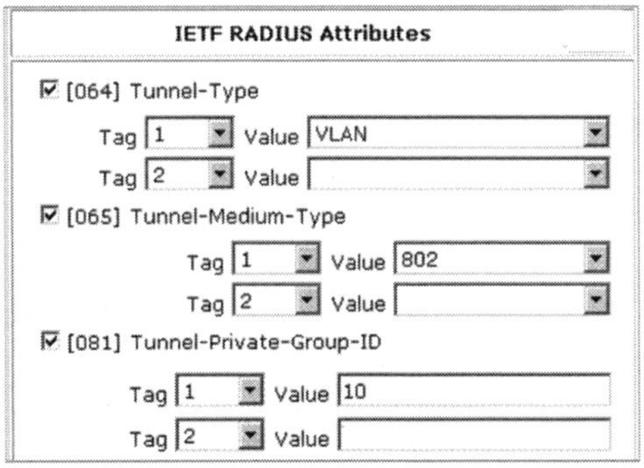

Role-Based Access Control

Using RADIUS attributes for user VLAN assignment has been a network design strategy for many years. However, RADIUS attributes can be further leveraged to assign different groups of users to all kinds of different user traffic settings, including VLANs, firewall policies, bandwidth policies, and much more.

Role-based access control (RBAC) is an approach to restricting system access to authorized users. The majority of enterprise WLAN vendor solutions have RBAC capabilities. The three main components of an RBAC approach are *users, roles,* and *permissions.* Separate roles can be created, such as the sales role or the marketing role. User traffic permissions can be defined as Layer 2 permissions (MAC filters), VLANs, Layer 3 permissions (access control lists), Layers 4–7 permissions (stateful firewall rules), and bandwidth permissions. All these permissions can also be time based. The user traffic permissions are mapped to the roles. Some WLAN vendors use the term *roles* whereas other vendors use the term *user profiles.*

When a user authenticates using 802.1X/EAP, RADIUS attributes can be used to assign users to a specific role automatically. All users can associate to the same SSID but be assigned to unique roles. This method is often used to assign users from certain Active Directory (AD) groups into predefined roles created on a WLAN controller or access point. Each role has unique access restrictions. Once users are assigned to roles, they inherit the user traffic permissions of whatever roles they have been assigned.

Figure 9.16 depicts a RADIUS server with three unique access policies mapped to three different Active Directory groups. For example, user-2 belongs to the marketing AD group. Based on the RADIUS access policy for that AD group, when user-2 authenticates, the RADIUS server will send the AP a RADIUS packet with an attribute that contains a value relevant to Role-B, which has been configured on the AP. The user-2 WLAN client will then be assigned to VLAN 20, firewall-policy-B, and a bandwidth policy of 4 Mbps.

FIGURE 9.16 RADIUS attributes for role assignment

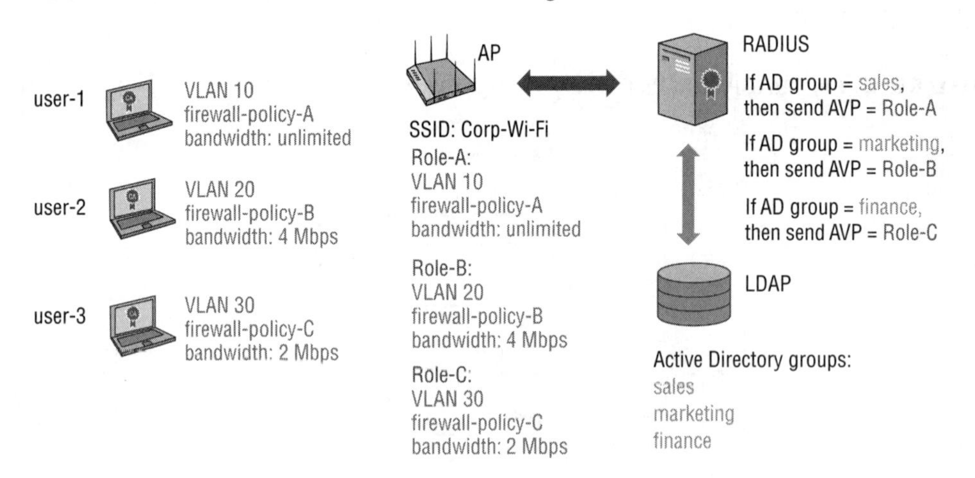

WLAN vendors often use VSAs for role assignment, but the standard IETF RADIUS attribute (11) `Filter-Id` is also often used for role assignment of user traffic permissions.

LDAP Attributes

As previously mentioned, WLAN vendors offer solutions where either a standalone AP or a WLAN controller can dual-function as a RADIUS server and perform direct LDAP queries, thus eliminating the need for an external RADIUS server. However, these dual-function devices most often do not offer extensive support for RADIUS attributes. You have just learned that RADIUS attributes can be leveraged for server-based role assignment of user traffic settings. The end result is that different groups of users and devices can be assigned to different VLANs, firewall policies, and so forth while connected to the same SSID. Can this be accomplished without RADIUS attributes? The answer is yes, because LDAP attributes can also be used for server-based role assignment of user traffic settings. For example, Active Directory queries return the `memberOf` attribute, which is a list of AD groups to which a user belongs. Access policies can be configured on the WLAN device to assign roles based on AD group values from the `memberOf` attribute.

Summary

The RADIUS protocol has been around since 1991 and predates 802.11 technology. However, RADIUS is a key component of 802.1X/EAP security that is widely used in enterprise WLAN deployments. RADIUS servers almost always function as the authentication server that validates the credentials of a WLAN supplicant. RADIUS attributes can be used to transport EAP frames and for server-based role assignment for user traffic settings.

RADIUS can fully integrate with any type of LDAP database. Because both RADIUS and LDAP are mature technologies, multiple deployment models exist for both scalability and redundancy. RADIUS can be used with 802.1X/EAP security as well as with captive web portal and MAC authentication.

Exam Essentials

Define LDAP and directory services. Understand that LDAP is an application protocol for providing directory services over an IP network. A directory service is an infrastructure used to share information about network resources.

Explain the relationship between RADIUS and LDAP. Understand that an LDAP-compliant database can be queried by the RADIUS authentication server. 802.1X supplicant credentials can reside in an LDAP database.

Understand how the RADIUS protocol is used for AAA. Explain how RADIUS provides authentication, authorization, and accounting (AAA) capabilities for computers to connect to and use network services.

Describe RADIUS and LDAP deployment models. Explain the many different ways RADIUS and LDAP can be deployed together to provide scalability and redundancy.

Understand RADIUS attributes. Describe how standard RADIUS attributes can be used to communicate AAA information between a RADIUS client and a RADIUS server. Describe how attributes can be used for server-based role assignment of user traffic settings.

Review Questions

1. The ACME Company has over 300 WLAN users communicating through 25 access points using Generic Routing Encapsulation (GRE) to tunnel all 802.11 user traffic back to a central WLAN controller capable of role-based access control (RBAC). What type of access restrictions can be placed on the users after authentication?
   - A. UDP
   - B. TCP
   - C. Bandwidth
   - D. Time-of-day
   - E. All of the above
2. What type of database can be integrated with a RADIUS server for proxy authentication?
   - A. eDirectory
   - B. Active Directory
   - C. SQL
   - D. Open Directory
   - E. All of the above
3. What ports can be used by a RADIUS server for LDAP queries to an LDAP database server? (Choose all that apply.)
   - A. UDP 1812
   - B. UDP 1813
   - C. TCP 369
   - D. TCP 636
   - E. TCP 2083
4. Kenny has been tasked with designing an 802.1X/EAP solution for the corporate WLAN. The company headquarters and datacenter reside in London. Employees need secure WLAN access at 15 remote offices in other European cities. Kenny needs a solution that is cost-efficient but should provide secure WLAN connectivity for the majority of users if a remote WAN link goes down. Which of these RADIUS deployment models best meets his requirements? (Choose the best answer.)
   - A. Single-site deployment
   - B. Distributed autonomous sites
   - C. Distributed sites, centralized RADIUS and LDAP
   - D. Distributed sites with RADIUS, centralized LDAP
   - E. Distributed sites with RADIUS proxy, centralized RADIUS and LDAP

5. What port needs to be open on a firewall to permit RadSec protocol authentication and accounting traffic?
   - A. UDP 1812
   - B. UDP 1813
   - C. TCP 369
   - D. TCP 636
   - E. TCP 2083
6. Which RADIUS attribute is used to protect encapsulated EAP frames within RADIUS packets?
   - A. `(11) Filter-Id`
   - B. `(26) Vendor-Specific`
   - C. `(40) Acct-Status-Type`
   - D. `(79) EAP-Message`
   - E. `(80) Message-Authenticator`
7. Which of these terms best describes the capability of a RADIUS server to forward the RADIUS requests across the Internet between different ISPs or different companies?
   - A. Machine authentication
   - B. LDAP authentication
   - C. User authentication
   - D. Realm-based authentication
   - E. Domain authentication
8. During an 802.1X/EAP authentication process, which LDAP attribute is used to assign users that belong to different Active Directory groups into distinctive roles with unique user traffic policies? (Choose all that apply.)
   - A. `objectClass`
   - B. `userPrincipalName`
   - C. `userCert`
   - D. `memberOf`
   - E. `objectCategory`
   - F. `userSharedFolder`
9. Which RADIUS attribute is used in RADIUS packets that traverse through a firewall via UDP port 1813?
   - A. `(11) Filter-Id`
   - B. `(26) Vendor-Specific`
   - C. `(40) Acct-Status-Type`
   - D. `(79) EAP-Message`
   - E. `(80) Message-Authenticator`
   - F. `memberOf`

10. Which of these enterprise RADIUS deployment models is the most cost-efficient but will result in all 802.1X supplicants not being able to connect to the WLAN should a remote WAN link goes down? (Choose the best answer.)

A. Single-site deployment

B. Distributed autonomous sites

C. Distributed sites, centralized RADIUS and LDAP

D. Distributed sites with RADIUS, centralized LDAP

11. Within an 802.1X infrastructure framework, what is the name of the device that communicates directly with a RADIUS server using the RADIUS protocol? (Choose all that apply.)

A. Authenticator

B. RADIUS ports

C. Network access server

D. LDAP integration

E. RADIUS client

F. Supplicant

12. Which RADIUS packets can be sent from a RADIUS server to an access point when 802.1X/EAP is the deployed WLAN security solution? (Choose all that apply.)

A. RADIUS Access-Request

B. RADIUS Access-Challenge

C. RADIUS Access-Accept

D. RADIUS Access-Reject

13. Bob has been tasked with designing an 802.1X/EAP solution for the corporate WLAN. The company headquarters and datacenter reside in Denver. Employees need secure WLAN access at 15 remote offices in other cities. Which of these RADIUS deployment models guarantees secure WLAN connectivity even if a remote WAN link goes down? (Choose the best answer.)

A. Single-site deployment

B. Distributed autonomous sites

C. Distributed sites, centralized RADIUS and LDAP

D. Distributed sites with RADIUS, centralized LDAP

E. Distributed sites with RADIUS proxy, centralized RADIUS and LDAP

14. Which RADIUS attribute is used to encapsulate EAP frames within RADIUS packets?

A. `(11) Filter-Id`

B. `(26) Vendor-Specific`

C. `(40) Acct-Status-Type`

D. `(79) EAP-Message`

E. `(80) Message-Authenticator`

15. Which of these enterprise RADIUS deployment models is the most cost-efficient but may negatively impact fast secure roaming of WLAN clients? (Choose the best answer.)
    A. Single-site deployment
    B. Distributed autonomous sites
    C. Distributed sites, centralized RADIUS and LDAP
    D. Distributed sites with RADIUS, centralized LDAP
    E. Distributed sites with RADIUS proxy, centralized RADIUS and LDAP

16. Which of these authentication methods are supported by RADIUS and can be used for WLAN security? (Choose all that apply.)
    A. Hologram authentication
    B. Captive web portal authentication
    C. MAC authentication
    D. TSA authentication
    E. 802.1X/EAP authentication

17. When configuring an 802.1X/EAP solution, what must be configured on the RADIUS server for RADIUS protocol communications with an access point? (Choose all that apply.)
    A. NAS IP addresses
    B. Digital certificates
    C. EAP protocols
    D. LDAP integration settings
    E. Authentication and authorization ports
    F. Shared secret

18. Which of these authentication methods does not require any WLAN client configuration and is sometimes bound together with other authentication methods?
    A. Hologram authentication
    B. Captive web portal authentication
    C. MAC authentication
    D. TSA authentication
    E. 802.1X/EAP authentication

19. In which of these enterprise RADIUS deployments do the EAP communications not traverse across a WAN link? (Choose all that apply.)
    A. Single-site deployment
    B. Distributed autonomous sites
    C. Distributed sites, centralized RADIUS and LDAP
    D. Distributed sites with RADIUS, centralized LDAP
    E. Distributed sites with RADIUS proxy, centralized RADIUS and LDAP

20. During an 802.1X/EAP authentication process, which RADIUS attributes might be used to assign users that are members of different Active Directory groups to distinctive roles with unique user traffic policies? (Choose all that apply.)

A. `(11) Filter-Id`

B. `(26) Vendor-Specific`

C. `(40) Acct-Status-Type`

D. `(79) EAP-Message`

E. `(80) Message-Authenticator`

F. `memberOf`

Bring Your Own Device (BYOD) and Guest Access

IN THIS CHAPTER, YOU WILL LEARN ABOUT THE FOLLOWING:

✓ **Mobile Device Management**

- Company-issued devices vs. personal devices
- MDM architecture
- MDM enrollment
- MDM profiles
- MDM agent software
- Over-the-air management
- Application management

✓ **Self-service device onboarding for employees**

- Dual SSID onboarding
- Single SSID onboarding
- MDM vs. Self-service onboarding

✓ **Guest WLAN access**

- Guest SSID
- Guest VLAN
- Guest firewall policy
- Captive web portals
- Client isolation, rate limiting, and web content filtering
- Guest management
- Guest self-registration
- Employee sponsorship

- Social login
- Encrypted guest access

✓ **Network access control (NAC)**

- Posture
- NAC and BYOD
- OS fingerprinting
- AAA
- RADIUS change of authorization

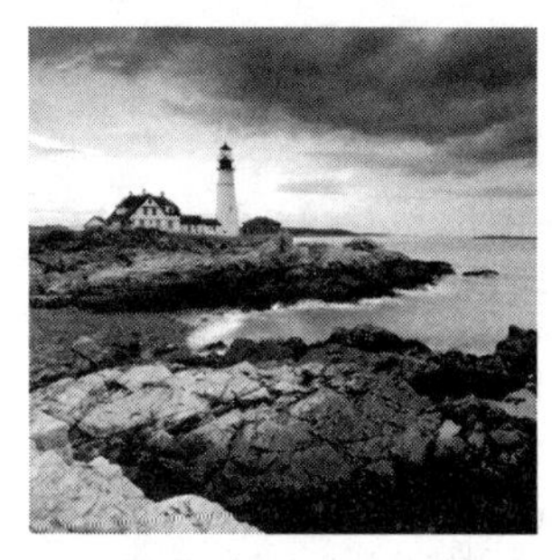

For many years, the primary purpose of enterprise WLANs was to provide wireless access for company-owned laptop computers used by employees. Some vertical markets, such as healthcare, retail, and manufacturing, also required WLAN access for company-owned mobile devices such as VoWiFi phones and wireless barcode scanners. However, in recent years there has been a massive population explosion of Wi-Fi personal mobile devices. Wi-Fi radios are now the primary communications component in smartphones, tablets, PCs, and many other mobile devices.

Although mobile devices initially were intended for personal use, organizations found ways of deploying corporate mobile devices with custom software to improve productivity or functionality. Employees also increasingly want to use their personal mobile devices in the workplace. Employees have expectations of being able to connect to a corporate WLAN with multiple personal mobile devices. The catchphrase of *bring your own device (BYOD)* refers to the policy of permitting employees to bring personally owned mobile devices such as smartphones, tablets, and laptops to their workplace. A BYOD policy dictates which corporate resources can or cannot be accessed when employees connect to the company WLAN with their personal devices.

The main focus of this chapter is how security is used to control and monitor BYOD access to a WLAN. *Mobile device management (MDM)* solutions can be used to remotely manage and control company-owned as well as personal mobile Wi-Fi devices. MDM solutions use server software or cloud services to configure client settings, along with client applications to control and monitor what the user can do. Additionally there is a growing trend to use self-service BYOD solutions where employees can securely provision their personal WLAN devices. *Network access control (NAC)* integrates different security technologies, such as AAA, RADIUS, client health check, guest services, and client self-registration and enrollment. Using these technologies, NAC can control and monitor client access on the network. NAC can be used to provide authentication and access control of MDM-managed devices, corporate Wi-Fi devices, or BYOD and guest devices.

This chapter will also cover the many components of WLAN guest access and how it has evolved over the years. Guest access technology includes support for visitor devices along with employee BYOD devices.

Mobile Device Management

Consumerization of IT is a phrase used to describe a shift in information technology (IT) that begins in the consumer market and moves into business and government facilities. It has become common for employees to introduce consumer market devices into the workplace after already embracing new technology at home. In the early days of Wi-Fi, most businesses did not provide wireless network access to the corporate network. Due to the limited wireless security options available at that time, along with a general mistrust of the unknown, it was common for companies to avoid implementing WLANs. However, because employees enjoyed the flexibility of Wi-Fi at home, they began to bring small office/home office (SOHO) wireless routers into the office and install them despite the objections of the IT department. Eventually, businesses and government agencies realized that they needed to deploy WLANs to take advantage of the technology as well as manage the technology.

Personal mobile Wi-Fi devices, such as smartphones and tablets, have been around for quite a few years. The Apple iPhone was first introduced in June 2007, and the first iPad debuted in April 2010. HTC introduced the first Android smartphone in October 2008. These devices were originally meant for personal use, but in a very short time, employees wanted to also use their personal devices on company WLANs. Additionally, software developers began to create enterprise mobile business applications for smartphones and tablets. Businesses began to purchase and deploy tablets and smartphones to take advantage of these mobile enterprise applications. Tablets and smartphones provided the true mobility that employees and businesses desired, and within a few years, the number of mobile devices connecting to corporate WLANs surpassed the number of laptop connections. This trend is continuing, with many, if not most, devices shipping with Wi-Fi as the primary networking adapter. Many laptop computers now ship without an Ethernet adapter because the laptop Wi-Fi radio is used for network access.

Because of the proliferation of personal mobile devices, a BYOD policy is needed to define how employees' personal devices may access the corporate WLAN. A mobile device management (MDM) solution might be needed for onboarding personal mobile devices as well as *company-issued devices (CIDs)* to the WLAN. Corporate IT departments can deploy MDM servers to manage, secure, and monitor the mobile devices. An MDM solution can manage devices across multiple mobile operating systems and across multiple mobile service providers. Most MDM solutions are used to manage iOS and Android mobile devices. However, mobile devices that use other operating systems such as BlackBerry OS and Windows Phone can also be managed by MDM solutions. Although the main focus of an MDM solution is the management of smartphones and tablets, some MDM solutions can also be used to onboard personal Mac OS and Chromebook laptops. A few of the devices that can be managed by an MDM solution are shown in Figure 10.1.

Some of the WLAN infrastructure vendors have developed small-scale MDM solutions that are specific to their WLAN controller and/or access point solution. However, the bigger MDM companies sell overlay solutions that can be used with any WLAN vendor's solution.

FIGURE 10.1 Personal mobile devices with Wi-Fi radios

These are some of the major vendors selling overlay MDM solutions:

VMware AirWatch—`www.air-watch.com`

Citrix—`www.citrix.com`

IBM—`www.maaS360.com`

JAMF Software—`www.jamfsoftware.com`

MobileIron—`www.mobileiron.com`

Company-Issued Devices vs. Personal Devices

An MDM solution can be used to manage both company-issued devices and personal devices. However, the management of CID and BYOD is quite different. A company mobile device was purchased by the company with the intent of enhancing employee performance. A tablet or smartphone might be issued to an individual employee or shared by employees on different shifts. Commercial business applications, and very often industry-specific applications, are deployed on these devices. Many companies even develop in-house applications unique to their own business needs. Very often company mobile devices are deployed to replace older hardware. For example, inventory control software running on a tablet might replace legacy handheld bar code scanners. A software Voice over Internet Protocol (VoIP) application running on a smartphone might be used to replace WLAN VoWiFi handsets. The IT department will usually choose one model of mobile device that runs the same operating system.

The management strategy for company mobile devices usually entails more in-depth security because very often the CIDs have company documents and information stored on them. When company devices are provisioned with an MDM solution, many configuration settings such as virtual private network (VPN) client access, email account settings, Wi-Fi profile settings, passwords, and encryption settings are enabled. The ability for employees to remove MDM profiles from a CID is disabled and the MDM administrator can remotely wipe company mobile devices if they are lost or stolen. The MDM solution is also used for hardware and software inventory control. Because these devices are not personal devices, the IT department can also dictate which applications can or cannot be installed on tablets and/or smartphones.

The concept of BYOD emerged because personally owned mobile devices are difficult to control and manage, while allowing access to the enterprise network. Access and control may be managed using an MDM or NAC solution, but BYOD needs are different than CID needs. Employees, visitors, vendors, contractors, and consultants bring a wide range of personal devices—different makes and models loaded with a variety of operating systems and applications—to the workplace. Therefore, a different management strategy is needed for BYOD. Every company should have its own unique BYOD containment strategy while still allowing access to the corporate WLAN. For example, when the personal devices are provisioned with an MDM solution, the camera may be disabled so that pictures cannot be taken within the building. As shown in Figure 10.2, many restrictions can be enforced on a CID or BYOD device after it has been enrolled in the MDM solution.

FIGURE 10.2 Device restrictions

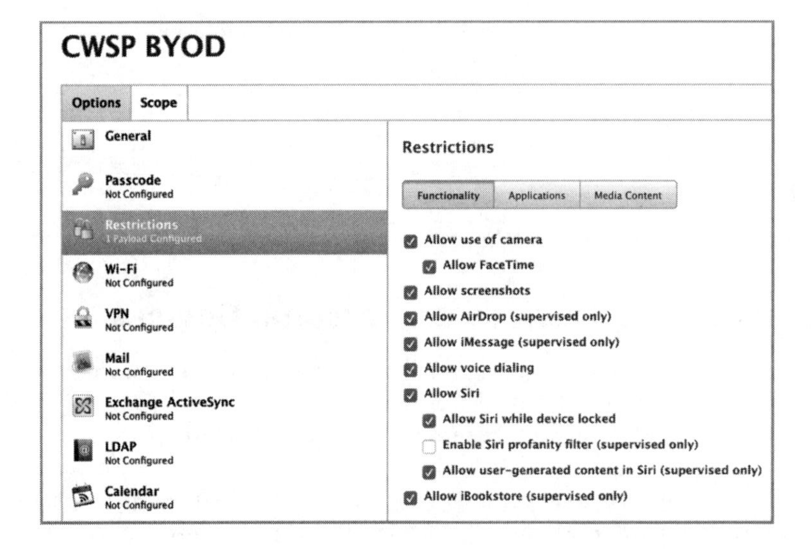

As an alternative to MDM solutions, NAC solutions can be used to authenticate BYOD devices along with controlling access to the enterprise network without having to necessarily install software on the client device. NAC solutions do not provide control of the individual device but can provide extensive control over the level of access the BYOD device has on the network.

MDM Architecture

The basic architecture of any MDM solution consists of four main components:

Mobile Device The mobile Wi-Fi device requires access to the corporate WLAN. The mobile device can be either a company-owned or an employee-owned device. Depending on the MDM vendor, multiple operating systems may be supported, including iOS, Android, Chrome, and Mac OS, among others. The mobile devices are not allowed onto the corporate network until an enrollment process has been completed and an MDM profile has been installed.

AP/WLAN Controller All Wi-Fi communications are between the mobile devices and the access point to which they connected. If the devices have not been enrolled via the MDM server, the AP or WLAN controller quarantines the mobile devices within a restricted area of the network known as a walled garden. Mobile devices that have been taken through the enrollment process are allowed outside of the walled garden.

MDM Server The MDM server is responsible for enrolling client devices. The MDM server provisions the mobile devices with MDM profiles that define client device restrictions as well as configuration settings. Certificates can be provisioned from the MDM server. MDM servers can also be configured for either enrollment whitelisting or blacklisting. Whitelisting policies restrict enrollment to a list of specific devices and operating systems. Blacklisting policies allow all devices and operating systems to enroll except for those that are specifically prohibited by the blacklist. Although the initial role of an MDM server is to provision and onboard mobile devices to the WLAN, the server is also used for client device monitoring. Device inventory control and configuration are key components of any MDM solution. The MDM server usually is available as either a cloud-based service or as an on-premises server that is deployed in the company datacenter. On-premise MDM servers can be in the form of a hardware appliance or can run as software in a virtualized server environment.

Push Notification Servers The MDM server communicates with push notification servers such as *Apple Push Notification service (APNs)* and *Google Cloud Messaging (GCM)* for over-the-air management of mobile Wi-Fi devices. Over-the-air management will be discussed in greater detail later in this chapter.

There are other key components to an MDM architecture deployment. MDM servers can be configured to query *Lightweight Directory Access Protocol (LDAP)* databases, such as Active Directory. Typically, a corporate firewall also will be in place. Proper outbound ports need to be open to allow for communications between all of the various components of the MDM architecture. For example, Transmission Control Protocol (TCP) port 443 needs to be open for encrypted SSL communications between the AP and the MDM server as well as SSL communications between the mobile device and the MDM server. TCP port 5223 needs to be open so that mobile devices can communicate with APNs. TCP ports 2195 and 2196 are needed for traffic between the MDM server and APNs. TCP ports 443, 5223, 5229, and 5330 are required for communication between mobile devices and GCM. Communications between the MDM server and GCM require TCP port 443 to be open.

MDM Enrollment

When MDM architecture is in place, mobile devices must go through an enrollment process in order to access network resources. The enrollment process can be used to onboard both company-issued devices and personal devices. Figure 10.3 illustrates the initial steps of the MDM enrollment process.

FIGURE 10.3 MDM enrollment—initial steps

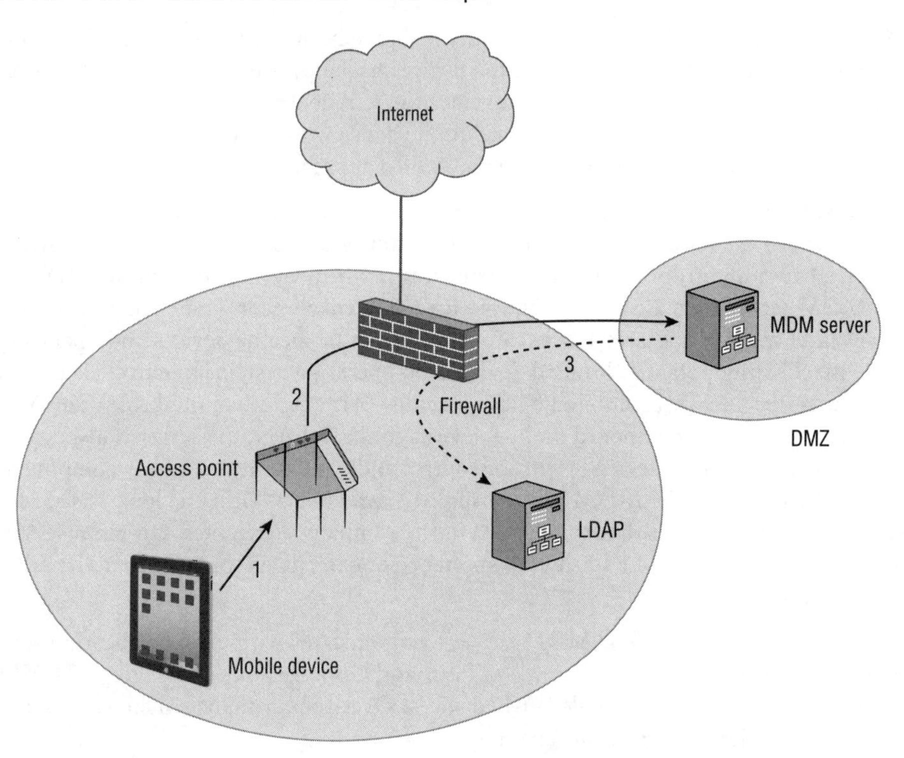

Step 1: The mobile device connects with the access point. The mobile device must first establish an association with an AP. The Wi-Fi security could be open, but usually the CID or personal devices are trying to establish a connection with a secure corporate SSID that is using 802.1X or preshared key (PSK) security. At this point, the AP holds the mobile client device inside a *walled garden*. Within a network deployment, a walled garden is a closed environment that restricts access to web content and network resources while still allowing access to some resources. A walled garden is a closed platform of network services provided for devices and/or users. While inside the walled garden designated by the AP, the only services that the mobile device can access are Dynamic Host Configuration Protocol (DHCP), Domain Name System (DNS), push notification services, and the MDM server. To escape from the walled garden, the mobile device must find the proper exit point, much like in a real walled garden. The designated exit point for a mobile device is the MDM enrollment process.

Step 2: The AP checks whether the device is enrolled. The next step is to determine if the mobile device has been enrolled. Depending on the WLAN vendor, the AP or a WLAN controller queries the MDM server to determine the enrollment status of the mobile device. If the MDM is provided as a cloud-based service, the enrollment query crosses a WAN link. (An on-premise MDM server typically will be deployed in a DMZ.) If the mobile device is already enrolled, the MDM server will send a message to the AP to release the device from the walled garden. Unenrolled devices will remain quarantined inside the walled garden.

Step 3: The MDM server queries LDAP. Although an open enrollment process can be deployed, administrators often require authentication. The MDM server queries an existing LDAP database, such as Active Directory. The LDAP server responds to the query, and then the MDM enrollment can proceed.

Step 4: The device is redirected to the MDM server. Although the unenrolled device has access to DNS services, the quarantined device cannot access any web service other than the MDM server. When the user opens a browser on the mobile device, it is redirected to the captive web portal for the MDM server, as shown in Figure 10.4. The enrollment process can then proceed. For legal and privacy reasons, captive web portals contain a legal disclaimer agreement that gives the MDM administrator the ability to restrict settings and remotely change the capabilities of the mobile device. The legal disclaimer is particularly important for a BYOD situation where employees are onboarding their own personal devices. If the user does not agree to the legal disclaimer, they cannot proceed with the enrollment process and will not be released from the walled garden.

FIGURE 10.4 MDM server—enrollment captive web portal—step 4

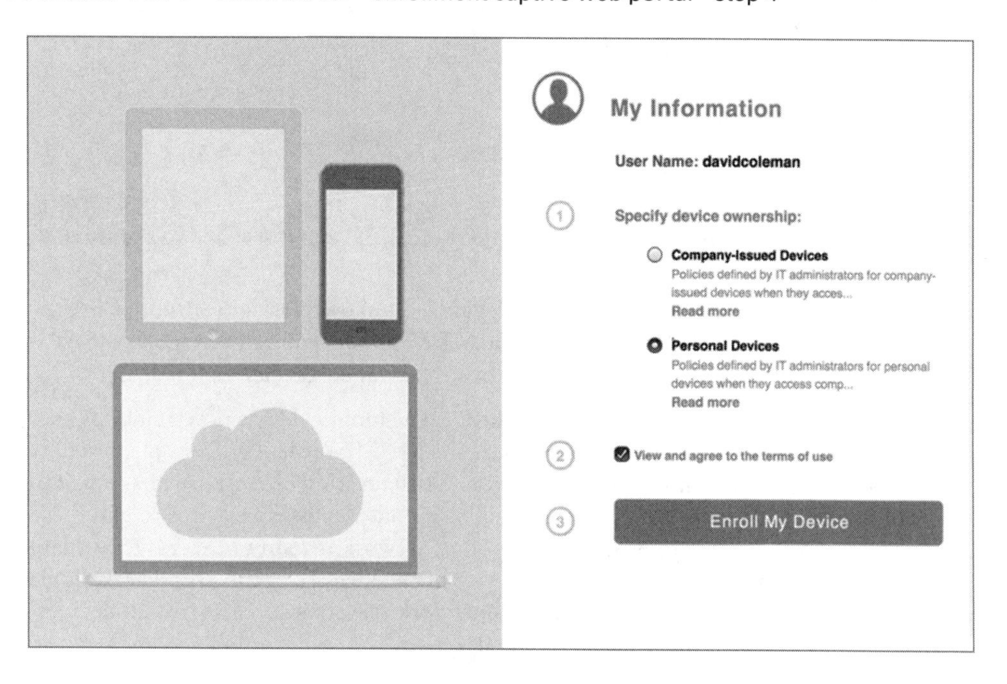

Step 5: The device installs the certificate and MDM profile. Once enrollment begins, a secure *over-the-air provisioning* process for installing the MDM profile is needed. Over-the-air provisioning differs between different device operating systems, but using trusted certificates and SSL encryption is the norm. For this example we will describe how iOS devices are provisioned. For iOS devices, the *Simple Certificate Enrollment Protocol (SCEP)* uses certificates and Secure Sockets Layer (SSL) encryption to protect the MDM profiles. The user of the mobile device accepts an initial profile that is installed on the device. After installation of the initial profile, device-specific identity information can be sent to the MDM server. The MDM server then sends an SCEP payload that instructs the mobile device about how to download a trusted certificate from the MDM certificate authority (CA) or a third-party CA. Once the certificate is installed on the mobile device, the encrypted MDM profile with the device configuration and restrictions payload is sent securely to the mobile device and installed. Figure 10.5 depicts the installation of MDM profiles using SCEP on an iOS device.

FIGURE 10.5 Certificate and MDM profile installation—step 5

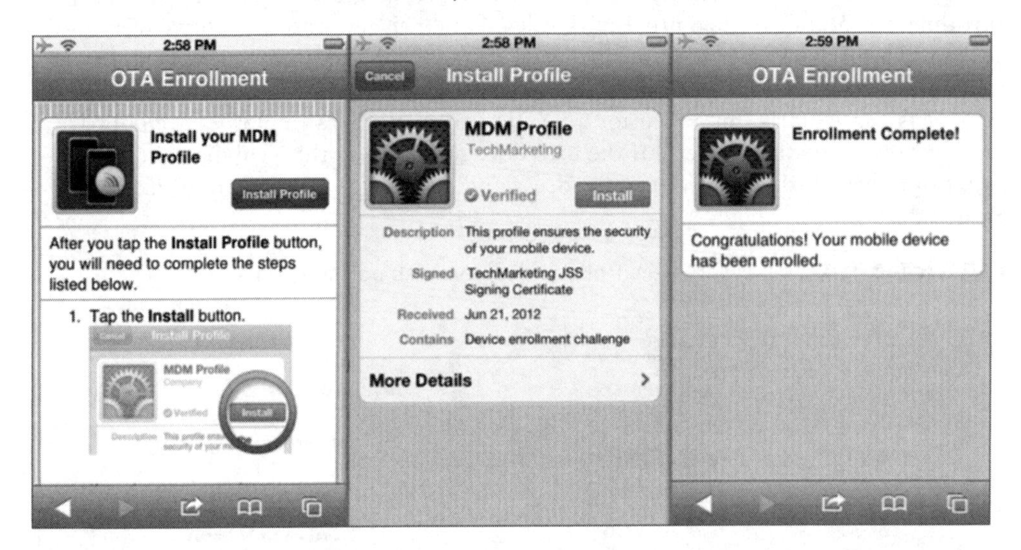

Step 6: The MDM server releases the mobile device. As shown in Figure 10.6, once the device has completed the MDM enrollment, the MDM server sends a message to the AP or WLAN controller to release the mobile device from the walled garden.

Step 7: The mobile device exits the walled garden. The mobile device now abides by the restrictions and configuration settings defined by the MDM profile. For example, use of the mobile device's camera may no longer be allowed. Configuration settings, such as email or VPN settings, also may have been provisioned. The mobile device is now free to exit the walled garden and access the Internet and corporate network resources. Access to available network resources is dictated by the type of device or the identity of the user. For example, company-owned devices may have access to all network servers whereas personal devices may only access specific servers such as the email server.

FIGURE 10.6 The mobile device exits the walled garden in the final step.

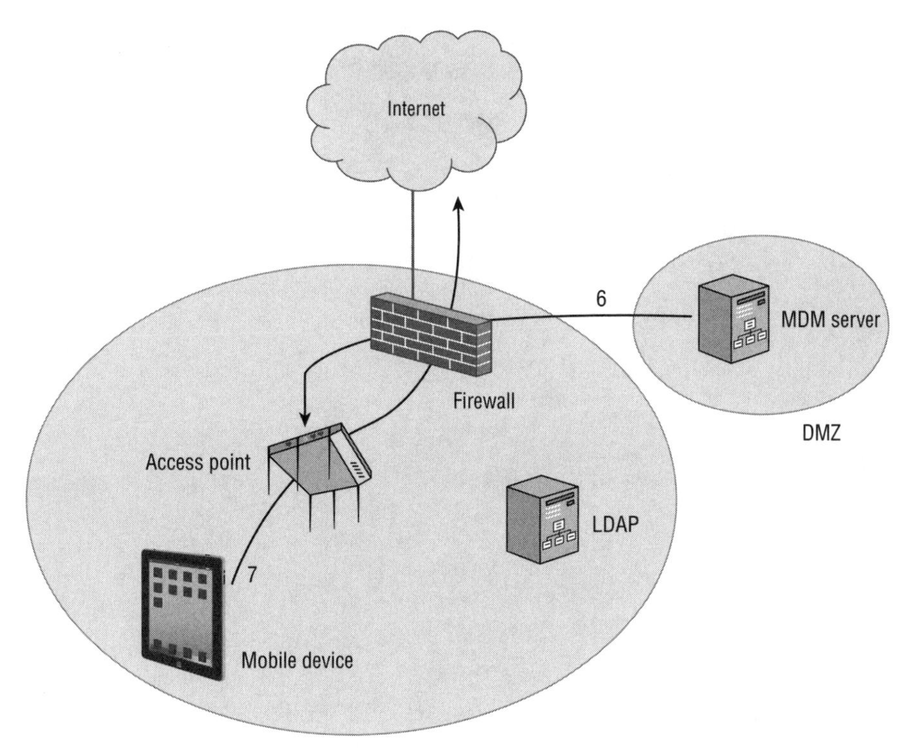

MDM Profiles

We have already learned that MDM profiles are used for mobile device restrictions. The MDM profiles can also be used to globally configure various components of a mobile device. MDM profiles are essentially configuration settings for a mobile device. As the example in Figure 10.7 shows, MDM profiles can include device restrictions, email settings, VPN settings, LDAP directory service settings, and Wi-Fi settings. MDM profiles can also include *webclips*, which are browser shortcuts that point to specific URLs. A webclip icon is automatically installed on the desktop screen of the mobile device. For example, a company-issued device could be provisioned with a webclip link to the company's internal intranet.

The configuration profiles used by Mac OS and iOS devices are *Extensible Markup Language (XML)* files. Apple has several tools to create profiles, including the Apple Configurator and the iPhone Configuration Utility. For manual installations, the XML profiles can be delivered via email or through a website. Manual installation and configuration is fine for a single device, but what about in an enterprise where thousands of devices might need to be configured? In the enterprise, a method is needed to automate the delivery of configuration profiles, and that is where an MDM solution comes into play. MDM configuration profiles are created on the MDM server and installed onto the mobile devices during the enrollment process.

FIGURE 10.7 MDM profile settings

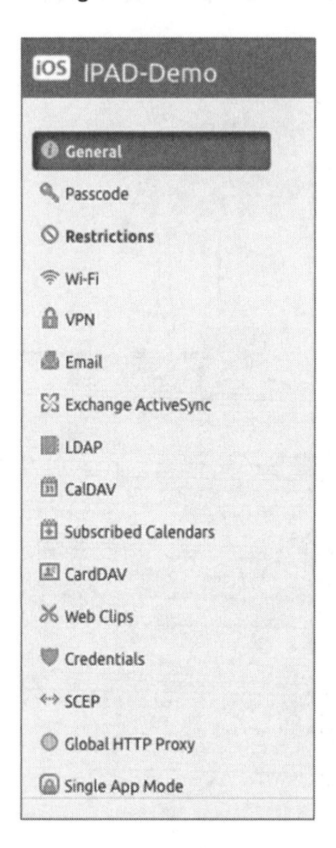

As mentioned, one aspect of an MDM profile is that the Wi-Fi settings can be provisioned. Company-owned devices can be locked down with a specific Wi-Fi profile that designates the corporate SSID and proper security settings. An MDM profile can also be used to deploy Wi-Fi settings to an employee's personal device. If 802.1X/EAP is deployed, a root CA certificate must be installed on the supplicant mobile device. An MDM solution is the easiest way to provision root CA certificates on mobile devices. Client certificates can also be provisioned if EAP-TLS is the chosen 802.1X security protocol.

MDM profiles can be removed from the device locally or can be removed remotely through the Internet via the MDM server.

Can Employees Remove the MDM Profiles from the Mobile Device?

Once a mobile device has gone through an enrollment process, the MDM configuration profiles and related certificates are installed on the mobile device. The following figure shows the settings screen of an iPad with installed MDM profiles.

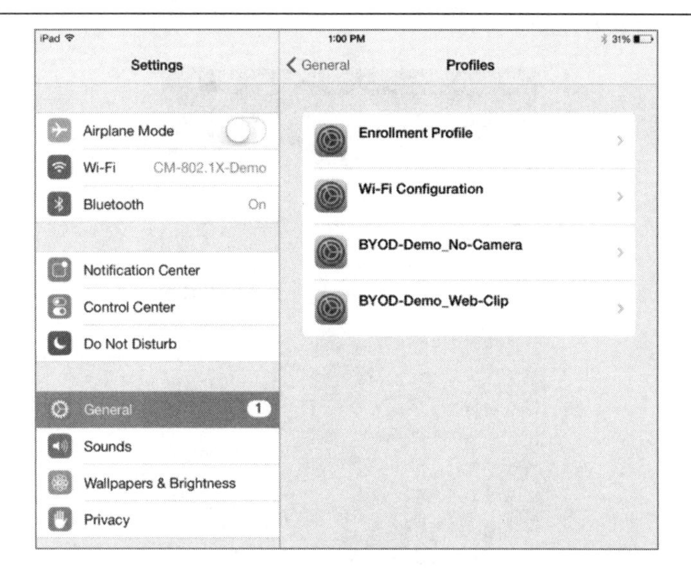

Can an employee remove the MDM profiles? The answer to this question is a matter of company policy. Company-owned mobile devices usually have the MDM profiles locked and they cannot be removed. This prevents the employee from making unauthorized changes to the device. If the mobile device is stolen and sensitive information resides on the device, the MDM administrator can remotely wipe the mobile device if it is connected to the Internet. The BYOD policy of personal devices is usually less restrictive. When employees enroll their personal devices through the corporate MDM solution, typically the employee retains the ability to remove the MDM profiles because they own the device. If the employee removes the MDM profiles, the device is no longer managed by the corporate MDM solution. The next time the employee tries to connect to the company's WLAN with the mobile device, the employee will have to once again go through the MDM enrollment process.

MDM Agent Software

The operating systems of some mobile devices require *MDM agent* application software. For example, Android devices require an MDM agent application like the one shown in Figure 10.8. The Android OS is an open source operating system that can be customized by the various mobile device manufacturers. Although this provides much more flexibility, managing and administering Android devices in the enterprise can be challenging due to the sheer number of hardware manufacturers. An MDM agent application can report unique information about the Android device back to an MDM server, which can later be used in MDM restriction and configuration policies. An MDM agent must support multiple Android device manufacturers.

FIGURE 10.8 MDM agent application

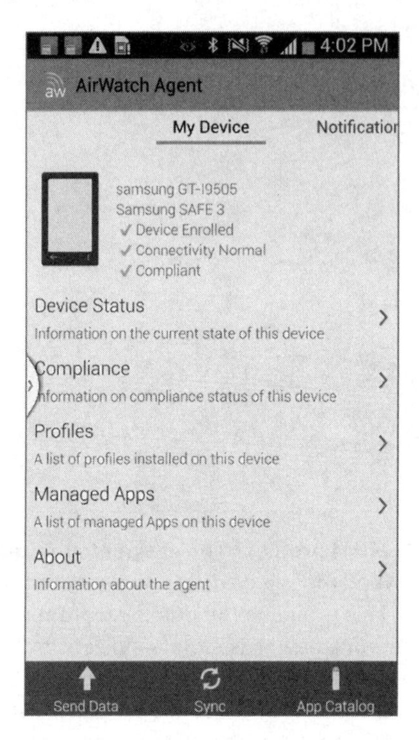

An employee downloads the MDM agent from a public website or company website and installs it on their Android device. The MDM agent contacts the MDM server over the WLAN and is typically required to authenticate to the server. The MDM agent must give the MDM server permission to make changes to the device and function as the administrator of the device. Once this secure relationship has been established, the MDM agent software enforces the device restriction and configuration changes. MDM administration on an Android device is handled by the agent application on the device. Changes can, however, be sent to the MDM agent application from the MDM server via the Google Cloud Messaging (GCM) service.

Although iOS devices do not require MDM agent software, some MDM solutions do offer iOS MDM agents. The MDM agent on the iOS device could potentially send information back to the MDM server that is not defined by the Apple APIs.

Over-the-Air Management

Once a device has been provisioned and enrolled with an MDM server, a permanent management relationship exists between the MDM server and the mobile device. As shown in Figure 10.9, the MDM server can monitor such device information as its name, serial number, capacity, battery life, and the applications that are installed on the device. Information that cannot be seen includes SMS messages, personal emails, calendars, and browser history.

FIGURE 10.9 Device information

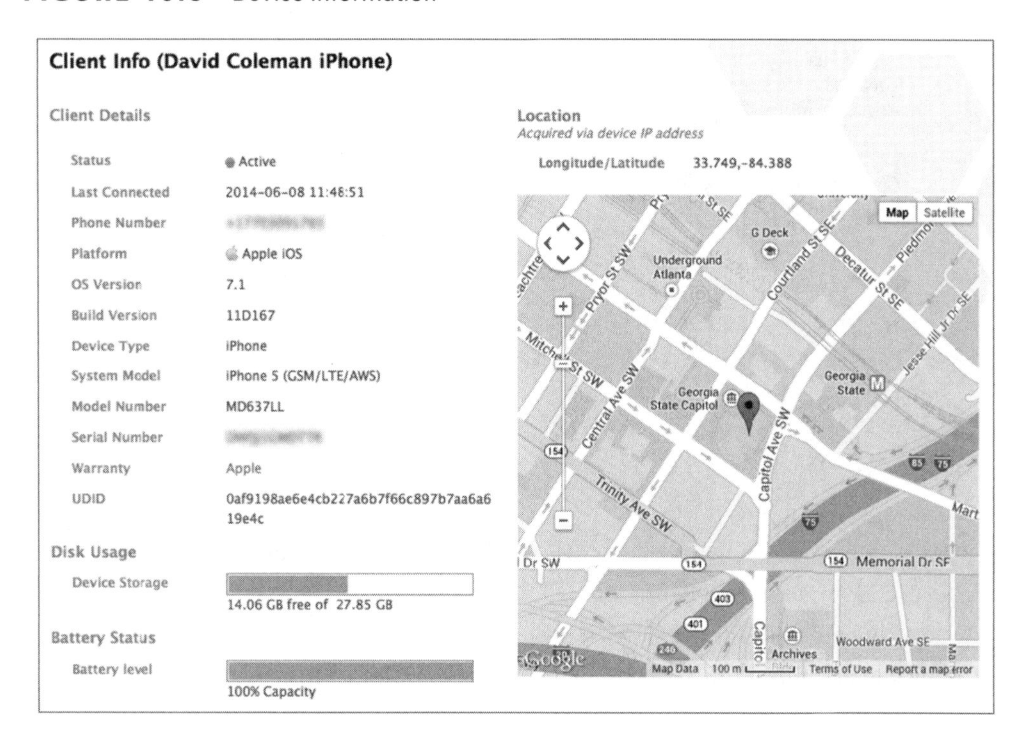

The mobile device can still be managed remotely, even if the mobile device is no longer connected to the corporate WLAN. The MDM server can still manage the device as long as the device is connected to the Internet from any location. The communication between the MDM server and the mobile devices requires push notifications from a third-party service. Both Google and Apple have APIs that allow applications to send push notifications to mobile devices. iOS applications communicate with the Apple Push Notification service (APNs) servers and Android applications communicate with the Google Cloud Messaging (GCM) servers.

As seen in Figure 10.10, the first step is for the MDM administrator to make changes to the MDM configuration profile on the MDM server. The MDM server then contacts push notification servers. A previously established secure connection already exists between the push notification servers and the mobile device. The push notification service then sends a message to the mobile device telling the device to contact the MDM server over the Internet. Once the mobile device contacts the MDM server, the MDM server sends the configuration changes and/or messages to the mobile device.

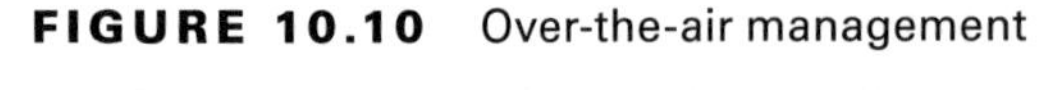

FIGURE 10.10 Over-the-air management

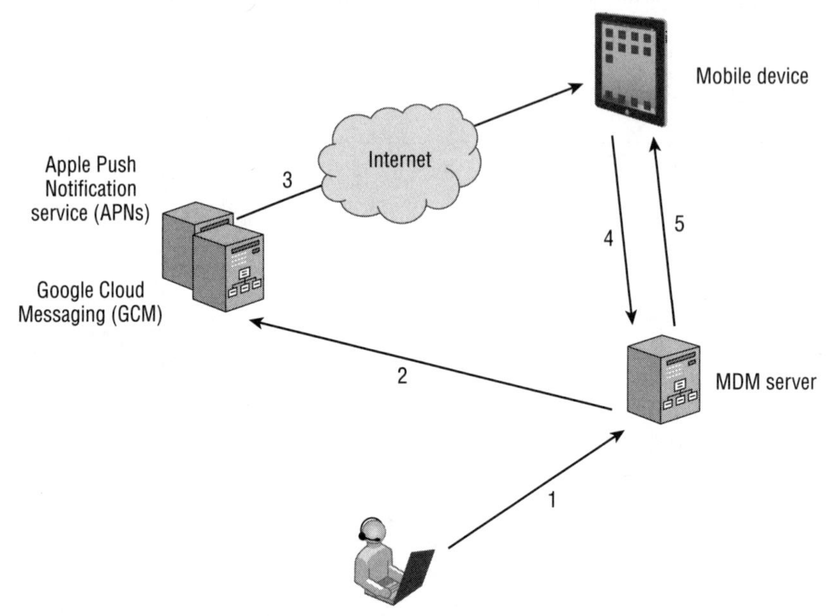

What kind of remote actions can an MDM administrator accomplish over the Internet?

- Make changes to the configuration.
- Make changes to the device restrictions.
- Deliver a message to the device.
- Lock the device.
- Wipe the device.
- Make application management changes.

Stop Thief!

A stolen company-owned device can be remotely wiped. MDM vendors implement different types of remote wipes.

Enterprise Wipe Wipes all corporate data from the selected device and removes the device from the MDM. All of the enterprise data contained on the device is removed, including MDM profiles, policies, and internal applications. The device will return to the state it was in prior to the enrollment with the MDM.

Device Wipe Wipes all data from the device, including all data, email, profiles, and MDM capabilities and returns the device to factory default settings.

Application Management

Enterprise MDM solutions also offer various levels of management of the applications that run on mobile devices. Once an MDM profile is installed, all of the applications installed on the device can be viewed from the MDM server, as shown in Figure 10.11. The MDM server can manage applications by whitelisting and/or blacklisting specific applications that can be used on the mobile devices. Managing applications on company-owned devices is commonplace; however, application management on employee's personal devices is not as prevalent.

FIGURE 10.11 Mobile device applications

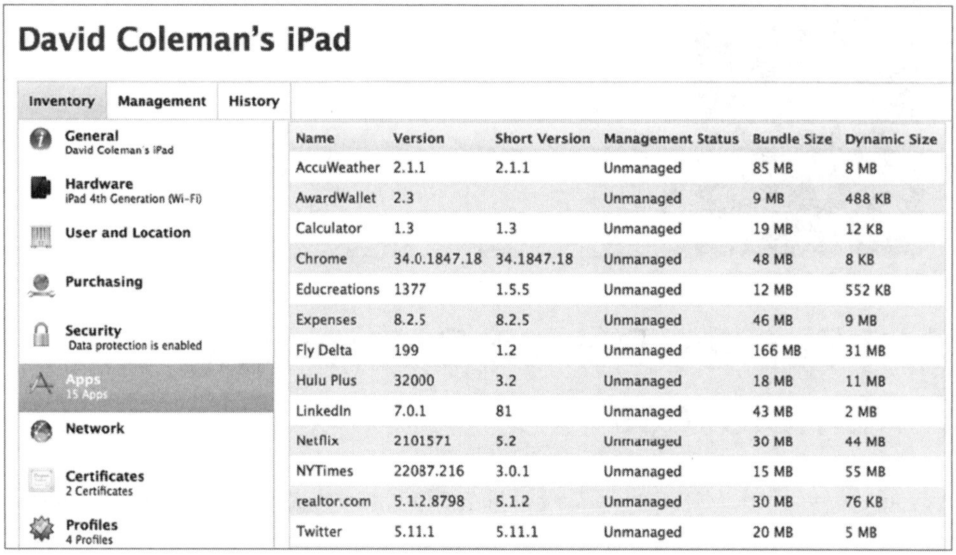

David Coleman's iPad

Inventory | Management | History

General — David Coleman's iPad
Hardware — iPad 4th Generation (Wi-Fi)
User and Location
Purchasing
Security — Data protection is enabled
Apps — 15 Apps
Network
Certificates — 2 Certificates
Profiles — 4 Profiles

Name	Version	Short Version	Management Status	Bundle Size	Dynamic Size
AccuWeather	2.1.1	2.1.1	Unmanaged	85 MB	8 MB
AwardWallet	2.3		Unmanaged	9 MB	488 KB
Calculator	1.3	1.3	Unmanaged	19 MB	12 KB
Chrome	34.0.1847.18	34.1847.18	Unmanaged	48 MB	8 KB
Educreations	1377	1.5.5	Unmanaged	12 MB	552 KB
Expenses	8.2.5	8.2.5	Unmanaged	46 MB	9 MB
Fly Delta	199	1.2	Unmanaged	166 MB	31 MB
Hulu Plus	32000	3.2	Unmanaged	18 MB	11 MB
LinkedIn	7.0.1	81	Unmanaged	43 MB	2 MB
Netflix	2101571	5.2	Unmanaged	30 MB	44 MB
NYTimes	22037.216	3.0.1	Unmanaged	15 MB	55 MB
realtor.com	5.1.2.8798	5.1.2	Unmanaged	30 MB	76 KB
Twitter	5.11.1	5.11.1	Unmanaged	20 MB	5 MB

MDM solutions integrate with public application stores, such as iTunes and Google Play, in order to allow access to public applications. The MDM server communicates with the push notification server, which then places an application icon on the mobile device. The mobile device user can then install the application. The Apple Volume Purchase Program (VPP) provides a way for businesses and educational institutions to purchase apps in bulk and distribute them across their organization. Applications can be purchased and pushed silently to the remote devices. An MDM server can also be configured to deliver custom in-house applications that might be unique to the company.

As shown in Figure 10.12, eBooks can also be managed and distributed to mobile devices via an MDM platform. We suggest that your company make a bulk purchase of the *CWSP Study Guide eBook*.

FIGURE 10.12 MDM distribution of the *CWSP Study Guide eBook*

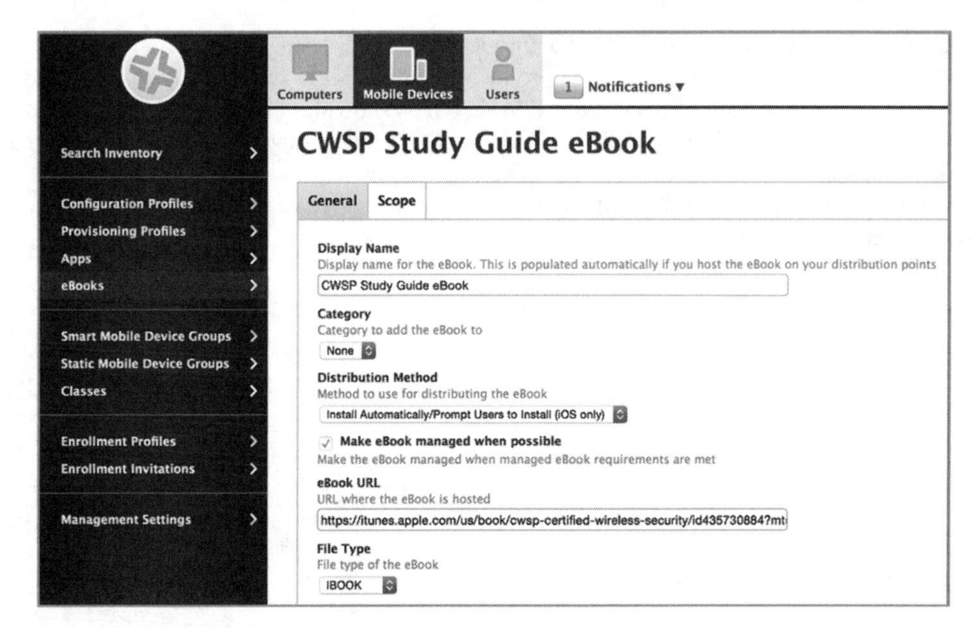

Self-Service Device Onboarding for Employees

As you have learned, MDM solutions can be used to manage and provision both company-owned devices as well as employee-owned WLAN devices. Typical configuration for BYOD management is *self-service device onboarding* solutions as opposed to a robust enterprise MDM. The main purpose of these onboarding solutions is to provide an inexpensive and simple way to provision employee personal WLAN devices onto a secure corporate SSID. A self-service device onboarding solution is not meant to offer all the monitoring and restriction aspects of a full-blown MDM. Instead, device onboarding provides a self-service method for an employee to configure the BYOD supplicant and install security credentials such as an 802.1X/EAP root CA certificate.

Consider this scenario: Jeff logs in to the enterprise network as an employee from his corporate computer using 802.1X/EAP. His username is verified in the LDAP database using RADIUS. In this scenario, Jeff is trusted as the user because his username and password are valid. His corporate laptop is trusted because his machine can be validated. However, Jeff using his corporate laptop is different from Jeff using his smartphone, Jeff using his personal laptop, or Jeff using his tablet.

In the world of authentication and encryption, 802.1X/EAP is typically the required method to provide secure access to corporate networks and data. However, configuration of the client supplicant is typically not a task that can be easily performed by a nontechnical user. Additionally, the root CA certificate needs to be securely transferred to the client device and installed. This can be a problem for corporations and employee BYOD users, since the corporation will not provide access without a properly configured device. As you can imagine, requiring a trained IT help desk person to configure employee personal devices is not practical. The solution is a process known as *onboarding*.

A properly configured and secured 802.1X/EAP network requires that a root CA certificate be installed on the supplicant. Installing the root certificate onto Windows laptops can be easily automated using a *Group Policy Object (GPO)* if the Windows laptop is part of the Active Directory (AD) domain. However, a GPO cannot be used for Mac OS, iOS, or Android mobile devices, or for personal Windows BYOD devices that are not joined to the AD domain. Manually installing certificates on mobile devices and employee-owned devices is an administrative nightmare.

The onboarding solution is most often used to install the root CA certificates on mobile devices to be used with an 802.1X/EAP-enabled SSID. Client certificates can also be provisioned with an onboarding solution. Some of the Wi-Fi vendors that offer dynamic PSK solutions also offer onboarding solutions that can provision mobile devices with Wi-Fi client profiles configured with unique individual PSKs.

Self-service onboarding solutions for personal employee devices can come in many different forms and are often WLAN vendor-specific. Onboarding solutions typically use an application that uses similar over-the-air provisioning aspects of MDM solutions to securely install certificates and Wi-Fi client profiles onto mobile devices. Self-service onboarding solutions may also use custom applications built using a WLAN vendor's *application programming interface (API)*. Regardless of the solution, the device onboarding normally requires an initially WLAN connection to complete the self-service process.

Dual-SSID Onboarding

Dual-SSID onboarding is performed using an open SSID and an 802.1X/EAP secure corporate SSID. The employee initially connects to the open SSID and will be prompted with a captive portal page. Depending on how onboarding is implemented, the employee may log in directly through the captive portal using their corporate username and password, or the employee may click on a link that would take them to an onboard login screen where they would enter their corporate username and password. The captive web portal authentication validates the employee's username and password via RADIUS and LDAP. The captive web portal authentication is protected via HTTPS. After the employee logs in to the network, an onboarding application is typically downloaded to the mobile device. The onboarding application is then executed to securely download the 802.1X root certificate and/or other security credentials as well as provision the supplicant on the mobile device. Custom onboarding applications are often distributed via the open SSID. An onboarding solution for different types of users might also be available as a web-based application. Figure 10.13 shows a custom onboarding web application used by the Calgary Board of Education used to provision different security credentials to guests, students, and staff.

FIGURE 10.13 BYOD onboarding application

Courtesy of the Calgary Board of Education

After the employee device is provisioned using the onboarding application, it is ready to connect to the secure network. Because the secure network is a separate SSID, the employee would have to manually disconnect from the open SSID and reconnect to the secure SSID.

Single-SSID Onboarding

Single-SSID onboarding uses a single SSID that is capable of authenticating 802.1X/EAP-PEAP clients and 802.1X/EAP-TLS clients. The client initially logs in to the SSID using an 802.1X/EAP-PEAP connection, using their corporate username and password. After the device is logged on to the network, the employee would bring up a captive portal page, requiring the user to log in again, this time to validate that they are allowed to perform the onboard process. As with the dual-SSID process, an onboarding program is downloaded to the device and executed, and then the application downloads the server certificate via SSL and provisions the supplicant on the device.

After the employee device is configured, the RADIUS server will initiate a change of authorization (CoA) to the employee device, disconnecting the device from the network. The device will immediately reconnect to the same SSID, using either 802.1X/EAP-PEAP or 802.1X/EAP-TLS, depending on the wireless profile that was installed on the employee device. This time, the client will also validate the server certificate.

MDM vs. Self-Service Onboarding

MDM solutions are often the preferred choice for large corporations. An enterprise MDM gives a corporation the ability to manage and monitor company WLAN devices as well as provide a provisioning solution for employee personal WLAN devices. However, an MDM solution is not always the best choice for a BYOD solution. Enterprise MDM deployments are often cost-prohibitive for medium-size and small businesses. As previously mentioned, employees often do not like to use MDM due to privacy issues.

Self-service device onboarding solutions are typically much cheaper and simpler to deploy as an employee BYOD solution. Self-service onboarding solutions are used primarily to provision employee WLAN devices and are not used to enforce device restrictions or for over-the-air management. The privacy concerns are no longer an issue for the employee personal devices.

Depending on a company's security requirements, MDM, self-service onboarding, or a combination of the two solutions can be chosen for a BYOD solution.

Guest WLAN Access

Although the primary purpose for enterprise WLANs has always been to provide employees wireless mobility, WLAN access for company guests can be just as important. Customers, consultants, vendors, and contractors often need access to the Internet to accomplish job-related duties. When they are more productive, employees will also be more productive. Guest access can also be a value-added service and often breeds customer loyalty. In today's world, business customers have come to expect guest WLAN access. Free guest access is often considered a value-added service. There is a chance that your customers will move toward your competitors if you do not provide guest WLAN access. Retail, restaurants, and hotel chains are all prime examples of environments where wireless Internet access is often expected by customers.

The primary purpose of a guest WLAN is to provide a wireless gateway to the Internet for company visitors and/or customers. Generally, guest users do not need access to company network resources. Therefore, the most important security aspect of a guest WLAN is to protect the company network infrastructure from the guest users. In the early days of Wi-Fi, guest networks were not very common because of fears that the guest users might access corporate resources. Guest access was often provided on a separate infrastructure. Another common strategy was to send all guest traffic to a separate gateway that was different from the Internet gateway for company employees. For example, a T1 or T3 line might have been used for the corporate gateway, whereas all guest traffic was segmented on a separate DSL phone line.

WLAN guest access has grown in popularity over the years, and the various types of WLAN guest solutions have evolved to meet the need. In the following sections, we will discuss the security aspects of guest WLANs. At a minimum, there should be a separate guest SSID, a unique guest VLAN, a guest firewall policy, and a captive web portal.

We will also discuss the many guest access options that are available, including guest self-registration.

Guest SSID

In the past, a common SSID strategy was to segment different types of users—even employees—on separate SSIDs; each SSID was mapped to an independent VLAN. For example, a hospital might have unique SSID/VLAN pairs for doctors, nurses, technicians, and administrators. That strategy is rarely recommended now because of the Layer 2 overhead created by having many SSIDs. Today, the more common method is to place all employees on the same SSID and leverage Remote Authentication Dial-In User Service (RADIUS) attributes to assign different groups of users to different VLANS. What has not changed over time is the recommendation that all guest user traffic be segmented onto a separate SSID. The guest SSID will always have different security parameters than the employee SSID, and therefore the necessity of a separate guest SSID continues. For example, employee SSIDs commonly use 802.1X/EAP security, whereas guest SSIDs are most often an open network that uses a captive web portal for authentication. Although encryption is not usually provided for guest users, some WLAN vendors have begun to offer encrypted guest access and provide data privacy using dynamic PSK credentials. Encrypted guest access can also be provided with 802.1X/EAP with Hotspot 2.0, which is discussed later.

Like all SSIDs, a guest SSID should never be hidden and should have a simple name, such as CWSP-Guest. In most cases, the guest SSID is prominently displayed on a sign in the lobby or entrance of the company offices.

Guest VLAN

Guest user traffic should be segmented into a unique VLAN tied to an IP subnet that does not mix with the employee VLANs. Segmenting your guest users into a unique VLAN is a security and management best practice. The main debate about the guest VLAN is whether or not the guest VLAN should be supported at the edge of the network. As shown in Figure 10.14, a frequent design scenario is that the guest VLAN does not exist at the edge of the network and instead is isolated in what is known as a *demilitarized zone (DMZ)*. As shown in Figure 10.14, the guest VLAN (VLAN 10) does not exist at the access layer, and therefore, all guest traffic must be tunneled from the AP back to the DMZ where the guest VLAN does exist. An IP tunnel, commonly using Generic Routing Encapsulation (GRE) protocol, transports the guest traffic from the edge of the network back to the isolated DMZ. Depending on the WLAN vendor solution, the tunnel destination in the DMZ can be either a WLAN controller or simply a Layer 2 server appliance.

Although isolating the guest VLAN in a DMZ has been a common practice for many years, it is no longer necessary if guest firewall policies are being enforced at the edge of the network. Various WLAN vendors are now building enterprise-class firewalls into access points. If the guest firewall policy can be enforced at the edge of the network, the guest VLAN can also reside at the access layer and no tunneling is needed.

FIGURE 10.14 GRE tunneling guest traffic to a DMZ

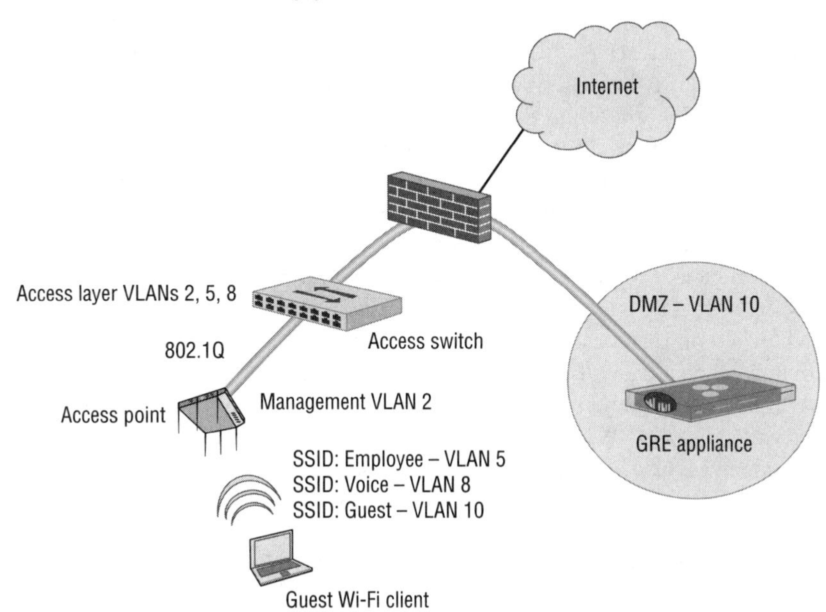

Guest Firewall Policy

The most important security component of a guest WLAN is the firewall policy. The guest WLAN firewall policy prevents guest user traffic from getting near the company network infrastructure and resources. Figure 10.15 shows a very simple guest firewall policy that allows DHCP and DNS but restricts access to private networks 10.0.0.0/8, 172.16.0.0/12, and 192.168.0.0/16. Guest users are not allowed on these private networks because corporate network servers and resources often reside on that private IP space. The guest firewall policy should simply route all guest traffic straight to an Internet gateway and away from the corporate network infrastructure.

FIGURE 10.15 Guest firewall policy

Source IP	Destination IP	Service	Action
Any	Any	DHCP-Server	PERMIT
Any	Any	DNS	PERMIT
Any	10.0.0.0/255.0.0.0	Any	DENY
Any	172.16.0.0/255.240.0.0	Any	DENY
Any	192.168.0.0/255.255.0.0	Any	DENY
Any	Any	Any	PERMIT

Firewall ports that should be permitted include the DHCP server (UDP port 67), DNS (UDP port 53), HTTP (TCP port 80), and HTTPS (TCP port 443). This allows the guest user's wireless device to receive an IP address, perform DNS queries, and browse the Web. Many companies require their employees to use a secure VPN connection when the employee is connected to an SSID other than the company SSID. Therefore, it is recommended that IPsec IKE (UDP port 500) and IPsec NAT-T (UDP port 4500) also be permitted.

The firewall policy shown in Figure 10.15 represents the minimum protection needed for a guest WLAN. The guest firewall policy can be much more restrictive. Depending on company policy, many more ports can be blocked. A good practice is to force the guest users to use webmail and block SMTP and other email ports so that users cannot "spam through" the guest WLAN. It is up to the security policy of the company to determine what ports need to be blocked on the guest VLAN. If the policy forbids the use of SSH on the guest WLAN, then TCP port 22 will need to be blocked. In addition to blocking UDP and TCP ports, several WLAN vendors now have the ability to block applications. In addition to stateful firewall capability, WLAN vendors have begun to build application-layer firewalls capable of *deep packet inspection (dpi)* into access points or WLAN controllers. An application-layer firewall can block specific applications or groups of applications. For example, some popular video streaming applications can be blocked on the guest SSID, as shown in Figure 10.16. The company security policy will also determine which applications should be blocked on a guest WLAN.

FIGURE 10.16 Application firewall policy

Source IP	Destination IP	Service	Action
Any	Any	YOUTUBE	DENY
Any	Any	NETFLIX VIDEO STREAM	DENY
Any	Any	FACETIME	DENY
Any	Any	GOOGLE VIDEO	DENY
Any	Any	INSTAGRAM VIDEO	DENY
Any	Any	Any	PERMIT

Captive Web Portals

Often, guest users must log in through a captive web portal page before they are given access to the Internet. One of the most important aspects of the captive web portal page is the legal disclaimer. A good legal disclaimer informs the guest users about acceptable behavior when using the guest WLAN. Businesses are more likely to be legally protected if something bad, such as being infected by a computer virus, should happen to a guest user's WLAN device while connected through the portal. A *captive portal* solution effectively turns a web browser into an authentication service. To authenticate, the user must first connect to the WLAN and launch a web browser. After the browser is launched and the user attempts to go to a website, no matter what web page the user attempts to browse to, the user is redirected to a different URL, which displays a captive portal logon page. Captive portals can redirect unauthenticated users to a logon page using an IP redirect, DNS redirection, or redirection by HTTP. As shown in Figure 10.17, many captive web portals are triggered by DNS redirection. The guest user attempts to browse to a web page but the DNS query redirects the browser to the IP address of the captive web portal.

FIGURE 10.17 Captive web portal—DNS redirect

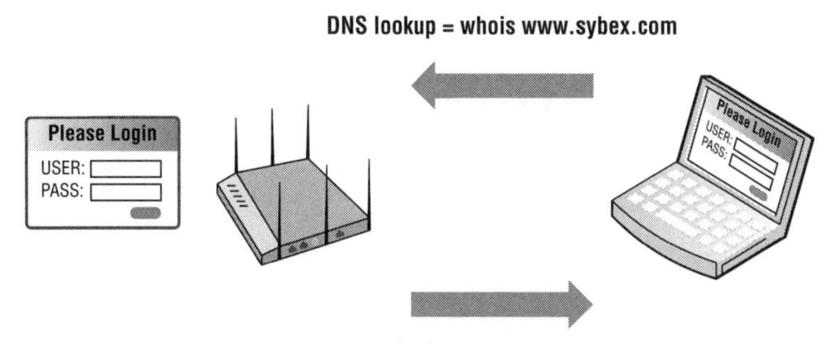

Captive portals are available as standalone software solutions, and most WLAN vendors offer integrated captive portal solutions. The captive portal may exist within a WLAN controller, or it may be deployed at the edge within an access point. WLAN vendors that support captive portals provide the ability to customize the captive portal page. You can typically personalize the page by adding graphics such as a company logo, inserting an acceptable use policy, or configuring the logon requirements. Depending on the chosen security of the guest WLAN, different types of captive web portal logon pages can be used. A user authentication logon page requires the AP or WLAN controller to query a RADIUS server with the guest user's name and password. If the guest user does not already have an account, the logon page may provide a link, allowing the user to create a guest account, as shown in Figure 10.18. The guest registration page allows the user to enter the necessary information for them to self-register, as shown in Figure 10.19. The guest user may also be connected to a captive portal web page requiring them to simply acknowledge a user policy acceptance agreement, as shown in Figure 10.20.

FIGURE 10.18 Captive web portal—guest logon

CWSP Guest Network

Please login to the network using your username and password.

If you do not have a login account, click here Create Account

CWSP Guest Network

Username: david@westcott-consulting.com

Password: ••••••••

Terms: I accept the terms of use

Log In

Contact a staff member if you are experiencing difficulty logging in.

FIGURE 10.19 Captive web portal—guest self-registration

Guest Registration

Please complete the form below to gain access to the network.

Visitor Registration

* Your Name:	David Westcott Please enter your full name.
Phone Number:	555-123-4567 Please enter your contact phone number.
* Company Name:	Westcott Consulting, Inc. Please enter your company name.
* Email Address:	david@westcott-consulting.com Please enter your email address. This will become your username to log into the network.
* Confirm:	I accept the terms of use

Register

* required field

Already have an account? Sign In

FIGURE 10.20 Captive web portal—policy acceptance

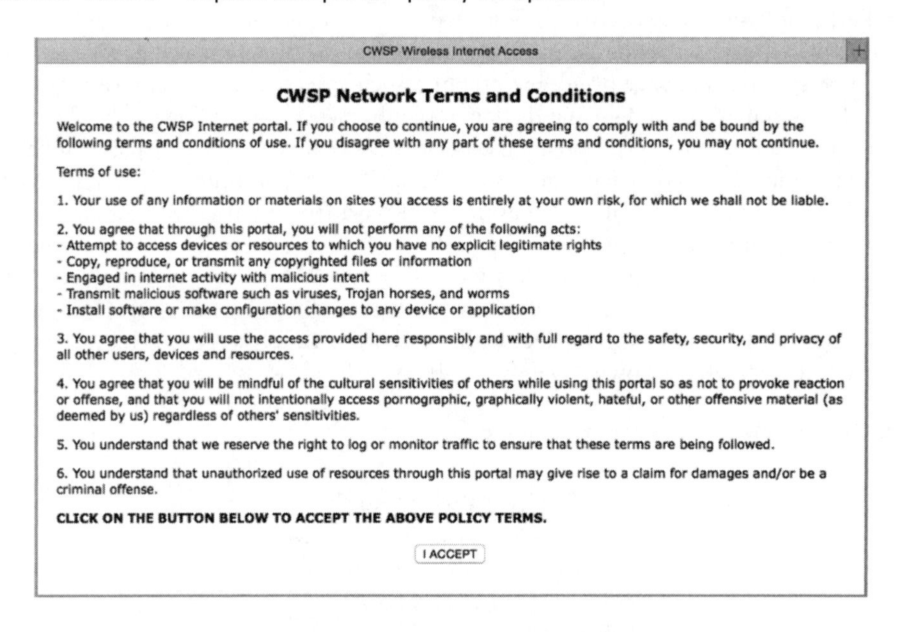

CWSP Wireless Internet Access

CWSP Network Terms and Conditions

Welcome to the CWSP Internet portal. If you choose to continue, you are agreeing to comply with and be bound by the following terms and conditions of use. If you disagree with any part of these terms and conditions, you may not continue.

Terms of use:

1. Your use of any information or materials on sites you access is entirely at your own risk, for which we shall not be liable.

2. You agree that through this portal, you will not perform any of the following acts:
- Attempt to access devices or resources to which you have no explicit legitimate rights
- Copy, reproduce, or transmit any copyrighted files or information
- Engaged in internet activity with malicious intent
- Transmit malicious software such as viruses, Trojan horses, and worms
- Install software or make configuration changes to any device or application

3. You agree that you will use the access provided here responsibly and with full regard to the safety, security, and privacy of all other users, devices and resources.

4. You agree that you will be mindful of the cultural sensitivities of others while using this portal so as not to provoke reaction or offense, and that you will not intentionally access pornographic, graphically violent, hateful, or other offensive material (as deemed by us) regardless of others' sensitivities.

5. You understand that we reserve the right to log or monitor traffic to ensure that these terms are being followed.

6. You understand that unauthorized use of resources through this portal may give rise to a claim for damages and/or be a criminal offense.

CLICK ON THE BUTTON BELOW TO ACCEPT THE ABOVE POLICY TERMS.

I ACCEPT

Client Isolation, Rate Limiting, and Web Content Filtering

When guest users are connected to the guest SSID, they are all in the same VLAN and the same IP subnet. Because they reside in the same VLAN, the guests can perform peer-to-peer attacks against each other. Client isolation is a feature that can be enabled on WLAN access points or controllers to block wireless clients from communicating directly with other wireless clients on the same wireless VLAN. *Client isolation* (or the various other terms used to describe this feature) usually means that packets arriving at the AP's wireless interface are not allowed to be forwarded back out of the wireless interface to other clients. This isolates each user on the wireless network to ensure that a wireless station cannot be used to gain Layer 3 or higher access to another wireless station. The client isolation feature is usually a configurable setting per SSID linked to a unique VLAN. Client isolation is highly recommended on guest WLANs to prevent peer-to-peer attacks.

Enterprise WLAN vendors also offer the capability to throttle bandwidth of user traffic. Bandwidth throttling, which is also known as *rate limiting*, can be used to curb traffic at either the SSID level or the user level. Rate limiting is recommended on guest WLANs. It can ensure that the majority of the wireless bandwidth is preserved for employees. Rate limiting the guest user traffic to 1024 Kbps is a common practice.

Enterprise companies often deploy *web content filter* solutions to restrict the type of websites that their employees can view while at the workplace. A web content filtering solution blocks employees from viewing websites based on content categories. Each category contains websites or web pages that have been assigned based on their primary web content. For example, the company might use a web content filter to block employees from viewing any websites that pertain to gambling or violence. Content filtering is most often used to block what employees can view on the Internet, but web content filtering can also be used to block certain types of websites from guest users. All guest traffic might be routed through the company's web content filter.

Guest Management

As Wi-Fi has evolved, so have WLAN guest management solutions. Most guest WLANs require a guest user to authenticate with credentials via a captive web portal. Therefore, a database of user credentials must be created. Unlike user accounts in a preexisting Active Directory database, guest user accounts are normally created on the fly and are created in a separate guest user database. Guest user information is usually collected when the guests arrive at company offices. Someone has to be in charge of managing the database and creating the guest user accounts. IT administrators are typically too busy to manage a guest database; therefore, the individual who manages the database is often a receptionist or the person who greets guests at the front door. This individual requires an administrative account to the guest management solution, which might be a RADIUS server or some type of other guest database server. The guest management administrators have the access rights to create guest user accounts in the guest database and issue the guest credentials, which are usually usernames and passwords.

A guest management server can be cloud based or can reside as an on-premise server in the company datacenter. Although most guest management systems are built around a RADIUS server, the guest management solution offers features in addition to providing RADIUS services. Modern WLAN guest management solutions offer robust report-generation capabilities for auditing and compliance requirements. As shown in Figure 10.21, a guest management solution can also be used as a 24/7 full-time monitoring solution. An IT administrator usually configures the guest management solution initially; however, a company receptionist will have limited access rights to provision guest users. Guest management solutions can also be integrated with LDAP for employee sponsorships and usually have some method for guest users to self-register. Most often, guest management solutions are used for wireless guests, but they might also be used to authenticate guests connected to wired ports.

FIGURE 10.21 Guest management and monitoring

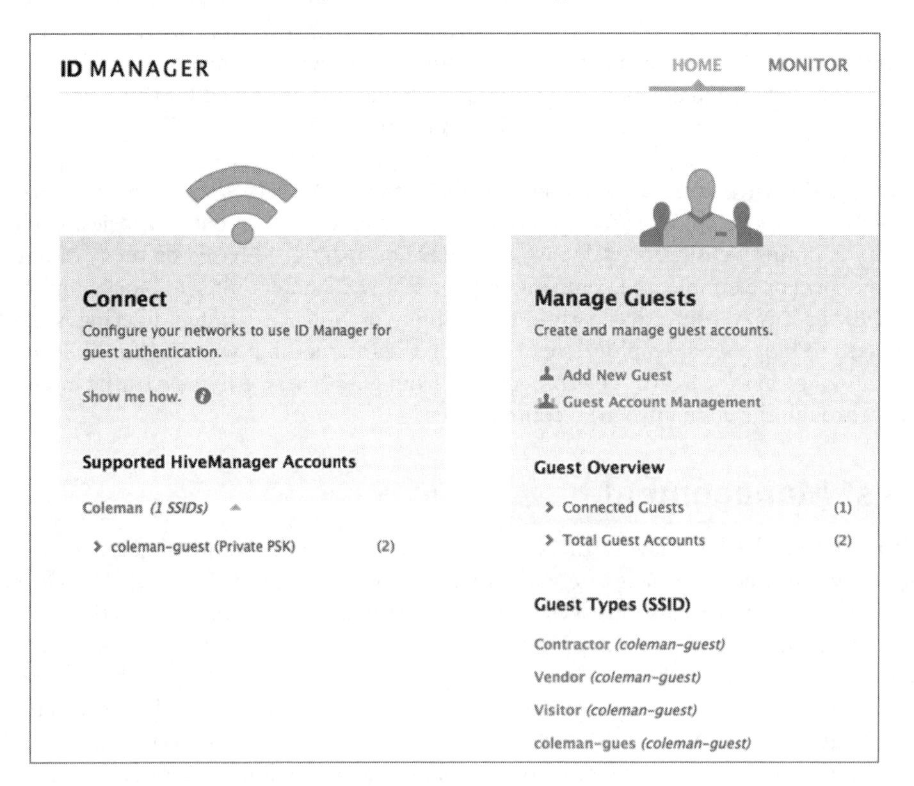

As you can see in Figure 10.22, there can be multiple ways to deliver the guest credentials to the guest user. The credentials can be delivered via an electronic wallet, SMS text message, an email message, or a printed receipt. The SMS, email, and receipt can also be customized with company information. The guest registration logon pages can all be customized with company logos and information.

FIGURE 10.22 Guest credential delivery methods

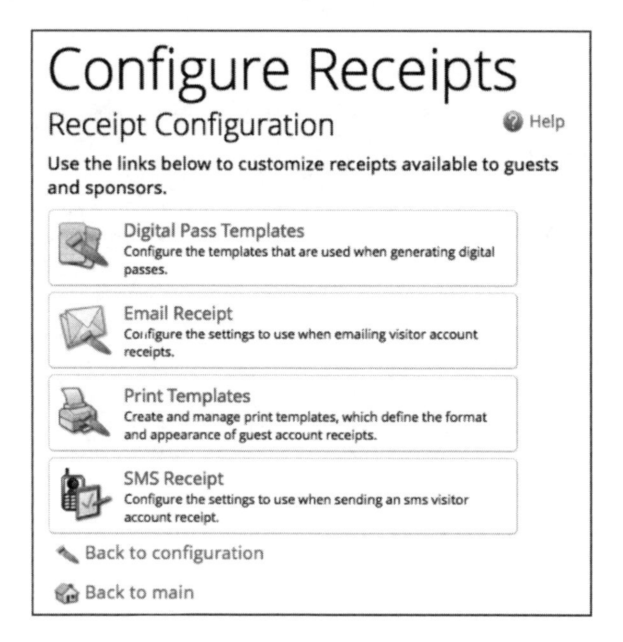

Guest Self-Registration

Guest management solutions have traditionally relied on a company receptionist or lobby ambassador to register the guest users. A good guest management solution allows the receptionist to register a single guest user or groups of users. Over the past few years, there has also been a greater push for guest users to create their own account, what is commonly referred to as self-registration. When the guest is redirected to the captive web portal, if they do not already have a guest account, a link on the logon web page redirects them to a self-registration page. Simple self-registration pages allow guests to fill out a form and their guest account is created and displayed or printed for them. More advanced self-registration pages require guests to enter an email or SMS address, which is then used by the registration system to send users their logon credentials.

As shown in Figure 10.23, some guest management solutions now offer a kiosk mode, where the self-registration logon page runs on a tablet that functions as the kiosk. Self-registration via a kiosk is quite useful when the kiosk is deployed in the main lobby or at the entrance to the company. An advantage of self-registration kiosks is that the receptionist does not have to assist the users and can concentrate on other work duties.

FIGURE 10.23 Kiosk mode

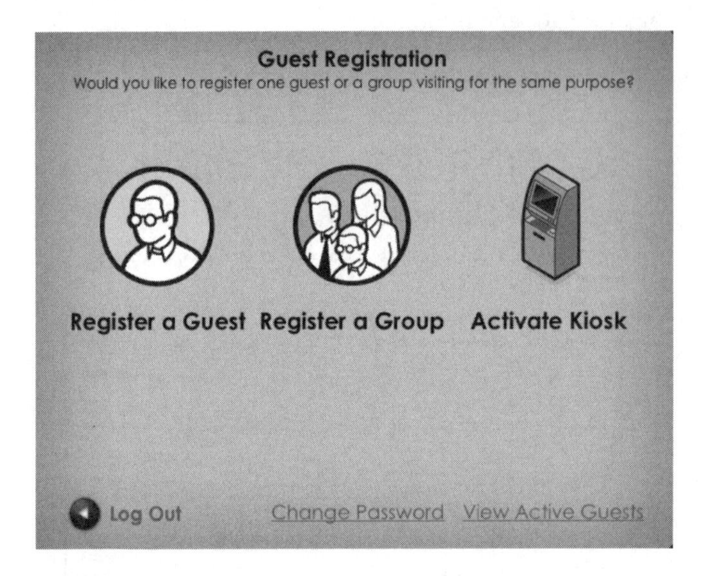

Employee Sponsorship

Guest users can also be required to enter the email address of an employee, who in turn must approve and sponsor the guest. The sponsor typically receives an email with a link that allows them to easily accept or reject the guest's request. Once the user is registered or sponsored, they can log on using their newly created credentials. A guest management solution with *employee sponsorship* capabilities can be integrated with an LDAP database, such as Active Directory.

As you already learned, a receptionist can register guest users or a company may choose to use a registration kiosk so that guests can self-register. For larger or distributed organizations, a central registration kiosk does not scale well. Self-registration with employee sponsorship is becoming popular for many organizations.

When guest users initially connect to the guest network, they are redirected to a captive portal page. The captive portal page prompts them to log on if they already have an account, or it allows them to click a link that allows them to create their own guest account. As shown in Figure 10.24, the guest must enter the email address of the employee who is sponsoring them. Typically, this is the person they are meeting with.

When the registration form is completed and submitted, the sponsor receives an email notifying them that the guest would like network access. As shown in Figure 10.25, the email typically contains a link that the sponsor must click to approve network access. Once the link is clicked, the guest account is approved and the guest receives confirmation either by email or SMS, and they will then be allowed to log on to the network. If the sponsor does not click the link, the guest account is never created and the guest is denied access to the network.

FIGURE 10.24 Employee sponsorship registration

CWSP Guest Registration

Please complete the form below to gain access to the network.

Visitor Registration

* Your Name: David Coleman
Please enter your full name.

Phone Number: 555-123-4567
Please enter your contact phone number.

* Email Address: dcoleman@aerohive.com
Please enter your email address.
This will become your username to log into the network.

* Sponsor's Name: Metka Dragos-Radanovic
Name of the person sponsoring this account.

* Sponsor's Email: mdragos@aerohive.com
Email of the person sponsoring this account.

* Confirm: I accept the terms of use

Register

* required field

Already have an account? Sign In

FIGURE 10.25 Employee sponsorship confirmation email

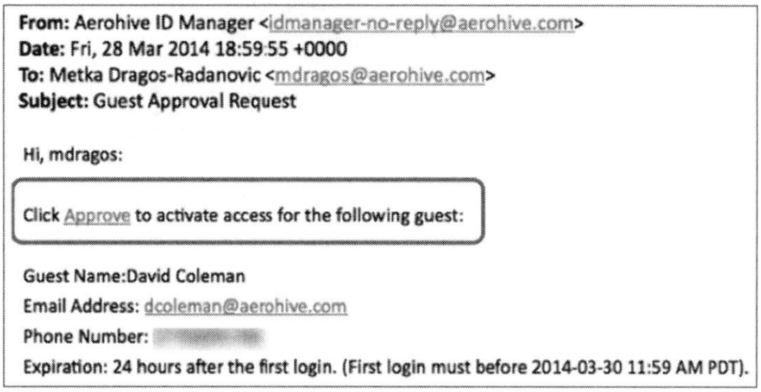

From: Aerohive ID Manager <idmanager-no-reply@aerohive.com>
Date: Fri, 28 Mar 2014 18:59:55 +0000
To: Metka Dragos-Radanovic <mdragos@aerohive.com>
Subject: Guest Approval Request

Hi, mdragos:

Click Approve to activate access for the following guest:

Guest Name:David Coleman
Email Address: dcoleman@aerohive.com
Phone Number:
Expiration: 24 hours after the first login. (First login must before 2014-03-30 11:59 AM PDT).

Employee sponsorship ensures that only authorized guest users are allowed onto the guest WLAN and that the company employees are actively involved in the guest user authorization process.

Social Login

A new trend in guest networks in retail and service industries is *social login*. Social login is a method of using existing logon credentials from a social networking service (such as Twitter, Facebook, or LinkedIn) to register on a third-party website. Social login allows a user to forgo the process of creating new registration credentials for the third-party website. Social login is often enabled using the *OAuth* protocol. OAuth (Open Standard for Authorization) is a secure authorization protocol that allows access tokens to be issued to third-party clients by an authorization server. As shown in Figure 10.26, the OAuth 2.0 authorization framework enables a third-party application to obtain limited access to an HTTP service and can be used for social login for Wi-Fi guest networks.

FIGURE 10.26 OAuth 2.0 application

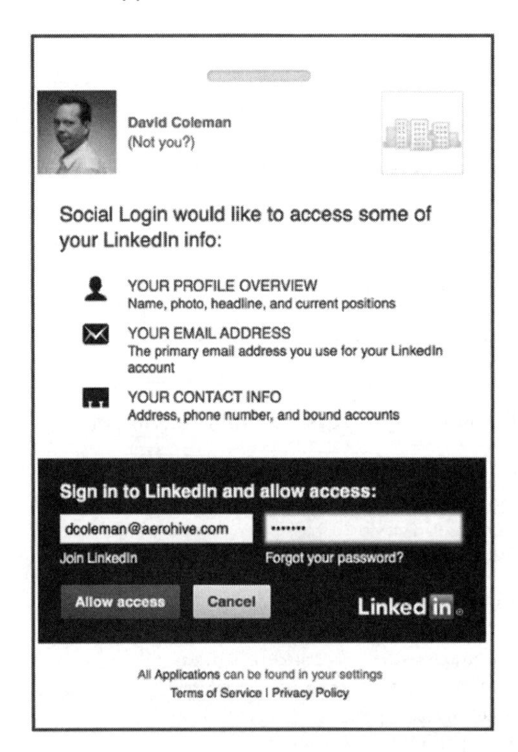

As shown in Figure 10.27, social login can be tied to an open guest SSID. Guest users are redirected to a captive web portal page and they can then log on to the guest WLAN using their existing social media logon credentials. Retail and service businesses like the idea of social login because it allows the business to obtain meaningful marketing information about the guest user from the social networking service. Businesses can then build a database of the type of customers who are using the guest Wi-Fi while shopping at the business. It should be noted that there are serious privacy concerns with social login, and the logon captive web portal always has a legal disclaimer stating that customer information might be gathered if the customer agrees to use the social login registration to the guest WLAN.

FIGURE 10.27 Social login

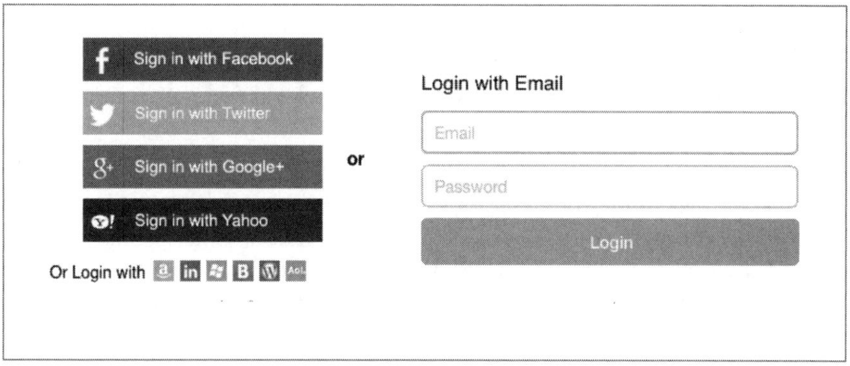

Encrypted Guest Access

Most guest networks are open networks that do not use encryption; thus, there is no data privacy for guest users. In Chapter 14, "Wireless Security Monitoring," you learned about the numerous wireless attacks that make unsecured Wi-Fi users vulnerable. Because no encryption is used on most guest WLANs, the guest users are low-hanging fruit and often targets of skilled hackers or attackers. For that reason, many corporations require their employees to use an IPsec VPN solution when connected to any kind of public or open guest SSID. Because the guest SSID does not provide data protection, the guest user must bring their own security in the form of a VPN connection that provides encryption and data privacy.

The problem is that many consumers and guest users are not savvy enough to know how to use a VPN solution when connected to an open guest WLAN. As a result, there is a recent trend to provide encryption and better authentication security for WLAN guest users. Protecting the company network infrastructure from attacks from a guest user still remains a top security priority. However, if a company can also provide encryption on the guest SSID, the protection provided to the guest user is a value-added service.

One simple way to provide encryption on a guest SSID is to use a static PSK. Although encryption is provided when using static PSK, this is not ideal because of brute-force dictionary attacks and social engineering attacks. Some WLAN vendors offer cloud-based servers to distribute secure guest credentials in the form of unique dynamic PSKs. A guest management solution that utilizes unique PSKs as credentials also provides data privacy for guest users with WPA2 encryption.

Another growing trend with public access networks is the use of 802.1X/EAP with Hotspot 2.0. Hotspot 2.0 is a Wi-Fi Alliance technical specification that is supported by the Passpoint certification program. With Hotspot 2.0, the client device is equipped by an authentication provider with one or more credentials, such as a SIM card, username/password pair, or X.509 certificate. Passpoint devices can query the network prior to connecting in order to discover the authentication providers supported by the network. Though open networks are still the norm today, growing interest in security and automated connectivity in public access networks will motivate adoption and use of Hotspot 2.0.

Network Access Control (NAC)

Network access control (NAC) evaluates the capability or state of a computer to determine the potential risk of the computer on the network and to determine the level of access to allow. NAC has changed over the years from an environment that primarily assessed the virus and spyware health risk to an environment where checks and fingerprinting are performed on a computer, extensively identifying its capabilities and configuration. These checks are integrated with 802.1X/EAP and RADIUS to authenticate and authorize network access for the user and the computer.

Posture

NAC began as a response to computer viruses, worms, and malware that appeared in the early 2000s. The early NAC products date back to around 2003 and provided what is known as *posture assessment.* Posture is a process that applies a set of rules to check the health and configuration of a computer and determine whether it should be allowed access to the network. NAC products do not perform the health checks themselves but rather validate that the policy is adhered to. A key task of posture assessment is to verify that security software (antivirus, antispyware, and a firewall) is installed, up-to-date, and operational. Figure 10.28 shows an example of some of the antivirus settings that can be checked. Essentially, posture assessment "checks the checkers." In addition to checking security software status, posture assessment can check the state of the operating system. Posture policy can be configured to make sure that specific patches or updates are installed, verify that certain processes are running or not running, or even check to determine whether or not specific hardware (such as USB ports) is active.

A posture check is performed by a *persistent agent* (software that is permanently installed on the computer) or by a *dissolvable agent* (software that is temporarily installed). If a company deploys posture software, a persistent agent will likely be installed on all of the corporate laptops to make sure that they are healthy. The company may also want to check guest computers that are trying to connect to the network; however, the guest is not likely to allow your company to install software on their computer. When the guest connects to the captive portal, a posture assessment process can temporarily run and check the guest computer for compliance.

After the posture check is performed, if a computer is considered unhealthy, the ideal scenario would be for the posture agent to automatically fix or remediate the problem so that the computer can pass the check and gain network access. Since the persistent agent is installed on the corporate computer and typically has permissions to make changes, automatic remediation can be performed. Computers that are running dissolvable agents typically cannot be automatically updated. The guest user must resolve the problems before network access will be allowed.

FIGURE 10.28 Antivirus posture settings

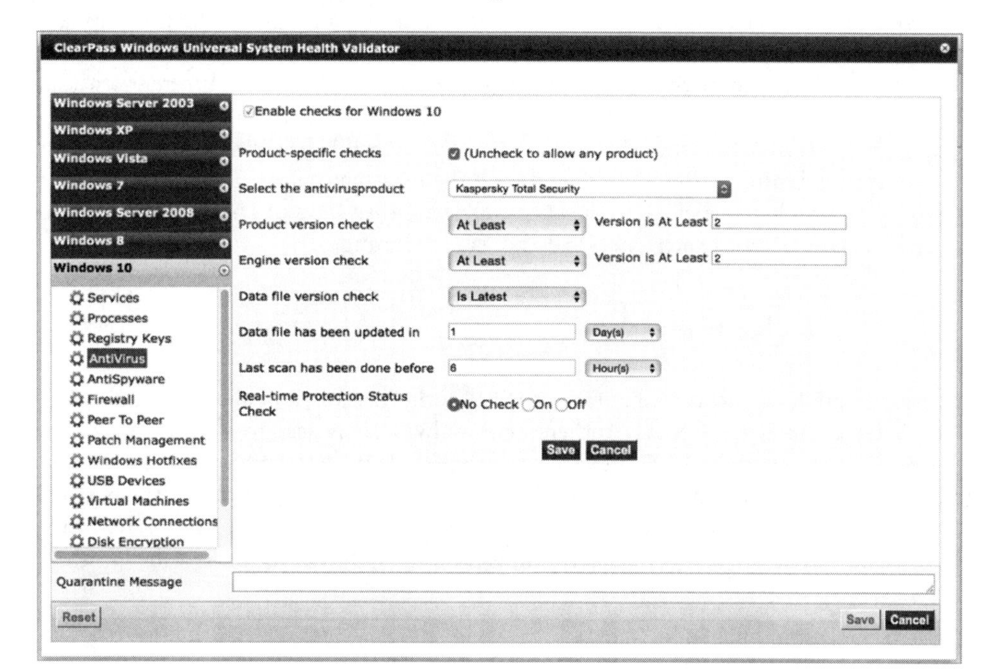

OS Fingerprinting

The operating system of WLAN client devices can be determined by a variety of fingerprinting methods, including DHCP and HTTP snooping. After a client successfully establishes a Layer 2 connection, the next action is to send a DHCP request to obtain an IP address. As part of that request, the client device includes DHCP option information and requests a list of DHCP parameters or options from the DHCP server. These options may include subnet mask, domain name, default gateway, and the like. When a client sends DHCP discover and request messages, each type of client requests different parameters under the DHCP option 55 portion of the request. The parameters within DHCP option 55 create a fingerprint that can be used to identify the operating system of the client.

For example, iOS devices include a common set of parameters when performing a DHCP request, thus making it possible to identify that the device is most likely an iOS device. DHCP fingerprinting is not perfect, and it is often not possible to discern the difference between similar devices such as an iPod, an iPhone, or an iPad. Depending on the NAC vendor, the DHCP fingerprint is referenced as an ASCII list of parameter request options, such as 1, 3, 6, 15, 119, 252. Or it might be display as a hexadecimal string, such as 370103060F77FC. In the string, the first two hex digits are equal to ASCII 55 (option 55), and each of the following two digits pairs are the hex values of each option.

You will find an extensive list of DHCP fingerprints at `www.fingerbank.org`. Although the parameter request list is not guaranteed to be unique, it can typically be used along with other fingerprinting techniques to identify devices.

Another OS detection method is *HTTP fingerprinting*. The user-agent header within an HTTP packet identifies the client operating system. During captive portal authentication, NAC solutions are able to inspect HTTP/HTTPS frames while handling the client requests. This fingerprinted information is combined with the information obtained through other methods to paint a better picture of the client device. Other ways of obtaining client information are active methods such as SNMP and TCP scanning.

AAA

Earlier in the book we mentioned authentication, authorization, and accounting (AAA). AAA is a key component of NAC. Authentication obviously is used to identify the user who is connecting to the network. We often refer to this as identifying "who you are." Although "who you are" is a very important piece of the process for allowing access to the network, an equally important component of the connection is authorization. We often refer to this as identifying "what you are." Authorization is used to analyze information such as the following:

- User type (admin, help desk, staff)
- Location, connection type (wireless, wired, VPN)
- Time of day
- Device type (smartphone, tablet, computer)
- Operating system
- Posture (system health or status)

When configuring AAA for authentication, one of the configuration tasks is to define or specify the database that will be used to verify the user's identity. Historically we have referred to this as the user database; however, the user's identity could be verified by something other than a user account and password, such as a MAC address or a certificate. If you are not sure of the identity type that is being used, or if you want to maintain a more neutral stance, the term *identity store* is a good one to use.

By utilizing both authentication and authorization, a NAC can distinguish between Jeff using his smartphone and Jeff using his personal laptop. From this information, the NAC can control what Jeff can do with each device on the network.

To use an analogy to explain authorization, say that George is a member of a country club. As he drives onto the property, the guard at the entrance checks his identification card and verifies his membership. He has been authenticated. After parking his car, he decides to go to the restaurant since it is 6:30 p.m. and he is hungry. When he arrives at the restaurant, he is told that he is not allowed in the restaurant. Confused, George questions why since he has already verified that he is a member of the club at the entrance.

The hostess then explains to him that after 6 p.m., the restaurant has a policy that all male guests must be wearing slacks and a sports jacket or suit jacket. Unfortunately George was wearing shorts and did not have a jacket; therefore he was not authorized to eat at the restaurant. The hostess was polite, and did tell George that he was authorized to go to the lounge and have a meal there, since the requirements for the lounge were not as strict.

As you can see from the analogy, authentication is about who you are, whereas authorization is about other parameters, such as what, where, when, and how. Also, unlike authentication where you are or are not authenticated, authorization varies depending on the parameters and the situation.

RADIUS Change of Authorization

Prior to RADIUS *Change of Authorization (CoA)*, if a client was authenticated and assigned a set of permissions on the network, the client authorization would not change until the client logged out and logged back in. This only allowed the authorization decision to be made during the initial connection of the client.

RADIUS accounting (the final *A* in *AAA*) is used to monitor the user connection. In the early days of AAA, it typically tracked client connection activity: logging on and logging off events, which in some environments may be all you want or need to track. Enhancements to accounting allow the AAA server to also provide interim accounting. Interim accounting can track resource activity such as time and bytes used for the connection. If the user exceeds or violates the allowed limits of resources, RADIUS CoA can be used to dynamically change the permissions that the user has on the network.

To use an analogy to explain RADIUS CoA, Jack is going to a club with some friends to enjoy some cocktails and dancing. When they arrive, a bouncer at the door admits them into the club but tells them that they are not allowed to become drunk or cause trouble in the club. While telling them this, the bouncer checks to make sure that they are not already drunk or causing trouble. Unfortunately, the bouncer must stand at the door and monitor the guests only as they enter the club. The bouncer cannot monitor the guests once they are inside the club. After a few nights of experiencing some problems in the club, the manager decides to hire additional bouncers, who walk around the club and monitor the guests who are already in the club. Anyone who is found to be drunk or causing problems is either restricted within the club (maybe they are no longer allowed to purchase alcoholic drinks), or the guest may be removed from the club. Once the guest is outside the club, the bouncer at the door can reevaluate the status of the guest, possibly denying reentry into the club, allowing the guest back in the club, or allowing the guest to reenter, but with a different set of permissions.

RADIUS CoA was originally defined by RFC 3576 and later updated in RFC 5176. Before you begin to worry, no, you do not need to know this for the CWSP exam. We are mentioning it because many of the AAA servers, NAC servers, and enterprise wireless equipment reference "RADIUS RFC 3576" on configuration menus without referring to CoA. Therefore, from a practical perspective, you should be aware that if you see RFC 3576 on any configuration menu, that is the section where RADIUS CoA is configured.

Single Sign-On

In the early days of networking, users had to log on to the file or print server in order to get access to the network resources. The user accounts were managed and stored on each server. Initially this was rarely a problem since networks were smaller, but as the number of internal servers and server types increased, logging in to multiple servers became a hassle. To simplify the process, companies began to implement *single sign-on (SSO)* within the organization, allowing users to access many if not all of the internal resources using a single network logon. Not only did this simplify the logon process for the user, it also simplified network management by consolidating user accounts into one central user database.

Within the organization, single sign-on worked well for many years until corporate resources began migrating to Internet- and cloud-based servers and services. User logons now had to extend outside the corporate network, and many cloud-servers were actually services provided by other companies, such as CRM systems, office applications, knowledge bases, and file-sharing servers. Authentication and authorization across organizational boundaries introduce more complexity and a greater security risk.

Two technologies, *Security Assertion Markup Language (SAML)* and OAuth can be used to provide the access security needed to expand outside the organization's network. The following sections will briefly explain the components of these technologies and how they work.

SAML

SAML provides a secure method of exchanging user security information between your organization and an external service provider, such as third-party cloud-based *customer relationship management (CRM)* platform. When a user attempts to connect to the CRM platform, instead of requiring the user to log on, a trust relationship between your authentication server and the CRM server will validate the user's identity and provide access to the application or service. This allows users to log on once to the enterprise network and then seamlessly and securely access external services and resources without having to revalidate their identities.

The SAML specification defines three roles that participate in the SSO process; the identity provider (IdP), which is the asserting party; the service provider (SP), which is the relying party; and the user. This section will briefly explain two scenarios in which SAML can be used to provide SSO.

The first scenario is the service provider–initiated login, as illustrated in Figure 10.29. Here, the user attempts to access a resource on the CRM server (SP). In this scenario, if the user has not been authenticated, the user is redirected to the enterprise authentication server (IdP) using a SAML request. After successful authentication, the user is then redirected to the CRM server using a SAML assertion, at which time the user will have access to the requested resources.

FIGURE 10.29 Service provider–initiated login

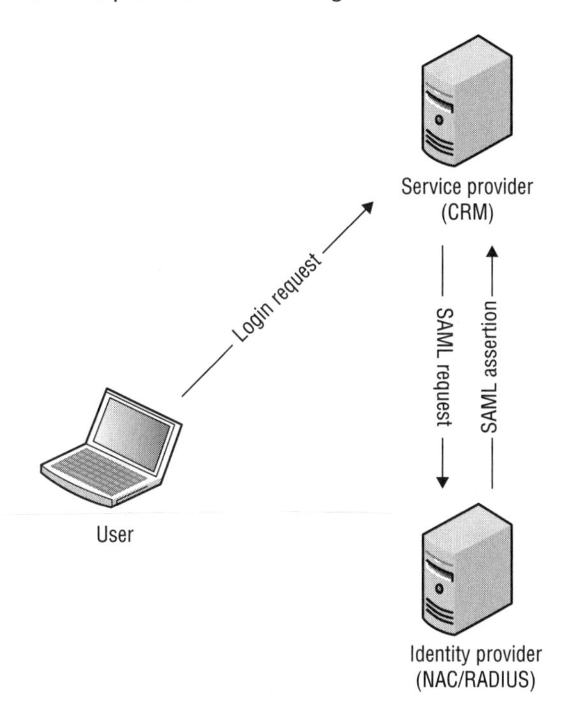

The second scenario is the identity provider–initiated login, as illustrated in Figure 10.30. Here, the user logs on to the enterprise authentication server (IdP) first. Once logged on, the user is redirected to the CRM server and a SAML assertion verifies the user's access.

There are many ways of configuring SAML. The key concept is that it provides access to resources outside of the enterprise network, using the user's enterprise credentials, and without requiring the user to log on multiple times.

OAuth

OAuth is different from SAML; it is an authorization standard and not an authentication standard. With OAuth, a user logs on to the authenticating application. Once logged on, the user (resource owner) can authorize a third-party application to access specific user information or resources, providing an authorization flow for web and desktop applications, along with mobile devices. NACs can use OAuth to communicate with external resources and systems.

FIGURE 10.30 Identity provider–initiated login

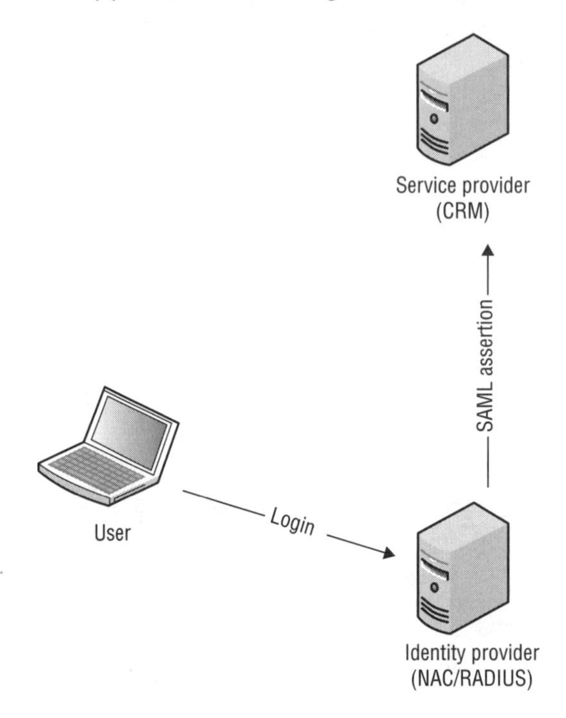

Summary

In this chapter, we discussed BYOD policy and the MDM solutions that are needed to manage company-issued mobile devices as well as employee personal mobile devices. We examined the differences between CID and BYOD devices and the MDM policy considerations for both. We discussed the various components of an MDM architecture and how the components interact. We explained the MDM enrollment process and over-the-air provisioning. We reviewed the types of mobile devices that use MDM profiles and those that use MDM agent software. We also discussed over-the-air management and application management when using an MDM solution for mobile devices.

We discussed self-service device onboarding solutions for employees. Self-service onboarding for employee devices is the fastest growing trend for BYOD provisioning.

We reviewed guest WLAN access and the key security components needed to protect corporate network infrastructure from guest users. We examined the various methods of guest management, including employee sponsorship, self-registration, and social login. Finally, we discussed how NAC can be used to provide access control by monitoring posture and by fingerprinting the client device prior to it connecting to the network. AAA services can authenticate the user connecting to the network and can authorize the device onto the network. RADIUS CoA can be used to modify the authorization of a user if a new set of permissions needs to be assigned.

Although MDM, self-service device onboarding, WLAN guest management, and NAC are separate components of a WLAN, we chose to write about all four together in this chapter because several WLAN vendors package these security solutions together as one application suite. MDM, device onboarding, WLAN guest management, and NAC can be deployed as separate components or can be deployed in unison to provide mobile device security management, guest user security, and network access security.

Exam Essentials

Define the differences between company-issued devices and personal mobile devices. Be able to explain the MDM policy concerns for both CID and BYOD devices.

Describe the four main components of an MDM architecture. Define the roles of a mobile device, an MDM server, an AP, and push notification servers. Explain how they interact.

Explain how MDM profiles and MDM agents are used within an MDM solution. Describe how MDM profiles can be used for restrictions and mobile device configurations. Describe the role of MDM agents and which mobile devices require MDM agent software.

Discuss MDM over-the-air management and MDM application management. Be able to explain how push notification servers are used to manage mobile devices across the Internet. Explain how an MDM can manage mobile device applications.

Explain self-service device management. Be able to discuss both dual- and single-SSID methods used to provision employee devices. Explain the advantages and differences between self-service device management versus MDM.

Define the four main security objectives of a guest WLAN. Discuss the importance of guest SSIDs, guest VLANs, guest firewall policies, and captive web portals.

Explain the many components and methods of WLAN guest management. Be able to explain self-registration, employee sponsorship, social login, and other ingredients of guest management.

Explain NAC and how it is used to control access to the network. Describe how posture, RADIUS attributes, and DHCP fingerprinting are used along with AAA to authenticate and authorize a user and device onto the network. Describe how RADIUS CoA can be used to modify the authorization of the user.

Review Questions

1. In a guest firewall policy, what are some of the ports that are recommended to be permitted? (Choose all that apply.)
   A. TCP 22
   B. UDP 53
   C. TCP 443
   D. TCP 110
   E. UDP 4500
2. In a guest firewall policy, which IP networks should be restricted? (Choose all that apply.)
   A. 172.16.0.0/12
   B. 20.0.0.0/8
   C. 192.16.0.0/16
   D. 172.10.0.0/24
   E. 10.0.0.0/8
3. What are some the components within an MDM architecture? (Choose all that apply.)
   A. AP
   B. RADIUS
   C. BYOD
   D. APNs
   E. GCM
4. What are some the methods that can be used to provision a root certificate onto Wi-Fi clients that function as 802.1X supplicants? (Choose all that apply.)
   A. GPO
   B. RADIUS
   C. MDM
   D. APNs
   E. GCM
5. What type of files are used by the MDM profiles for Apple Mac OS and iOS devices? (Choose all that apply.)
   A. HTTP
   B. XML
   C. JAVA
   D. PHP
   E. Python

6. What type of information can be seen on a mobile device that is monitored by an MDM server? (Choose all that apply.)
   - A. SMS messages
   - B. Battery life
   - C. Web browsing history
   - D. Installed applications
   - E. Device capacity

7. Which of these is used to report back to an MDM server unique information about mobile devices that can later be used in MDM restriction and configuration policies?
   - A. MDM profile
   - B. Push notification service
   - C. Captive web portal
   - D. MDM agent
   - E. Access point

8. What are some of the methods that can be used by a captive web portal to redirect a user to the captive portal logon page? (Choose all that apply.)
   - A. HTTP redirection
   - B. IP redirection
   - C. UDP redirection
   - D. TCP redirection
   - E. DNS redirection

9. During the MDM enrollment process, what resources can a mobile client reach while quarantined inside a walled garden? (Choose all that apply.)
   - A. SMTP
   - B. DHCP
   - C. DNS
   - D. MDM server
   - E. Exchange server

10. What is the protocol used by iOS and Mac OS devices for over-the-air provisioning of MDM profiles using certificates and SSL encryption?
    - A. OAuth
    - B. GRE
    - C. SCEP
    - D. XML
    - E. HTTPS

11. What mechanism can be used if the guest VLAN is not supported at the edge of the network and only resides in a DMZ?

 A. GRE

 B. VPN

 C. STP

 D. RTSP

 E. IGMP

12. Which type of guest management solution needs to integrate with LDAP?

 A. Social login

 B. Kiosk mode

 C. Receptionist registration

 D. Self-registration

 E. Employee sponsorship

13. An employee has enrolled a personal device with an MDM server over the corporate WLAN. The employee removes the MDM profile while at home. What will happen with the employee's personal device the next time the employee tries to connect to the company SSID?

 A. The MDM server will reprovision the MDM profile over the air.

 B. The push notification service will reprovision the MDM profile over the air.

 C. The device will be quarantined in the walled garden and will have to reenroll.

 D. The device will be free to access all resources because the certificate is still on the mobile device.

14. Which phrase best describes a policy of permitting employees to bring personally owned mobile devices such as smartphones, tablets, and laptops to their workplace?

 A. MDM

 B. NAC

 C. DMZ

 D. BYOD

15. Which method of guest management can be used by a company to gather valuable personal information about guest users?

 A. Social login

 B. Kiosk mode

 C. Receptionist registration

 D. Self-registration

 E. Employee sponsorship

16. What kind of remote actions can an MDM administrator send to the mobile device over the Internet?
    A. Configuration changes
    B. Restrictions changes
    C. Locking the device
    D. Wiping the device
    E. Application changes
    F. All of the above

17. What are some extra restrictions that can be placed on a guest user other than those defined by the guest firewall policy? (Choose all that apply.)
    A. Encryption
    B. Web content filtering
    C. DHCP snooping
    D. Rate limiting
    E. Client isolation

18. With a WLAN infrastructure, where can the guest captive web portal operate? (Choose the best answer.)
    A. AP
    B. WLAN controller
    C. Third-party server
    D. All of the above

19. When an MDM solution is deployed, after a mobile device connects to an access point, where does the mobile device remain until the MDM enrollment process is complete?
    A. DMZ
    B. Walled garden
    C. Quarantine VLAN
    D. IT sandbox
    E. None of the above

20. To calculate the capability Jeff should have on the network, which of the following can the NAC server use to initially identify and set his permission? (Choose all that apply.)
    A. Posture
    B. DHCP fingerprinting
    C. RADIUS attributes
    D. RADIUS CoA
    E. MDM profiles

Wireless Security Troubleshooting

IN THIS CHAPTER, YOU WILL LEARN ABOUT THE FOLLOWING:

✓ **Five Tenets of WLAN Troubleshooting**

- Troubleshooting best practices
- Troubleshoot the OSI model
- Most Wi-Fi problems are client issues
- Proper WLAN design reduces problems
- WLAN always gets the blame

✓ **PSK troubleshooting**

✓ **802.1X/EAP troubleshooting**

- 802.1X/EAP troubleshooting zones
- Zone 1 - Backend communication problems
- Zone 2 - Supplicant certificate problems
- Zone 2 - Supplicant credential problems

✓ **Roaming troubleshooting**

✓ **VPN troubleshooting**

Throughout this book, you have learned about the building blocks of WLAN security. We've focused on protecting the Layer 3–7 MSDU payload of 802.11 data frames as well as on protecting the WLAN portal. You have learned about the Layer 2 dynamic encryption that is used to provide data privacy and the secure authentication methods used prior to authorizing WLAN access for users and devices. However, as with any type of communications network, problems with WLAN networks arise that might require attention from an administrator. Client connectivity issues often arise that might be the result of improper implementation of WLAN security. In this chapter, you will learn how to troubleshoot PSK and 802.1X/EAP authentication that might be the root cause of connectivity and roaming problems. You will also learn other WLAN troubleshooting strategies from a security perspective.

Five Tenets of WLAN Troubleshooting

Before we discuss specific WLAN security troubleshooting strategies, you should understand five basic tenets for troubleshooting any type of WLAN problem:

- Implement troubleshooting best practices.
- Troubleshoot the OSI model.
- Most problems are client side.
- Proper WLAN design/planning is important.
- The WLAN will always get the blame.

We will now review these WLAN troubleshooting doctrines in greater detail.

Troubleshooting Best Practices

The fundamentals of troubleshooting best practices are to ask questions and collect information. When troubleshooting any type of computer network, you must ask the correct questions to collect information that is relevant to the problem. It is easy to get sidetracked when troubleshooting, so asking the proper questions will help an IT administrator focus on the pertinent data with a goal of isolating the root cause of the problem. WLAN security problems often result in WLAN client connectivity issues; asking the appropriate questions will point you in the right direction toward solving the problem. Some of the basic questions that need to be asked include the following:

- When is the problem happening?

  At what time did the problem occur? Did this problem happen during a very specific time period? This information can be easily determined by looking at the log files of APs, WLAN controllers, and applicable servers such as RADIUS. Best practices mandate that all *Network Time Protocol (NTP)* and time zone settings be correctly configured on all network hardware.

- Where is the problem happening?

  Is the problem widespread or does it only exist in one physical area? Is the problem occurring on a single floor or in the entire building? Does the problem affect just one access point or a group of access points? Determining the location of problem will help you gather better information toward solving the problem.

- Does the problem affect one client or numerous clients?

  If the problem is only affecting a single client, you may have a simple driver issue or an incorrectly configured supplicant. If the issue is affecting numerous clients, then the problem is obviously of greater concern. Most connectivity problems are client side whether they are detrimental to a single client or multiple clients.

- Does the problem reoccur or did it just happen once?

  Troubleshooting a problem that only happens one time or only a few times can be difficult. Collecting data is much easier with recurring problems. You may have to enable debug commands on APs or WLAN controllers to hopefully capture the problem again in a log file.

- Did you make any changes recently?

  This is a question that the support personnel of WLAN vendors always ask their customers. And the answer is almost always no despite the fact that changes to the network indeed take place. Best practices dictate that any network configuration changes be planned and scheduled. WLAN infrastructure security audit logs will always leave a paper trail of which administrator made which changes at any specific time.

Once you have asked numerous questions, you can begin the process of solving the problem. Troubleshooting best practices include the following:

1. Identify the issue.

   Because the WLAN always seems to get the blame, it is even more important to correctly identify the problem. Determine that a problem actually exists. Asking questions and collecting information will help you identify the true issue.

2. Re-create the problem.

   Having the ability to duplicate the problem either onsite or in a remote lab gives you the ability to collect more information to diagnose the problem. If you cannot re-create a problem, you may need to ask more questions.

3. Locate and isolate the cause.

   The whole point of asking the pointed questions and gathering data is so that you can isolate the root cause of the problem. Troubleshooting up the OSI model will also help you identify the culprit.

4. Solve the problem.

   Formulate and implement a plan to solve the problem. This may require network changes, firmware updates, and so forth.

5. Test to verify the problem is solved.

   Always be sure to test in different areas during different times and with multiple devices. Extensive testing will ensure that the problem is indeed resolved.

6. Document the problem and the solution.

   Troubleshooting best practices dictate that you document all problems, diagnostics, and resolutions. A reference help desk database will assist you in solving problems in a timely fashion should any problem reoccur.

7. Provide feedback.

   As a professional courtesy, always be sure to follow up with the individual(s) who first alerted you to the problem.

WLAN security problems usually result in WLAN client connection failures. Many WLAN vendors offer Layer 2 diagnostic tools to troubleshoot client device authentication and association. These diagnostic tools may be accessible directly from an AP, a WLAN controller, or a cloud-based network management system (NMS). Better diagnostic tools may even offer suggested remediation for detected problems. Security and AAA log files from the WLAN hardware and the RADIUS server are also a great place to start when troubleshooting either PSK or 802.1X/EAP authentication problems. Log files may also be gathered from individual WLAN supplicants.

Third-party tools are also available for diagnostics. One example is the handheld AirCheck G2 wireless tester tool from NetScout, shown in Figure 11.1. Another example is a protocol analyzer, which can be used to capture 802.11 frames relevant to RSN security associations.

FIGURE 11.1 Handheld diagnostic tool

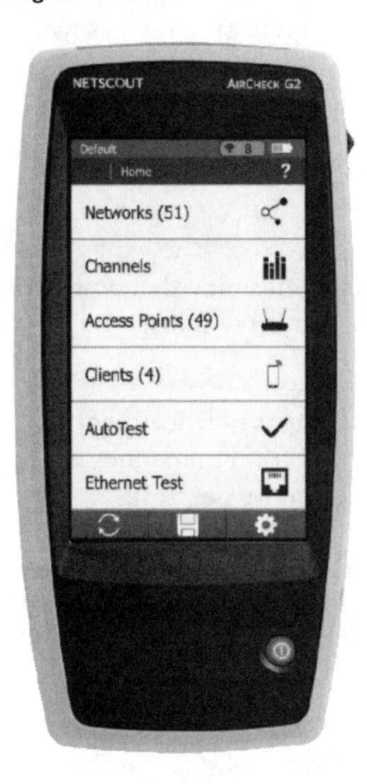

Troubleshoot the OSI Model

The diagnostic approach that is used to troubleshoot wired 802.3 networks should also be applied when troubleshooting a wireless local area network (WLAN). A bottoms-up approach to analyzing the OSI reference model layers also applies to wireless networking. Remember that 802.11 technology is similar to 802.3 in that it operates at the first two layers of the OSI model. For that reason, a WLAN administrator should always try to first determine whether problems exist at Layer 1 and Layer 2. If the first two layers of the OSI model have been eliminated as the cause of the problem, the problem is not a Wi-Fi problem and the higher layers of the OSI model should be investigated.

As with most networking technologies, most problems usually exist at the Physical layer. Simple Layer 1 problems, such as nonpowered access points or client radio driver problems, are often the root cause of connectivity or performance issues. Disruption of RF signal propagation and RF interference will affect both the performance and coverage of your WLAN. But what about Physical layer problems that are actually security related? The most likely culprit is improperly configured supplicant security settings. Later in this chapter, we will discuss troubleshooting the problems that can occur at the Physical layer due to misconfigured supplicants.

After eliminating Layer 1 as the source of the problem, a WLAN administrator should try to determine whether the problem exists at the Data-Link layer. As shown in Figure 11.2, WLAN security mechanisms operate at Layer 2. You have already learned that modern-day 802.11 radios use CCMP encryption that provides data privacy for Layers 3–7. The chosen encryption method must match on both the AP and client radios. For example, if an AP has disabled backward compatibility for TKIP encryption, a legacy client that only supports TKIP will not be able to connect. Remember that only CCMP encryption can be used for 802.11n (HT) and 802.11ac (VHT) data rates. An access point might be configured to transmit an SSID that supports both TKIP and CCMP encryption. In this situation, a common support call may be that the legacy TKIP clients seem slow because of the lack of support for higher data rates. The simple solution is to replace the legacy clients with modern-day clients that support CCMP.

FIGURE 11.2 OSI model

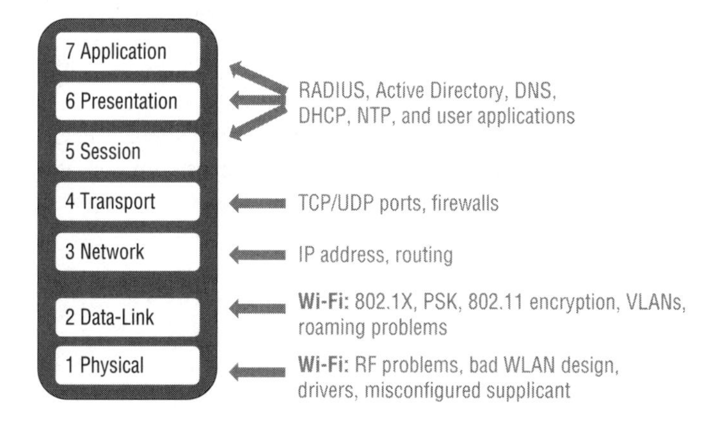

Also remember that there is a symbiotic relationship between the creation of dynamic encryption keys and authentication. A pairwise master key (PMK) is used to seed the 4-Way Handshake that generates the unique dynamic encryption keys employed by any two 802.11 radios. The PMK is generated as a byproduct of either PSK or 802.1X/EAP authentication. Therefore, if authentication fails, no encryption keys are generated. We will discuss troubleshooting both 802.11 authentication methods later in this chapter.

As stated earlier, if the first two layers of the OSI model have been eliminated, the problem is not a Wi-Fi problem and therefore the problem exists within Layers 3–7. It is likely the problem is either a TCP/IP networking issue or an application issue. As shown in Figure 11.1, TCP/IP problems should be investigated at Layers 3–4, whereas most application issues exist between Layers 5 and 7.

Most Wi-Fi Problems Are Client Issues

As previously mentioned, whenever you troubleshoot a WLAN, you should start at the Physical layer and 70 percent of the time the problem will reside on the WLAN client. If there are any client connectively problems, WLAN Troubleshooting 101 dictates that you disable and re-enable the WLAN network adapter. The driver for the WLAN network interface card (NIC) is the interface between the 802.11 radio and the operating system (OS) of the client device. For whatever reason, the WLAN driver and the OS of the device may not be communicating properly. A simple disable/re-enable of the WLAN NIC will reset the driver. Always eliminate this potential problem before investigating anything else. Additionally, first-generation radio drivers and firmware are notorious for possible bugs. Always make sure the WLAN client population has the latest available drivers installed.

Another change that is quick and easy to make is to reconfigure the client configuration profile. Most client supplicants allow the user to define a WLAN configuration profile or connection parameters. Sometimes troubleshooting a problem is as easy as deleting the old profile and configuring a new profile.

As mentioned earlier, client-side security issues usually evolve around improperly configured supplicant settings. This could be something as simple as a mistyped WPA2-Personal passphrase or as complex as 802.1X/EAP digital certificate problems. Many roaming problems are also a direct result of lack of support for fast secure roaming (FSR) mechanisms on the client. Most businesses and corporations can eliminate many of the client connectivity and performance problems by simply upgrading company-owned client devices before updating the WLAN infrastructure. Sadly, the opposite is often more common, with companies spending many thousands of dollars on new access point technology upgrades while still deploying legacy clients.

Real World Scenario

Is There a Master Database of Wi-Fi Client Capabilities?

The short answer is that there is not any official IEEE database of 802.11 client devices and their capabilities. There are, however, a few resources including the Wi-Fi Alliance, which maintains a *Wi-Fi CERTIFIED Product Finder* database at `www.wi-fi.org/product-finder`. Although most WLAN infrastructure vendors submit their access points for certification, it should be understood that many manufacturers of WLAN client devices do not go through the certification process. As shown in Figure 11.3, Wi-Fi expert Mike Albano (CWNE #150) maintains a free public listing of WLAN client capabilities at `clients.mikealbano.com`. Mike has put together a good database of many of the modern-day popular WLAN client devices. You can also download 802.11 frame captures of the client devices as well as submit WLAN client information. Often, a laptop or mobile device manufacturer will list the radio model in the specification sheet for the laptop or mobile device. However, some manufacturers may not list detailed radio specifications and capabilities. Another method of identifying the Wi-Fi radio in your device is from the FCC ID. In the United States, all Wi-Fi radios must be certified by the Federal Communication Commission (FCC) government agency. The FCC maintains a searchable equipment authorization database at `www.fcc.gov/fccid`. You can enter the FCC ID of your device into the database search engine and find documentation and pictures submitted by the manufacturer to the FCC. The FCC database is very useful in helping identity Wi-Fi radio models and specifications if the information is not available on the manufacturer's website.

FIGURE 11.3 WLAN client database

Device/Chipset	CC	Version	36	40	44	48	52	56	60	64	100	104	108	112	116	120	124	128	132	136	140	144	149	153	157	161	165	SS
Amazon Echo	US		Y	Y	Y	Y	Y	Y	Y	Y	Y	Y	Y	Y	Y	Y	Y	Y	Y	Y	Y	Y	Y	Y	Y	Y	Y	2
Amazon Fire Phone	US		Y	Y	Y	Y	N	N	N	N	N	N	N	N	N	N	N	N	N	N	N	N	Y	Y	Y	Y	Y	1
Amazon Fire TV	US		Y	Y	Y	Y	Y	Y	Y	Y	Y	Y	Y	Y	Y	Y	Y	Y	Y	Y	Y	Y	Y	Y	Y	Y	Y	2
Amazon Kindle Fire HD	US	3rd Gen	Y	Y	Y	Y	Y	Y	Y	Y	Y	Y	Y	Y	Y	Y	Y	Y	N	N	N	N	Y	Y	Y	Y	Y	1
Apple TV 3rd Gen	US		Y	Y	Y	Y	Y	Y	Y	Y	Y	Y	Y	Y	Y	Y	Y	Y	Y	Y	Y	Y	Y	Y	Y	Y	Y	
Centrino7260AC	EU	Windows	Y	Y	Y	Y	Y	Y	Y	Y	Y	Y	Y	Y	Y	Y	Y	Y	Y	Y	Y	/	Y	Y	Y	Y	Y	2
Chromebook - Acer C7	US	C710-2847	Y	Y	Y	Y	Y	Y	Y	Y	Y	Y	Y	Y	Y	Y	Y	Y	Y	Y	Y	N	Y	Y	Y	Y	Y	2

Proper WLAN Design Reduces Problems

Poor WLAN performance is often a problem that must be addressed, and often the performance issues are a result of improper WLAN design. A huge percentage of WLAN support phone calls are a symptom of a lack of WLAN design. Proper capacity and coverage planning, spectrum analysis, and a validation site-survey will eliminate the majority of WLAN support tickets in regard to performance. Additionally, many WLAN security holes can be eliminated in advance with proper WLAN security planning. If 802.1X/EAP is deployed, one of the biggest challenges is how to provision the root CA certificates for mobile devices such as smart phones and tablets. A well-thought-out security strategy for employee WLAN devices, BYOD devices, and guest WLAN access is essential. Proper WLAN security planning and design in advance will reduce time spent troubleshooting WLAN security problems at a later juncture.

WLAN Always Gets the Blame

Despite all your best WLAN troubleshooting practices and best efforts, you should resign yourself to the fact the WLAN will always get the blame. Experienced WLAN administrators know the WLAN will be blamed for problems that have nothing to do with the WLAN. This is another reason that troubleshooting up the OSI stack is important. If the problem is not a Layer 1 or Layer 2 problem, then Wi-Fi is not the culprit. However, put yourself in the shoes of the end user who is connected to the WLAN. 802.11 technology exists at the access layer. The whole point of an AP is to provide a wireless portal to a preexisting network infrastructure. Your employees and guests who connect to the WLAN expect seamless wireless mobility, and they have no concept of problems that exist at Layers 3–7. A WLAN end user is not aware that the DHCP server is out of leases. A WLAN end user is not aware the Internet service provider (ISP) is experiencing difficulty and the WAN link is down. The WLAN end user just knows that they cannot access `www.facebook.com` through the WLAN and therefore they point the finger at the Wi-Fi network.

PSK Troubleshooting

Troubleshooting PSK authentication is relatively easy. WLAN vendor diagnostic tools, log files, or a protocol analyzer can all be used to observe the 4-Way Handshake process between a WLAN client and an access point. Let's first take a look at a successful PSK authentication. In Figure 11.4, you can see the client associate with the AP and then PSK authentication begins. Because the PSK credentials matched on both the access point and the client, a pairwise master key (PMK) is created to seed the 4-Way Handshake. The 4-Way Handshake process is used to create the dynamically generated unicast encryption key that is unique to the AP radio and the client radio.

FIGURE 11.4 Successful PSK authentication

Device Name	Device BSSID	Event Type	Description
12-A-3BD500	08EA443BD514	Basic	Rx assoc req (rssi 40dB)
12-A-3BD500	08EA443BD514	Basic	Tx assoc resp <accept> (status 0, pwr 3dBm)
12-A-3BD500	08EA443BD514	Info	WPA-PSK auth is starting (at if=wifi0.1)
12-A-3BD500	08EA443BD514	Info	Sending 1/4 msg of 4-Way Handshake (at if=wifi0.1)
12-A-3BD500	08EA443BD514	Info	Sending 1/4 msg of 4-Way Handshake (at if=wifi0.1)
12-A-3BD500	08EA443BD514	Info	Received 2/4 msg of 4-Way Handshake (at if=wifi0.1)
12-A-3BD500	08EA443BD514	Info	Sending 3/4 msg of 4-Way Handshake (at if=wifi0.1)
12-A-3BD500	08EA443BD514	Info	Received 4/4 msg of 4-Way Handshake (at if=wifi0.1)
12-A-3BD500	08EA443BD514	Info	PTK is set (at if=wifi0.1)
12-A-3BD500	08EA443BD514	Basic	Authentication is successfully finished (at if=wifi0.1)
12-A-3BD500	08EA443BD514	Info	station sent out DHCP DISCOVER message
12-A-3BD500	08EA443BD514	Info	DHCP server sent out DHCP OFFER message to station
12-A-3BD500	08EA443BD514	Info	DHCP server sent out DHCP OFFER message to station
12-A-3BD500	08EA443BD514	Info	station sent out DHCP REQUEST message
12-A-3BD500	08EA443BD514	Info	DHCP server sent out DHCP ACKNOWLEDGE message to station
12-A-3BD500	08EA443BD514	Basic	DHCP session completed for station
12-A-3BD500	08EA443BD514	Basic	IP 10.5.1.162 assigned for station

Figure 11.4 shows that the 4-Way Handshake process was successful and that the *pairwise transient key (PTK)* is installed on the AP and the client. The Layer 2 negotiations are now complete, and it is time for the client to move on to higher layers. So of course the next step is that the client obtains an IP address via DHCP. If the client does not get an IP address, there is a networking issue and therefore the problem is not a Wi-Fi issue.

Perhaps a Wi-Fi administrator receives a phone call from an end user who cannot get connected using WPA2-Personal. The majority of problems are at the Physical layer; therefore, Wi-Fi Troubleshooting 101 dictates that the end user first enable and disable the Wi-Fi network card. This should ensure the Wi-Fi NIC drivers are communicating properly with the operating system. If the connectivity problem persists, the problem exists at Layer 2. You can then use diagnostic tools, log files, or a protocol analyzer to observe the failed PSK authentication of the WLAN client.

In Figure 11.5, you can see the client associate and then start PSK authentication. However, the 4-Way Handshake process fails. Notice that only two frames of the 4-Way Handshake complete.

FIGURE 11.5 Unsuccessful PSK authentication

2016-02-22 16:06:48	05-A-764fc0	08EA44764FD4	Info	WPA-PSK auth is starting (at if=wifi0.1)
2016-02-22 16:06:48	05-A-764fc0	08EA44764FD4	Info	Sending 1/4 msg of 4-Way Handshake (at if=wifi0.1)
2016-02-22 16:06:49	05-A-764fc0	08EA44764FD4	Info	Received 2/4 msg of 4-Way Handshake (at if=wifi0.1)
2016-02-22 16:06:52	05-A-764fc0	08EA44764FD4	Info	Sending 1/4 msg of 4-Way Handshake (at if=wifi0.1)
2016-02-22 16:06:52	05-A-764fc0	08EA44764FD4	Info	Received 2/4 msg of 4-Way Handshake (at if=wifi0.1)

The problem is almost always a mismatch of the PSK credentials. If the PSK credentials do not match, a *pairwise master key (PMK)* seed is not properly created and therefore the 4-Way Handshake fails entirely. The final pairwise transient key (PTK) is never created. Remember a symbiotic relationship exists between authentication and the creation of dynamic encryption keys. If PSK authentication fails, so does the 4-Way Handshake that is used to create the dynamic encryption keys. There is no attempt by the client to get an IP address because the Layer 2 process did not complete.

Remember that an 8–63 character case-sensitive passphrase is entered by the user or administrator. This passphrase is then used to create the PSK. The passphrase could possibly be improperly configured on the access point; however, the majority of the time, the problem is simple: the end user is incorrectly typing in the passphrase. The administrator should make a polite request to the end user to retype the passphrase slowly and carefully, which is a well-known cure for what is known as fat-fingering.

Another possible cause of the failure of PSK authentication could be a mismatch of the chosen encryption methods. An access point might be configured to support only WPA2 (CCMP-AES), which a legacy WPA (TKIP) client does not support. A similar failure of the 4-Way Handshake would occur.

802.1X/EAP Troubleshooting

PSK authentication (also known as WPA2-Personal) is simple to troubleshoot because the authentication method was designed to be uncomplicated. However, troubleshooting the more complex 802.1X/EAP authentication (also known as WPA2-Enterprise) is a bigger challenge because multiple points of failure exist.

As you learned in Chapter 4, "802.1X/EAP Authentication," 802.1X is a port-based access control standard that defines the mechanisms necessary to authenticate and authorize devices to network resources. The 802.1X authorization framework consists of three main components, each with a specific role. These three 802.1X components work together to make sure only properly validated users and devices are authorized to access network resources. The three 802.1X components are known as the supplicant, authenticator, and authentication server. The supplicant is the user or device that is requesting access to network resources. The authentication server's job is to validate the supplicant's credentials.

The authenticator is a gateway device that sits in the middle between the supplicant and authentication server, controlling or regulating the supplicant's access to the network.

802.1X/EAP Troubleshooting Zones

In the example shown in Figure 11.6, the supplicant is a Wi-Fi client, an AP is the authenticator, and an external RADIUS server functions as the authentication server. The RADIUS server can maintain an internal user database or query an external database, such as an LDAP database. Extensible Authentication Protocol (EAP) is used within the 802.1X framework to validate users at Layer 2. The supplicant will use an EAP protocol to communicate with the authentication server at Layer 2. The Wi-Fi client will not be allowed to communicate at the upper layers of 3–7 until the RADIUS server has validated the supplicant's identity at Layer 2.

FIGURE 11.6 802.1X/EAP

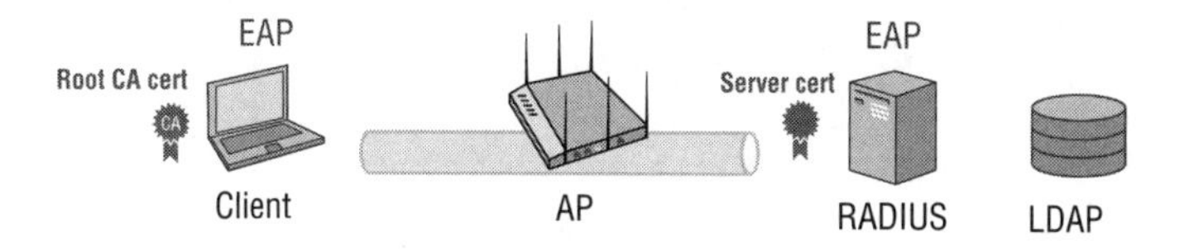

The AP blocks all of the supplicant's higher-layer communications until the supplicant is validated. When the supplicant is validated, higher layer communications are allowed through a virtual "controlled port" on the AP (the authenticator). Layer 2 EAP authentication traffic is encapsulated in RADIUS packets between the authenticator and the authentication server. The authenticator and the authenticator server also validate each other with a "shared secret."

Better versions of EAP such as EAP-PEAP and EAP-TTLS use "tunneled authentication" to protect the supplicant credentials from offline dictionary attacks. Certificates are used within the EAP process to create an encrypted SSL/TLS tunnel and ensure a secure authentication exchange. As illustrated in Figure 11.6, a server certificate resides on the RADIUS server and the root CA public certificate must be installed on the supplicant. As mentioned earlier, there are many points of failure in an 802.1X/EAP process. However, as depicted in Figure 11.7, there are effectively two troubleshooting zones within the 802.1X/EAP framework where failures will occur. Troubleshooting zone 1 consists of the backend communications between the authenticator, the authentication server, and the LDAP database. Troubleshooting zone 2 resides solely on the supplicant device that is requesting access.

FIGURE 11.7 802.1X/EAP troubleshooting zones

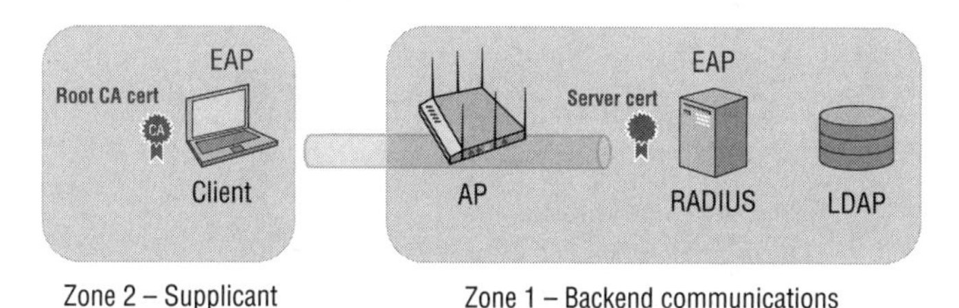

Zone 1: Backend Communication Problems

Zone 1 should always be investigated first. If an AP and a RADIUS server cannot communicate with each other, the entire authentication process will fail. If the RADIUS server and the LDAP database cannot communicate, the entire authentication process will also fail.

Figure 11.8 shows a capture of a supplicant (Wi-Fi client) trying to contact a RADIUS server. The authenticator forwards the request to the RADIUS server, but the RADIUS server never responds. The AP (authenticator) then sends a deauthentication frame to the Wi-Fi client because the process failed. This is an indication that there is a backend communication problem in the first troubleshooting zone.

FIGURE 11.8 The RADIUS server does not respond.

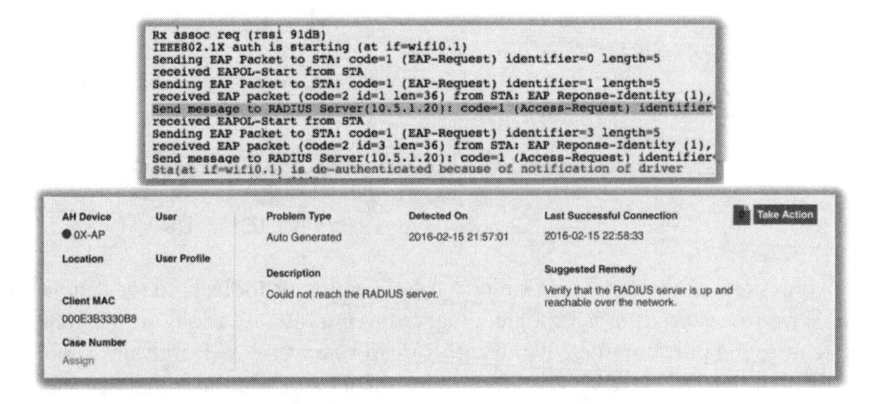

As shown in Figure 11.9, if the RADIUS server never responds to the supplicant, there are four possible points of failure in the first troubleshooting zone:

- Shared secret mismatch
- Incorrect IP settings on the AP or the RADIUS server
- Authentication port mismatch
- LDAP query failure

FIGURE 11.9 Points of failure – 802.1X/EAP troubleshooting zone 1

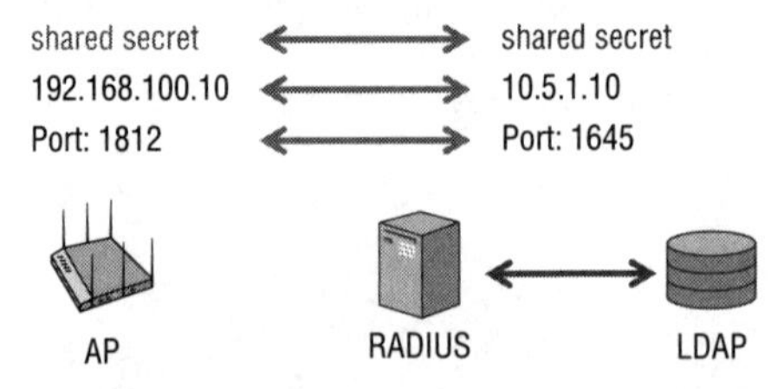

The first three possible points of failure are between the authenticator and the RADIUS server. The authenticator and the authentication server validate each other with a *shared secret*. The most common failure in RADIUS communications is that the shared secret has been typed in wrong on either the RADIUS server or the AP functioning as the authenticator.

The second most common failure in RADIUS communications is simply misconfigured IP networking settings. The AP must know the correct IP address of the RADIUS server. Likewise the RADIUS server must be configured with the IP addresses of any APs or WLAN controllers functioning as authenticators. Incorrect IP settings will result in miscommunications.

The third point of failure between an authenticator and an authentication server is a mismatch of RADIUS authentication ports. UDP ports 1812 and 1813 are defined as the industry standard ports used for RADIUS authentication and accounting. However, some older RADIUS servers may be using UDP ports 1645 and 1646. UDP ports 1645 and 1646 are rarely used anymore but do occasionally show up on older RADIUS servers. Although not a common point of failure, if the authentication ports do not match between a RADIUS server and the AP, the authentication process will fail.

The final point of failure on the backside is a failure of the LDAP query between a RADIUS server and the LDAP database. A standard domain account can be used for LDAP queries; however, if the account has expired or if there is a networking issue between the RADIUS server and the LDAP server, the entire 802.1X/EAP authentication process will fail.

Real World Scenario

What Tools Can Be Used to Troubleshoot 802.1X/EAP Backend Communications?

The good news is that multiple troubleshooting resources are available to troubleshoot Zone 1. Several WLAN vendors offer built-in diagnostic tools to test the communications between an authenticator and a RADIUS server as well as LDAP communications. Depending on the WLAN vendor and architecture, the authenticator may be either an access point or a WLAN controller. As shown in Figure 11.10, a standard domain account and password can be used to test the RADIUS and EAP communications.

Several software utilities are also available to test backend 802.1X/EAP communications. EAPTest is a commercial test utility available for the Mac OS. EAPTest is available from the Mac App Store. More information can be found at `www.ermitacode.com/eaptest.html`. RADLogin is a free test utility for the Windows and Linux platforms. More information can be found at `www.iea-software.com/products/radlogin4.cfm`. RADIUS server and LDAP database logs are also a great resource for troubleshooting 802.1X/EAP backend communication problems. In worst-case scenarios, a wired protocol analyzer may be needed to capture RADIUS packets. In Chapter 9, "RADIUS and LDAP," you learned that RADIUS attributes can be leveraged during 802.1X/EAP authentication for role-based access control (RBAC). Many of these test tools can also be used to troubleshoot issues with RADIUS attributes.

FIGURE 11.10 802.1X/EAP backend diagnostic tool

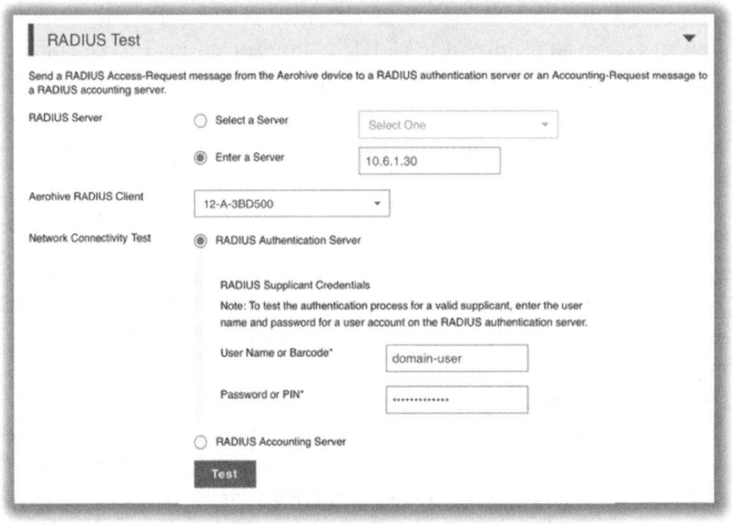

RADIUS Test

Send a RADIUS Access-Request message from the Aerohive device to a RADIUS authentication server or an Accounting-Request message to a RADIUS accounting server.

RADIUS Server
Select a Server
Select One
Enter a Server
10.6.1.30
Aerohive RADIUS Client
12-A-3BD500
Network Connectivity Test
RADIUS Authentication Server
RADIUS Supplicant Credentials
Note: To test the authentication process for a valid supplicant, enter the user name and password for a user account on the RADIUS authentication server.
User Name or Barcode*
domain-user
Password or PIN*
RADIUS Accounting Server
Test

Zone 2: Supplicant Certificate Problems

If all backend communications between the authenticator and the RADIUS server are functioning properly, then the 802.1X/EAP troubleshooting focus should now be redirected to Zone 2. In simpler words, the culprit is the supplicant. Problems with the supplicant usually either revolve around certificate issues or client credential issues. Let's take a look at Figure 11.11. Note that the RADIUS server is responding and therefore verifying that the backend communications are good. Also notice an SSL tunnel negotiation starts and finishes successfully. This 802.1X/EAP diagnostic log confirms that the certificate exchange was successful and that an SSL/TLS tunnel was successfully created to protect the supplicant credentials.

FIGURE 11.11 Successful SSL/TLS tunnel creation

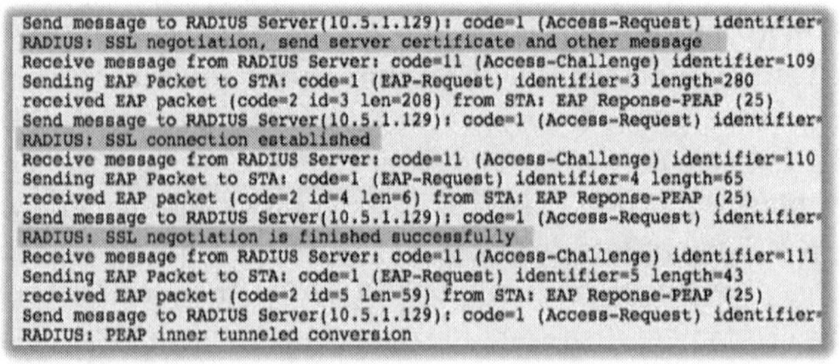

```
Send message to RADIUS Server(10.5.1.129): code=1 (Access-Request) identifier=
RADIUS: SSL negotiation, send server certificate and other message
Receive message from RADIUS Server: code=11 (Access-Challenge) identifier=109
Sending EAP Packet to STA: code=1 (EAP-Request) identifier=3 length=280
received EAP packet (code=2 id=3 len=208) from STA: EAP Reponse-PEAP (25)
Send message to RADIUS Server(10.5.1.129): code=1 (Access-Request) identifier=
RADIUS: SSL connection established
Receive message from RADIUS Server: code=11 (Access-Challenge) identifier=110
Sending EAP Packet to STA: code=1 (EAP-Request) identifier=4 length=65
received EAP packet (code=2 id=4 len=6) from STA: EAP Reponse-PEAP (25)
Send message to RADIUS Server(10.5.1.129): code=1 (Access-Request) identifier=
RADIUS: SSL negotiation is finished successfully
Receive message from RADIUS Server: code=11 (Access-Challenge) identifier=111
Sending EAP Packet to STA: code=1 (EAP-Request) identifier=5 length=43
received EAP packet (code=2 id=5 len=59) from STA: EAP Reponse-PEAP (25)
Send message to RADIUS Server(10.5.1.129): code=1 (Access-Request) identifier=
RADIUS: PEAP inner tunneled conversion
```

Figure 11.12 displays an 802.1X/EAP diagnostic log where you can see the SSL negotiation begin and the server certificate sent from the RADIUS server to the supplicant. However, the SSL/TLS tunnel is never created, and EAP authentication fails. If the SSL/TLS tunnel cannot be established, this is an indication that there is some sort of certificate problem.

FIGURE 11.12 Unsuccessful SSL/TLS tunnel creation

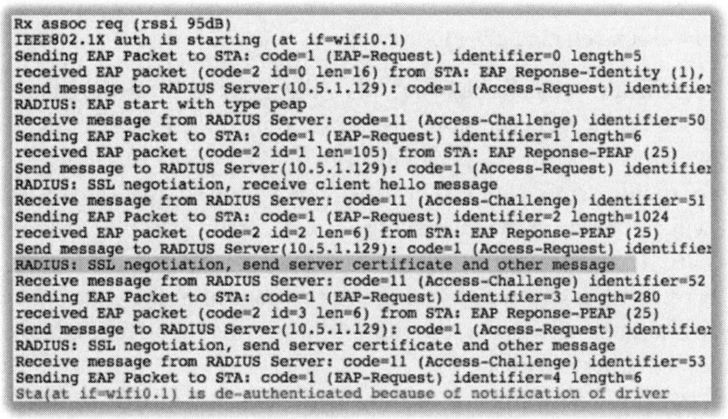

```
Rx assoc req (rssi 95dB)
IEEE802.1X auth is starting (at if=wifi0.1)
Sending EAP Packet to STA: code=1 (EAP-Request) identifier=0 length=5
received EAP packet (code=2 id=0 len=16) from STA: EAP Reponse-Identity (1),
Send message to RADIUS Server(10.5.1.129): code=1 (Access-Request) identifie
RADIUS: EAP start with type peap
Receive message from RADIUS Server: code=11 (Access-Challenge) identifier=50
Sending EAP Packet to STA: code=1 (EAP-Request) identifier=1 length=6
received EAP packet (code=2 id=1 len=105) from STA: EAP Reponse-PEAP (25)
Send message to RADIUS Server(10.5.1.129): code=1 (Access-Request) identifie
RADIUS: SSL negotiation, receive client hello message
Receive message from RADIUS Server: code=11 (Access-Challenge) identifier=51
Sending EAP Packet to STA: code=1 (EAP-Request) identifier=2 length=1024
received EAP packet (code=2 id=2 len=6) from STA: EAP Reponse-PEAP (25)
Send message to RADIUS Server(10.5.1.129): code=1 (Access-Request) identifie
RADIUS: SSL negotiation, send server certificate and other message
Receive message from RADIUS Server: code=11 (Access-Challenge) identifier=52
Sending EAP Packet to STA: code=1 (EAP-Request) identifier=3 length=280
received EAP packet (code=2 id=3 len=6) from STA: EAP Reponse-PEAP (25)
Send message to RADIUS Server(10.5.1.129): code=1 (Access-Request) identifie
RADIUS: SSL negotiation, send server certificate and other message
Receive message from RADIUS Server: code=11 (Access-Challenge) identifier=53
Sending EAP Packet to STA: code=1 (EAP-Request) identifier=4 length=6
Sta(at if=wifi0.1) is de-authenticated because of notification of driver
```

You can usually verify that there is a certificate problem by editing the supplicant client software settings and temporarily disabling the validation of the server certificate, as shown in Figure 11.13. If EAP authentication is successful after you temporarily disable the validation of the server certificate, then you have confirmed there is a problem with the implementation of the certificates within the 802.1X framework. Please note that this is not a fix but an easy way to verify that some sort of certificate issue exists.

FIGURE 11.13 Server certificate validation

Protected EAP Properties
When connecting:
Validate server certificate
Connect to these servers:
Trusted Root Certification Authorities:
ah-lab-AD2008-CA
ah-lab-AD2008-CA
Class 3 Public Primary Certification Authority
Equifax Secure Certificate Authority
GTE CyberTrust Global Root
HiveManager
Microsoft Root Authority
Do not prompt user to authorize new servers or trusted certification authorities.

A whole range of certificate problems could be causing the SSL/TLS tunnel not to be successfully created. The most common certificate issues are

- The root CA certificate is installed in the incorrect certificate store.
- The incorrect root certificate is chosen.
- The server certificate has expired.
- The root CA certificate has expired.
- The supplicant clock settings are incorrect.

The Root CA certificate needs to be installed in the Trusted Root Certificate Authorities store of the supplicant device. A common mistake is to install the root CA certificate in the default location, which is typically the personal store of a Windows machine. Another common mistake is to select the incorrect root CA certificate with the supplicant configuration. The SSL/TLS tunnel will fail because the incorrect root CA certificate will not be able to validate the server certificate. Digital certificates are also time-based, and a common problem is that the server certificate has expired. Although not as common, the root CA certificate can also have expired. The clock settings on the supplicant may be incorrect and might possibly predate the creation of either certificate.

Because of all the possible points of failure involving certificates, troubleshooting 802.1X/EAP certificate problems in Zone 2 can be difficult. Additionally, there are more potential problems with certificates. The server certificate configuration may be incorrect on the RADIUS server. In other words, the certificate problem exists back in troubleshooting Zone 1. What if EAP-TLS is the deployed authentication protocol? EAP-TLS requires the provisioning of client-side certificates in addition to server certificates. Client certificates add an additional layer of possible certificate troubleshooting on the supplicant as well as within the private PKI infrastructure that has been deployed.

There is one final complication that might result in the failure of tunneled authentication. The chosen Layer 2 EAP protocols must match on both the supplicant and the authentication server. For example, the authentication will fail if PEAPv0 (EAP-MSCHAPv2) is selected on the supplicant while PEAPv1 (EAP-GTC) is configured on the RADIUS sever. Although the SSL/TLS tunnel might still be created, the inner tunnel authentication protocol does not match and authentication will fail. Although it is possible for multiple flavors of EAP to operate simultaneously over the same 802.1X framework, the EAP protocols must match on both the supplicant and the authentication server.

Zone 2: Supplicant Credential Problems

If you can verify that you do not have any certificate issues and the SSL/TLS tunnel is indeed established, the supplicant problems are credential failures. Figure 11.14 displays an 802.1X/EAP diagnostic log where the RADIUS server is rejecting the supplicant credentials. Possible supplicant credential problems include

- Expired password or user account
- Wrong password
- User account does not exist in LDAP
- Machine account has not been joined to the Windows domain

FIGURE 11.14 RADIUS server rejects supplicant credentials

```
RADIUS: SSL connection established
Receive message from RADIUS Server: code=11 (Access-Challenge) identifier=127 length=123
Sending EAP Packet to STA: code=1 (EAP-Request) identifier=5 length=65
received EAP packet (code=2 id=5 len=6) from STA: EAP Reponse-PEAP (25)
Send message to RADIUS Server(10.5.1.129): code=1 (Access-Request) identifier=128 length=176
RADIUS: SSL negotiation is finished successfully
Receive message from RADIUS Server: code=11 (Access-Challenge) identifier=128 length=101
Sending EAP Packet to STA: code=1 (EAP-Request) identifier=6 length=43
received EAP packet (code=2 id=6 len=43) from STA: EAP Reponse-PEAP (25)
Send message to RADIUS Server(10.5.1.129): code=1 (Access-Request) identifier=129 length=213
RADIUS: PEAP inner tunneled conversion
Receive message from RADIUS Server: code=11 (Access-Challenge) identifier=129 length=117
Sending EAP Packet to STA: code=1 (EAP-Request) identifier=7 length=59
received EAP packet (code=2 id=7 len=91) from STA: EAP Reponse-PEAP (25)
Send message to RADIUS Server(10.5.1.129): code=1 (Access-Request) identifier=130 length=261
RADIUS: PEAP Tunneled authentication was rejected. NTLM_auth failed for Logon failure (0xc00
Receive message from RADIUS Server: code=11 (Access-Challenge) identifier=130 length=101
Sending EAP Packet to STA: code=1 (EAP-Request) identifier=8 length=43
received EAP packet (code=2 id=8 len=43) from STA: EAP Reponse-PEAP (25)
Send message to RADIUS Server(10.5.1.129): code=1 (Access-Request) identifier=131 length=213
RADIUS: rejected user 'user' through the NAS at 10.5.1.129.
Authentication is terminated (at if=wifi0.1) because it is rejected by RADIUS server
Sending EAP Packet to STA: code=4 (EAP-Failure) identifier=8 length=4
Sta(at if=wifi0.1) is de-authenticated because of notification of driver
```

If the user credentials do not exist in the LDAP database or the credentials have expired, authentication will fail. Unless single sign-on capabilities have been implemented on the supplicant, there is always the possibility that the domain user password can be incorrectly typed by the end user.

Another common error is that the Wi-Fi supplicant has been improperly configured for machine authentication and the RADIUS server has only been configured for user authentication. In Figure 11.15 we see a diagnostic log that clearly shows the machine credentials being sent to the RADIUS server and not the user credentials. The RADIUS server was expecting a user account and therefore rejected the machine credentials because no machine accounts had been set up for validation. In the case of Windows, the machine credentials are based on a *System Identifier (SID)* value that is stored on a Windows domain computer after being joined to a Windows domain with Active Directory.

FIGURE 11.15 Machine authentication failure

```
Send message to RADIUS Server(10.5.1.129): code=1 (Access-Request) identifier=151 length=203,
RADIUS: SSL negotiation, send server certificate and other message
Receive message from RADIUS Server: code=11 (Access-Challenge) identifier=151 length=340
Sending EAP Packet to STA: code=1 (EAP-Request) identifier=4 length=280
received EAP packet (code=2 id=4 len=17) from STA: EAP Reponse-PEAP (25)
Send message to RADIUS Server(10.5.1.129): code=1 (Access-Request) identifier=152 length=214,
RADIUS:
RADIUS: rejected user 'host/TRAINING-PC16.ah-lab.local' through the NAS at 10.5.1.129.
Authentication is terminated (at if=wifi0.1) because it is rejected by RADIUS server
Sending EAP Packet to STA: code=4 (EAP-Failure) identifier=4 length=4
Sta(at if=wifi0.1) is de-authenticated because of notification of driver
```

Of course, a WLAN administrator can always verify that all is well with an 802.1X/EAP client session. Always remember that a byproduct of the EAP process is the generation of the pairwise master key (PMK) that seeds the 4-Way Handshake exchange. Figure 11.16 shows the EAP process completing; the pairwise master key (PMK) is sent to the AP from the RADIUS server. The 4-Way Handshake process then begins to dynamically generate the pairwise transient key (PTK) that is unique between the radios of the AP and the client device. When the 4-Way Handshake completes, the encryption keys are installed and the Layer 2 connection is completed. The virtual controlled port on the authenticator opens up for this Wi-Fi client. The supplicant can now proceed to higher layers and get an IP address. If the client does not get an IP address, there is a networking issue and therefore the problem is not a Wi-Fi issue.

FIGURE 11.16 4-Way Handshake

```
Receive message from RADIUS Server: code=2 (Access-Accept) identifier=125
PMK is got from RADIUS server (at if=wifi0.1)
Sending EAP Packet to STA: code=3 (EAP-Success) identifier=5 length=4
Sending 1/4 msg of 4-Way Handshake (at if=wifi0.1)
Received 2/4 msg of 4-Way Handshake (at if=wifi0.1)
Sending 3/4 msg of 4-Way Handshake (at if=wifi0.1)
Received 4/4 msg of 4-Way Handshake (at if=wifi0.1)
PTK is set (at if=wifi0.1)
Authentication is successfully finished (at if=wifi0.1)
IP 10.5.10.100 assigned for station
station sent out DHCP REQUEST message
DHCP server sent out DHCP ACKNOWLEDGE message to station
DHCP session completed for station
```

One final consideration when troubleshooting 802.1X/EAP is RADIUS attributes. RADIUS attributes can be leveraged during 802.1X/EAP authentication for role-based access control, providing custom settings for different groups of users or devices. For example, different groups of users may be assigned to different VLANs even though they are connected to the same 802.1X/EAP SSID. If the RADIUS attribute configuration does not match on the authenticator and the RADUS server, users might be assigned to default role or VLAN assignments. In worst-case scenarios, a RADIUS attribute mismatch might result in authentication failure.

Roaming Troubleshooting

Mobility is the whole point behind wireless network access. 802.11 clients need the ability to seamlessly roam between access points without any interruption of service or degradation of performance. As shown in Figure 11.17, seamless roaming has become even more important in recent years because of the proliferation of handheld personal Wi-Fi devices such as smart phones and tablets.

FIGURE 11.17 Seamless roaming

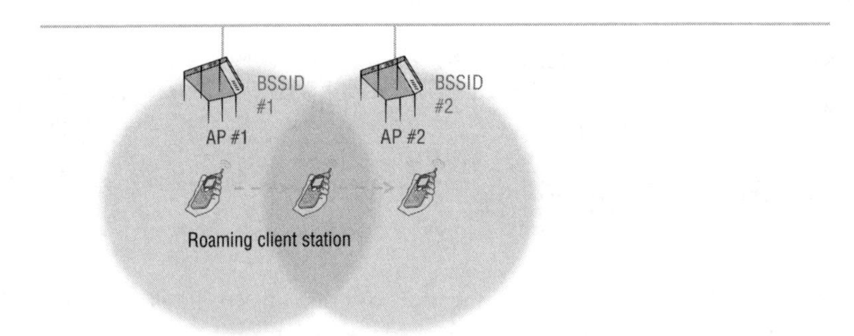

The most common roaming problems are a result of either bad client drivers or bad WLAN design. The very common *sticky client problem* is when client stations stay connected to their original AP and do not roam to a new AP of closer vicinity and stronger signal. The sticky client problem and other roaming performance issues can usually be avoided with proper WLAN design and site surveys. Good roaming design entails defining primary coverage and secondary coverage zones.

Roaming performance also has a direct relationship to WLAN security. Every time a client station roams, new encryption keys must be generated between the AP and the client station radios via the 4-Way Handshake. When using 802.1X/EAP security, roaming can be especially troublesome for VoWiFi and other time-sensitive applications. Due to the multiple frame exchanges between the authentication server and the supplicant, an 802.1X/EAP authentication can take 700 milliseconds (ms) or longer for the client to authenticate. VoWiFi requires a handoff of 150 ms or less to avoid a degradation of the quality of the call, or even worse, a loss of connection. Therefore, faster, secure roaming handoffs are required.

In Chapter 7, "802.11 Fast Secure Roaming," you learned about *opportunistic key caching (OKC)* and *fast BSS transition (FT)*, both of which produce roaming handoffs of closer to 50 ms even when 802.1X/EAP is the chosen security solution. Both OKC and FT use key distribution mechanisms so that roaming clients do not have to reauthenticate every time they roam. OKC is now considered a legacy method of fast secure roaming. The FT roaming mechanisms defined in both 802.11r and Voice Enterprise are considered the standard. Many WLAN enterprise vendor APs are now certified for Voice Enterprise by the Wi-Fi Alliance. However, client-side support for Voice Enterprise is not widespread. Any client devices that were manufactured before 2012 simply will not support 802.11r/k/v operations. Therefore, the bulk of client devices do not support Voice Enterprise capabilities. However, client-side support is growing.

Most security-related roaming problems are based on the fact that many clients simply do not support either OKC or fast BSS transition (FT). Client-side support for any device that will be using voice applications and 802.1X/EAP is critical. Proper planning and verification of client-side and AP support for OKC or FT will be necessary. Figure 11.18 shows the results of a diagnostic command that displays the roaming cache of an access point. This type of diagnostic command can verify if PMKs are being forwarded between access points. In this situation FT is enabled on the AP and supported on the client radio. You can verify the MAC address of the supplicant and the authenticator as well as the PMKR0 and the PMKRO holder. Always remember that the supplicant must also support FT; otherwise, the suppliant will reauthenticate every time the client roams.

FIGURE 11.18 Roaming cache

```
sh roam cache mac  b844:d90e:006e
Supplicant Address(SPA): b844:d90e:006e
PMK(1st 2 bytes): n/a
PMKID(1st 2 bytes): n/a
Session time: -1 seconds
(-1 means infinite)
PMK Time left in cache: 3581
PMK age: 1040
Roaming cache update interval: 60
last time logout: 1221 seconds ago
Authenticator Address: MAC=9c5d:122e:c124, IP=172.16.255.93
Roaming entry is got from neighbor AP: 9c5d:122e:c124
PMK is got(Flag): Locally
Station IP address: 172.16.255.90 (from DHCP)
Station hostname: Davids-iPhone
Station default gateway: 172.16.255.1
Station DNS server: 172.16.255.1
Station DHCP lease time: 85349 seconds
Hops: 0
WPA key mgmt: 64
R0KH: 9c5d:1263:6464
R0KH IP: 172.16.255.94
PMKR0 Name: 19D2*
```

You also learned in Chapter 7 that enabling Voice Enterprise mechanisms on an access point may actually create connectivity problems for legacy clients. When FT is configured on an access point, the AP will broadcast management frames with new information elements. For example, the mobility domain information element (MDIE) will be in all beacon and probe response frames. Unfortunately, the drivers of some older legacy client radios may not be able to process the new information in these management frames. The result is that legacy clients may have connectivity problems when an AP is configured for FT. Always test the legacy client population when configuring APs for fast BSS transition. If connectivity problems arise, consider using a separate SSID solely for fast BSS transition devices. However, please remember that every SSID consumes airtime due to the Layer 2 management overhead. Additionally, as more devices begin to support FT, upgrade your client devices.

Because 802.11 wireless networks are usually integrated into preexisting wired topologies, crossing Layer 3 boundaries is often a necessity, especially in large deployments. The only way to maintain upper-layer communications when crossing Layer 3 subnets is to provide a *Layer 3 roaming* solution. When clients roam to a new subnet, a GRE tunnel must be created to the original subnet so that the WLAN client can maintain its original IP address. As shown in Figure 11.19, the major WLAN vendors offer diagnostic tools and commands to verify that Layer 3 roaming tunnels are being successfully created.

FIGURE 11.19 Layer 3 roaming

```
show amrp dnxp cache 0022:4368:3aa2
owner:    10.5.2.153(0019:7704:c000)
age:      00:36:29
TTL:      00:00:41
homeLAN:      10.5.2.155(0019:7703:2c40)
tunnel:   10.5.2.161
Vlan:     1
upid:     10
unroam:   2000/60 sec
flag:     0x0
```

VPN Troubleshooting

VPNs are rarely used anymore as the primary method of security for WLANs. Occasionally, a VPN may be used to provide data privacy across a point-to-point 802.11 wireless bridge link. IPsec VPNs are still commonly used to connect remote branch offices with corporate offices across WAN links. Although a site-to-site VPN link is not necessarily a WLAN security solution, the wireless user traffic that originated at the remote location may be required to traverse through a VPN tunnel. Most WLAN vendors also offer VPN capabilities within their solution portfolio.

The creation of an IPsec tunnel involves two phases, called *Internet Key Exchange (IKE)* phases:

- IKE Phase 1

  The two VPN endpoints authenticate one another and negotiate keying material. The result is an encrypted tunnel used by Phase 2 for negotiating the *Encapsulating Security Payload (ESP)* security associations.

- IKE Phase 2

  The two VPN endpoints use the secure tunnel created in Phase 1 to negotiate ESP *security associations (SAs)*. The ESP SAs are used to encrypt user traffic that traverses between the endpoints.

The good news is that any quality VPN solution offers diagnostic tools and commands to troubleshoot both IKE phases. Some of the common problems that can occur if IKE Phase 1 fails are

- Certificate problems
- Incorrect networking settings
- Incorrect NAT settings on the external firewall

In Figure 11.20 you see the results of an IKE Phase 1 diagnostic command executed on a VPN server. IPsec uses digital certificates during Phase 1. If IKE Phase 1 fails due to a certificate problem, ensure that you have the correct certificates installed properly on the VPN endpoints. Also remember that certificates are time based. Very often, a certificate problem during IKE Phase 1 is simply an incorrect clock setting on either VPN endpoint.

FIGURE 11.20 IPsec Phase 1 – certificate failure

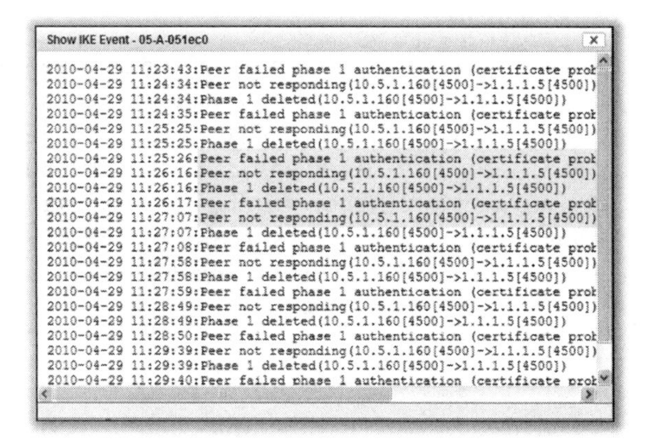

In Figure 11.21 you see the results of an IKE Phase 1 diagnostic command executed on a VPN server that indicates a possible networking error due to incorrect configuration. IPsec uses private IP addresses for tunnel communications and also uses external IP addresses, which are normally the public IP address of a firewall. If an IKE Phase 1 failure occurs as shown in Figure 11.19, check the internal and external IP settings on the VPN devices. If an external firewall is being used, also check the *Network Address Translation (NAT)* settings. Another common networking problem that causes VPNs to fail is that needed firewall ports are blocked. Ensure that the following ports are open on any firewall that the VPN tunnel may traverse:

- UDP 500 (IPsec)
- UDP 4500 (NAT Transversal)

FIGURE 11.21 IPsec Phase 1 – networking failure

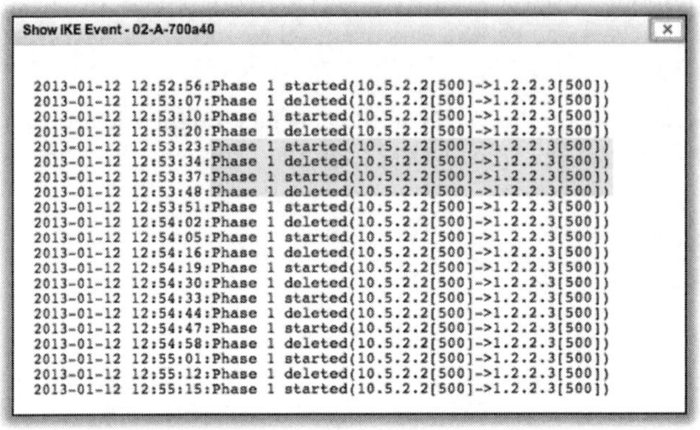

```
Show IKE Event - 02-A-700a40

2013-01-12 12:52:56:Phase 1 started(10.5.2.2[500]->1.2.2.3[500])
2013-01-12 12:53:07:Phase 1 deleted(10.5.2.2[500]->1.2.2.3[500])
2013-01-12 12:53:10:Phase 1 started(10.5.2.2[500]->1.2.2.3[500])
2013-01-12 12:53:20:Phase 1 deleted(10.5.2.2[500]->1.2.2.3[500])
2013-01-12 12:53:23:Phase 1 started(10.5.2.2[500]->1.2.2.3[500])
2013-01-12 12:53:34:Phase 1 deleted(10.5.2.2[500]->1.2.2.3[500])
2013-01-12 12:53:37:Phase 1 started(10.5.2.2[500]->1.2.2.3[500])
2013-01-12 12:53:48:Phase 1 deleted(10.5.2.2[500]->1.2.2.3[500])
2013-01-12 12:53:51:Phase 1 started(10.5.2.2[500]->1.2.2.3[500])
2013-01-12 12:54:02:Phase 1 deleted(10.5.2.2[500]->1.2.2.3[500])
2013-01-12 12:54:05:Phase 1 started(10.5.2.2[500]->1.2.2.3[500])
2013-01-12 12:54:16:Phase 1 deleted(10.5.2.2[500]->1.2.2.3[500])
2013-01-12 12:54:19:Phase 1 started(10.5.2.2[500]->1.2.2.3[500])
2013-01-12 12:54:30:Phase 1 deleted(10.5.2.2[500]->1.2.2.3[500])
2013-01-12 12:54:33:Phase 1 started(10.5.2.2[500]->1.2.2.3[500])
2013-01-12 12:54:44:Phase 1 deleted(10.5.2.2[500]->1.2.2.3[500])
2013-01-12 12:54:47:Phase 1 started(10.5.2.2[500]->1.2.2.3[500])
2013-01-12 12:54:58:Phase 1 deleted(10.5.2.2[500]->1.2.2.3[500])
2013-01-12 12:55:01:Phase 1 started(10.5.2.2[500]->1.2.2.3[500])
2013-01-12 12:55:12:Phase 1 deleted(10.5.2.2[500]->1.2.2.3[500])
2013-01-12 12:55:15:Phase 1 started(10.5.2.2[500]->1.2.2.3[500])
```

If you can confirm that IKE Phase 1 is successful yet the VPN is still failing, then IKE Phase 2 is the likely culprit. Some of the common problems if IKE Phase 2 fails are

- Mismatched transform sets between the client and server (encryption algorithm, hash algorithm, etc.)
- Mixing different vendor solutions

In Figure 11.22 you see the successful results of an IKE Phase 2 diagnostic command executed on a VPN server. If this command had indicated a failure, be sure to check both encryption and hash settings on the VPN endpoints. Check other IPsec settings such as *tunnel mode*. You will need to verify that all settings match on both ends. IKE Phase 2 problems often occur when different VPN vendors are used on opposite sides of the intended VPN tunnel. Although IPsec is a standards-based suite of protocols, mixing different VPN vendor solutions often results in more troubleshooting.

FIGURE 11.22 IPsec Phase 2 – Success

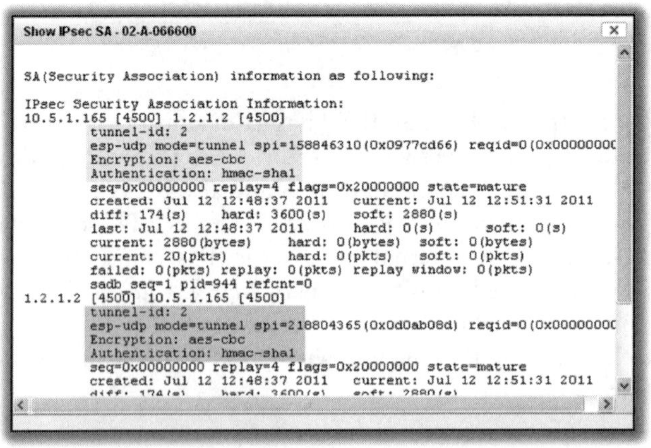

```
Show IPsec SA - 02-A-066600

SA(Security Association) information as following:

IPsec Security Association Information:
10.5.1.165 [4500] 1.2.1.2 [4500]
        tunnel-id: 2
        esp-udp mode=tunnel spi=158846310(0x0977cd66) reqid=0(0x0000000C
        Encryption: aes-cbc
        Authentication: hmac-sha1
        seq=0x00000000 replay=4 flags=0x20000000 state=mature
        created: Jul 12 12:48:37 2011   current: Jul 12 12:51:31 2011
        diff: 174(s)    hard: 3600(s)   soft: 2880(s)
        last: Jul 12 12:48:37 2011      hard: 0(s)      soft: 0(s)
        current: 2880(bytes)    hard: 0(bytes)  soft: 0(bytes)
        current: 20(pkts)       hard: 0(pkts)   soft: 0(pkts)
        failed: 0(pkts) replay: 0(pkts) replay window: 0(pkts)
        sadb_seq=1 pid=944 refcnt=0
1.2.1.2 [4500] 10.5.1.165 [4500]
        tunnel-id: 2
        esp-udp mode=tunnel spi=218804365(0x0d0ab08d) reqid=0(0x0000000C
        Encryption: aes-cbc
        Authentication: hmac-sha1
        seq=0x00000000 replay=4 flags=0x20000000 state=mature
        created: Jul 12 12:48:37 2011   current: Jul 12 12:51:31 2011
```

Summary

Troubleshooting WLANs can be very challenging. Much of WLAN troubleshooting revolves around performance issues that are a result of improper WLAN design. However, WLAN troubleshooting can also revolve around the 802.11 security that is implemented. If you have a deep understanding of PSK authentication, 802.1X/EAP authentication, and the 4-Way Handshake mechanisms, you will be better prepared to troubleshoot potential WLAN security problems. Always remember to also use troubleshooting best practices, analyze the problems at the different layers of the OSI model, and utilize all diagnostic tools that might be available.

Exam Essentials

Understand troubleshooting basics. Recognize the importance of asking the correct questions and gathering the proper information to determine the root cause of the problem.

Explain where in the OSI model various WLAN problems occur. Remember that troubleshooting up the OSI model is a recommended strategy. WLAN security issues almost always reside at Layers 1 and 2. Remember that most WLAN connectivity problems also exist on the client devices as opposed to the WLAN infrastructure.

Explain how to troubleshoot PSK authentication. Understand that the usual causes of failed PSK authentication are client driver issues and mismatched passphrase credentials. The 4-Way Handshake will fail if PSK authentication fails.

Define the multiple points of failure of 802.1X/EAP authentication. Explain all the potential backend communications points of failure and possible supplicant failures. Understand how to analyze the 802.1X/EAP process to pinpoint the exact point of failure.

Explain potential WLAN security problems with roaming. Understand that both the WLAN infrastructure and the WLAN clients must support fast secure roaming mechanisms such as OKC or Voice Enterprise.

Define troubleshooting strategies for an IPsec VPN. Recognize that IPsec establishes a VPN tunnel through two IKE phases. Explain how to troubleshoot each independent IKE phase and how to rectify the problem.

Review Questions

1. What can cause PSK authentication to fail? (Choose all that apply.)
   - **A.** Passphrase mismatch
   - **B.** Expired root CA certificate
   - **C.** WLAN client driver problem
   - **D.** Expired LDAP user account
   - **E.** Encryption mismatch
2. When the Wi-Fi network is the actual source of either a connectivity, security, or performance problem, which WLAN device is usually where the problem resides?
   - **A.** WLAN controller
   - **B.** Access point
   - **C.** WLAN client
   - **D.** Wireless network management server
3. When you are troubleshooting client connectivity problems with a client using 802.1X/EAP security, what is the first action you should take to investigate a potential Layer 1 problem?
   - **A.** Reboot the WLAN client.
   - **B.** Verify the root CA certificate.
   - **C.** Verify the EAP protocol.
   - **D.** Disable and re-enable the client radio network interface.
   - **E.** Verify the server certificate.
4. Proper implementation of 802.1X/EAP security requires the exact same EAP protocol on which of these two devices?
   - **A.** Supplicant and authenticator
   - **B.** Supplicant and authentication server
   - **C.** Authenticator and authentication server
   - **D.** Authentication server and LDAP server
   - **E.** Supplicant and LDAP server
5. Bob the WLAN administrator is troubleshooting an IPsec VPN problem that has been deployed as the security solution over a point-to-point 802.11 wireless bridge link between two buildings. Bob cannot get the VPN tunnel to establish and notices that there is a certificate error during the IKE Phase 1 exchange. What are the possible causes of this problem? (Choose all that apply.)
   - **A.** The VPN server behind the root bridge is using AES-256 encryption, and the VPN endpoint device behind the nonroot bridge is using AES-192 encryption.

B. The VPN server behind the root bridge is using SHA-1 hash for data integrity, and the VPN endpoint device behind the nonroot bridge is using MD-5 for data integrity.

C. The root CA certificate installed on the VPN device behind the nonroot bridge was not used to sign the server certificate on the VPN server behind the root bridge.

D. The clock settings of the VPN server that is deployed behind the root bridge predate the creation of the server certificate.

E. The public/private IP address settings are misconfigured on the VPN device behind the nonroot bridge.

6. Andrew Garcia, the WLAN administrator, is trying to explain to his boss that the WLAN is not the reason that Andrew's boss cannot post on Facebook. Andrew has determined that the problem does not exist at Layer 1 or Layer 2 of the OSI model. What should Andrew say to his boss? (Choose the best answer.)

A. Wi-Fi only operates at Layer 1 and Layer 2 of the OSI model. The WLAN is not the problem.

B. The problem is most likely a networking problem or an application problem.

C. Don't worry, boss; I will fix it.

D. Why are you looking at Facebook during business hours?

7. You have been tasked with troubleshooting a client connectivity problem at your company's headquarters. All the APs and employee iPads are configured for PSK authentication. An employee notices that he cannot connect his iPad to the AP in the reception area of the main building but can connect to other APs. View the following graphic and describe the cause of the problem.

```
BASIC   Rx assoc req (rssi 93dB)
INFO    WPA-PSK auth is starting (at if=wifi0.1)
INFO    Sending 1/4 msg of 4-Way Handshake (at if=wifi0.1)
INFO    Received 2/4 msg of 4-Way Handshake (at if=wifi0.1)
INFO    Sending 1/4 msg of 4-Way Handshake (at if=wifi0.1)
INFO    Received 2/4 msg of 4-Way Handshake (at if=wifi0.1)
BASIC   Sta(at if=wifi0.1) is de-authenticated because of notification of driver
```

A. The WLAN client driver is not communicating properly with the device's OS.

B. The APs are configured for CCMP encryption only. The client only supports TKIP.

C. The client has been configured with the wrong WPA2 Personal passphrase.

D. The AP in the reception area has been configured with the wrong WPA2 Personal passphrase.

8. You have been tasked with configuring a secure WLAN for 300 APs at the corporate offices. All the APs and employee Windows laptops have been configured for 802.1X using PEAPv0 (EAP-MSCHAPv2). The domain user accounts are failing authentication with every attempt. After viewing the graphic shown here, determine the possible causes of the problem. (Choose all that apply.)

```
Rx assoc req (rssi 91dB)
IEEE802.1X auth is starting (at if=wifi0.1)
Sending EAP Packet to STA: code=1 (EAP-Request) identifier=0 length=5
received EAPOL-Start from STA
Sending EAP Packet to STA: code=1 (EAP-Request) identifier=1 length=5
received EAP packet (code=2 id=1 len=36) from STA: EAP Reponse-Identity (1),
Send message to RADIUS Server(10.5.1.20): code=1 (Access-Request) identifier
received EAPOL-Start from STA
Sending EAP Packet to STA: code=1 (EAP-Request) identifier=3 length=5
received EAP packet (code=2 id=3 len=36) from STA: EAP Reponse-Identity (1),
Send message to RADIUS Server(10.5.1.20): code=1 (Access-Request) identifier
Sta(at if=wifi0.1) is de-authenticated because of notification of driver
```

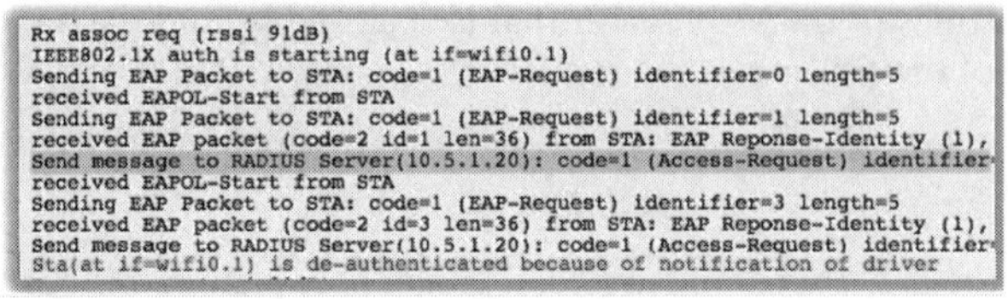

A. Windows OS laptops have the root certificate installed in the incorrect store.

B. Windows OS laptops' supplicant has been configured for machine authentication.

C. The shared secret does not match between the AP and the RADIUS server.

D. The RADIUS cannot query LDAP.

E. The Windows OS laptops have been configured for PEAPv1 (EAP-GTC).

F. The server certificate has expired.

9. You have been tasked with configuring a secure WLAN for 500 APs at the corporate offices. All the APs and employee Windows laptops have been configured for 802.1X using PEAPv1 (EAP-GTC). The domain user accounts are failing authentication with every attempt. After viewing the graphic shown here, determine the possible causes of the problem. (Choose all that apply.)

```
Rx assoc req (rssi 95dB)
IEEE802.1X auth is starting (at if=wifi0.1)
Sending EAP Packet to STA: code=1 (EAP-Request) identifier=0 length=5
received EAP packet (code=2 id=0 len=16) from STA: EAP Reponse-Identity (1),
Send message to RADIUS Server(10.5.1.129): code=1 (Access-Request) identifier
RADIUS: EAP start with type peap
Receive message from RADIUS Server: code=11 (Access-Challenge) identifier=50
Sending EAP Packet to STA: code=1 (EAP-Request) identifier=1 length=6
received EAP packet (code=2 id=1 len=105) from STA: EAP Reponse-PEAP (25)
Send message to RADIUS Server(10.5.1.129): code=1 (Access-Request) identifier
RADIUS: SSL negotiation, receive client hello message
Receive message from RADIUS Server: code=11 (Access-Challenge) identifier=51
Sending EAP Packet to STA: code=1 (EAP-Request) identifier=2 length=1024
received EAP packet (code=2 id=2 len=6) from STA: EAP Reponse-PEAP (25)
Send message to RADIUS Server(10.5.1.129): code=1 (Access-Request) identifier
RADIUS: SSL negotiation, send server certificate and other message
Receive message from RADIUS Server: code=11 (Access-Challenge) identifier=52
Sending EAP Packet to STA: code=1 (EAP-Request) identifier=3 length=280
received EAP packet (code=2 id=3 len=6) from STA: EAP Reponse-PEAP (25)
Send message to RADIUS Server(10.5.1.129): code=1 (Access-Request) identifier
RADIUS: SSL negotiation, send server certificate and other message
Receive message from RADIUS Server: code=11 (Access-Challenge) identifier=53
Sending EAP Packet to STA: code=1 (EAP-Request) identifier=4 length=6
Sta(at if=wifi0.1) is de-authenticated because of notification of driver
```

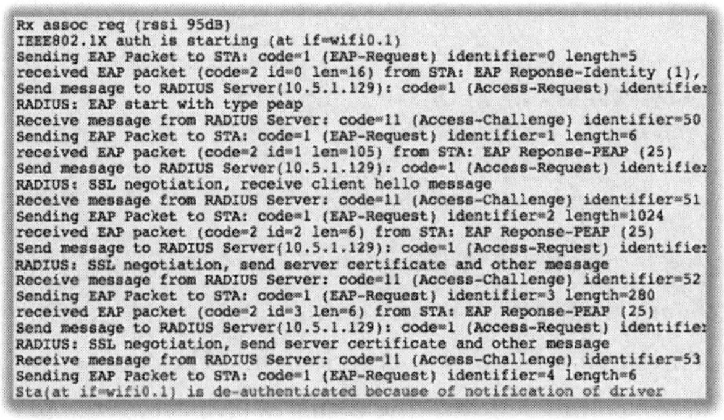

A. The Windows OS laptops have the root certificate installed in the incorrect store.

B. The Windows OS laptops' supplicant has been configured for machine authentication.

C. The shared secret does not match between the AP and the RADIUS server.

D. The RADIUS cannot query LDAP.

E. The Windows OS laptops have been configured for PEAPv0 (EAP-MSCHAPv2).

F. The server certificate has expired.

10. The corporate IT administrators, Hunter, Rion, and Liam, are huddled together to try to solve an issue with the newly deployed VoWiFi phones. The chosen security solution is PEAPv0 (EAP-MSCHAPv2) for the voice SSID that also has Voice Enterprise enabled on the access points. The VoWiFi phones are authenticating flawlessly and voice calls are stable when the employees use the devices from their desk. However, there seem to be gaps in the audio and sometimes disconnects when the employees are talking on the VoWiFi phones and move to other areas of the building. What are the possible causes of the interruption of service for the voice calls while the employees are mobile? (Choose all that apply.)

A. VoWiFi phones should only be configured for PSK authentication when roaming is a requirement.

B. VoWiFi phones are reauthenticating every time they roam to a new AP.

C. VoWiFi phones do not use opportunistic key caching.

D. VoWiFi phones do not support fast BSS transition.

11. You have been tasked with troubleshooting a client connectivity problem at your company's headquarters. All the APs and employee iPads are configured for PSK authentication. An employee notices that he cannot connect to any of the APs with his iPad; however, all the other corporate iPads are connecting. After viewing the graphic shown here, determine the cause of the problem.

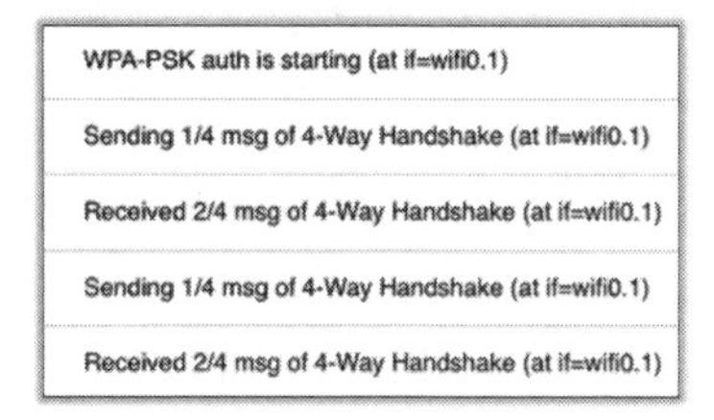
WPA-PSK auth is starting (at if=wifi0.1)
Sending 1/4 msg of 4-Way Handshake (at if=wifi0.1)
Received 2/4 msg of 4-Way Handshake (at if=wifi0.1)
Sending 1/4 msg of 4-Way Handshake (at if=wifi0.1)
Received 2/4 msg of 4-Way Handshake (at if=wifi0.1)

A. The WLAN client driver is not communicating properly with the device OS.

B. The APs are configured for CCMP encryption only. The client only supports TKIP.

C. The APs have been configured with the WPA2 Personal passphrase.

D. The APs have been configured for WPA2 Enterprise.

12. You have been tasked with configuring a secure WLAN for 400 APs at the corporate offices. All the APs and employee Windows laptops have been configured for 802.1X using EAP-MSCHAPv2. The domain user accounts are failing authentication with every attempt. After viewing the graphic shown here, determine the possible causes of the problem. (Choose all that apply.)

```
Rx assoc req (rssi 95dB)
IEEE802.1X auth is starting (at if=wifi0.1)
Sending EAP Packet to STA: code=1 (EAP-Request) identifier=0 length=5
received EAP packet (code=2 id=0 len=16) from STA: EAP Reponse-Identity (1),
Send message to RADIUS Server(10.5.1.129): code=1 (Access-Request) identifier
RADIUS: EAP start with type peap
Receive message from RADIUS Server: code=11 (Access-Challenge) identifier=50
Sending EAP Packet to STA: code=1 (EAP-Request) identifier=1 length=6
received EAP packet (code=2 id=1 len=105) from STA: EAP Reponse-PEAP (25)
Send message to RADIUS Server(10.5.1.129): code=1 (Access-Request) identifier
RADIUS: SSL negotiation, receive client hello message
Receive message from RADIUS Server: code=11 (Access-Challenge) identifier=51
Sending EAP Packet to STA: code=1 (EAP-Request) identifier=2 length=1024
received EAP packet (code=2 id=2 len=6) from STA: EAP Reponse-PEAP (25)
Send message to RADIUS Server(10.5.1.129): code=1 (Access-Request) identifier
RADIUS: SSL negotiation, send server certificate and other message
Receive message from RADIUS Server: code=11 (Access-Challenge) identifier=52
Sending EAP Packet to STA: code=1 (EAP-Request) identifier=3 length=280
received EAP packet (code=2 id=3 len=6) from STA: EAP Reponse-PEAP (25)
Send message to RADIUS Server(10.5.1.129): code=1 (Access-Request) identifier
RADIUS: SSL negotiation, send server certificate and other message
Receive message from RADIUS Server: code=11 (Access-Challenge) identifier=53
Sending EAP Packet to STA: code=1 (EAP-Request) identifier=4 length=6
Sta(at if=wifi0.1) is de-authenticated because of notification of driver
```

A. The networking settings on the AP are incorrect.

B. The Windows OS laptops' supplicant has been configured for machine authentication.

C. The supplicant clock settings are incorrect.

D. An authentication port mismatch exists between the AP and the RADIUS server.

E. The networking settings on the RADIUS server are incorrect.

F. The incorrect root certificate is selected in the supplicant.

13. You have been tasked with configuring a secure WLAN for 600 APs at the corporate offices. All the APs and employee Windows laptops have been configured for 802.1X/EAP. The domain user accounts are failing authentication with every attempt. After looking at some packet captures of the authentication failures, you have determined that an SSL/TLS tunnel is never created. After viewing the graphic shown here, determine the possible causes of the problem. (Choose all that apply.)

```
Version: 802.1X-2004 (2)
Type: EAP Packet (0)
Length: 26
▽ Extensible Authentication Protocol
  Code: Request (1)
  Id: 137
  Length: 26
  Type: Flexible Authentication via Secure Tunneling EAP (EAP-FAST) (43)
 ▽ EAP-TLS Flags: 0x21
    0... .... = Length Included: False
    .0.. .... = More Fragments: False
    ..1. .... = Start: True
    .... .001 = Version: 1
 ▽ Secure Sockets Layer
    Ignored Unknown Record
```

A. The Windows laptops are missing a client certificate.

B. The incorrect root certificate is selected in the supplicant.

C. The server certificate has expired.

D. PACs have not been provisioned properly.

E. The root certificate has expired.

14. The network administrator of the WonderPuppy Coffee Company calls up the support hotline for his WLAN vendor and informs the support personnel that the WLAN is broken. The support personnel ask the customer a series of questions so that they can isolate and identify the cause of a potential problem. What are some common Troubleshooting 101 questions? (Choose all that apply.)

A. When is the problem happening?

B. What is your favorite color?

C. What is your quest?

D. Does the problem reoccur or did it just happen once?

E. Did you make any changes recently?

15. You have been tasked with configuring a secure WLAN for 900 APs at the corporate offices. All the APs and employee Windows laptops have been configured for EAP-MSCHAPv2. You are required to provide both machine and user authentication as part of the security solution. You have verified that the backend communications between the RADIUS server and the AP are working. After viewing the graphic shown here, determine the possible causes of the problem. (Choose all that apply.)

```
Send message to RADIUS Server(10.5.1.129): code=1 (Access-Request) identifier=151 length=203,
RADIUS: SSL negotiation, send server certificate and other message
Receive message from RADIUS Server: code=11 (Access-Challenge) identifier=151 length=340
Sending EAP Packet to STA: code=1 (EAP-Request) identifier=4 length=280
received EAP packet (code=2 id=4 len=17) from STA: EAP Reponse-PEAP (25)
Send message to RADIUS Server(10.5.1.129): code=1 (Access-Request) identifier=152 length=214,
RADIUS:
RADIUS: rejected user 'host/TRAINING-PC16.ah-lab.local' through the NAS at 10.5.1.129.
Authentication is terminated (at if=wifi0.1) because it is rejected by RADIUS server
Sending EAP Packet to STA: code=4 (EAP-Failure) identifier=4 length=4
Sta(at if=wifi0.1) is de-authenticated because of notification of driver
```

A. The domain account has expired.

B. The machine accounts were not joined to the domain.

C. The server certificate has expired.

D. The supplicant has only been configured for user authentication.

E. The root certificate has expired.

F. The incorrect root certificate is selected in the supplicant.

16. The network administrator of the Holy Grail Corporation calls up the support hotline for his WLAN vendor and informs the support personnel that the WLAN bridge link is no longer working. The support personnel ask the customer a series of questions so that they can isolate and identify the cause of a potential problem. What are some common Troubleshooting 101 questions? (Choose all that apply.)

A. When is the problem happening?

B. Where is the problem happening?

C. Does the problem affect one client or numerous clients?

D. What is the airspeed velocity of an unladen swallow?

17. WLAN administrator Marko Tisler is troubleshooting an IPsec VPN problem that has been deployed as the security solution over a point-to-point 802.11 wireless bridge link between two buildings. Marko cannot get the VPN tunnel to establish and notices that the IKE Phase 1 exchange is successful; however, IKE Phase 2 is failing. What are the possible causes of this problem? (Choose all that apply.)

A. The VPN server behind the root bridge is using AES-256 encryption and the VPN endpoint device behind the nonroot bridge is using AES-192 encryption.

B. The VPN server behind the root bridge is using SHA-1 hash for data integrity and the VPN endpoint device behind the nonroot bridge is using MD-5 for data integrity.

C. The root CA certificate installed on the VPN device behind the nonroot bridge was not used to sign the server certificate on the VPN server behind the root bridge.

D. The clock settings of the VPN server that sits behind the root bridge predate the creation of the server certificate.

E. The public/private IP address settings are misconfigured on the VPN device behind the nonroot bridge.

18. You have been tasked with configuring a secure WLAN for 900 APs at the corporate offices. All the APs and employee Windows laptops have been configured for EAP-MSCHAPv2. The WLAN clients are never able to connect to the WLAN. After viewing the graphic shown here, determine the possible causes of the problem. (Choose all that apply.)

```
Receive message from RADIUS Server: code=2 (Access-Accept) identifier=125
PMK is got from RADIUS server (at if=wifi0.1)
(63)Sending 1/4 msg of 4-Way Handshake (at if=wifi0.1)
(64)Received 2/4 msg of 4-Way Handshake (at if=wifi0.1)
(65)Sending 3/4 msg of 4-Way Handshake (at if=wifi0.1)
(66)Received 4/4 msg of 4-Way Handshake (at if=wifi0.1)
(67)PTK is set (at if=wifi0.1)
(68)Authentication is successfully finished (at if=wifi0.1)
(69)station sent out DHCP REQUEST message
(70)station sent out DHCP REQUEST message
(71)station sent out DHCP REQUEST message
```

A. The VLAN on the access layer switch is incorrectly configured.

B. The machine accounts were not joined to the domain.

C. The server certificate has expired.

D. The supplicant has only been configured for user authentication.

E. The root certificate has expired.

F. The DHCP server has run out of leases.

19. You have been tasked with configuring a secure WLAN for 600 APs at the corporate offices. All the APs and employee Windows laptops have been configured for EAP-MSCHAPv2. User authentication is failing for one of the employee laptops. After viewing the graphic shown here, determine the possible causes of the problem. (Choose all that apply.)

```
RADIUS: SSL connection established
Receive message from RADIUS Server: code=11 (Access-Challenge) identifier=127 length=123
Sending EAP Packet to STA: code=1 (EAP-Request) identifier=5 length=65
received EAP packet (code=2 id=5 len=6) from STA: EAP Reponse-PEAP (25)
Send message to RADIUS Server(10.5.1.129): code=1 (Access-Request) identifier=128 length=176
RADIUS: SSL negotiation is finished successfully
Receive message from RADIUS Server: code=11 (Access-Challenge) identifier=128 length=101
Sending EAP Packet to STA: code=1 (EAP-Request) identifier=6 length=43
received EAP packet (code=2 id=6 len=43) from STA: EAP Reponse-PEAP (25)
Send message to RADIUS Server(10.5.1.129): code=1 (Access-Request) identifier=129 length=213
RADIUS: PEAP inner tunneled conversion
Receive message from RADIUS Server: code=11 (Access-Challenge) identifier=129 length=117
Sending EAP Packet to STA: code=1 (EAP-Request) identifier=7 length=59
received EAP packet (code=2 id=7 len=91) from STA: EAP Reponse-PEAP (25)
Send message to RADIUS Server(10.5.1.129): code=1 (Access-Request) identifier=130 length=261
RADIUS: PEAP Tunneled authentication was rejected. NTLM auth failed for Logon failure (0xc00
Receive message from RADIUS Server: code=11 (Access-Challenge) identifier=130 length=101
Sending EAP Packet to STA: code=1 (EAP-Request) identifier=8 length=43
received EAP packet (code=2 id=8 len=43) from STA: EAP Reponse-PEAP (25)
Send message to RADIUS Server(10.5.1.129): code=1 (Access-Request) identifier=131 length=213
RADIUS: rejected user 'user' through the NAS at 10.5.1.129.
Authentication is terminated (at if=wifi0.1) because it is rejected by RADIUS server
Sending EAP Packet to STA: code=4 (EAP-Failure) identifier=8 length=4
Sta(at if=wifi0.1) is de-authenticated because of notification of driver
```

A. There is an incorrect shared secret on the RADIUS server.

B. The machine accounts were not joined to the domain.

C. The server certificate has expired.

D. The user account does not exist.

E. The user password has expired.

F. The DHCP server has run out of leases.

20. At what layer of the OSI model do most networking problems occur?

A. Physical

B. DataLink

C. Network

D. Transport

E. Session

F. Presentation

G. Application

Wireless Security Risks

IN THIS CHAPTER, YOU WILL LEARN ABOUT THE FOLLOWING:

✓ **Unauthorized rogue access**

- Rogue devices
- Rogue prevention

✓ **Eavesdropping**

- Casual eavesdropping
- Malicious eavesdropping
- Eavesdropping risks
- Eavesdropping prevention

✓ **Authentication attacks**

✓ **Denial of Service (DoS) attacks**

- Layer 1 DoS
- Layer 2 DoS

✓ **MAC spoofing**

✓ **Wireless hijacking**

✓ **Encryption cracking**

✓ **Peer-to-Peer attacks**

✓ **Management interface exploits**

✓ **Vendor proprietary attacks**

✓ **Social engineering**

✓ **Physical damage and theft**

✓ **Guest access and WLAN hotspots**

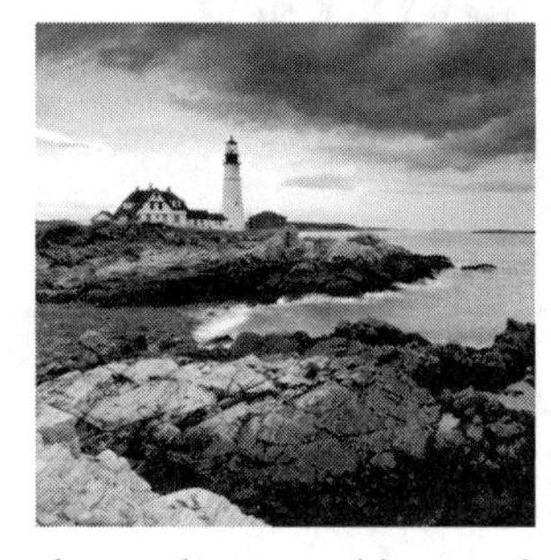

802.11 wireless networking is inherently insecure due to the use of a shared and unbounded medium—radio frequency (RF) signals. Unlike traditional bounded, or wired, networks, the medium used by wireless networks extends beyond the confines of your office or home and even beyond your campus and property as far as the signal can propagate. Essentially, your network is shared with not only the computers and other devices on your cables but also with any other 802.11 devices on the same frequency channel. 802.11 transmissions can be monitored by any third party not participating in the WLAN conversations. This lack of signal containment makes wireless networking inherently insecure. Wi-Fi technology is now an everyday part of our global society. New networks are increasingly wireless by default and wired by exception. Access points and client devices have become less expensive and are sold by popular retail outlets and even in vending machines. The increased use of 802.11 wireless technology has made it easier for workers to be more productive in the office and on the road. As the popularity of WLANs continues to grow, so does the potential for WLAN attacks and security risks. In this chapter, we will examine the risks involved with the use of 802.11 wireless networking and discuss some of the ways to mitigate these risks.

Unauthorized Rogue Access

The corporate WLAN is an authorized wireless portal to network resources. In Chapter 4, "802.1X/EAP Authentication" we discussed the proper 802.1X/EAP mechanisms that should be used to authenticate users before they are authorized through the corporate WLAN portal. However, more often than not, there is nothing to prevent a trusted individual or intruder from installing their own unauthorized wireless portal into the network backbone. A big buzz phrase in Wi-Fi security has always been the *rogue access point* or *rogue device*. A rogue access device is any WLAN radio that is connected to the wired infrastructure but is not under the management of proper network administrators. A rogue device is any unauthorized WLAN portal to network resources.

Rogue Devices

It is not uncommon for a company to have a wireless network installed within their facilities and not even know about its existence. The individuals usually responsible for installing rogue access points (APs) are not hackers; they are employees not realizing the consequences of their actions. Wi-Fi networking has become ingrained in our society, and the average employee has become accustomed to the convenience and mobility that Wi-Fi offers. As a result, employees often install their own wireless devices in the workplace because the company for which they work has yet to deploy an enterprise wireless network or they are not aware of the corporate policy forbidding installation of WLAN devices. The problem is that, while these self-installed access points might provide the wireless access that the employees desire, they are rarely secured. Any consumer-grade Wi-Fi router or access point can be plugged into a live data port. The rogue access point is a potential open and unsecured gateway straight into the wired infrastructure that the company wants to protect (see Figure 12.1).

FIGURE 12.1 Rogue access point

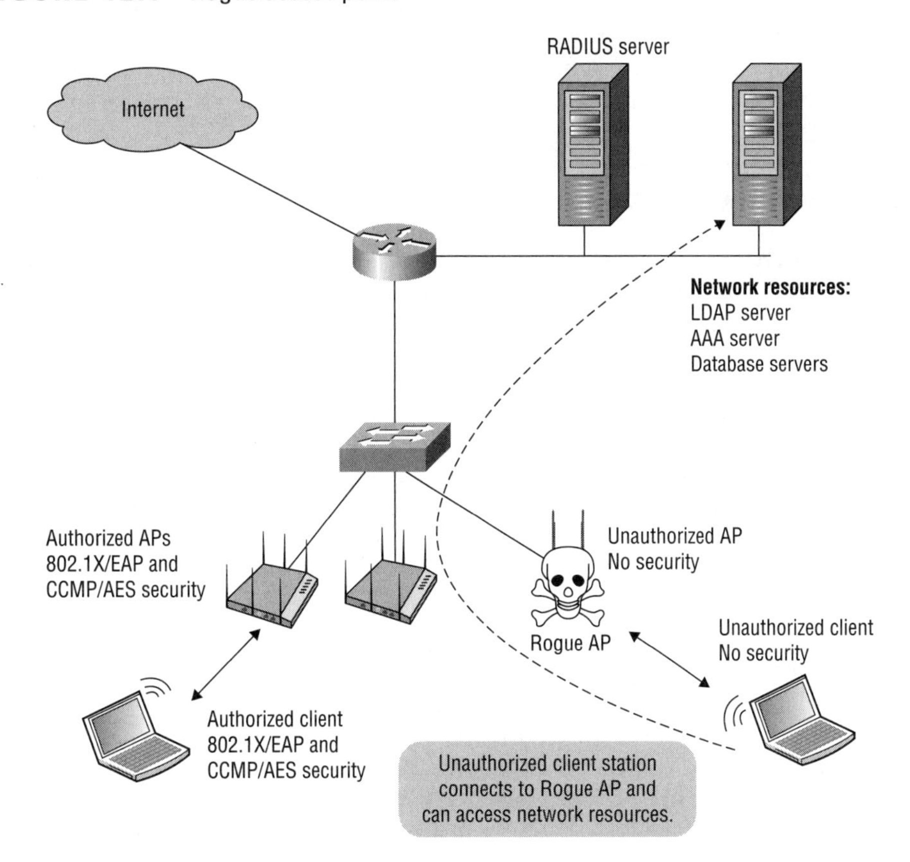

Although it is true that in some industries corporate espionage exists and in some government deployments dedicated attackers can be found, the vast majority of rogue devices are not installed for these malicious purposes. The majority of rogue devices are placed on networks by approved network users, employees, contractors, and visitors. Employees, contractors, and visitors are granted physical access to the buildings on a daily basis, something a dedicated attacker is not given. These "trusted" individuals rarely place rogue devices for malicious purposes. They place them to extend their wireless coverage or provide wireless coverage in areas that they feel it should exist without organizational permission. Some of the "trusted" individuals do not know that they are doing anything wrong due to the lack of an enforced security policy or training covering the use of wireless devices. They are simply uninformed about the risks of such device use.

Other members of this "trusted" group know that what they are doing is against policy but believe their need for wireless networking is more important than written policy. This subgroup will often take measures to hide the rogue access points or laptops under their desks, in boxes, or behind furniture. Rogue devices are even found in server rooms, having been placed there by the IT staff against policy. Although only a single open portal is needed to expose network resources, many large companies have discovered literally dozens of rogue access points installed by employees and/or contractors. Most of the rogue devices that are installed by employees are actually not access points but are instead consumer-grade Wi-Fi routers that can be purchased inexpensively and brought into the workplace.

Don't Have Wi-Fi at Work? Surprise!

A United States WLAN services company, Netrepid, performed a wireless survey for a hospital in 2007 that was found to have over 75 rogue access points throughout the main building. The IT department was certain that, prior to the survey, there were no rogue devices in the building. The rogue devices were almost exclusively consumer-grade Wi-Fi routers.

Probably the most overlooked rogue device is the ad hoc wireless network. The technical term for an 802.11 ad hoc WLAN is an *independent basic service set (IBSS)*. The 802.11 radios communicating within an IBSS network consist solely of client stations, and no access point is deployed. An IBSS network that consists of just two stations (STAs) is analogous to a wired crossover cable. An IBSS, however, can have multiple client stations in one physical area communicating in an ad hoc fashion. Unfortunately, ad hoc networks also have the potential of providing rogue access into the corporate network. Very often an employee will have a laptop or desktop plugged into the wired network via an Ethernet network adapter. On that same computer, the employee has a Wi-Fi radio and has set up an ad hoc Wi-Fi connection with another employee. As shown in Figure 12.2, the Ethernet connection and the Wi-Fi can be bridged together—an intruder might access the ad hoc wireless network and then potentially route their way to the Ethernet connection and get onto the wired network.

Another common rogue type is the wireless printer. Many printers now have 802.11 radios with a default configuration of ad hoc mode. Attackers can connect to these printers using the printer manufacturer's administrative tools, downloadable from the company's website. Then, using these tools, attackers can upload their own firmware to your printer, thus allowing them to bridge the wired and wireless connections of your printer to gain access to your wired network, without the use of an access point. Many 802.11 wireless camera security systems can be breached in a similar manner.

FIGURE 12.2 Bridged ad hoc WLAN

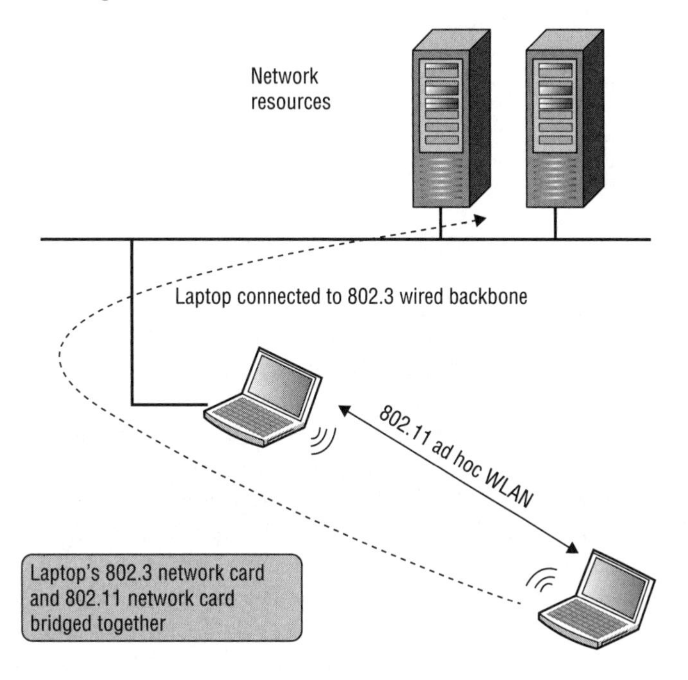

As previously mentioned, the individuals most often responsible for installing rogue access points are not hackers but instead are employees. The employees leave an open wireless portal for anyone to pass through. Because rogue devices are unauthorized WLAN portals, all of your network resources are potentially exposed. If network resources are exposed, the following risks exist:

Data Theft Corporate data on database servers can be compromised. Credit card information, corporate trade secrets, personnel information, and medical data can all be stolen if exposed via a rogue device. Any data stored on network servers or desktop workstations is entirely at risk. Data theft is usually the most common risk associated with rogue access.

Data Destruction Destruction of data can also occur. Databases can be erased and drives can be reformatted.

Loss of Services Network services can also be disabled. Even if no data was stolen or destroyed, imagine the loss of productivity and the potential losses if email services were disabled by an attacker through a rogue AP.

Malicious Data Insertion An attacker can use the unauthorized portal to upload viruses and pornography. Remote control applications and keystroke loggers can also be uploaded to network resources and used to gather information at a later date. Attackers have been known to upload illegally copied software and set up illegal FTP servers to distribute the illegal software.

Third-Party Attacks Once an attacker has accomplished rogue access, your wired network can be used as a launching pad for third-party attacks against other networks across the Internet. Distributed denial-of-service (DDoS) attacks against other corporate networks can be launched from your network infrastructure. Spammers long ago figured out that they can use a rogue AP as the originating source to send spam.

Rogue Prevention

In Chapter 15, "Wireless Security Policies," we will discuss corporate policies that ban employees from installing unauthorized APs. In addition to unauthorized APs, many government agencies and corporations ban the use of ad hoc networks. The ability to configure an ad hoc network can be disabled on most enterprise laptops. Endpoint WLAN security software can also be installed on WLAN client devices to prevent bridging between 802.11 client radios and 802.3 Ethernet radios.

Policy is a great start. However, beyond physical security or wired port control, there is nothing to prevent an intruder from connecting their own rogue AP via an Ethernet cable into any live data port provided in a wall plate. The best way of preventing rogue access is wired port control. The main focus of Chapter 4 was how 802.1X/EAP security is used for authentication and authorization via the WLAN. It should be noted that 802.1X/EAP can also be used to authorize access through wired ports on an access layer switch. EAP-PEAP and EAP-TLS can be used for wired 802.1X/EAP authentication to control wired side access. When 802.1X/EAP is used for port control on an access layer switch, desktop clients function as the supplicant requesting access. Some WLAN vendor APs can also function as a supplicant and cannot forward user traffic unless the approved AP is authenticated. As depicted in Figure 12.3, unless proper credentials are presented at Layer 2, upper-layer communications are not possible through the wired port. A rogue device cannot act as a wireless portal to network resources if the rogue device is plugged into a managed port that is blocking upper-layer traffic. Therefore, a wired 802.1X/EAP solution is an excellent method for preventing rogue access.

It should be noted that many businesses do not use a wired 802.1X/EAP solution for wired port control. Therefore a WLAN monitoring solution known as a *wireless intrusion detection system (WIDS)* is always needed to detect potential rogue devices. Most WIDS vendors prefer to call their products *wireless intrusion prevention systems (WIPSs)*. The reason that they refer to their products as prevention systems is that they are all now capable of mitigating attacks from rogue access points and rogue clients.

In Chapter 14, "Wireless Security Monitoring," you will be introduced to the concept of *classification*, which is used by wireless intrusion detection systems (WIDSs) to differentiate between authorized devices and rogue devices. A more detailed discussion of methods of rogue detection, classification, and mitigation is also found in that chapter.

FIGURE 12.3 Wired 802.1X/EAP prevents rogue access.

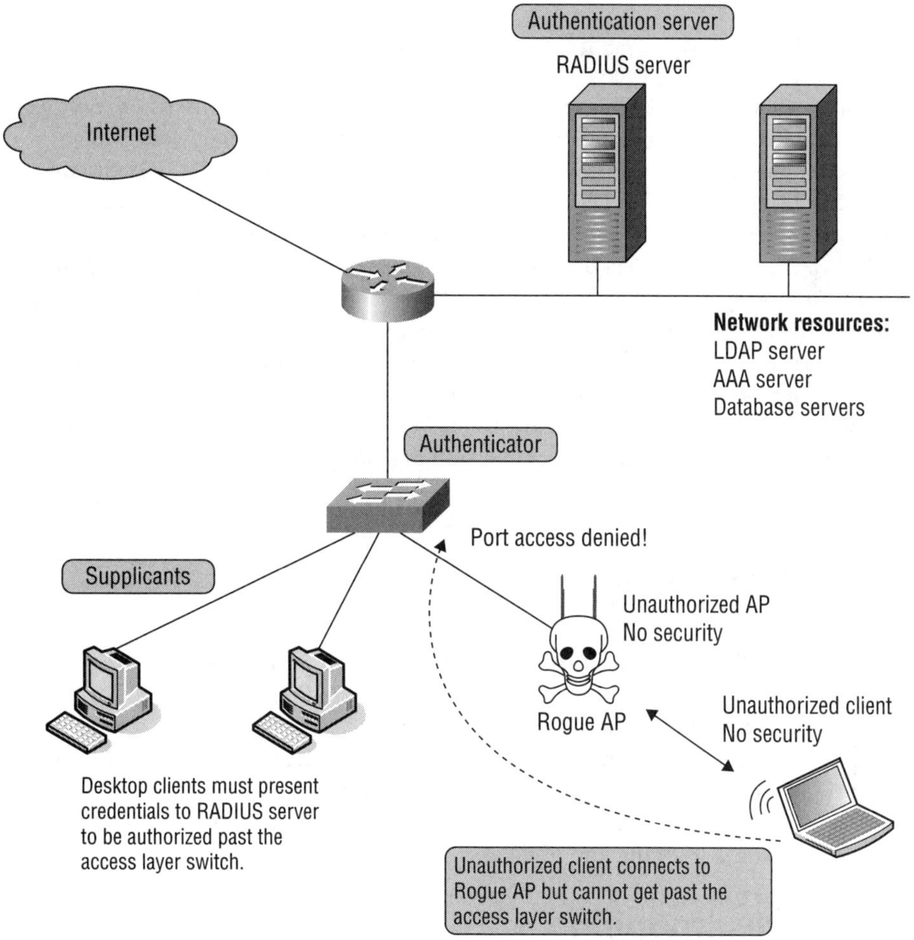

WIPS solutions use several methods to effectively terminate, suppress, and contain communications from rogue devices. The most common method of rogue containment uses a known Layer 2 DoS attack against the rogue device as a countermeasure. Rogue APs and ad hoc clients can be effectively contained until they are located and removed. Another method of rogue mitigation uses the Simple Network Management Protocol (SNMP). Most WIPSs can determine that the rogue AP is connected to the wired infrastructure and may be able to use SNMP to disable the managed switch port that is connected to the rogue AP. If the switch port is closed, the attacker cannot attack network resources that are behind the rogue AP. This method of rogue AP mitigation is known as *port suppression*.

Will a WIPS Protect Against All Known Devices?

The simple answer is "no." Although wireless intrusion prevention systems are outstanding products that can mitigate most rogue attacks, some rogue devices will go undetected. The radios inside the WIPS sensors typically monitor the 2.4 GHz ISM band and the 5 GHz UNII frequencies. Older legacy wireless networking equipment exists that transmits in the 900 MHz ISM band, and these devices will not be detected. The radios inside the WIPS sensors also use only *direct sequencing spread spectrum (DSSS)* and *orthogonal frequency division multiplexing (OFDM)* technologies. Wireless networking equipment exists that uses *frequency hopping spread spectrum (FHSS)* transmissions in the 2.4 GHz ISM band and will go undetected. The only tool that will detect with 100 percent certainty either a 900 MHz or a frequency hopping rogue AP is a spectrum analyzer capable of operating in those frequencies. Some WIPSs offer *Distributed Spectrum Analysis Systems (DSAS)*. A more detailed discussion of DSAS can be found in Chapter 14.

Eavesdropping

Just as human conversations can be overheard by any third party within hearing range of the speakers' voices, WLAN communications between two 802.11 radios can be overheard by any third-party 802.11 station on the same frequency channel. Because the RF medium is half-duplex, and therefore a shared medium, only one 802.11 station can transmit at any given time. However, any 802.11 radio within listening range can monitor any active 802.11 transmissions. WLAN communications can be monitored via two eavesdropping methods: *casual eavesdropping* and *malicious eavesdropping*.

Casual Eavesdropping

Casual eavesdropping is sometimes referred to as WLAN discovery. Casual eavesdropping is accomplished by simply exploiting the 802.11 frame exchange methods that are clearly defined in the 802.11-2012 standard. As we discussed in Chapter 2, "Legacy 802.11 Security," in order for an 802.11 client station to be able to connect to an access point, it must first discover the access point. A station discovers an access point by either listening for an AP (passive scanning) or searching for an AP (active scanning). In *passive scanning*, the client station listens for 802.11 beacon management frames that are continuously sent by the access points.

A casual eavesdropper can simply use any 802.11 client radio to listen for 802.11 beacon management frames and to discover Layer 2 information about the WLAN. Some of the information found in beacon frames includes the service set identifier (SSID), MAC addressing, supported data rates, and other basic service set (BSS) capabilities. All of this Layer 2 information is in cleartext and can be seen by any 802.11 radio.

In addition to scanning passively for APs, client stations can actively scan for them. In *active scanning*, the client station transmits management frames known as probe requests. The access point then answers back with a probe response frame that basically contains all of the same Layer 2 information that can be found in a beacon frame. A probe request without the SSID information is known as a *null probe request*. If a directed probe request is sent, all APs that support that specific SSID and hear the request should reply by sending a probe response. If a null probe request is heard, all APs, regardless of their SSID, should reply with a probe response.

Casual eavesdroppers can discover 802.11 networks using software tools that send null probe requests. Casual eavesdropping is typically considered harmless and is also often referred to as *wardriving*. Wardriving is strictly the act of looking for wireless networks, usually while in a moving vehicle. The term wardriving was derived from *wardialing* from the 1983 film *WarGames*. Wardialing was an old technique employed by hackers using computer modems to scan thousands of telephone numbers automatically to search for other computers with which they could connect.

Wardiving is now considered an outdated term and concept. In the very early days of Wi-Fi, wardriving was a hobby and sport for techno-geeks and hackers looking to find WLANs. Wardriving competitions were often held at hacker conventions to see who could find the most WLANs. While the sport of wardriving has faded into the past, millions of individuals now use WLAN discovery tools to still find available Wi-Fi networks. A more current term would be *WLAN discovery*. In the early days of Wi-Fi, the original WLAN discovery software tool was a freeware program called NetStumbler. Although still available as a free download, NetStumbler has not been updated in many years. However, many newer WLAN discovery tools exist that operate on a variety of operating systems. Figure 12.4 depicts a very popular WLAN discovery tool, inSSIDer, which is available from `www.metageek.net`.

FIGURE 12.4 inSSIDer WLAN discovery tool

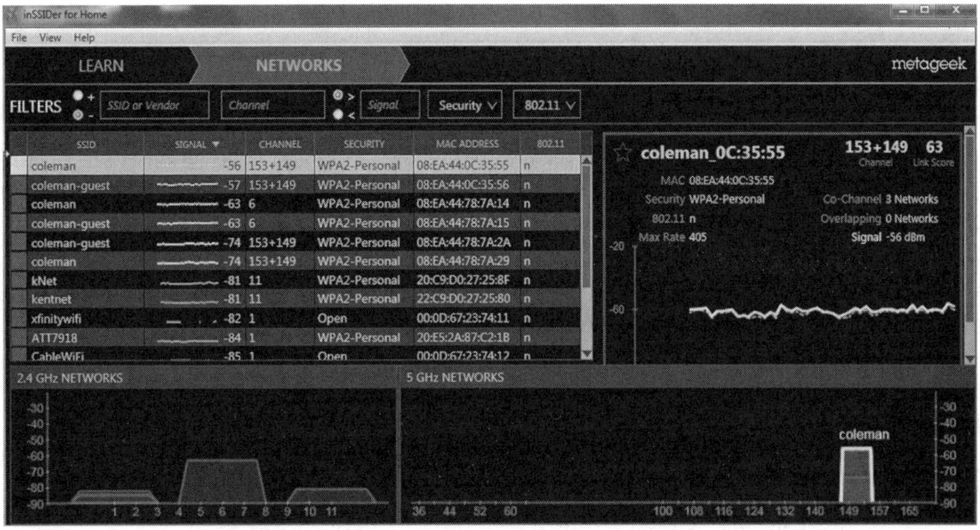

The original intent of wardriving was to find an open WLAN to gain free wireless access to the Internet. WLAN discovery tools send out null probe requests across all license-free 802.11 channels with the hope of receiving probe response frames containing wireless network information, such as SSID, channel, encryption, and so on. By design, the very nature of 802.11 passive and active scanning is to provide the identifying network information that is accessible to anyone with an 802.11 radio. Because this is an inherent and necessary function of 802.11, wardriving is not a crime. The legality of using someone else's wireless network without permission is often unclear, but be warned that people have been arrested and prosecuted as a result of these actions. An alarming decision about the use of networks owned by others was reached in March 2011. The Hague Court ruled that they were no longer going to prosecute as a criminal offense the unauthorized use of networks belonging to others if the person using the network without permission only used it to access the Internet, even if the access was gained by extraordinary means. The ruling does leave the opportunity for civil action. Every nation has its own laws covering such actions.

We do not encourage or support the efforts of using wireless networks that you are not authorized to use. We recommend that you connect only to 802.11 wireless networks that you are authorized to access.

Real World Scenario

What Tools Are Needed for WLAN Discovery?

To start finding WLANs, you will need an 802.11 client NIC and a WLAN discovery application. Numerous freeware-based discovery tools exist, including inSSIDer for Windows, WiFi Explorer for the Mac OS, and WiFiFoFum for Android and iOS. You can download inSSIDer from `www.metageek.net`, WiFi Explorer from `www.adriangranados.com`, and WiFiFoFum from `www.wififofum.net`.

Global positioning system (GPS) devices in conjunction with WLAN discovery tools can be used to pinpoint longitude and latitude coordinates of the signal from APs that are discovered. WLAN discovery capture files with GPS coordinates can be uploaded to large dynamic mapping databases on the Internet. The Wireless Geographic Logging Engine (WIGLE) maintains a searchable database of more than 273 million Wi-Fi networks. Go to `www.wigle.net` and type in your address to see whether any wireless access points have already been discovered in your neighborhood.

Malicious Eavesdropping

Malicious eavesdropping is the unauthorized use of protocol analyzers to capture wireless communications and is typically considered illegal. Most countries have some type of wiretapping law that makes it a crime to listen in on someone else's phone conversation. Additionally, most countries have laws making it illegal to listen in on any type of electromagnetic communications, including 802.11 wireless transmissions. Protocol analysis and packet analysis are used to diagnose problems in network communications, identify traffic patterns, and find bottlenecks. Many commercial and freeware 802.11 protocol analyzers exist that allow wireless network administrators to capture 802.11 traffic, for the purpose of analyzing and troubleshooting their own wireless networks. A *protocol analyzer* is a passive device that operates in an RF monitoring mode to capture any 802.11 frame transmissions within their range. Commercial and freeware WLAN protocol analyzers are widely available. In earlier chapter exercises, you used a popular freeware protocol analyzer, *Wireshark,* to view 802.11 frame captures.

A WLAN protocol analyzer is meant to be used as a diagnostic tool. However, an attacker can use a WLAN protocol analyzer as a malicious listening device for unauthorized monitoring of 802.11 frame exchanges. Encryption is the best protection against unauthorized monitoring of the WLAN. Although all Layer 2 information is always available, WPA2 encryption provides data privacy for all the Layer 3–7 information.

Eavesdropping Risks

Because malicious eavesdropping is a passive attack, it should be understood that a WIDS/WIPS solution will not be able to detect a protocol analyzer because it is not a transmitting device. The WLAN protocol analyzer is only a listening device and will go undetected. Because protocol analyzers capture 802.11 frames passively, a wireless intrusion detection system (WIDS) cannot detect malicious eavesdropping, and the attacker cannot be located.

802.11 frames can be passively monitored and data can be captured from great distances well beyond the limits of corporate buildings and property lines. An attacker does not need physical access to buildings or property to perform malicious eavesdropping. WLANs cannot be hidden from the outside world when the RF signal propagates beyond property lines. Even if a strong RF signal does not propagate from beyond your own walls, an eavesdropper could use a high gain antenna to amplify a weak signal and still be able to monitor 802.11 frame transmissions passively from locations beyond your physical control and well out of sight. 802.11 frames can be passively captured with a protocol analyzer from a distance of many miles if the attacker has a clear RF line of sight, and the attacker will remain undetected.

Postal Analogy

Think of WLAN communications as sending postcards, letters, or any other package. If someone sees the package in transit, they can always see its origin and destination as well as the means by which it is being conveyed—mail, courier, express delivery service, freight, and so on. Despite the fact that the package is well wrapped or in an envelope, the addressing is still exposed. In wireless terms, no matter how the information is encrypted, Layers 1 and 2 are exposed to allow proper transmission and reception. Therefore, all Layer 2 information, such as MAC addresses, will always be exposed in a WLAN environment. When a postcard is sent, not only will the addressing and means of conveyance be seen, but the actual postcard message can also be read by anyone while the post is in transit. If no encryption is used in a WLAN environment, the Layer 3–7 data payload is exposed to anyone who happens to be listening.

Many people believe that if their data is encrypted, they have nothing else that an attacker may wish to collect. That belief brings with it a false sense of security. All Layer 2 information is still seen in cleartext, and this information can be gathered passively using a WLAN protocol analyzer. MAC addresses and Layer 2 discovery protocols can be seen in the clear.

As discussed in Chapter 2, a legacy security measure is MAC filtering. MAC filtering is the blocking of all MAC addresses from connecting unless they are specifically allowed or, alternately, allowing all MAC addresses unless they are specifically denied access. As stated earlier, the MAC addresses of WLAN devices are visible each time a device transmits any type of frame. Since the MAC addresses can be seen by any device on the same channel, an attacker can document their use. Anyone with a WLAN protocol analyzer can capture 802.11 frame exchanges between an AP and a client and see the MAC addressing that is used for the Layer 2 communications. Simply put, MAC filters do not offer any real measure of security for wireless transmissions.

Wired leakage is also a security risk and a type of information that an attacker can use to gain access to your network or data. Wired leakage often occurs when wired stations, servers, or infrastructure devices use a broadcast protocol to communicate or to find other devices with which to replicate. Access points may forward broadcast and multicast traffic from the wired infrastructure. Layer 2 discovery protocols, such as the *Link Layer Discovery Protocol (LLDP)* or the *Cisco Discovery Protocol (CDP),* will reveal information about the wired network as well as what can be seen wirelessly. Again, passively using protocol and packet analysis will not deter a potential attacker from gaining valuable information about your infrastructure.

If encryption is not being used, the Layer 3–7 payload of any 802.11 data frame will also be exposed. Any cleartext communications such as email and Telnet passwords can be captured if no encryption is provided. Furthermore, any unencrypted 802.11 frame transmissions can be reassembled at the upper layers of the OSI model. For example, email messages can be reassembled and therefore read by an eavesdropper. Web pages and instant messages can also be reassembled. VoIP packets can be reassembled and saved as a WAV sound file. Malicious eavesdropping of this nature is highly illegal. Because of the passive and undetectable nature of this attack, encryption must always be implemented to provide data privacy.

Eavesdropping Prevention

How can you stop attackers and others from gaining access to your exposed information? The easiest and most important method of protection against malicious eavesdropping attacks is to use encryption. Encryption provides the data privacy necessary to protect the *MAC Service Data Unit (MSDU)* upper layer payload of 802.11 data frames. A strong, dynamic encryption solution—such as CCMP/AES—is a mandatory requirement to protect the Layer 3–7 payload.

To prevent anyone other than intended recipients from hearing your transmissions, you can use RF shielding to stop transmissions from exiting or entering your building. Mylar films can be placed on all of your windows, stopping signals from escaping through them. Special paint or wallpapers can be used to do the same for your walls, essentially making your building a *Faraday cage*. A Faraday cage, also known as a Faraday shield, is an enclosure made of a wired mesh or other conductive material to contain electric fields such as RF signals. Faraday shields can be built into the walls of buildings, but the construction costs are very high. Usually only well-funded and extremely security-conscious organizations, such as government offices and military institutions, go through the time and expense and take these measures to contain RF transmissions from exposure to the outside world.

The bottom line is that because 802.11 technology operates at Layers 1 and 2 of the OSI model, there is virtually no way to protect those two layers from eavesdropping. To prevent some Layer 2 wired leakage, we highly recommended that you disable Layer 2 discovery protocols such as CDP.

However, the number one priority should always be to protect the MAC Service Data Unit (MSDU) upper layer payload of 802.11 data frames. Strong, dynamic encryption solutions, such as CCMP/AES, should always be considered mandatory requirements to protect the Layer 3–7 payload and provide data privacy.

Authentication Attacks

As you learned earlier, the usual purpose of an 802.11 wireless network is to act as a portal into an 802.3 wired network. It is therefore necessary to protect that portal with very strong authentication methods so that only legitimate users with the proper credentials will be authorized to access network resources.

Authentication is the method of verifying the presented identity and credentials. Once the method of authentication has been determined by an attacker, they can begin to try to break the authentication process. Some forms of authentication are stronger than others. There are some forms of authentication that are very easy to break and should not be used in secure environments. There are others that are very complex, which are better suited to more secure environments. The type of authentication used is often dictated by things other than the security requirements of the transmissions and environments, such as ease of use, cost, device types, firmware used, regulations policy, and legacy deployments.

In Chapter 2, you learned about Open System authentication, which essentially validates all clients. As you have also already learned, stronger authorization to access network resources can be achieved by either an 802.1X/EAP authentication solution or the use of PSK authentication. The 802.11-2012 standard does not define which type of EAP authentication method to use, and all flavors of EAP are not equal. Some types of EAP authentication methods are more secure than others. As a matter of fact, Cisco's *Lightweight Extensible Authentication Protocol (LEAP)*, once one of the most commonly deployed 802.1X/EAP solutions, is susceptible to an *offline dictionary attack*. The hashed password response during the LEAP authentication process is crackable. An attacker merely has to capture a frame exchange when a LEAP user authenticates and then run the capture file through an offline dictionary attack tool, as shown in Figure 12.5. The password can be derived in a matter of seconds. The username is also seen in cleartext during the LEAP authentication process. After the attacker gets the username and password, they are free to impersonate the user by authenticating onto the WLAN and then accessing any network resources that are available to that user. It should be noted that weaker VPN solutions such as PPTP using MS-CHAPv2 authentication are also susceptible to offline dictionary attacks. Stronger EAP authentication protocols that use "tunneled authentication" are not susceptible to offline dictionary attacks.

FIGURE 12.5 Offline dictionary attack

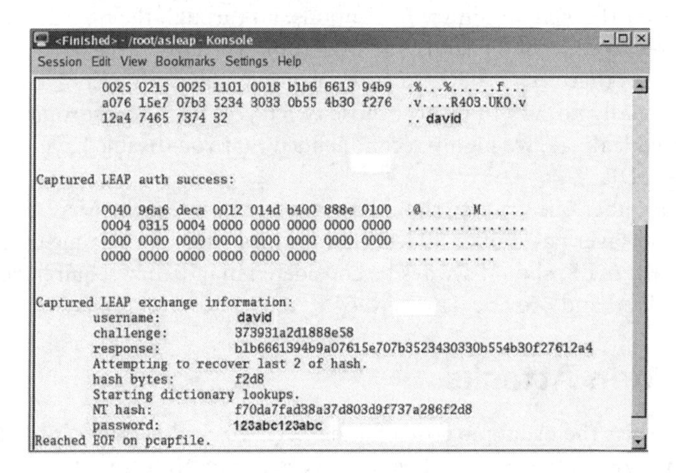

The biggest risk with any authentication attack is that all network resources become vulnerable if the authentication credentials are compromised. The risks of authentication attacks are similar to rogue access points. If an authorized WLAN portal can be compromised and the authentication credentials can be obtained, the following risks also apply:

- Data theft
- Data destruction
- Loss of services
- Malicious data
- Third-party attacks

Due to the existence of these severe risks, it is necessary to secure the corporate WLAN infrastructure properly with an 802.1X/EAP solution that uses a RADIUS server and the tunneled authentication EAP protocols discussed in Chapter 4. Multifactor authentication, also known as two-factor authentication, also increases the difficulty of cracking security immensely by adding another set of required credentials. WPA-Enterprise and WPA2-Enterprise certified solutions are almost always a necessity for the strong authentication security that is required in the workplace.

Because most home users do not have a RADIUS server in their house, weaker WPA/WPA2-Personal authentication methods are normally used. PSK authentication using a static passphrase was never intended to be used in the enterprise.

WPA/WPA2-Personal, using *preshared keys*, is a weak authentication method that is vulnerable to an offline *brute-force dictionary attack*. Shared keys or passphrases are also easily obtained through social engineering techniques. Social engineering is the act of manipulating people into performing actions or divulging confidential information. Hacking utilities are available that can derive the WPA/WPA2 passphrase by using an offline brute-force dictionary attack. An attacker who obtains the passphrase can associate with the WPA/WPA2 access point and access network resources. The biggest risk with any authentication attack is that all network resources could become vulnerable if the authentication credentials are compromised.

Even worse is that after obtaining the passphrase, the hacker can begin to decrypt the dynamically generated TKIP/RC4 or CCMP/AES encryption key. In Chapter 6, "PSK Authentication," you learned that the passphrase is used to derive the pairwise master key (PMK), which is used with the 4-Way Handshake to create the final dynamic encryption keys. If a hacker has the passphrase and captures the 4-Way Handshake, they can re-create the dynamic encryption keys and decrypt traffic. WPA/WPA2-Personal is not considered a strong security solution for the enterprise because if the passphrase is compromised, the attacker can not only access network resources, they can also decrypt traffic. Because of these risks, a static PSK authentication solution should never be used in the enterprise.

CloudCracker is an online password cracking tool for deriving the keys used in PSK networks. CloudCracker can also be used to crack different password hashes. An administrator can simply upload the corresponding 4-Way Handshake file, enter the SSID, and start the tool. CloudCracker uses an extremely large dictionary of nearly 300 million words in cracking the passphrases. CloudCracker can be found at `www.cloudcracker.com`. A policy mandating very strong passphrases of 20 characters or more should always be in place whenever a WPA/WPA2-Personal solution is deployed. Furthermore, because passphrases are static, they are susceptible to social engineering attacks. To prevent social engineering attacks, policy must dictate that only the administrator have knowledge of any static passphrases and that the passphrases are never shared with end users.

Denial-of-Service Attacks

Denial-of-service (DoS) attacks are not by themselves an attempt to gain access to your information or data. A DoS attack against a WLAN is an attack that effectively disables the WLAN. With the proper tools, any individual with ill intent can temporarily disable a Wi-Fi network by preventing legitimate users from accessing network resources. For mission-critical systems, this is a serious security concern. If the WLAN goes down, any application or network resource being accessed through the WLAN is no longer available. The wireless VoIP phone conversation comes to an abrupt end, communications with your database server are no longer possible, and wireless access to an Internet gateway has been closed. DoS attacks can be either malicious attempts to disrupt your WLAN or of an accidental nature. DoS attacks can also be used to jump-start other attacks such as wireless hijacking and Wi-Fi phishing.

The good news is that monitoring systems exist that can detect and identify DoS attacks immediately. The bad news is that usually nothing can be done to prevent DoS attacks other than locating and removing the source of the attack. DoS attacks can be targeted against the entire WLAN or can be targeted against individual access points or individual WLAN clients.

Layer 1 DoS Attacks

A DoS attack to a WLAN is most easily accomplished at Layer 1 in the RF environment. Layer 1 DoS attacks are a result of radio frequency interference. What can cause Layer 1 DoS? Layer 1 DoS can result from either intentional interference or unintentional interference.

Denial of service at Layer 1 usually occurs as an unintentional result of transmissions from non-802.11 devices. All sorts of devices transmit in the very crowded 2.4 GHz ISM band. RF video cameras, baby monitors, cordless phones, and microwave ovens are all potential sources of interference. The whole point of a spectrum analysis site survey is to identify and eliminate these sources of interference. But what if an employee forgets about corporate policy and employs a leaky microwave oven or a 2.4 GHz cordless phone after the original site survey was performed? Microwave ovens typically operate at 800 to 1,000 watts. Although microwave ovens are shielded, they can become leaky over time. A received signal of –60 dBm is about 1 millionth of 1 milliwatt and is considered a very strong signal for normal 802.11 communications. If a 1,000 watt microwave oven is even 0.000000001 percent leaky, the oven will interfere with the 802.11 radio.

Unintentional interference may cause continuous DoS; however, the disruption of service is often sporadic. This disruption of service will upset the performance of Wi-Fi networks used for data applications and can completely disrupt VoWiFi communications within a WLAN. At the very least, unintentional interference will result in retransmissions that negatively affect WLAN performance. The majority of unintentional interfering devices transmit in the 2.4 GHz ISM frequency band, and 2.4 GHz cordless phones, Bluetooth devices, medical equipment, and many other devices can cause unintentional interference in the 2.4 GHz ISM band. The 5 GHz UNII bands are less susceptible to unintentional interference. 5 GHz cordless phones often cause interference with the five channels of the 5 GHz UNII-3 band.

Now let's examine deliberate attempts to disrupt wireless networking at Layer 1. If the RF medium can be accidentally interfered with, it can also be purposely jammed at Layer 1. *Intentional interference* can be accomplished using a wide-band jamming device or a narrow-band jamming device. A wide-band jammer transmits a signal that raises the noise floor for most of the entire frequency band and therefore disrupts communications across multiple channels. As shown in Figure 12.6, there are companies that sell wide-band jammers as security tools for enforcing no Wi-Fi zones. Most of these jamming devices transmit in the 2.4 GHz frequency range, but 5 GHz jammers exist as well. The use of such devices is usually illegal in most countries.

FIGURE 12.6 RF jamming device

Narrow-band or single-channel jamming can also be done with commercially available devices. Figure 12.7 shows a signal generator that is normally used for legitimate testing purposes, such as to provide a power source to measure coax cable loss with a wattmeter. However, what is to prevent a villainous individual from transmitting a 1 watt (+30 dBm) signal via an ordinary antenna? The signal generator would then be transformed into a jamming device that will overtake most 802.11 radios that transmit at a maximum of 100 mw (+20 dBm). Higher-gain antennas can be combined with the signal generator to achieve more radiated power and extend the range of the DoS attack. Unidirectional antennas can be used to focus a Layer 1 DoS jamming attack.

FIGURE 12.7 RF signal generator and wattmeter

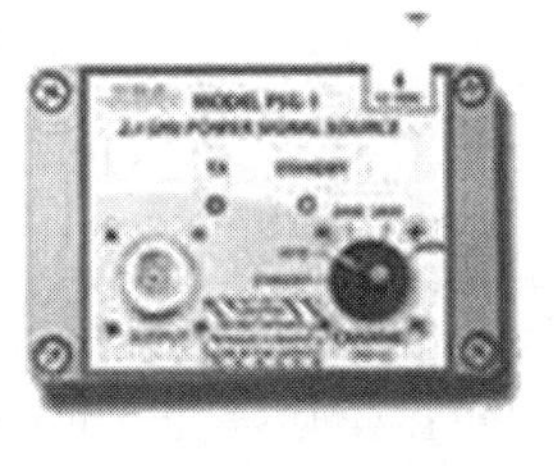

For much less money, an attacker could use the *Queensland Attack* to disrupt an 802.11 WLAN. What if an 802.11 radio could be placed in a "continuous transmit" state? In this scenario, the radio would not actually be sending data or modulating data, but would be sending out a constant RF signal much like a narrow-band signal generator. Other 802.11 radios never get to access the medium because whenever they perform a clear channel assessment, the medium is occupied by the continuous transmitter. Researchers at Queensland University in Australia discovered that this attack is indeed feasible. As shown in Figure 12.8, a major chipset manufacturer of 802.11b radios produced a software utility that placed the radios in a continuous transmit state for testing purposes. This utility can also be used for malicious purposes and is often referred to as the Queensland Attack. An 802.11b radio operating in a continuous transmit state at 30 mW may not be as large a threat as a 1 watt jammer; however, any 2.4 GHz 802.11 radios within range of the malicious radio will be affected.

FIGURE 12.8 Testing utility used for Queensland Attack

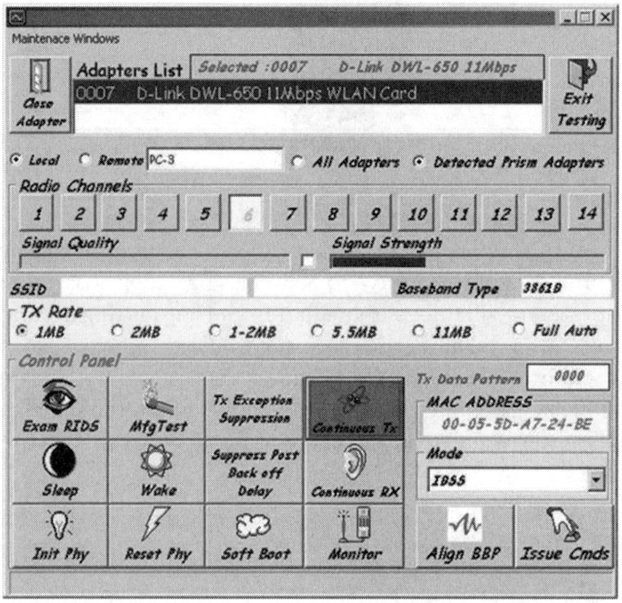

Why do jamming attacks, both accidental and intentional, cause a denial of service? Because of the half-duplex nature of the RF medium, it is necessary to ensure that at any given time only one 802.11 radio has control of the medium. *Carrier Sense Multiple Access with Collision Avoidance (CSMA/CA)* is a process used to ensure that only one 802.11 radio is transmitting at a time on the medium. One major component of the CSMA/CA method of medium contention is *physical carrier sense.*

Physical carrier-sensing is performed constantly by all stations that are not transmitting or receiving. When a station performs a physical carrier sense, it is actually listening to the channel to see whether any other transmitters are taking up the channel. Physical carrier sense has two purposes. The first purpose is to determine whether a frame transmission is inbound for a station to receive. If the medium is busy, the radio will attempt to synchronize with the transmission. The second purpose is to determine whether the medium is busy before transmitting. This is known as the *clear channel assessment (CCA)*. As shown in Figure 12.9, the CCA involves listening for 802.11 RF transmissions at the Physical layer. The medium must be clear before a station can transmit. However, if the medium is not clear (based on sensing RF transmissions that exceed predefined energy thresholds), the 802.11 radio will defer for a defined amount of time and then perform the CCA once again to listen for a clear medium before transmitting. However, if there is a "continuous" RF transmission that is constantly heard during the CCA intervals, 802.11 transmissions will completely cease until the signal is no longer present. If 802.11 transmissions cease due to an interfering RF signal, the result is a denial of service to the WLAN.

FIGURE 12.9 Clear channel assessment (CCA)

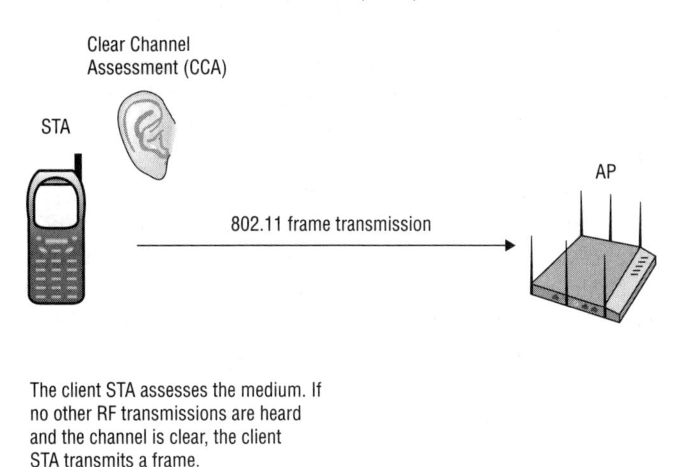

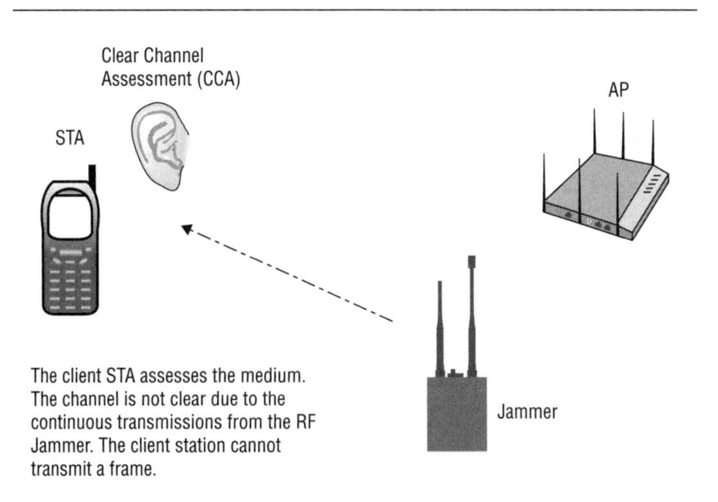

Whether intentional or unintentional, a Layer 1 attack may also result in a partial DoS attack. Every time an 802.11 radio transmits a unicast frame, if the frame is received properly, the 802.11 radio that received the frame will reply with an acknowledgment (ACK) frame. If the ACK is received, the original station knows that the frame transfer was successful. All unicast 802.11 frames must be acknowledged. Broadcast and multicast frames do not require an acknowledgment. If any portion of a unicast frame is corrupted, the cyclic redundancy check (CRC) will fail and the receiving 802.11 radio will not send an ACK frame to the transmitting 802.11 radio. If an ACK frame is not received by the original transmitting radio, the unicast frame is not acknowledged and will have to be retransmitted. RF devices that just transmit intermittently can disrupt with 802.11 transmissions. The intermittent RF interferer will cause corruption of 802.11 unicast frames that are being transmitted and result in Layer 2 retransmissions. An increase in Layer 2 retransmissions will result in decreased throughput and increased latency. While this might not be as traumatic as a continuous transmitting device causing a complete denial of service, the WLAN performance is still adversely affected.

Taking out the entire band or a single channel for only a few seconds breaks all the communications of any upper-layer applications being used over the WLAN. When the attack is stopped, client stations must locate the AP again, authenticate/associate, get an IP address, and reestablish the application session.

Jamming attacks are often used to kick-start other types of attacks. Jamming can force client stations to reauthenticate. A protocol analyzer can then be used to capture the authentication process of clients using a weak method of authentication, such as LEAP or WPA/WPA2-Personal. The information needed to proceed with an offline dictionary attack has been captured. Narrow-band jamming can also be used to jump-start wireless hijacking, man-in-the-middle, and Wi-Fi phishing attacks described later in this chapter.

If you suspect that sources of interference are causing problems for your network, you can find them using a *spectrum analyzer*. A spectrum analyzer is a frequency domain measurement and troubleshooting tool. A spectrum analyzer can help identify and locate an interfering transmitter. Spectrum analyzers will be discussed in greater detail in Chapter 14.

Layer 2 DoS Attacks

The more common type of DoS attacks that originate from hackers are Layer 2 DoS attacks. A wide variety of Layer 2 DoS attacks exist that are a result of tampering with 802.11 frames and retransmitting them into the air. The most common involves spoofing *disassociation* or *deauthentication* management frames. Let's examine some of the more common Layer 2 intentional DoS attacks.

Many of the intentional DoS attacks found here at Layer 2 use combinations of basic wireless networking requirements and the manipulation of what is required by wireless transmissions for them to succeed. For a client STA to pass data within the basic service set (BSS), it must be authenticated and associated. Without authentication, there is neither association nor Layer 2 connection.

An 802.11 management frame, called a deauthentication frame, is sometimes used by client stations and APs to sever communications at Layer 2. An 802.11 deauthentication frame is a notification and not a request. If a station wants to deauthenticate from an AP, or an AP wants to deauthenticate from stations, either device can send a deauthentication frame. Because authentication is a prerequisite for association, a deauthentication frame will automatically cause a disassociation to occur.

Sadly, deauthentication frames can easily be spoofed, and a deauthentication attack can be launched against a single device or the entire BSS. An attacker simply observes the MAC addresses of client stations and access points using a protocol analyzer. The attacker then uses a hex editor to edit a previously captured deauthentication frame. As shown in Figure 12.10, the attacker can edit the 802.11 header and spoof the MAC address of an access point or a client in either the transmitter address (TA) field or the receiver address (RA) field. The attacker then retransmits the spoofed deauthentication frame repeatedly. The station that receives the spoofed deauthentication frame thinks it is coming from another legitimate station and disconnects at Layer 2. Unicast deauthentication frames can be used as an attack against a single client or multiple clients can be deauthenticated if the destination address is a broadcast address.

FIGURE 12.10 Deauthentication attack

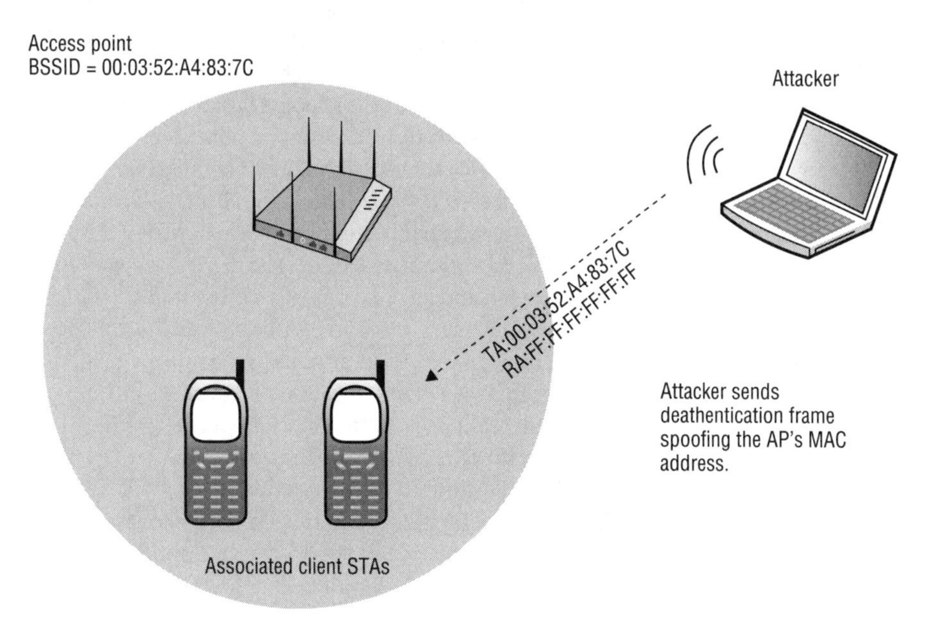

Disassociation attacks work in the same manner and are equally effective for the attacker. Just like deauthentication, disassociation is a notification, not a negotiation. Disassociation management frames can also be spoofed and thereby accomplish the same result as a deauthentication attack.

802.11w-2009 Amendment

The 802.11w-2009 amendment is intended to help stop some DoS attacks that use spoofed management frames. Many more types of Layer 2 DoS attacks exist, including association floods, reassociation floods, authentication floods, EAPOL floods, PS-Poll floods, and virtual carrier attacks. Luckily, any good wireless intrusion detection system will be able to alert an administrator immediately to a Layer 2 DoS attack. The 802.11w-2009 amendment is the *management frame protection (MFP)* amendment with a goal of delivering certain types of management frames in a secure manner. These 802.11w frames are referred to as *robust management frames*. Robust management frames can be protected by the management frame protection service and include disassociation, deauthentication, and robust action frames.The end result will hopefully prevent some of the Layer 2 DoS attacks that currently exist, but many Layer 2 DoS attacks will never be circumvented. Although most new access points support MFP, most client devices, particularly legacy clients, do not support protected management frame protection. More information about management frame protection can be found in Chapter 14.

Let's discuss some other Layer 2 DoS attacks that can be accomplished by simply editing 802.11 frames and retransmitting them into the air. Once such attack is called *illegal channel beaconing*. This attack uses a spoofed beacon frame transmitted on the same real channel as the legitimate AP, so that the associated client stations will hear the spoofed beacon. The spoofed beacon uses the same SSID, but the channel field has been edited to display a nonexistent, or illegal, channel. For example, there are 14 channels available for use in the 2.4 GHz range, none of which are used in the spoofed beacon. The attacker's beacon could be telling the client STAs that the AP is on channel 0 or channel 432 or some other unused channel number, when in reality the attacker is beaconing on the same real channel used by the legitimate devices. The drivers of some WLAN vendor radios cannot interpret the illegal channel, and a denial of service is the result.

Probe requests and responses are management frames and can be spoofed just as beacons can be spoofed to disrupt network connectivity. If an attacker sends probe response frames to a victim station, even if it is already associated with a real AP, that station will assume that it should try to connect to that AP. The stations that fall victim to this attack did not send a probe request frame looking for this AP but will try to connect to it anyway. This attack is called a *probe response flood*.

Many Layer 2 DoS attacks are flooding attacks that use management or control frames to overwhelm an access point or client station. An example of a Layer 2 flooding involves an attack on an AP's association table. The 802.11-2012 standard defines the maximum number of client associations to an access point radio as 2,007. In reality, no access point radio would ever want 2,007 clients associations, but in theory it is possible. Most vendors offer settings on APs or WLAN controllers to limit the number of active client associations to an AP for capacity purposes. For example, an administrator might set the maximum number of client associations to 75 per access point radio. An attacker can flood an AP with bogus association request frames and fill up the AP's association table. Then, when any legitimate client attempts to associate, the legitimate clients are denied association because the maximum has already been reached. This attack is called an *association flood*.

Another Layer 2 DoS attack is called *FakeAP*. This software tool generates thousands of counterfeit management advertising fake SSIDs and BSSIDs. The original intent was to hide a real access point in plain sight among all the fake APs to confuse any wardrivers, script kiddies, or other undesirables. The problem is that the FakeAP tool can also be used for a DoS attack. Legitimate client stations may spend time attempting to associate to the APs that do not exist. Additionally, the extra airtime consumption within the service set area will degrade the performance of the WLAN.

As mentioned earlier is this chapter, Carrier Sense Multiple Access with Collision Avoidance (CSMA/CA) is the process used to ensure that only one 802.11 radio is transmitting at a time on the medium. Another major component of the CSMA/CA method of medium contention is *virtual carrier sense*. Virtual carrier sense uses a timer mechanism known as the *network allocation vector (NAV)* or the NAV timer. The NAV timer maintains a prediction of future traffic on the medium based on Duration value information seen in a previous frame transmission. When an 802.11 radio is not transmitting, it is listening. As shown in Figure 12.11, when the listening radio hears a frame transmission from another station, it looks at the Layer 2 header of the frame and determines whether the *Duration/ID* field contains a Duration value or an ID value. If the field contains a Duration value, the listening station will set its NAV timer to this value. The listening station will then use the NAV as a countdown timer, knowing that the RF medium should be busy until the countdown reaches 0. This process essentially allows the transmitting 802.11 radio to notify the other stations that the medium will be busy for a period of time (the Duration/ID value). The stations that are not transmitting listen and hear the Duration/ID value, set a countdown timer (NAV), and wait until their timer hits 0 before they can contend for the medium and eventually transmit on the medium. A station cannot contend for the medium until its NAV timer is 0, nor can a station transmit on the medium if the NAV timer is set to a nonzero value.

FIGURE 12.11 Virtual carrier sense

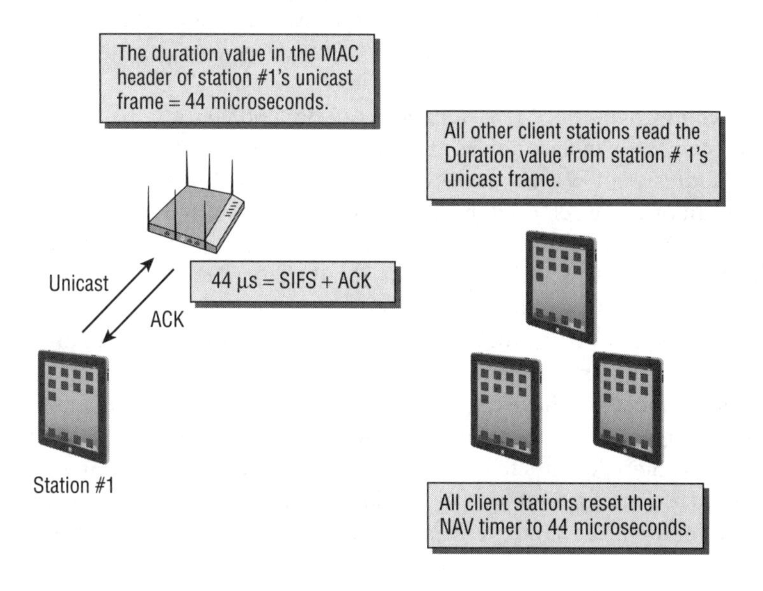

The Duration value can be set from 0 to 32767. Most frames do not approach the limit. If an attacker is not contending for the medium but simply transmitting frames with spoofed Duration/ID field values set near the upper limit, an attack known as the *virtual-carrier attack* will disrupt the WLAN. By selecting a value near the limit such as 29000, the attacker knows that stations are not likely to ignore the spoofed frame. Some stations will ignore values over 30000. The high value used by the attacker will cause stations in the area hearing it to set their NAV timer to that high number. The victims then must count down from that number to zero. Once they reach zero, the stations listen to the medium again, and they hear another spoofed frame from the attacker and reset their NAV timers, thus starting the process all over. In this case, the victim stations would never be allowed access to the medium. The attacker is simply winning the contention for the medium by using spoofed Duration values and causing everyone else to remain idle during the attacker's transmissions.

Most Layer 2 DoS attacks are not doing anything "special." The attackers are merely exploiting the way that 802.11 communications function at Layer 2 to disrupt legitimate traffic. As mentioned earlier, numerous Layer 2 DoS attacks exist such as association floods, reassociation floods, authentication floods, EAPOL floods, and PS-Poll floods. All of these attacks are accomplished by simply editing 802.11 frames and retransmitting them to disrupt Layer 2 communications.

MAC Spoofing

All 802.11 radios have a physical address known as a *MAC address*. This address is a 12-digit hexadecimal number that is seen in cleartext in the Layer 2 header of 802.11 frames. Wi-Fi vendors provide MAC filtering capabilities on their access points and WLAN controllers. Usually, MAC filters are configured to apply restrictions that will allow traffic only from specific client stations to pass through. These restrictions are based on their unique MAC addresses. All other client stations whose MAC addresses are not on the allowed list will not be able to pass traffic through the virtual port of the access point and onto the distribution system medium.

As you learned in Chapter 2, MAC addresses can be spoofed, or impersonated, and any amateur hacker can easily bypass any MAC filter by spoofing an allowed client station's address. MAC addresses of WLAN devices can be spoofed within any operating system. MAC spoofing can often be achieved in the Windows operating system by simply editing the wireless radio's MAC address in Device Manager or by performing a simple edit in the Registry. Third-party software utilities, such as SMAC (described in a moment), can also be used to accomplish MAC spoofing. MAC spoofing renders MAC filters useless as a form of security on wireless networks, since MAC addresses are always visible to anyone on the same channel and in the same area.

No two devices should ever have the same MAC address configured on them from the manufacturer. The organizationally unique identifier (OUI) address is the first three octets of the MAC address that identifies the manufacturer of the radio. The remaining octets of the MAC address are unique and are used to identify the individual radio. The existence of two radios with the same MAC address should not happen because vendors are very careful to avoid this in the manufacturing process to prevent address conflicts. Attackers, on the other hand, use duplicate MAC addressing to their advantage. By cloning a MAC address an attacker can bypass MAC filters. The filter cannot distinguish between a legitimate device and a spoofed device, thereby allowing them access to spoofed devices. The MAC filter has no way of knowing that more than one device is using the same addressing. Frames from either device—the real one or the attackers—are accepted by the network. Traffic leaving the network bound for either device is transmitted into the air. Both devices receive the frames but only the one expecting it processes the data. This happens because the devices are using different sockets. If the attacker is in close proximity to the cloned device, the attack may go undetected by some WIPS solutions that only use sensor location to trigger a MAC spoof–based alarm. To be able to detect spoofing more efficiently, better WIPS products also look at the sequence numbers of the frames being transmitted by the devices. If they are out of sequence, an alarm can be triggered, alerting a WLAN administrator to the MAC spoofing attack.

One place where a MAC spoofing attack is still used with great effect is at public-access WLAN hotspots. A *MAC piggy-backing* attack is used to circumvent the hotspot captive portal login requirements. The attacker is not trying to break into a network to steal data but rather to exploit the way the hotspot's *captive portal* works to gain free Internet access. Captive portal authentication solutions are usually the only security provided for guest WLANs and public-access hotspots. Once a station connects to the hotspot SSID and gets an IP address, the user opens their browser. Rather than going to their normal home page, the user is redirected to the captive portal web login page. This page may have a simple terms-of-use agreement or may require credentials to access the Internet beyond the captive portal page. The required credentials could be a username and password or credit card information. When the captive portal is satisfied that the user has the correct information to gain Internet access—password, credit card, and so on—the captive portal authentication allows the user access to a gateway to the Internet. The only security beyond this point is a common MAC filter allowing access for users who have authenticated via the captive portal login page. An attacker uses a WLAN protocol analyzer to determine which stations are passing data frames through the AP, indicating the captive portal has approved their MAC addresses to do so. Then the attacker clones the MAC address of a station passing data through the AP onto their wireless radio. The attacker can then connect to the AP and pass data as well because the AP and its captive portal believe the attacker is an approved device. MAC piggy-backing is not normally considered a malicious attack, but the attack could be considered as theft of services if the hotspot requires payment for access.

If MAC addresses are designated by the manufacturer of the radios and no two MAC addresses should ever be the same, how are attackers able to change them? Attackers and users alike can change MAC addresses very easily using utilities within the operating system or programs designed to allow this change. As shown in Figure 12.12, changing the MAC address through the Windows OS can be done by directly editing the Registry value for the radio's MAC address and rebooting the system. When the system comes back up, the radio will be using the new cloned MAC address.

FIGURE 12.12 MAC address Registry settings

Moderate	REG_SZ	1
NetCfgInstanceId	REG_SZ	{5D683157-AC8D-4E21-959B-7CF790DB8B37}
NetworkAddress	REG_SZ	001A733D673C
ProviderName	REG_SZ	Marvell
SGMapRegisters...	REG_SZ	64
TagHeaderSuppo...	REG_SZ	0
TcpLargeSend	REG_SZ	1
WaitForRxResou...	REG_SZ	1
WakeFromShutd...	REG_SZ	1
WakeFromShutd...	REG_SZ	17
WakeUpModeCa...	REG_SZ	27
YKVS	REG_SZ	762751

As shown in Figure 12.13, many WLAN radios will allow a change of the MAC address to be made in the GUI without a system reboot. This option is typically found in the Advanced settings of the networking properties of the radio's MAC address.

FIGURE 12.13 MAC address: networking settings

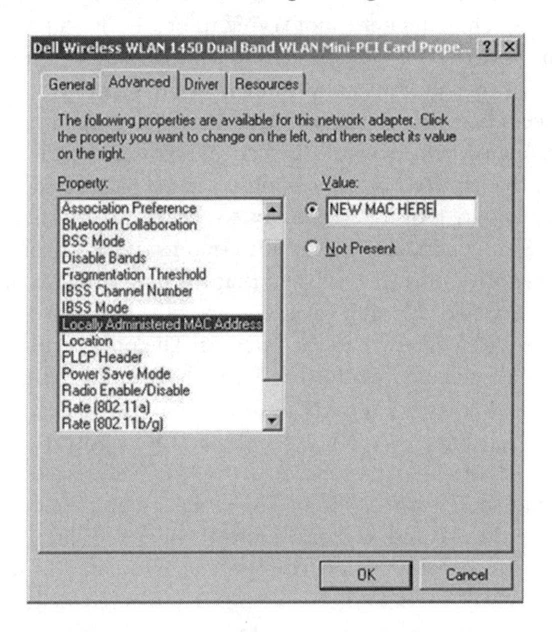

If the attacker does not wish to edit the Registry directly and the radio they are using does not have the option to configure a locally assigned MAC address, programs are available that will allow the user to change the MAC address despite the other limitations. One such program is SMAC from KLC Consulting. As shown in Figure 12.14, SMAC has the ability to set specific MAC addresses as well as the ability to generate MAC addresses based on known OUI structures from several vendors. It also keeps track of recently used MAC addresses for future use.

FIGURE 12.14 SMAC

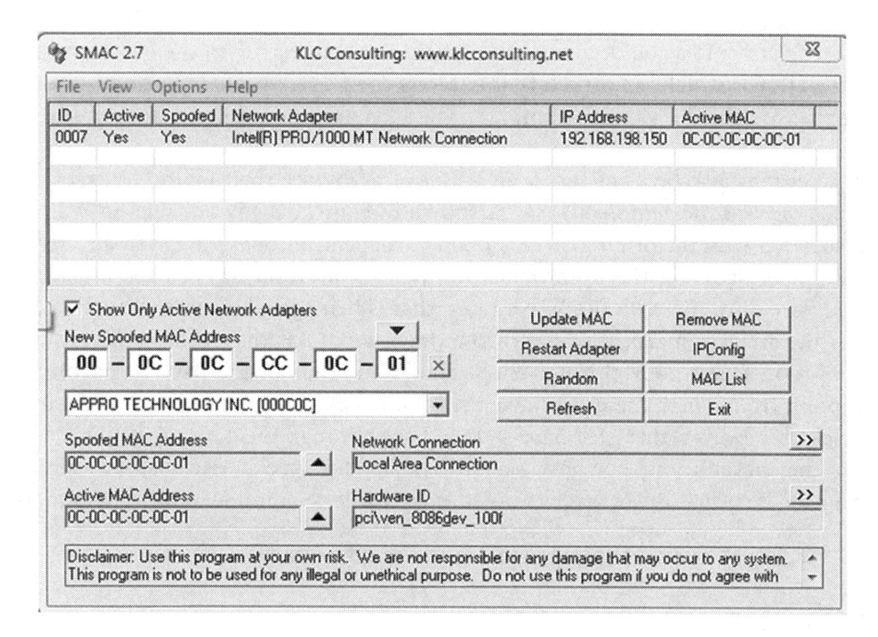

Because of spoofing and because of all the administrative work that is involved with setting up MAC filters, MAC filtering is not considered a reliable means of security for wireless enterprise networks and should be implemented only if stronger security is not available or in addition to some stronger form of security as part of a multifactor security plan.

Wireless Hijacking

An attack that often generates a lot of press is *wireless hijacking*, also known as the *evil twin attack*. The attacker configures access point software on a laptop, effectively turning a Wi-Fi client radio into an access point. Some small Wi-Fi USB devices also have the ability to operate as an AP. The access point software on the attacker's laptop is configured with the same SSID that is used by a public-access hotspot. The attacker's access point is now functioning as an evil twin AP with the same SSID but is transmitting on a different channel. The attacker then sends spoofed disassociation or deauthentication frames, forcing client stations associated with the hotspot access point to roam to the evil twin access point. At this point, the attacker has effectively hijacked wireless clients at Layer 2 from the original access point. Although deauthentication frames are usually used as one way to start a hijacking attack, RF jammers can also be used to force any clients to roam to an evil twin AP.

The evil twin AP will typically be configured with a Dynamic Host Configuration Protocol (DHCP) server available to issue IP addresses to the clients. At this point, the attacker will have hijacked the client stations at Layer 3. The attacker has a private WLAN network and is free to perform peer-to-peer attacks on any of the hijacked clients. The user's computer could, during the process of connecting to the evil twin, fall victim to the DHCP attack, an attack that exploits the DHCP process to dump root kits or other malware onto the victim's computer in addition to giving them an IP address as expected.

The attacker may also be using a second wireless radio with their laptop to execute what is known as a *man-in-the-middle attack*, as shown in Figure 12.15. The second WLAN radio is associated with the original access point as a client. In operating systems, network adapters can be bridged together to provide routing. The attacker has bridged together their second wireless radio with the Wi-Fi radio that is being used as the evil twin access point. After the attacker hijacks the users from the original AP, the traffic is then routed from the evil twin access point through the second Wi-Fi radio, right back to the original access point from which the users have just been hijacked. The result is that the users remain hijacked; however, they still have a route back through the gateway to their original network, so they never know they have been hijacked. The attacker can therefore sit in the middle and execute peer-to-peer attacks indefinitely while remaining completely unnoticed.

FIGURE 12.15 Wireless hijacking/man-in-the-middle attack

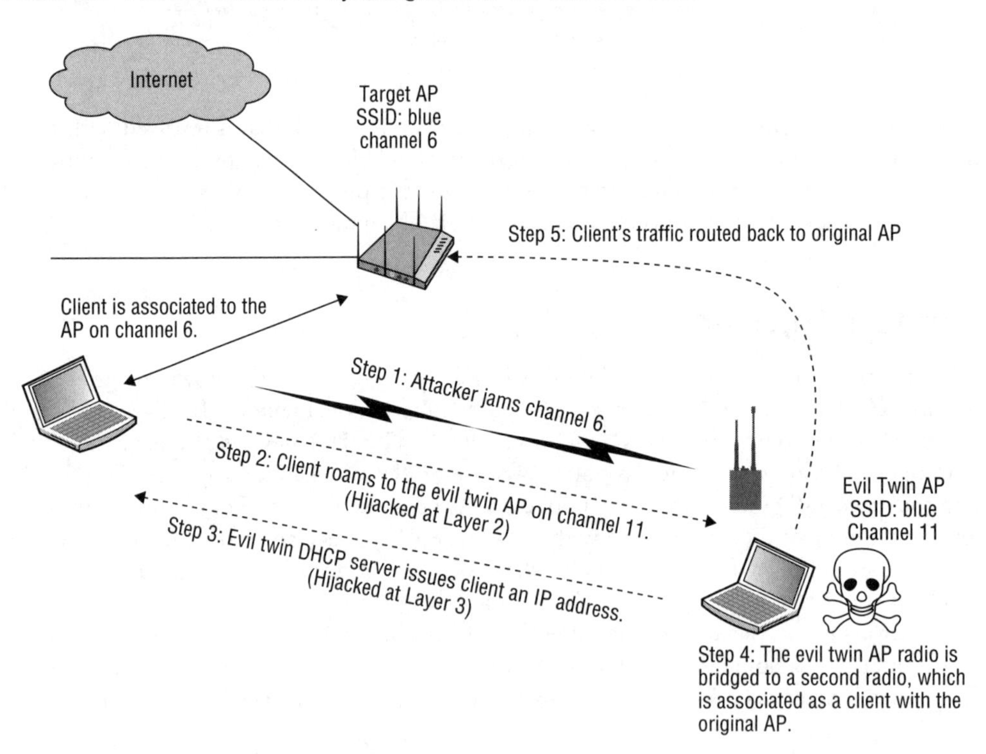

These attacks can take another form in what is known as the *Wi-Fi phishing attack*. The attacker may also have web server software and captive portal software. After the users have been hijacked to the evil twin access point, they will be redirected to a login web page that looks exactly like the hotspot's login page. Then the attacker's fake login page may request a credit card number from the hijacked user. Phishing attacks are common on the Internet and are now appearing at your local hotspot.

The only way to prevent a hijacking, man-in-the-middle, or Wi-Fi phishing attack is to use a mutual authentication solution. Mutual authentication solutions not only validate the user connecting to the network, but they also validate the network to which the user is connecting. 802.1X/EAP authentication solutions require that mutual authentication credentials be exchanged before a user can be authorized. A user cannot get an IP address unless authorized; therefore, users cannot be hijacked.

Encryption Cracking

Encrypting transmitted information is extremely important in wireless communications, since the medium is shared and unbounded. If any degree of data privacy is to be expected, some form of encryption should be used—the stronger, the better. Earlier wireless networks used no encryption at all or they used Wired Equivalent Privacy (WEP) to encrypt the transmissions. As you learned in Chapter 2, WEP is a Layer 2 encryption method that does indeed provide data privacy when encrypting the Layer 3–7 data payload known as the MAC Service Data Unit (MSDU). Unfortunately, WEP has been cracked, and software tools exist that can derive the static WEP key from 802.11 data frame traffic that has been captured using a protocol analyzer. Older WEP cracking software could take days to run because the number of Initialization Vectors (IVs) needed to crack the key is relatively high. WEP uses either a 40-bit secret key or a 104-bit secret key to protect the data. No matter which secret key length is used, 40-bit or 104-bit, WEP only uses a 24-bit IV. If an attacker gathers enough IVs they can crack 64-bit and 128-bit WEP in minutes. An attacker would need to capture about 500,000 IVs to be able to crack the WEP key. In a SOHO environment, this could take weeks of captures. In an enterprise, with a larger traffic volume than SOHO networks, this could still take days.

As shown in Figure 12.16, modern freeware cracking tools can now break through WEP protected frames in a matter of minutes and obtain the static WEP key. Modern cracking tools now use an injection attack that forces devices to generate the IVs much faster than regular traffic requires. The injection attacks often use Address Resolution Protocol (ARP) flooding to force this to happen. Now cracking 128-bit WEP keys can be done in a few minutes versus the days and weeks of older tools that employed large capture files and brute-force attacks. Once an attacker has obtained the WEP key, they can connect to the WLAN and decode the data captured offline or decode data frames in real time.

FIGURE 12.16 Cracked WEP key

```
                [00:00:03] Tested 2 keys (got 1040384 IVs)

KB    depth   byte(vote)
 0    0/  1   D7(  93) 59(  15) D2(  13) 6C(  12) EE(  10) 5A(   5)
 1    0/  1   57( 227) AE(  40) F7(  2/) 65(  25) 62(  22) 91(  22)
 2    0/  1   B7( 933) 9B(  27) 01(  25) 39(  25) F0(  23) 06(  20)
 3    0/  1   C9( 330) 62(  39) E8(  38) F6(  38) 66(  37) 0F(  35)
 4    0/  1   A8( 475) 25(  69) 0F(  60) 56(  50) 26(  48) 82(  44)
 5    0/  1   EB( 519) 75(  59) E2(  46) C4(  44) 66(  43) 74(  39)
 6    0/  2   60( 171) 81( 135) 7F(  44) 82(  44) EA(  37) C4(  35)
 7    0/  2   7E( 358) 17( 150) 16(  36) 92(  34) BE(  32) E6(  31)
 8    0/  3   DB( 196) 8E( 101) BF(  68) 8D(  39) DC(  35) 5C(  33)
 9    0/  1   86( 496) A7(  87) A8(  48) 16(  45) A6(  41) 23(  40)
10    0/  2   07( 283) 14( 120) 0E(  45) 91(  42) 10(  41) 15(  38)
11    0/  1   A4( 340) 19(  77) FE(  72) 3E(  46) 3C(  44) 4E(  44)
12    0/  2   A4( 328) 4C( 187) 53(  65) 48(  55) A5(  45) 9A(  42)

           KEY FOUND! [ D7-57-B7-C9-A8-EB-60-7E-DB-86-07-A4-A4 ]
```

Network managers abandoned WEP for newer, more secure dynamic encryption methods. To improve on the security offered by WEP, the Temporal Key Integrity Protocol (TKIP) was developed. TKIP uses 48-bit IVs, time-bound keys, and the RC4 algorithm for an improvement over the security offered by WEP. TKIP encryption is no longer considered secure and is not supported for any 802.11n or 802.11ac data rates. Modern-day WLAN radios use dynamic CCMP encryption, which uses the 128-bit AES cipher.

Peer-to-Peer Attacks

A commonly overlooked risk is the *peer-to-peer attack*. As you learned in earlier chapters, an 802.11 client station can be configured in either infrastructure mode or ad hoc mode. When configured in ad hoc mode, the wireless network is known as an independent basic service set (IBSS) and all communications are peer-to-peer without the need for an access point. Because an IBSS is by nature a peer-to-peer connection, any user who can connect wirelessly with another user can gain access to any resource available on either computer. A common use of ad hoc networks is to share files on the fly. If shared access is provided, files and other assets can accidentally be exposed. A personal firewall is often used to mitigate peer-to peer attacks. Some client devices can also disable this feature so that the device will connect only to certain networks and will not associate to a peer-to-peer without approval.

Users who are associated to the same access point are potentially just as vulnerable to peer-to-peer attacks as IBSS users. Properly securing your wireless network often involves protecting authorized users from each other, because hacking at companies is often performed internally by employees. Any users associated to the same AP who are members of the same basic service set (BSS) and are in the same VLAN are susceptible to peer-to-peer attacks because they reside in the same Layer 2 and Layer 3 domains. In most WLAN deployments, Wi-Fi clients communicate only with devices on the wired network, such as email or web servers, and peer-to-peer communications are not needed. Therefore, most enterprise AP vendors provide some proprietary method of preventing users from inadvertently sharing files with other users or bridging traffic between the devices. If connections are required to other wireless peers, the traffic is routed through a Layer 3 switch or other network device before passing to the desired destination station.

Client isolation is a feature that can often be enabled on WLAN access points or controllers to block wireless clients from communicating with other wireless clients on the same wireless VLAN. Client isolation, or the various other terms used to describe this feature, usually means that packets arriving at the AP wireless interface are not forwarded back out of the wireless interface to other clients. This isolates each user on the wireless network to ensure that a wireless station cannot be used to gain Layer 3 or higher access to another wireless station. The client isolation feature is usually a configurable setting per SSID linked to a unique VLAN. With client isolation enabled, client devices cannot communicate directly with other client devices on the wireless network, as shown in Figure 12.17.

FIGURE 12.17 Client isolation

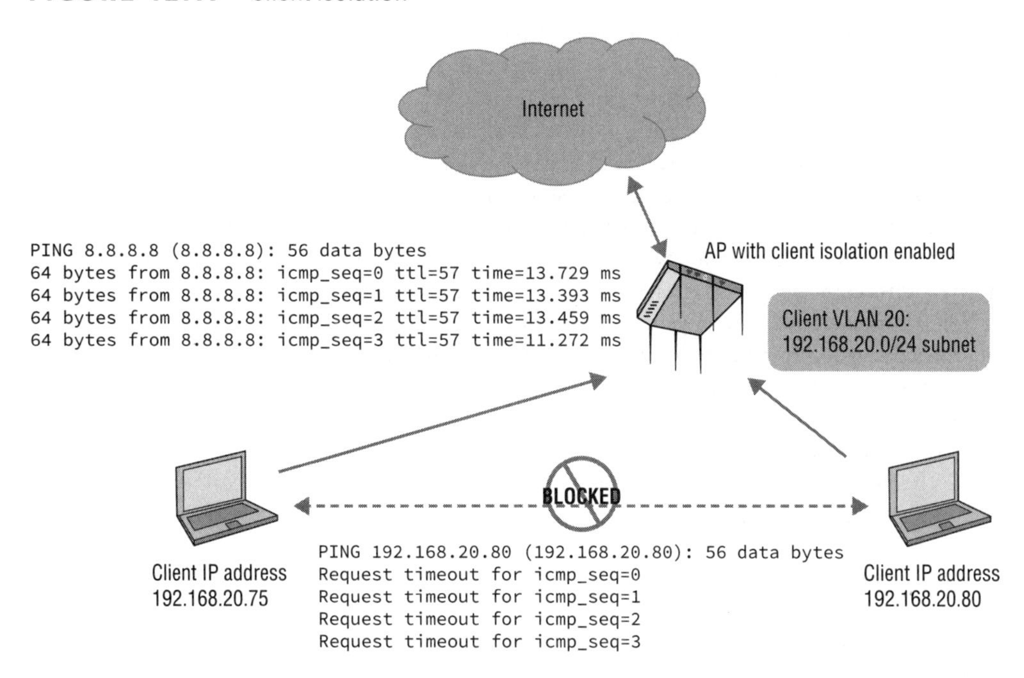

Although *client isolation* is the most commonly used term, some vendors instead use the terms *peer-to-peer blocking* or *public secure packet forwarding (PSPF)*. Not all vendors implement client isolation in the same fashion. Some WLAN vendors can only implement client isolation on an SSID/VLAN pair on a single access point, whereas others can enforce the peer-blocking capabilities across multiple APs.

Management Interface Exploits

One of the main goals of attackers is to gain access to administrative accounts or root privilege. Once they gain that access, they can run several attacks against networks and individual devices. On wired networks these attacks are launched against firewalls, servers, and infrastructure devices. In wireless attacks, these are first launched against access points or WLAN controllers and subsequently against the same targets as in wired attacks. Wireless infrastructure hardware such as autonomous access points and WLAN controllers can be managed by administrators via a variety of interfaces, much like managing wired infrastructure hardware. Devices can typically be accessed via a web interface, a command-line interface, a serial port, a console connection, and/or Simple Network Management Protocol (SNMP). It is imperative that these interfaces be protected. Interfaces that are not used should be disabled. Strong passwords should be used, and encrypted login capabilities such as Hypertext Transfer Protocol Secure (HTTPS) should be used if available.

Lists of all the default settings of every major manufacturer's access points exist on the Internet and are often used for security exploits by hackers. It is not uncommon for intruders to use security holes left in management interfaces to reconfigure access points. Legitimate users and administrators can find themselves locked out of their own wireless networking equipment. After gaining access via a management interface, an attacker might even be able to initiate a firmware upgrade of the wireless hardware and, while the upgrade is being performed, disable the equipment's power source. This attack could likely render the hardware useless, requiring it to be returned to the manufacturer for repair.

Many WLAN devices often have settings that allow for remote administration via the Internet. Although these settings are intended for legitimate administrators, an attacker may use remote access to management interfaces to perform the same attacks just described. Remote access should either be turned off or locked down tight.

Policy often dictates that all WLAN infrastructure devices be configured from only the wired side of the network. If an administrator attempts to configure a WLAN device while connected wirelessly, the administrator could lose connectivity due to configuration changes being made. Some WLAN vendors offer secure wireless console connectivity capabilities for troubleshooting and configuration.

Vendor Proprietary Attacks

Hackers often find holes in the firmware code used by specific WLAN access points and WLAN controller vendors. New WLAN vulnerabilities and attacks are discovered on a regular basis, including vendor proprietary attacks. Many of these vendor-specific exploits are in the form of buffer overflow attacks. When these vendor-specific attacks become known, the WLAN vendor usually makes a firmware fix available in a timely manner. Once the exploits are discovered, the affected WLAN vendor will make safeguard recommendations on how to avoid the exploit. In most cases the WLAN vendor will release a patch that can fix the problem. These attacks can be best avoided by staying informed through your WLAN vendor's support services.

Physical Damage and Theft

An important aspect of the installation of wireless equipment is the "pretty factor." The majority of businesses prefer that all wireless hardware remain completely out of sight. Aesthetics is extremely important in retail environments and in the hospitality industry (restaurants and hotels). Any business that is dealing with the public will require that the Wi-Fi hardware be hidden or at least secured. Many vendors are designing better-looking access points and antennas. Some vendors have even camouflaged access points to resemble smoke detectors. Indoor enclosures that are mounted in place of ceiling tiles are also often used to conceal access points. It should also be noted that most enclosure units can be locked to help prevent physical damage or theft of expensive Wi-Fi hardware.

Client devices such as VoWiFi phones and WLAN barcode scanners are often easily stolen. Several companies such as AeroScout and Ekahau provide a WLAN *real-time location system (RTLS)*, which can track the location of any 802.11 radio device as well as active Wi-Fi RFID tags with great accuracy. The components of an overlay WLAN RTLS solution include the preexisting WLAN infrastructure, preexisting WLAN clients, Wi-Fi RFID tags, and an RTLS server. Additional RTLS WLAN sensors can be added to supplement the preexisting WLAN APs.

Active RFID tags and/or standard Wi-Fi devices transmit a brief signal at a regular interval, adding status or sensor data if appropriate. Figure 12.18 shows an active RFID tag attached to heathcare monitoring equipment. The signal is received by standard wireless APs (or RTLS sensors), without any infrastructure changes needed, and is sent to a processing engine that resides in the RTLS server at the core of the network. The RTLS server uses signal strength and/or time-of-arrival algorithms to determine location coordinates.

FIGURE 12.18 Active 802.11 RFID tag

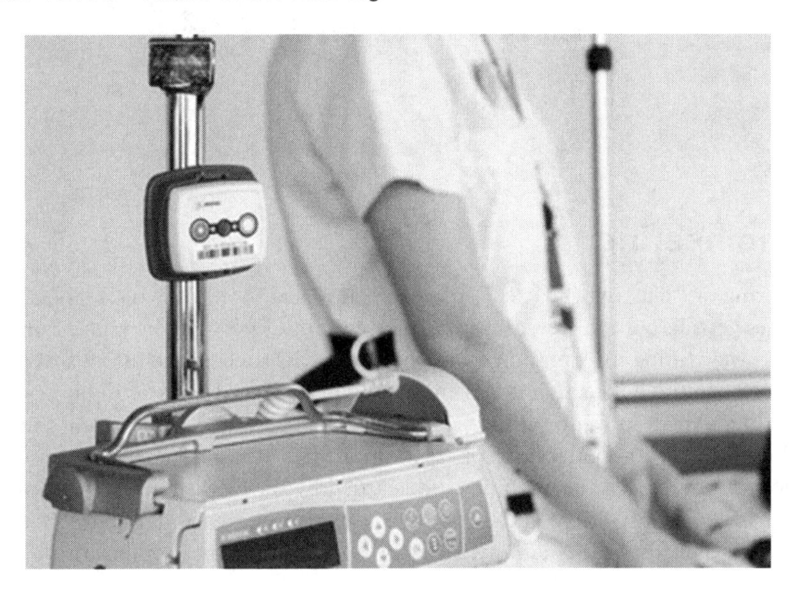

Courtesy of Ekahau

As pictured in Figure 12.19, a software application interface is then used to see location and status data on a display map of the building's floor plan. The RTLS application can be used to define zones on the floor plan that will trigger alarms if client devices or RFID tags leave a certain area or zone. An RTLS solution can greatly reduce theft of WLAN devices as well as any company assets with an attached RFID tag.

FIGURE 12.19 RTLS application

Courtesy of Ekahau

Social Engineering

Social engineering or "hacking the user" is a very real threat to all networks. Social engineering involves getting users to reveal information without knowing they have done so. It also may involve gaining information from public sources such as the target's website or corporate reports. Another effective social engineering attack is dumpster diving—going through the target's trash looking for information. This once was a big problem for credit cards. Retailers used carbon papers and physical imprints of the customer's card as part of the transaction. The carbon papers would be discarded by the clerks and later deposited in common garbage dumpsters. People wanting to use the credit card numbers of others would go through the trash in the dumpster to find the carbon papers and thus the credit card numbers.

Phishing is also a social engineering attack. Phishing is a criminal process used to acquire sensitive information fraudulently, such as usernames, passwords, and credit card information, by masquerading as a valid authority during electronic communication. Many of us have gotten the infamous email "I am in another country and need help getting my millions out, and I will share the money with you for your help." Perhaps you received the email from a bank or government office asking for your personal information or bank account numbers. Why do these obvious junk emails, phone calls, and letters continue to be sent? These scams continue because some unfortunate people respond to them, giving the criminals the information or money they desire. The Wi-Fi phishing attack discussed earlier in this chapter uses wireless hijacking techniques and a bogus captive portal to perpetrate a similar scam.

Computer network users sometimes are even more careless with corporate information and device security. People are often more concerned with ease of use than they are with security. Users will often keep the same simple passwords for months or years. Their passwords can be found written on sticky notes under their keyboards or stuck to their monitors. No matter how much money is spent on security equipment and software or even on security guards, users still may carelessly or unwittingly give attackers information that will compromise security measures. For example, a call can be placed to users stating that the caller is from the IT security team. The caller may tell the user that some unapproved software or freeware that is not allowed has been detected on their machine. The caller tells the user they want to help the user become compliant with corporate policy and, that by giving them their username and password, the caller will be able to remove the suspicious software from their machine without the caller having to report it as a policy violation to the user's manager. Desperate to avoid trouble, users may comply with the attacker's requests.

Users cannot all be trained security professionals. How can we train our users to be more security conscious as they travel with company laptops? We can train them to make sure their personal firewalls are always on and that their antivirus software is up to date and running, and to recognize social engineering attacks and report them to corporate security teams as soon as possible.

A variant of these attacks is called reverse social engineering. This is a process by which an attacker poses as someone in authority and gives the user bogus information rather than trying to get information from the user. The users are almost always the weakest link in network security.

The best defense against social engineering attacks are strictly enforced policies to prevent confidential information from being shared. Any information that is static is extremely susceptible to social engineering attacks. WEP encryption uses a static key, and WPA/WPA2-Personal requires the use of a static PSK or passphrase. You should avoid both of these security methods because of their static nature.

Social Engineering and Chocolate

In 2004 a survey was conducted in London in the Liverpool Street Station. The survey (`http://news.bbc.co.uk/2/hi/technology/3639679.stm`) found that commuters were willing to share their passwords with strangers conducting the survey in exchange for a piece of candy. A whopping 34 percent revealed their passwords when asked if it had something to do with a pet or child's name. A second survey revealed that 79 percent of those questioned gave away personal information without even knowing what they were exposing when questioned by "survey takers."

Guest Access and WLAN Hotspots

When you are using the WLAN at an Internet cafe, coffee shop, airport, hotel, and other public WLAN hotspots, you are not as protected as you are as when you are using your organization's WLAN where security measures have been taken to protect your wireless communications. Hotspots are often breeding grounds for peer-to-peer attacks, viruses, hijacking, data theft or manipulation, eavesdropping attacks, and other malicious events. So what makes these convenient connections so vulnerable to attacks? There is no real security provided by the hotspot's host. The hotspots may be using a captive portal requiring users to agree to the terms of use statement prior to granting the connection. Captive portal authentication is considered to be a weak authentication method but is normally adequate for simply authorizing users to access a gateway to the Internet. However, public-access WLANs do not offer an encryption solution. Additionally, most companies now offer guest WLAN access at their place of business and encryption is rarely used. All 802.11 data frames are therefore unencrypted and susceptible to malicious eavesdropping attacks.

Hotspots do not deploy wireless intrusion prevention systems (WIPS) to protect users either. Thus there is also no security monitoring of the insecure WLAN connectivity. The vast majority of hotspots users are not security conscious as they travel, making them vulnerable to attack as well.

Knowing that hotspots are the "wild west of Wi-Fi," we can implement a remote access policy for all corporate employees who access hotspots and other non-corporate guest Wi-Fi networks. End users will be taking their laptops and handheld devices off site and away from company grounds. Most users will likely use wireless networks at home and at wireless hotspots to access the Internet. By design, many of these remote wireless networks have absolutely no security in place, and it is imperative that a remote access WLAN policy be strictly enforced. This policy should include the required use of an IPsec VPN solution to provide device authentication, user authentication, and strong encryption of all wireless data traffic. Hotspots are prime targets for malicious eavesdropping attacks.

Personal firewalls should also be installed on all remote computers to prevent peer-to-peer attacks. Personal firewalls will not prevent hijacking attacks or peer-to-peer attacks, but will prevent attackers from accessing most critical information. Endpoint WLAN policy enforcement software solutions exist that force end users to use VPN and firewall security when accessing any wireless network other than the corporate WLAN. The remote access policy is mandatory because the most likely and vulnerable location for an attack to occur is at a public-access hotspot.

The problem is that many consumers and guest users are not savvy enough to know how to use a VPN solution when connected to an open guest WLAN. As a result, there is a recent trend to provide encryption and better authentication security for WLAN guest users. Encrypted guest access is discussed in more detail in Chapter 10, "BYOD and Guest Access."

Summary

Due to the unbounded functionality of wireless networking exposing Layers 1 and 2 to anyone within listening range and on the same frequency, there are many more risks involved in its use than are found in traditional bounded networking. Strong security measures exist for WLAN communications that should be employed to mitigate the risks involved in 802.11 communications. WLAN administrators should have a thorough understanding of all of the WLAN security risks and potential attacks. Thorough comprehension of all WLAN threats will ultimately result in the better decisions when choosing 802.11 security solutions for the enterprise.

Exam Essentials

Understand the risk of the rogue access point. Be able to explain why the rogue AP and other rogue devices provide a portal into network resources. Understand that hackers are usually not the source of rogue devices.

Define peer-to-peer attacks. Understand that peer-to-peer attacks can happen via an AP or through an ad hoc network. Explain how to defend against the attack.

Know the risks of eavesdropping. Explain the difference between casual and malicious eavesdropping. Explain why encryption is needed for protection.

Define authentication and hijacking attacks. Explain the risks behind these attacks. Understand that a strong 802.1X/EAP solution is needed to mitigate these attacks.

Explain wireless denial-of-service attacks. Know the difference between Layer 1 and Layer 2 DoS attacks. Explain why these attacks cannot be mitigated and can only be monitored.

Understand management interface exploits. Explain the reasons for hardening the WLAN infrastructure for authorized and secure administration only.

Understand the concept of social engineering. Know that end users are always the weakest link in WLAN security. Explain why end users need to be trained to recognize social engineering attacks. Understand that any type of static password, passphrase, or shared encryption key is always susceptible to a social engineering attack.

Review Questions

1. When an attacker passively captures and examines wireless frames from a victim's network, what type of attack is taking place?
   - **A.** Injection
   - **B.** Data destruction
   - **C.** Frame manipulation
   - **D.** Man in the middle
   - **E.** Eavesdropping
2. Which of these attacks are considered denial-of-service attacks? (Choose all that apply.)
   - **A.** Man-in-the-middle
   - **B.** Jamming
   - **C.** Deauthentication spoofing
   - **D.** MAC spoofing
   - **E.** Peer-to-peer
3. The majority of rogue devices are placed by whom? (Choose all that apply.)
   - **A.** Attackers
   - **B.** Wardrivers
   - **C.** Employees
   - **D.** Contractors
   - **E.** Visitors
4. Which of these attacks would be typically associated with malicious eavesdropping? (Choose all that apply.)
   - **A.** NetStumbler
   - **B.** Peer-to-peer
   - **C.** Protocol analyzer capture
   - **D.** Packet reconstruction
   - **E.** PS polling attack
5. What are some of the risks if a rogue device goes undetected?
   - **A.** Data theft
   - **B.** Data destruction
   - **C.** Loss of network services
   - **D.** Data insertion
   - **E.** Third-party attacks
   - **F.** All of the above

6. Which of these can cause unintentional RF jamming attacks against an 802.11 wireless network? (Choose all that apply.)
   - **A.** Microwave oven
   - **B.** Signal generator
   - **C.** 2.4 GHz cordless phones
   - **D.** 900 MHz cordless phones
   - **E.** Deauthentication transmitter

7. Which of these attacks are wireless users susceptible to at a public-access hotspot? (Choose all that apply.)
   - **A.** Wi-Fi phishing
   - **B.** Happy AP attack
   - **C.** Peer-to-peer attack
   - **D.** Malicious eavesdropping
   - **E.** 802.11 reverse ARP attack
   - **F.** Man-in-the-middle
   - **G.** Wireless hijacking

8. Which of these encryption technologies have been cracked? (Choose all that apply.)
   - **A.** 64-bit static WEP
   - **B.** 128-bit Dynamic WEP
   - **C.** CCMP/AES
   - **D.** 128-bit static WEP

9. Which of these attacks are considered Layer 2 denial-of-service attacks? (Choose all that apply.)
   - **A.** Deauthentication spoofing
   - **B.** Jamming
   - **C.** Virtual carrier attacks
   - **D.** PS-Poll floods
   - **E.** Authentication floods

10. What type of security solution can be used to prevent rogue WLAN devices from becoming an unauthorized portal to a wired network infrastructure? (Choose all that apply.)
    - **A.** 802.1X/EAP
    - **B.** Port control
    - **C.** WIPS
    - **D.** CCMP/AES
    - **E.** WIDS

11. What can happen when an intruder compromises the preshared key used during WPA/WPA2-Personal authentication? (Choose all that apply.)

    A. Decryption
    B. Eavesdropping
    C. Spoofing
    D. Encryption cracking
    E. Access to network resources

12. What is another name for a wireless hijacking attack?

    A. Wi-Fi phishing
    B. Man-in-the-middle
    C. Fake AP
    D. Evil twin
    E. AirSpy

13. MAC filters are typically considered to be a weak security implementation because of what type of attack?

    A. Spamming
    B. Spoofing
    C. Phishing
    D. Cracking
    E. Eavesdropping

14. Which components of Carrier Sense Multiple Access with Collision Avoidance (CSMA/CA) can be compromised by a denial-of-service attack? (Choose all that apply.)

    A. NAV timer
    B. Interframe spacing
    C. Clear channel assessment
    D. Random backoff timer

15. What security can be used to stop attackers from seeing the MAC addresses used by your legitimate 802.11 WLAN devices?

    A. MAC spoofing
    B. VPN tunneling
    C. CCMP/AES encryption
    D. Use 802.1X authentication
    E. None of the above

16. Wired leakage occurs under which of the following circumstances?

    A. When weak wireless encryption is used
    B. When weak wireless authentication is used

C. When wired broadcast traffic is passed through an AP

D. When wired unicast traffic is passed through an AP

E. When the protection mode is disabled on an AP

17. Which of these attacks can be mitigated with a mutual authentication solution? (Choose all that apply.)

A. Malicious eavesdropping

B. Deauthentication

C. Man-in-the-middle

D. Wireless hijacking

E. Authentication flood

18. Wireless intrusion prevention systems (WIPSs) are unable to detect which of the following attacks?

A. Association flood

B. Malicious eavesdropping

C. Management frame fuzzing

D. Injection attacks

E. Null probe attacks

19. Name two types of rogue devices that cannot be detected by a Layer 2 wireless intrusion prevention system (WIPS).

A. 900 MHz radio

B. 802.11h-compliant device

C. FHSS radio

D. 802.11b routers

E. 802.11g mixed-mode device

20. What type of solution can be used to perform countermeasures against a rogue access point?

A. WPA-Enterprise

B. 802.1X/EAP

C. WIPS

D. TKIP/RC4

E. WINS

Wireless LAN Security Auditing

IN THIS CHAPTER, YOU WILL LEARN ABOUT THE FOLLOWING:

✓ **WLAN security audit**

- OSI Layer 1 audit
- OSI Layer 2 audit
- Penetration testing
- Wired infrastructure audit
- Social engineering audit
- WIPS audit
- Documenting the audit
- Audit recommendations

✓ **WLAN security auditing tools**

- Linux-based tools

The diligent practice of WLAN security auditing is just as important to a healthy network as good planning and performance tuning but is often overlooked due to budgetary and time constraints. When building an 802.11 WLAN, you focus a lot of attention on the users, the intended use of the network, the devices used, channels, application support, security measures, signal strength, and bandwidth utilization. Although often overlooked in the past, proper planning and deployment of 802.11 WLAN security infrastructure should be considered mandatory. Over time, the network utilization and coverage can change, along with the number of WLAN devices and the applications being used. In addition, the equipment deployed as well as the industry standards and government regulations that apply to the use of wireless communications may evolve.

These changes may directly affect the security posture of the WLAN. The users of a WLAN will often complain if the network becomes "slow" or "unavailable" but will not complain about security holes because they know neither of their existence nor how to look for them. Users only care about their ability to access resources and do their jobs. Security holes, such as weak keys, unencrypted traffic, wired leakage, or rogue devices, do not concern the average user as long as their work is not impeded. Improperly configured security settings may reduce the usability of the network or allow attackers access to private information. While end users serve as a network performance monitor of sorts, there is no such built-in user monitoring for WLAN security mechanisms. Security is sometimes considered a luxury or an undesired expense until there is a breach that costs an organization a lot of time or money. Furthermore, if a WLAN security breach becomes public news, the organization faces embarrassment, potential loss of stock values, and potential legal liabilities.

In this chapter, you will learn about recommended WLAN audit procedures as well as the hardware and software tools needed to carry out a successful audit with the goal of creating a more secure wireless network.

WLAN Security Audit

Security does not make a computer network function, nor does it create a profit for an organization. However, without correctly implemented security, profitability and confidential information are in grave danger. WLAN security audits must be conducted on a regular basis to ensure compliance and aid in the early detection of vulnerabilities. Audits should also be conducted after any change is made to the WLAN infrastructure in order to ensure that the WLAN is not vulnerable to attacks due to the changes that were made. Regularly scheduled internal audits are a recommended practice and should be included as part of the organization's security policy. Larger organizations should also consider hiring a third party for an outside WLAN security audit. When a different set of eyes examines a WLAN during an audit, potential vulnerabilities missed by the internal auditor are often exposed. A security audit should also be used to verify that the WLAN is still meeting security requirements that are often set by industry standards, government regulations, and organizational policies.

In most countries, there are mandated regulations on how to protect and secure data communications within all government agencies. Legislation also often exists for protecting information and communications in certain industries. Various industry standards and U.S. government regulations will be discussed in greater detail in Chapter 15, "Wireless Security Policies."

A series of evaluation procedures will normally comprise a typical WLAN security audit. These auditing best practices include the following:

- Layer 1 audit
- Layer 2 audit
- Penetration testing
- Wired infrastructure audit
- Social engineering audit
- WIPS audit

The physical security and inventory of the deployed WLAN devices should also be audited and documented. If you do not have physical security for the devices, what real security do you have? Wired infrastructure devices are usually locked in server rooms. However, WLAN devices, such as access points and antennas, are often exposed to the naked eye. Physical inspection of devices and cabling is part of a complete WLAN audit. Improperly secured APs are susceptible to theft or manipulation. It is not uncommon for expensive WLAN devices to have been replaced with low-cost clones purchased from Internet auction sites. Some APs have a console or serial port. An attacker may access the exposed ports of an unsecured AP to extract information about WLAN configuration settings or to make changes to the WLAN configuration.

A good audit also includes proper documentation as well as the final recommendations to make the WLAN more secure. Security auditing is a method of *threat assessment* with the eventual goal of *risk mitigation*. Any security threats and risks that are found will be presented to the WLAN network owner together with an assessment of their impact and often with a proposal for mitigation. Recommended technical and nontechnical solutions will be part of a final proposal.

Which Term Is Correct? Packets or Frames? Analyzer or Sniffer?

The terminology used during WLAN security audits can often be confusing. A packet and a frame are both packages of data that traverse through a computer network. A packet exists at Layer 3 of the OSI model, whereas a frame exists at Layer 2 of the model. As mentioned earlier in this book, the Layer 3–7 payload, known as the MAC Service Data Unit (MSDU), is essentially an IP packet encapsulated in the body of an 802.11 data frame. Although packets operate at Layer 3 and frames operate at Layer 2 of the OSI model, the terms are often used interchangeably. For example, the term 802.11 packet generator actually refers to a software tool used to generate and transmit 802.11 frames. WLAN protocol analyzers are often referred to as WLAN packet analyzers. The phrase "wireless packet capture" is more commonly used as opposed to the technically correct phrase "wireless frame capture" or "802.11 frame capture." In reality a WLAN protocol analyzer captures 802.11 frames, and most of the troubleshooting and analysis is that of Layer 2 frame exchanges. The IP packet payload can only be analyzed if an 802.11 data frame can be decrypted. WLAN protocol analyzers are also often referred to as "wireless sniffers." Although the term sniffer is commonplace, it should be noted that Sniffer® is a registered trademark of Network General Corporation much like Band-Aid® is a registered trademark for the Johnson & Johnson Consumer Companies.

OSI Layer 1 Audit

WLAN site surveys have changed dramatically over the years. When most individuals are asked to define a wireless site survey, the usual response is that a site survey is for determining RF coverage and capacity planning. Although that definition is absolutely correct, the site survey encompasses so much more, including looking for potential sources of RF interference. Before conducting the coverage analysis site survey, a *spectrum analysis* site survey should be considered mandatory for locating sources of potential interference.

Unfortunately, many site surveys completely ignore spectrum analysis because of the high cost generally associated with purchasing the necessary spectrum analyzer hardware. Spectrum analyzers are frequency domain measurement devices that can measure the amplitude and frequency space of electromagnetic signals. Spectrum analyzer hardware can cost upward of $40,000 (U.S. dollars), thereby making them cost-prohibitive for many smaller and medium-sized businesses. The good news is that several companies have solutions, both hardware and software based, that are designed specifically for 802.11 site survey spectrum analysis and are drastically less expensive. Figure 13.1 depicts an example of an affordable spectrum analyzer that can be used to monitor the 2.4 GHz and 5 GHz bands. Wi-Spy DBx is a custom USB spectrum analyzer from MetaGeek (`www.metageek.com`). The USB spectrum analyzer can be run from a laptop together with analysis software called Chanalyzer.

FIGURE 13.1 Wi-Spy DBx 2.4 GHz and 5 GHz spectrum analyzer

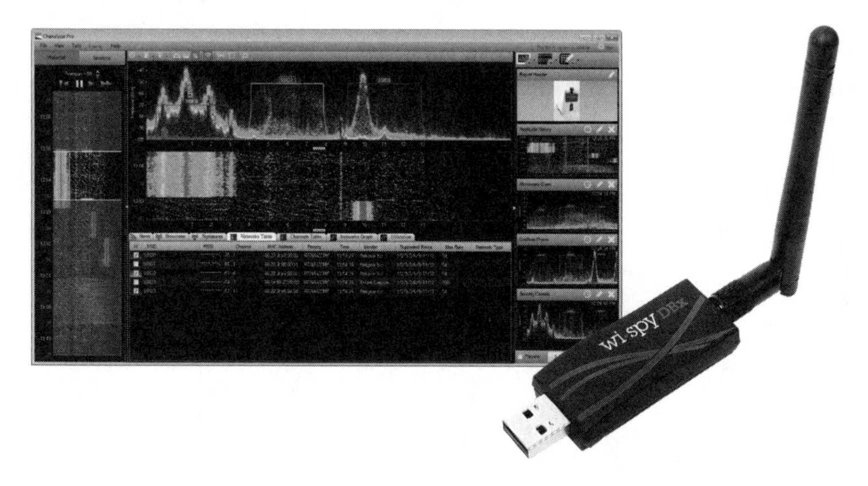

So what does the spectrum analysis during a site survey have to do with a Layer 1 security audit? Effectively the methods and tools used during both procedures are exactly the same. Spectrum analysis during a site survey is usually done for performance reasons, whereas spectrum analysis during a security audit is done to identify potential devices that will cause a Layer 1 *denial of service (DoS)*. The original site survey is executed prior to deployment and installation of the WLAN infrastructure. The spectrum analysis portion of the survey is performed only one time, and it can be for both performance and security evaluation. Layer 1 security audits are enacted on a regularly scheduled basis after the WLAN infrastructure is already operational.

The main purpose of spectrum analysis during a WLAN site survey is to locate the potential sources of interference that may negatively impact the performance of the WLAN. As you learned in Chapter 12, "Wireless Security Risks," most RF interference is unintentional. Unintentional interference will result in data corruption and Layer 2 retransmissions that negatively affect WLAN performance.

The main purpose of spectrum analysis during a security audit is to identify any devices that can cause a DoS at Layer 1. Any continuous transmitter will cause a DoS. RF jamming devices can be used by an attacker to cause an intentional Layer 1 DoS attack. Denial of service at Layer 1 usually occurs as an unintentional result of transmissions from non-802.11 devices. Video cameras, baby monitors, cordless phones, and microwave ovens are all potential sources of unintentional interference. Unintentional interference may cause a continuous DoS; however, the disruption of service is often sporadic. This disruption of service will upset the performance of Wi-Fi networks used for data applications but can completely disrupt VoWiFi communications within a WLAN. The majority of unintentional interfering devices transmit in the 2.4 GHz ISM frequency band. The 5 GHz UNII bands are less susceptible to unintentional interference.

The 2.4 to 2.4835 GHz ISM band is an extremely crowded frequency space. The following are potential sources of unintentional interference in the 2.4 GHz ISM band:

- Microwave ovens
- 2.4 GHz cordless phones: both direct sequence spread spectrum (DSSS) and frequency hopping spread spectrum (FHSS) phones
- Halogen gas lights
- 2.4 GHz video cameras
- Elevator motors
- Cauterizing devices
- Plasma cutters
- Bluetooth radios
- 2-way radios

During an audit, any source of interference must be identified, documented, and classified as either an intentional or unintentional source of interference that may cause a DoS. As you learned in Chapter 12, 802.11 radios use a *clear channel assessment (CCA)* to determine if the RF medium is busy or clear. Any "continuous" RF transmission that is constantly heard during the clear channel evaluation will cause 802.11 transmissions to cease completely until the signal is no longer present. Interfering devices may prevent an 802.11 radio from transmitting, thereby causing a DoS. There are multiple types of RF interference:

Narrow-band Interference A narrow-band RF signal occupies a smaller and finite frequency space and will not cause a DoS for an entire band such as the 2.4 GHz ISM band. A narrow-band signal is usually very high amplitude and will absolutely disrupt communications in the frequency space in which it is being transmitted. Narrow-band signals can disrupt one or several 802.11 channels. Narrow-band RF interference can also result in corrupted frames and Layer 2 retransmissions. The only way to eliminate *narrow-band interference* is to locate the source of the interfering device with a spectrum analyzer. Figure 13.2 shows a spectrum analyzer capture of a narrow-band signal close to channel 11 in the 2.4 GHz ISM band.

Wide-band Interference A source of interference is typically considered wide band if the transmitting signal has the capability to disrupt the communications of an entire frequency band. Wide-band jammers exist that can create a complete DoS for the 2.4 GHz ISM band. The only way to eliminate *wide-band interference* is to locate the source of the interfering device with a spectrum analyzer and remove the interfering device. Figure 13.3 shows a spectrum analyzer capture of a wide-band signal in the 2.4 GHz ISM band with average amplitude of –60 dBm.

FIGURE 13.2 Narrow-band RF interference

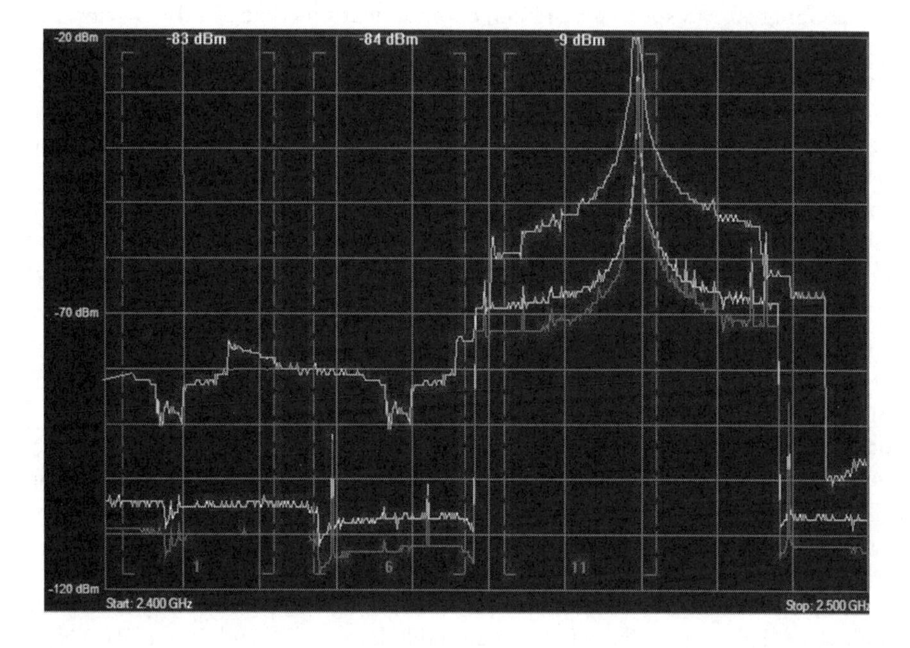

FIGURE 13.3 Wide-band RF interference

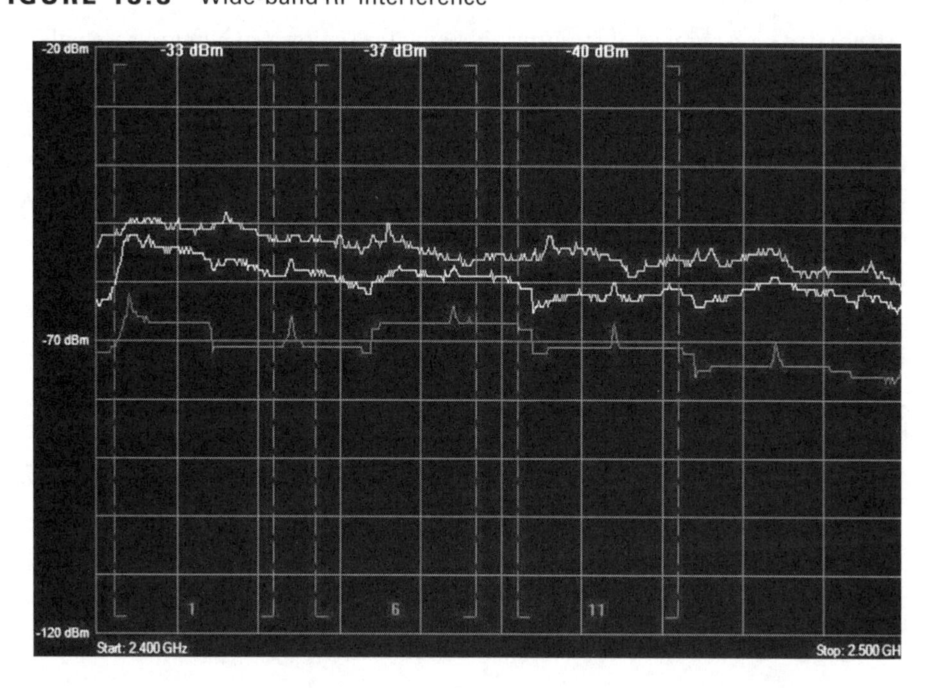

All-Band Interference The term *all-band interference* is typically associated with frequency hopping spread spectrum (FHSS) communications that usually disrupt the 802.11 communications at 2.4 GHz. FHSS constantly hops across an entire band, intermittingly transmitting on very small subcarriers of frequency space. A legacy 802.11 FHSS radio, for example, transmits on hops that are 1 MHz wide. 802.11b radios transmit in a stationary 22 MHz of frequency space and 802.11g/n radios transmit on fixed channels of 20 MHz of spectrum. While hopping and dwelling, an FHSS device will transmit in sections of the frequency space occupied by an 802.11b/g/n channel. Although an FHSS device will not typically cause a DoS, the frame transmissions from the 802.11b/g/n radios can be corrupted from the all-band transmissions of the FHSS interfering radio.

Bluetooth (BT) is a short-distance RF technology defined by the 802.15 standard. Bluetooth uses FHSS and hops across the 2.4 GHz ISM band at 1,600 hops per second. Older Bluetooth devices were known to cause severe all-band interference. Newer Bluetooth devices use adaptive mechanisms to avoid interfering with 802.11 WLANs. Bluetooth adaptive frequency hopping is most effective at avoiding interference, with a single AP transmitting on one 2.4 GHz channel. If multiple 2.4 GHz APs are transmitting on channels 1, 6, and 11 in the same physical area, it is impossible for the Bluetooth transmitters to avoid interfering with the WLAN. Digital Enhanced Cordless Telecommunications (DECT) cordless telephones also use frequency hopping transmissions. Some DECT phones transmit in the 2.4 GHz band. A now-defunct WLAN technology known as HomeRF also used FHSS; therefore, HomeRF devices can potentially cause all-band interference.

The existence of a high number of frequency-hopping transmitters in a finite space will result in some 802.11 data corruption and Layer 2 retransmissions. All-band interference may not cause a continuous DoS; however, the disruption of service due to Layer 2 retransmissions can be significant from a performance perspective. The only way to eliminate all-band interference is to locate the source of the interfering device with a spectrum analyzer and remove the interfering device. Figure 13.4 shows a spectrum analyzer capture of a frequency hopping transmission in the 2.4 GHz ISM band.

A Layer 1 security audit is normally accomplished using some sort of handheld spectrum analyzer or laptop spectrum analyzer. Some WLAN vendors offer 24/7 spectrum monitoring capabilities with a *distributed spectrum analysis system (DSAS)*. A more detailed discussion of DSAS can be found in Chapter 14, "Wireless Security Monitoring."

FIGURE 13.4 All-band RF interference

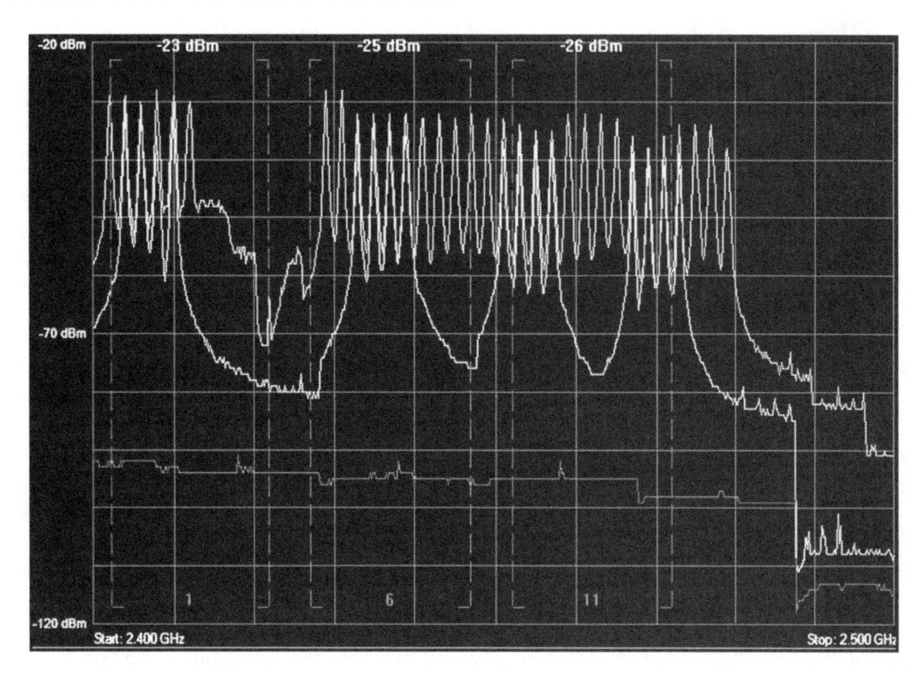

OSI Layer 2 Audit

The gathering and analysis of OSI Layer 2 information is a vital part of the wireless LAN security audit process and can reveal a great deal of information about both the WLAN functionality and the security posture of the network being examined. One of the main purposes of a Layer 2 audit is initially to detect any rogue devices or unauthorized 802.11 devices. Identifying rogue devices during an initial security audit is critical, especially if a distributed WIPS monitoring solution has not been deployed. A proper Layer 2 WLAN security audit will also be used initially to identify all authorized 802.11 WLAN devices, including access points and authorized clients. As shown in Figure 13.5, a Layer 2 audit can also be used to validate WLAN security compliance. In other words, if the mandated security required the use of Protected Extensible Authentication Protocol (PEAP) authentication and CCMP/AES encryption, all authorized devices can be evaluated to verify the proper security configuration. Any authorized devices that have not been properly configured will be flagged.

FIGURE 13.5 Device configuration compliance

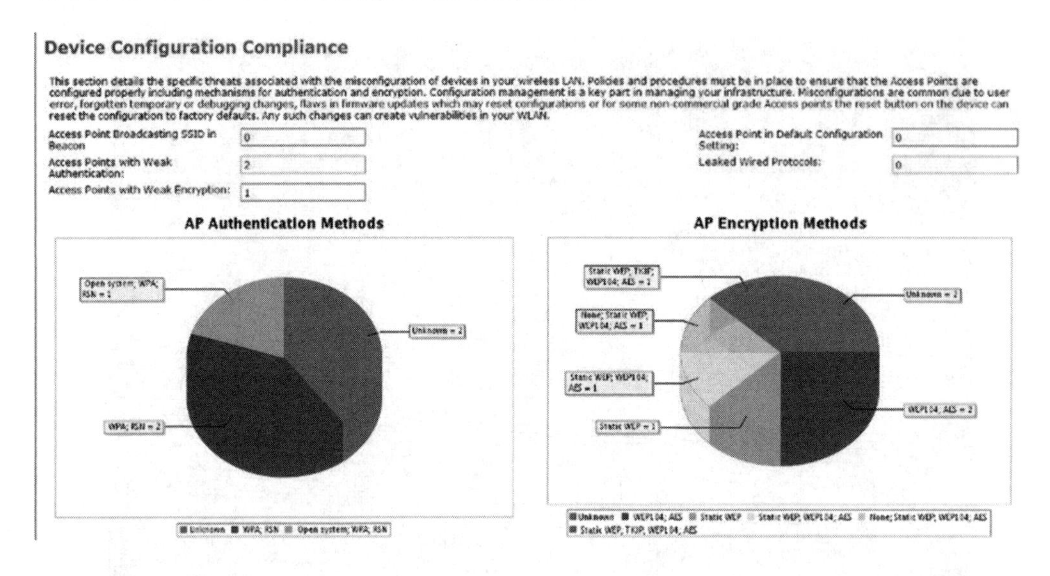

The payload of a wireless frame, meaning the Layer 3 and higher information, should normally be hidden due to encryption. The payload may be encrypted but the header and trailer information is always readable without any special decoding required. The header and trailer information must be in cleartext for the exchanges to work properly. Any 802.11 frame exchange can be captured and can reveal Layer 2 information about the devices directly involved in a frame exchange. A Layer 2 audit is necessary to ensure that no pertinent information is being exposed and that it is properly protected.

The following is a list (in no special order) of some of the more important things auditors should strive to find, identify, and classify at Layer 2 during audits:

- MAC addresses
- SSID
- BSSID
- Device types being used
- Authentication methods
- Encryption methods
- Traffic types
- Neighboring devices
- Channels in use
- Default configurations

- Active Layer 2 attacks
- Weak keys in use
- Ad hoc clients

As you can see in Figure 13.6, a simple WLAN protocol analyzer installed on a laptop is usually sufficient to perform a Layer 2 audit. WLAN protocol analyzers are typically used for Layer 2 troubleshooting and WLAN performance analysis. However, as Figure 13.6 shows, some software WLAN protocol analyzers provide a greater emphasis on security auditing and can be used effectively as a mobile wireless intrusion detection system (WIDS) solution. An example of such a professional tool, seen in Figure 13.6, is Fluke Networks' AirMagnet WiFi Analyzer PRO.

FIGURE 13.6 Layer 2 protocol analyzer

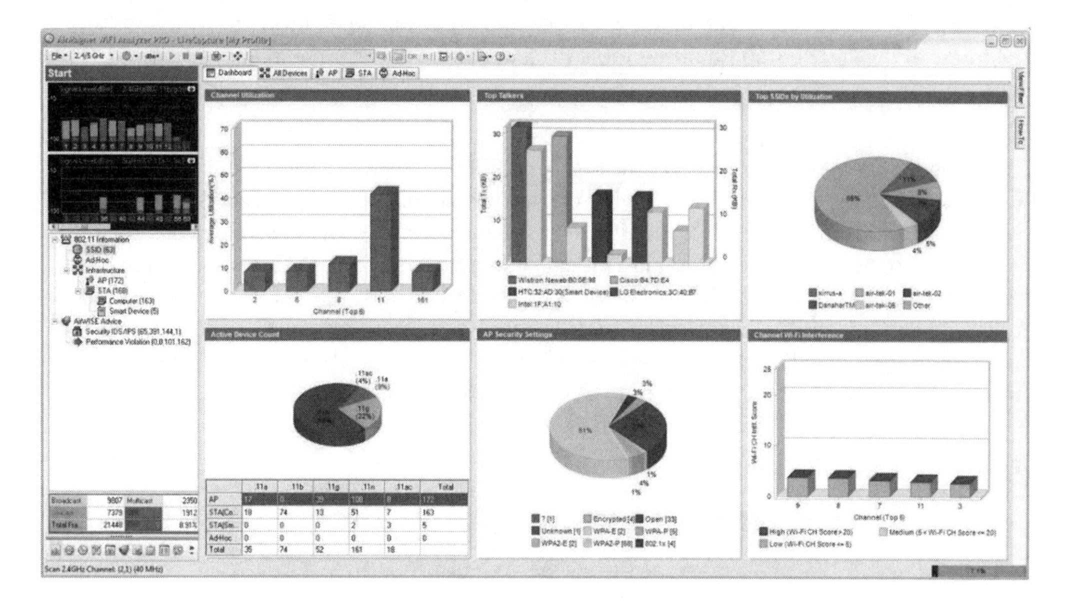

Penetration Testing

You have already learned that WLAN auditing is a process used to ensure that 802.11 communications and devices are secured and functioning as required by organizational policies, industry standards, and/or governmental regulations. Wireless LAN auditing may also include wireless *penetration testing* if desired as part of the scope of work agreement between the organization and the auditor. A WLAN penetration test is used to evaluate the security of the WLAN by simulating an attack from a malicious intruder. Many of the tools used during WLAN penetration testing are the same tools that hackers may use for malicious purposes. WLAN penetration testing tools are used to find security vulnerabilities due to hardware/software flaws, improper system configuration, and known technical weaknesses.

> **What Is a Rainbow Table?**
>
> A *rainbow table* is used to find the original plaintext for a hashed password. A rainbow table is a lookup table offering a time-memory trade-off used in recovering the plaintext password from a password hash generated by a hash function. With rainbow tables, time is initially spent precomputing the hashes and storing the data into a file. This file is later used to speed up the cracking process. Using brute-force methods to crack password hashes takes a long time, but it doesn't require much memory and requires no disk space. Rainbow tables require time to generate, take up disk space, and use more memory in the cracking process but are much faster when used to crack the hash. More information can be found at the Distributed Rainbow Table Project at `www.freerainbowtables.com`.

A good example of a penetration test is using known authentication cracking software tools to demonstrate the weakness of the chosen passwords or passphrases. Weaker authentication methods are often deployed due to cost concerns. A good penetration test would be an attempt to circumvent weaker authentication methods such as Lightweight Extensible Authentication Protocol (LEAP) or WPA/WPA-2-Personal. You learned in earlier chapters that LEAP is susceptible to an *offline dictionary attack*. As shown in Figure 13.7, a software auditing tool called *Asleap* can be used with a hashed rainbow table of adequate size to reveal the hashed LEAP password in a matter of seconds.

FIGURE 13.7 Asleap

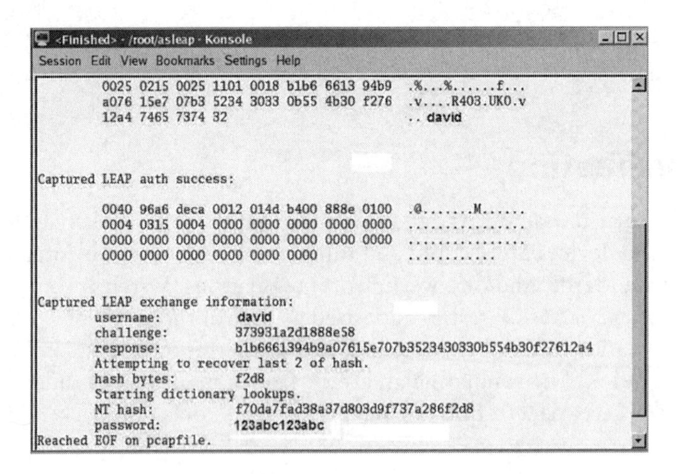

You also have already learned that WPA/WPA2-Personal, using preshared keys, is a weak authentication method that is vulnerable to an offline *brute-force dictionary attack*. There is no difference between cracking WPA or WPA2 preshared keys. The authentication methodology used in both formats is basically the same, so the technique used to obtain the passphrase is the same. As Figure 13.8 shows, a software auditing tool called coWPAtty can be used to derive weak passphrases.

FIGURE 13.8 coWPAtty

```
thallium cowpatty $ ./cowpatty -r wpa2psk-linksys.dump -d linksys.hash -s links
ys
cowpatty 4.0 - WPA-PSK dictionary attack. <jwright@hasborg.com>

Collected all necessary data to mount crack against WPA2/PSK passphrase.
Starting dictionary attack.  Please be patient.
key no. 10000: arrojadite
key no. 20000: calligraphical
key no. 30000: contestation

The PSK is "dictionary".

38333 passphrases tested in 0.97 seconds:  39655.26 passphrases/second
thallium cowpatty $
```

The Asleap and coWPAtty authentication cracking auditor tools were created by wireless security expert Joshua Wright. The tools can be downloaded from Joshua's website at `www.willhackforsushi.com`.

A great tool for auditing that you may have used in the past is BackTrack Linux. This older penetration testing tool has been replaced by a greatly improved successor, Kali Linux. As shown in Figure 13.9, Kali Linux is an open source project maintained and funded by Offensive Security, a provider of information security training and penetration testing services. Kali Linux comes preinstalled with over 600 penetration-testing programs. Kali Linux is a Debian-based distribution with a collection of security and forensics tools. These tools include Nmap (a port scanner), Wireshark (a packet analyzer), John the Ripper (a password cracker), Aircrack-ng (a software suite for WLAN penetration testing), and web application security scanners Burp Suite and OWASP Zed Attack Proxy (ZAP). Kali Linux can run natively when installed on a computer's hard disk, or you can boot it from a live CD or live USB, or run it from within a virtual machine.

This Study Guide is not a book on how to use the numerous WLAN penetration utilities that are available. A great instructional book that we recommend is *Kali Linux: Wireless Penetration Testing Beginner's Guide*, by Vivek Ramachandran and Cameron Buchanan (Packt Publishing, 2015).

FIGURE 13.9 Kali Linux

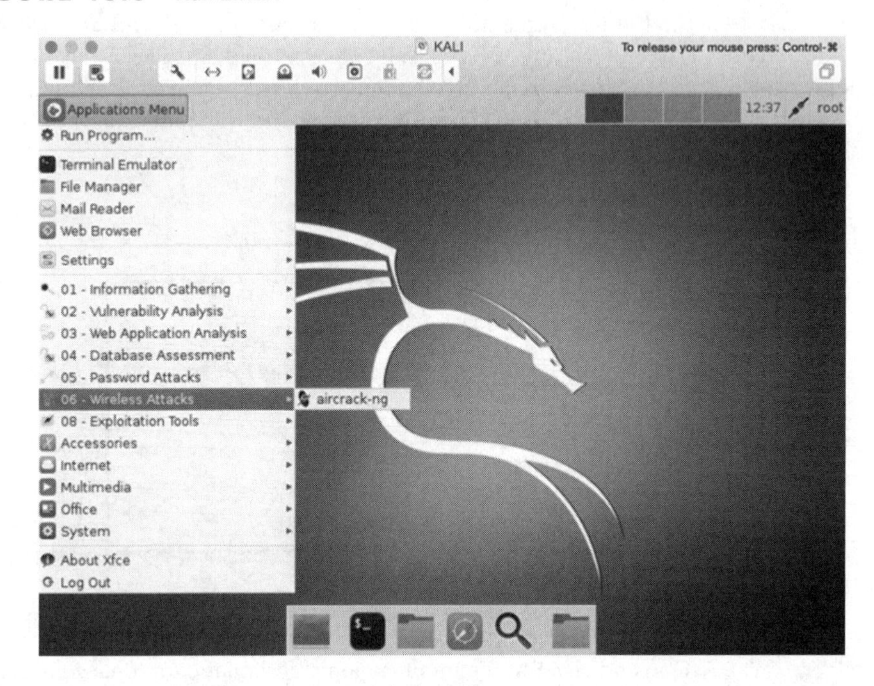

WLAN penetration testing tools are freely available on the Internet; however, there is often a learning curve as to how to properly use a wide variety of software programs. Several commercial vendors offer automated WLAN assessment solutions such as Immunity's SILICA solution, shown in Figure 13.10.

FIGURE 13.10 Immunity SILICA WLAN security assessment software

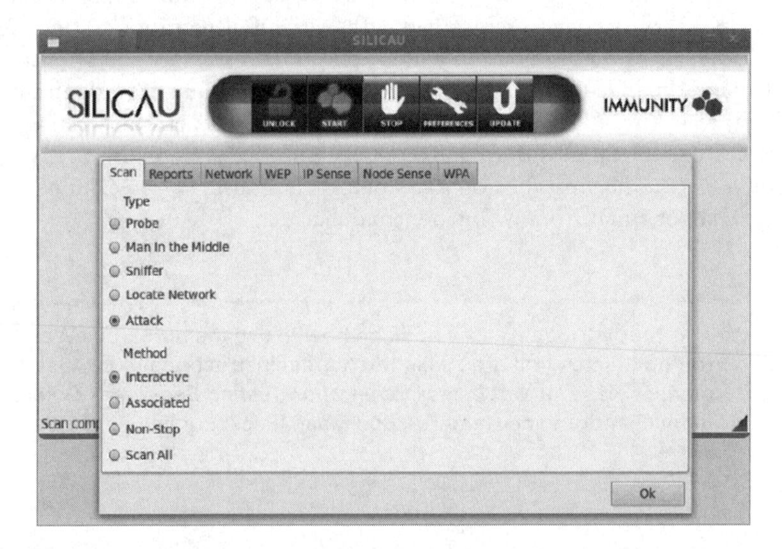

Another WLAN auditing tool that has become popular is Wi-Fi Pineapple from Hak5. WiFi Pineapple consists of custom, purpose-built hardware and software, enabling its users to quickly and easily deploy advanced attacks using an intuitive web interface. As shown in Figure 13.11, WiFi Pineapple features dual unlocked high-gain radios and SMA (SubMiniature version A) connectors to allow the user to install professional antennas. One appealing feature if you are conducting audits with little time to do so is the tool's auto-attack mode. It allows you to have customized boot-time payloads without the need to log in. All you need to do is set the switches to your attack mode of choice and power on.

FIGURE 13.11 WiFi Pineapple Mark V

Wired Infrastructure Audit

The normal purpose of a WLAN is to act as a wireless portal to network resources that reside on the wired network infrastructure. As you have already learned, penetration testing can be used to see if strong authentication solutions are deployed properly to protect the portal. Once authorized to access network resources, WLAN users can be further restricted as to what resources may be accessed and where they can go. *Role-based access control (RBAC)* security can be used to restrict certain groups of users to certain network resources. RBAC is usually accomplished with firewall policies and access control lists. Penetration testing of the firewall used to restrict WLAN user access to certain network resources should be a mandatory procedure during the security audit. All WLAN infrastructure interfaces should also be audited. Wireless infrastructure hardware, such as autonomous access points and WLAN controllers, can be managed by administrators via a web interface, a command-line interface, a serial port, a console connection, and/or Simple Network Management Protocol (SNMP). It is imperative that these interfaces be protected. Interfaces that are not used should be disabled. Strong passwords should be used, and encrypted login capabilities such as Hypertext Transfer Protocol Secure (HTTPS) and Secure Shell (SSH) should be utilized if available.

Social Engineering Audit

The weakest link in security for any type of computer network is usually the network end users. In Chapter 12, you learned how *social engineering* is the act of manipulating people into divulging confidential information for the purpose of obtaining information used to access a computer network. Therefore, social engineering techniques are often used by the security auditor to circumvent the WLAN and gain access to network resources. Social engineering performed by an auditor is a form of penetration testing and is usually the most successful method. An auditor will most likely find lapses in security due to improper employee training or employee policy enforcement. Although penetration testing using software tools is often successful, the auditor will probably have more luck with social engineering techniques used for penetration testing purposes.

Real World Scenario

Social Engineering Audit

A government agency once hired an outside company to perform a computer security audit. One of the auditors sent an email message introducing himself as Albert Kada, the new network administrator. Albert informed all of the government agency employees that he was updating the password database and needed everyone to email him their passwords. Over 40 percent of the employees sent their passwords to Al-Qaeda (Albert Kada). After the security audit, it was recommended that all employees be retrained in regard to password policy.

WIPS Audit

If a company has been hired to perform a WLAN security audit, chances are that the customer is not using a distributed *wireless intrusion detection system (WIDS)* or distributed *wireless intrusion prevention system (WIPS)*. Many of the WIDS/WIPS vendors will often perform an initial WLAN security audit for free. The WIPS/WIPS vendor will show the customer potential security holes, make recommendations, and then offer to sell the customer a distributed WIDS/WIPS solution that is capable of full-time WLAN security monitoring.

If a distributed WIDS/WIPS monitoring solution has already been deployed, the monitoring solution should also be audited. Hacker attacks, such as DoS, MAC spoofing, rogue devices, and so on, should be simulated to see if the proper WIDS/WIPS alarms are triggered for each attack. For example, a third-party access point can be connected to a wired port to test the rogue detection capabilities of a distributed WIPS monitoring solution. Another example is when Layer 2 DoS attacks are simulated to verify detection by the WIPS server and sensors. If certain alarms are not triggered, alarm thresholds may need to be adjusted on the WIDS/WIPS server.

WIDS and WIPS monitoring solutions will be discussed in greater detail in Chapter 14.

Documenting the Audit

The job is not finished until the paperwork is done. This is a common quotation but true nonetheless. Documenting the audit is of great importance to both the auditor and the customer. The auditor needs to be able to produce evidence of his or her findings to the customer and make recommendations for improving security and WLAN health based on these findings. The customer will need documentation to prove compliance with regulations. Documentation can also be used by the customer to have their own IT staff implement the security changes recommended by the auditor.

Documentation of an audit can be presented in several formats. An all-written presentation in PDF format is often used; some auditors use Microsoft PowerPoint slide shows. The actual deliverable can vary based on the *statement of work (SOW)* agreement and customer's requirements. The customer may also provide some documentation such as a network topology map and corporate policy documents prior to the audit. Here is some of the documentation required for the audit:

Statement of Work This document outlines the audit requirements, deliverables, and timeline that the auditor will execute for a customer. Pricing and detailed terms and conditions are specified in the SOW.

Liability Waiver It is very important that any auditor obtain signed permission in the form of a liability waiver or hold harmless agreement to perform the security audit. An audit should never take place without authorized permission.

Network Topology Map Understanding the layout of the customer's wired network infrastructure and WLAN infrastructure will speed up the audit process and allow for better planning of penetration testing. A computer network topology map will provide necessary information, such as the location of the wiring closets, access points, WLAN controllers, and firewalls.

Mutual Nondisclosure Agreement Some organizations may not wish to reveal their network topology for security reasons. It may be necessary to obtain security clearance and/or sign nondisclosure agreements to gain access to these documents.

Corporate Security Policy Obtaining a copy of the customer's written security policy document will be valuable for determining risk assessments. Very often the corporate security policy will be nonexistent or outdated.

Audits and or penetration tests should not go beyond the written corporate policy or the SOW agreement. Documentation in IT can be thought of as a type of unicorn, something you have heard about but never see. This does not mean it should not be requested prior to beginning the audit. Often an auditor is called upon to assist the customer in creating documentation such as policy documents.

By finding out what is visible in WLAN communications, the auditor is able to document the information being made available to attackers and to begin to secure the transmissions. If there is a written wireless security policy, auditors are able to determine if

WLAN communications are being conducted within the policy guidelines by comparing the information contained in captured traffic with the written policy.

Unfortunately, the existence of a written security policy that governs wireless communications is a rare thing. Most organizations rely on a generic security statement that focuses on inappropriate Internet and phone usage rather than on WLAN security. In the absence of written WLAN usage policy, auditors can compare captured traffic with industry compliance policies or governmental regulations such as PCI or SOX. If there is no organizational awareness of WLAN security, the comparison of current traffic and industry or governmental standards is a very useful tool in helping networks become and remain secure.

Audit Recommendations

The main purpose of the WLAN security audit is to expose potential security holes so that proper solutions and procedures can be implemented. After the audit is completed, recommendations will be made based on the audit findings on how to protect the WLAN and the network infrastructure more securely. Recommendations could include the following:

Stronger Dynamic Encryption If no encryption or weak encryption such as WEP is being used, the recommendations will be to upgrade to an available dynamic encryption method such as CCMP/AES.

Stringent Authentication Penetration testing may reveal weak authentication such as LEAP or WPA/WPA2-Personal. Upgrading to an 802.1X/EAP authentication solution using tunneled authentication will almost always be recommended.

Role-Based Access Control (RBAC) Recommendations can be made on how to segment groups of users and devices to certain network resources using firewall policies, access control lists, and VLANs.

Monitoring Recommendations will be made about the types of distributed WIDS/WIPS monitoring solutions that may be needed.

Corporate Policy An extra addendum to the security recommendations might be corporate WLAN policy recommendations. The auditor might assist the customer in drafting a wireless network security policy if they do not already have one.

Training One of the most overlooked areas of WLAN security is proper training. It is highly recommended that security training sessions be scheduled with the customer's network administration personnel. Additionally, condensed training WLAN security sessions should be scheduled with all end users.

Physical Security The installation of enclosure units to protect against theft and unauthorized physical access to access points may be a recommendation. Enclosure units are also often used for aesthetic purposes.

WLAN Security Auditing Tools

To conduct a successful WLAN security audit, you must use tools designed for that purpose. Although attackers may use some of the same tools to exploit poorly secured networks, WLAN security auditors use them to expose areas of the WLAN that are in need of increased protection or reconfiguration. The audit is performed in order to prevent attackers from gaining access to network resources and to assist the organization in reaching the required security levels demanded by industry and policy compliance. Auditors have an advantage over attackers in that they are typically allowed physical access to the premises that they are auditing. This access is much like the access granted during a WLAN site survey, and it may require an escort or special identification to traverse the area. A degree of overlap exists between a good WLAN site survey kit and a good WLAN auditing kit.

The successful WLAN auditor requires not only a thorough knowledge of 802.11 technology but also the proper software and hardware to perform the WLAN penetration testing. The hardware and software used during an audit vary based on the type of WLAN vulnerability to be identified. For example, if an auditor is looking for the source of noise that is causing a potential denial of service, the proper tool is a spectrum analyzer. If the auditor is conducting 802.11 traffic analysis, the proper tool would be a WLAN protocol analyzer.

The tools selected must match the job at hand. Typically a well-equipped auditor will have the following hardware devices and software to assist them in their work, either as part of their own kit or provided by the facility being audited:

- At least one laptop
- WLAN protocol analyzer and audit software
- Spectrum analyzer
- WLAN penetration testing software tools
- 802.11 packet generator software
- 2.4 GHz and 5 GHz signal generator
- Facility blueprints or floor plan
- WLAN cards for each frequency and protocol
- Omni-directional antennas
- Yagi or patch antennas
- Matching pigtail connectors
- Global positioning sensor (GPS)
- Access point
- Camera and/or video recorder
- Phone or walkie-talkie

- Ladders and or lifts
- Battery packs and power cables
- Wheeled cart

This is by no means an exhaustive listing of all the tools used to conduct a WLAN audit. The type and number of tools needed to conduct a successful audit vary based on the type of WLAN being audited and the security requirements of that WLAN. For example, security audits at a Wi-Fi hot spot may only require a visual inspection of the AP and making sure the captive portal is working, whereas large-scale enterprise WLANs may require several days of capturing 802.11 frames for evaluation as well as penetration testing.

The Layer 1 and Layer 2 auditing tools mentioned earlier in this chapter should all be considered mandatory. WLAN protocol analyzers are typically used for Layer 2 auditing, and spectrum analyzers are used for Layer 1 auditing. Many freeware Linux-based and Windows-based WLAN penetration testing software tools are also widely available. Penetration testing normally should be considered a mandatory part of the audit. Some of the other tools listed are obviously required, such as laptops and WLAN cards. Others require more explanation about their use, such as APs and battery packs. For example, an auditor may require several hours of analysis before rendering a security report; thus it's important that the laptop and other devices have adequate power the entire time they're gathering information. External battery packs to power the devices may be needed in areas where a wall outlet cannot be found. A third-party access point may be used to test the rogue detection capabilities of a distributed WIPS monitoring solution.

Physical inspection or building access may require identification or an escort during the audit. If conducting the audit from off the premises, the use of an antenna may be required to hear the wireless transmissions from inside the building(s). As discussed in Chapter 12, a high-gain unidirectional antenna can be used to hear RF signals from a great distance. Antennas are bidirectional passive amplifiers. In the case of an outside audit, proper antenna use often gives the auditor the ability to listen from a remote location. An outside auditor can emulate an attacker who might launch an attack from a location off company grounds.

Physical inspection of devices and cabling is part of a complete WLAN audit. To inspect such devices physically, ladders or lifts may be required. Carrying all of the equipment required for an audit about may become cumbersome and result in the auditor not covering the entire area. The use of a wheeled cart can reduce the wear and tear on the auditor (as will a comfortable pair of shoes), just like in a WLAN site survey. Pen and paper along with a camera and video recorder will give the auditor the ability to take notes and document what they find as part of the deliverable presented at the end of the audit. In a large outdoor area, the auditor may find the use of a GPS is required to record the location of outdoor 802.11 devices, such as bridges and mesh access points.

A larger deployment may require a team of auditors who need to communicate with one another during the audit. Having phones and or walkie-talkies will facilitate this communication. It is easy to see how the hardware and software used as part of a WLAN security audit resembles a kit used as part of a pre- or post-deployment WLAN site survey. Unlike a site survey, the audit will monitor channels, frequencies, and areas of coverage not required by the WLAN.

Many of the tools traditionally used for penetration testing are Linux based. Today, however, many of these tools are supported on a Windows platform. Which tool is used may depend on whether users being audited are Linux, Windows, or Mac OS users. Fewer tools are available for WLAN analysis for use on Mac OS than for Linux and Windows platforms. Table 13.1 contains a list of some of the more common WLAN security-auditing tools.

TABLE 13.1 Common tools and uses

Type of use	Possible audit/attack	Tools
Wireless discovery	Eavesdropping, discovery of rogue APs, ad hoc STAs, and open or misconfigured APs	inSSIDer, NetSurveyor, NetStumbler, Kismet, Wellenreiter, WiFiFoFum, WiFi Explorer, WiFi Hopper, WiFi Scanner, Win Sniffer, Wireshark, and commercial WLAN protocol analyzers
Encryption/authentication	WEP, WPA, LEAP cracking, dictionary attacks	Asleap, Aircrack-ng, CloudCracker, coWPAtty, AirSnort, Fern WiFi Cracker, WEPCrack, and WZCook
Masquerade	MAC spoofing, man-in-the-middle attacks, evil twin attacks, Wi-Fi phishing attacks	Airsnarf, Ettercap Karma, Ghost Fisher, Hotspotter, HostAP, and SMAC
Insertion	Multicast/broadcast injection, routing cache poisoning, man-in-the-middle attacks	Airpwn, CDPsniffer, chopchop, IrPass, VIPPR, and WiFitap
Denial-of-service	Layer 1 and Layer 2 DoS	AirJack, Void11, Bugtraq, IKECrack, FakeAP, and RF signal generators

Linux-Based Tools

The majority of effective wireless auditing tools run on Linux platforms, many of which can be accessed from a bootable CD such as Kali Linux. In this section, we will examine some of these tools and explain their use in conducting security audits.

There are several tools from which to choose to perform the same or similar tasks. The tools you choose will vary based on personal preference and the task at hand. The basics are the same no matter which tool you select. The devices and vulnerabilities must be discovered and documented. To find these vulnerabilities, you can use tools such as Kismet, AirSnort, and Aircrack-ng.

Where Can You Find Linux WLAN Penetration Software Tools?

BackTrack was probably the most popular Linux-based distribution of tools focused on penetration testing. It has been replaced by a new tool called Kali Linux which has 600+ tools included and is a Debian-based distribution with a collection of security and forensics tools. You can download Kali Linux at `www.kali.org`.

Commercially packaged WLAN penetration tools are also available. Information about WiFi Pineapple can be found at `www.wifipineapple.com`. Information about Immunity's WLAN security assessment solution, SILICA, can be found at `www.immunityinc.com`.

Blueprinting or enumerating network devices should be part of every audit. Looking for anything in use, not just the device traffic you would expect, will reveal rogue devices as well as incorrect configuration of authorized devices. Wardriving software tools are still used for simple WLAN discovery whereas other tools are used for packet capture. Since these tools are largely freeware and run on the Linux platform, there are lots of user groups and online tutorials and videos detailing their usage. *THC-wardrive* is a Linux-based wardriving tool that uses both an 802.11 radio and a GPS device to link the discovered 802.11 devices with latitude and longitude coordinates. As shown in Figure 13.12, *Kismet* is a Linux-based 802.11 Layer 2 wireless protocol analyzer and intrusion detection system. Kismet will work with any wireless card that supports raw monitoring (rfmon) mode, and it can sniff 802.11b, 802.11a, and 802.11g traffic.

Once the target has been identified, the auditor must determine what additional information might be extracted using penetration testing. For example, if the auditor has been tasked with determining the WEP key or the WPA preshared key (PSK), they may use a tool such as AirSnort or Aircrack-ng. These tools are used to derive keys. *AirSnort* passively gathers packets until enough containing the *Initialization Vector* (IV) are captured, and then it cracks the WEP key offline. It takes roughly 300,000 to 500,000 IVs in the packet capture to crack the WEP key based on the complexity of the WEP key used. As shown in Figure 13.13, Aircrack-ng is a set of tools used together for cracking WEP and WPA-PSK. The Aircrack-ng tools include Airodump (used for packet capturing), Aireplay (used to inject traffic), Aircrack (used to crack WEP and WPA-PSK), and AirPcap (used to decode WEP and WPA-PSK captures). Aircrack-ng enables WEP and WPA-PSK traffic to be cracked in a short time frame as compared to older methods.

FIGURE 13.12 Kismet

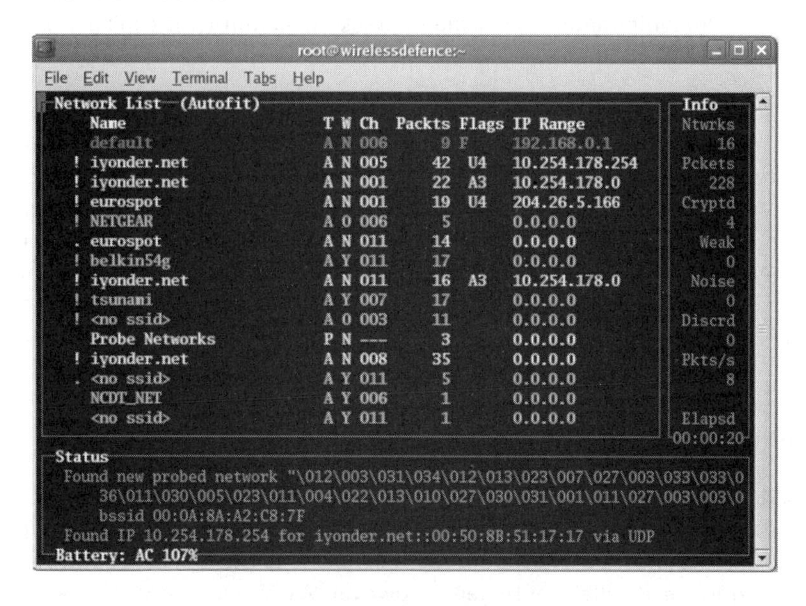

FIGURE 13.13 Aircrack-ng

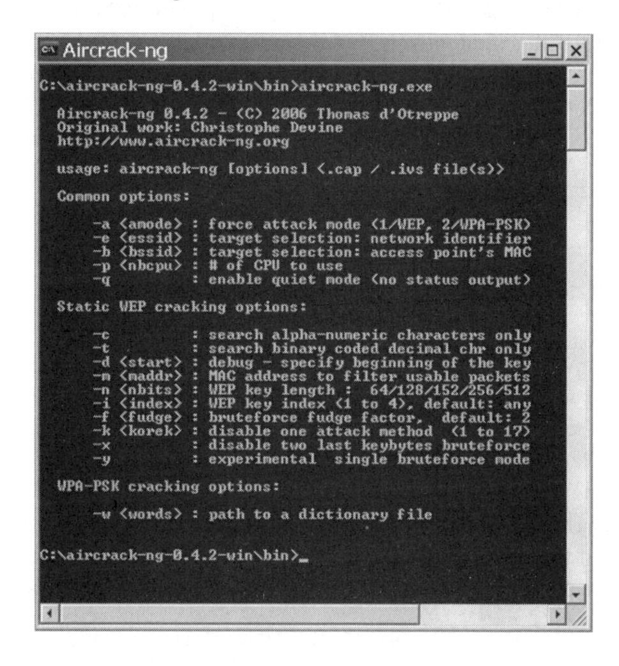

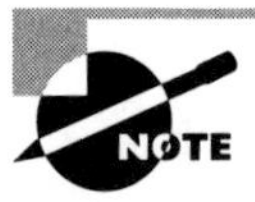

More information about the Aircrack-ng auditor tools can be found at `www.aircrack-ng.org`.

No matter which tools are used, the auditor's goals remain constant—find the devices, get on the network, and/or decode the data. Attackers use the same tools but for malicious reasons. Although there are many tools from which to choose, auditors and attackers alike find some tools they prefer more than others, with speed and ease of use weighing heavily in the decision. Most of the modern Linux command-line tools have been ported over to GUI-based tools that require less interaction and allow for quicker rendering of the desired information.

What about Windows and Mac WLAN Security Auditing Tools?

The bulk of WLAN security penetration tools remain Linux based although some have been ported over to Windows. If you require a full suite of WLAN penetration testing tools to run on a Windows PC or MacBook, your best bet is to run Kali Linux as a virtual machine using VMware, Parallels, or Hyper-V. However, WLAN discovery applications that run in either Windows or Mac OS can be used for some low-level security auditing. Several examples of WLAN discovery tools are discussed in Chapter 12. Additionally, packet-capturing tools, such as the freeware tool Wireshark or commercial 802.11 protocol analyzers such as WildPackets' OmniPeek, do a great job of gathering 802.11 frames and allowing the user to filter their view based on a vast number of items—from the MAC address to the frame subtype. Commercial WLAN protocol analyzers often have security monitoring capabilities and alarms. These tools let you see what each 802.11 device detected is doing, which is valuable information to auditors and attackers alike.

Summary

Diligently conducting WLAN security audits will reveal areas where improvements can be made in protecting the wireless network. The WLAN audit process involves inspection of Layer 1, Layer 2, the wired network, and the WIPS monitoring solution. The information gathered during these audits can be leveraged to make the proper recommendations for protecting both the WLAN as well as the wired network resources. Numerous tools are available to assist the auditor in their duties; some are commercial tools and others are freeware. Many of the tools used by hackers can also be used during a WLAN security audit for penetration testing. The best tool found in any auditor's toolbox is social engineering skills.

Exam Essentials

Explain the various components of a WLAN security audit. These include a Layer 1 and 2 audit, penetration testing, wired infrastructure audit, social engineering audit, and possibly a WIPS audit.

Describe social engineering techniques. Explain why social engineering skills are usually the most successful WLAN auditing tool.

Understand the various methods of WLAN security auditing. Explain the importance of auditing both Layers 1 and 2 of the OSI model. Define aspects of penetration testing, wired infrastructure auditing, and WIPS auditing.

Explain WLAN security auditing and penetration testing tools. Discuss how Layer 1 and Layer 2 tools are used during a WLAN security audit. Describe some of the methods and tools used during WLAN penetration testing.

Review Questions

1. Which of these devices are potential sources of all-band interference? (Choose all that apply.)
   - A. Bluetooth
   - B. Microwave oven
   - C. 2.4 GHz DSSS cordless phone
   - D. 802.11 FHSS access point
   - E. HomeRF access point
   - F. 2.4 GHz DECT phone
2. Which of these WLAN auditing tools has the most success in compromising network resources when penetration testing is performed?
   - A. WLAN protocol analyzer
   - B. Asleap
   - C. Aircrack-ng
   - D. coWPAtty
   - E. Social engineering
   - F. Kismet
3. A WLAN security auditor recently walked into the ACME Company corporate headquarters and presented documentation to ACME management based on the WLAN penetration testing that was performed during the audit. ACME management was not pleased and decided to call the police and have the WLAN security auditor arrested. What documentation did the auditor fail to obtain prior to the WLAN security audit?
   - A. Written corporate policy
   - B. Liability waiver
   - C. Statement of work
   - D. Nondisclosure agreement
   - E. Network topology map
4. What would be the intended purpose of simulating Layer 2 deauthentication attacks as part of a WLAN audit?
   - A. Audit Layer 1
   - B. Audit Layer 2
   - C. Audit the wired infrastructure
   - D. Audit the WIPS

5. Management has asked the WLAN administrator to perform a thorough WLAN security audit. The administrator explains that a wired-side audit is necessary to ensure a secure WLAN. What procedures should be followed during the wired-side portion of a WLAN audit? (Choose all that apply.)

   A. Audit firewall policies and rules

   B. Audit WLAN management interfaces

   C. Audit core Layer 3 switch

   D. Audit application services

6. What are some of the recommendations that might be made to a customer after a successful WLAN security audit? (Choose all that apply.)

   A. Physical security

   B. Employee training

   C. Dynamic RF configuration

   D. Monitoring capabilities

   E. AP and client power settings

7. Which of these tools are required for a proper WLAN security audit? (Choose all that apply.)

   A. Spectrum analyzer

   B. WLAN protocol analyzer

   C. WLAN penetration testing software tools

   D. Global positioning sensor (GPS)

   E. Cameras

8. The management at the ACME Corporation has asked Bob to perform a WLAN security audit. Bob informs management that he will need to purchase a Yagi antenna for the audit. What reasons should Bob give management to justify the purchase of the Yagi antenna? (Choose all that apply.)

   A. Attackers do not need physical access to the facility.

   B. Yagi antennas are used in high multipath areas.

   C. RF signals can be amplified from great distances.

   D. Yagi antennas are used for indoor audits.

   E. Spectrum analyzers require directional antennas.

9. As an auditor you have been asked to determine if the WLAN access points and client devices have been configured with the proper encryption. What should you use to answer this question for your customer? (Choose all that apply.)

   A. Written corporate security policy

   B. WLAN protocol analyzer

C. Aircrack-ng

D. coWPAtty

E. Asleap

10. What is some of the proper documentation needed prior to the WLAN security audit?

A. Statement of work

B. Liability waiver

C. Nondisclosure agreement

D. All of the above

11. A thorough security audit was conducted when a WLAN was deployed over 12 months ago and found no security issues. Recently, the organization failed to meet an industry compliance that the WLAN initially was able to meet due to security failures. What should have been done to help prevent the noncompliance issue the company now faces? (Choose all that apply.)

A. Update to the corporate security policy.

B. Remove all WLAN devices from the network.

C. Upgrade all firmware.

D. Perform a periodic WLAN audit.

12. As part of an audit and covered in the statement of work and nondisclosure agreements, you have been asked to determine if an outsider with no inside access or information would be able to gain access to the WLAN used in your client's WPA2-Personal protected warehouse. Which tools should you use to provide the answer? (Choose all that apply.)

A. WLAN protocol analyzer

B. Aircrack-ng

C. Dictionary file

D. NetStumbler

E. coWPAtty

F. Asleap

13. What would be the intended purpose of using a third-party AP as part of a WLAN audit?

A. Audit Layer 1.

B. Audit Layer 2.

C. Audit the wired infrastructure.

D. Audit the WIPS.

14. Users have recently been complaining about lost connections at various times of the day. An original site survey was conducted that initially confirmed connectivity in all areas of the facility. Which tools must you use as part of your audit to determine the unexplained cause of loss connectivity? (Choose all that apply.)

A. High-gain Yagi antenna

B. Low-gain dipole antenna

C. WLAN protocol analyzer

D. Spectrum analyzer

E. Rainbow table

15. While conducting a WLAN security audit, you find an access point being used by employees on the network configured with all of the correct corporate security settings. This AP is not on the authorized AP list of the company's WIPS but is configured securely and according to corporate written security policy. What should you do about this AP?

    A. Unplug it from the wired LAN immediately.

    B. Include it in your report to the company.

    C. Nothing; it is secured by company standard.

    D. Crack its security and decode the data.

16. What is the main purpose of using a WLAN protocol analyzer during the Layer 2 analysis of a WLAN security audit? (Choose all that apply.)

    A. Identifying unauthorized devices

    B. Auditing the wired infrastructure

    C. Performing penetration testing

    D. Validating security compliance of authorized devices

    E. Auditing the WIPS

17. During a WLAN audit, you see an AP deployed in a common hallway of a multitenant building. This AP provides coverage to a small meeting room used by your customer. It was deployed there to keep the small meeting room from looking cluttered and is using the appropriate authentication and encryption as defined by the written company security policy. Is this AP a risk to the company's network?

    A. Yes, it is not on company property.

    B. No, it is using required security settings.

18. When conducting a WLAN security audit, which of the following items would be of the least amount of use to the auditor?

    A. Physical access to the building

    B. Floor plan

    C. Security escort

    D. Ladder or lift

    E. All of these are useful.

19. As an auditor, you have been asked to determine if the WLAN access points and client devices have been configured with EAP-TTLS authentication. What should you use to answer this question for your customer?

    A. WLAN discovery tool

    B. WLAN protocol analyzer

C. Aircrack-ng

D. coWPAtty

E. Asleap

20. Another auditor tells you that they use the same toolkit to conduct audits as they use to conduct WLAN site survey work. Why would they do this since they are performing two different types of work?

A. Many auditors cannot afford a proper auditing kit.

B. The two types of work are similar enough to use the same tools.

C. They have never been shown how to conduct an audit.

D. Their scope of work document limits them to passive auditing.

Wireless Security Monitoring

IN THIS CHAPTER, YOU WILL LEARN ABOUT THE FOLLOWING:

✓ **Wireless intrusion detection and prevention systems (WIDS and WIPS)**

- WIDS/WIPS infrastructure components
- WIDS/WIPS architecture models
- Multiple radio sensors
- Sensor placement

✓ **Device classification**

- Rogue detection
- Rogue mitigation
- Device tracking

✓ **WIDS/WIPS analysis**

- Signature analysis
- Behavioral analysis
- Protocol analysis
- Spectrum analysis
- Forensic analysis
- Performance analysis

✓ **Monitoring**

- Policy enforcement
- Alarms and notification
- False positives
- Reports

✓ **802.11n/ac**

✓ **802.11w**

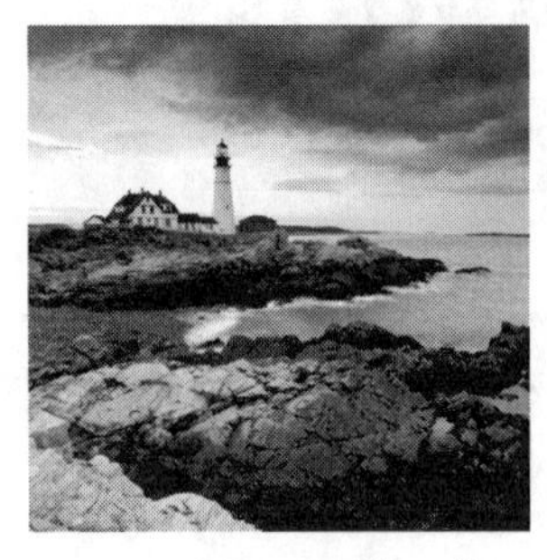

Wireless intrusion monitoring has evolved since its creation in 2006. Today most systems have methods to prevent and mitigate several of the better known wireless attacks. While most systems are distributed for scalability across a large enterprise, single laptop versions of intrusion monitoring systems also exist. Initially, wireless intrusion monitoring existed at Layer 2. Now Layer 1 wireless intrusion monitoring systems are also available to scan for potential Layer 1 attacks. A large-scale enterprise WLAN requires more than spot checks with handheld scanners and laptop-based tools. Although many of these tools have state-of-the-art detection and mitigation, by themselves they are not going to offer the needed level of 24-hour daily monitoring or security. Spot checks or mobile scans can be a vital part of a security plan and can even help you to meet the bare minimum of some security standards. However, by themselves they leave much to be desired as a total solution. Large enterprises require a distributed solution with centralized management capabilities and 24/7/365 monitoring capabilities. Distributed monitoring solutions may also have prevention capabilities, including mitigating rogue APs and clients. The use of distributed WLAN monitoring and rogue prevention reduces the time and expense required to maintain a healthy and secure wireless network.

Wireless Intrusion Detection and Prevention Systems (WIDS and WIPS)

When people think of wireless networking, they tend to think only in terms of access and not in terms of attacks or intrusions. However, it has become increasingly necessary to monitor constantly for many types of WLAN attacks because of the potential damage they can cause. Businesses of all sizes deploy 802.11 wireless networks for mobility and access. Many of these networks are running a *wireless intrusion detection system (WIDS)* to monitor for attacks. Because many organizations are worried about the potential damage that results from rogue access points, it is not unusual for a company to deploy a WIDS prior to deploying the WLAN. Most WIDS vendors prefer to call their product a *wireless intrusion prevention system (WIPS)*. The reason that they refer to their products as prevention systems is that all of them are now capable of mitigating attacks from rogue access points and or rogue clients.

Wireless intrusion detection systems and wireless intrusion prevention systems share many common features that help administrators and security staff members alike in the maintenance and protection of WLAN traffic. WIDS and WIPS both use a combination of sensors and management systems in the gathering and analysis of wireless traffic. Some use information gained from integration with managed switches and WLAN controllers as well as information reported from access points. The key thing to understand about WIDS and WIPS is that they use the same information that an auditor or an administrator would gather using laptop-based tools or handheld devices to conduct an audit, but they do so 24 hours per day using a distributed methodology. A WIDS solution gathers information from the 802.11 radio transmissions detected by multiple sensors and then correlates the captured information. A distributed WIDS solution offers the scalability that cannot be provided by an auditor with a single protocol analyzer.

WIPS act in a similar manner, with the additional capability of being able to keep rogue stations off legitimate WLANs, and some even help keep rogue APs off the wired network. Some WIPS are also able to enforce WLAN policy to stop authorized stations from engaging in unauthorized behaviors, such as connecting to unauthorized APs, forming ad hoc connections, and communicating without using approved authentication and encryption methods. Their names describe their functions; a WIDS detects and notifies about potential attacks or vulnerabilities, whereas a WIPS functions as a WIDS and additionally protects the WLAN using various methods beyond simple detection and notification.

WIDS/WIPS Infrastructure Components

The components of a WIDS/WIPS and their abilities vary from manufacturer to manufacturer. However, their core functions and hardware are similar. The typical WIDS/WIPS solution is a distributed client/server model that consists of two primary components:

- WIDS/WIPS server
- Sensors

A *WIDS/WIPS server* acts as a central point for monitoring security and performance data collection. A WIDS/WIPS server is often a standalone device that is available as either a hardware appliance or as a software appliance that can run as a virtual machine. The WIDS/WIPS server might also be integrated in a WLAN controller. WIDS/WIPS server monitoring capabilities can also be unified with a *network management server (NMS)* solution that is used to monitor all aspects of a WLAN. These WLAN security servers may be deployed in a datacenter as an on-premises solution or exist in a cloud-based environment.

The server can use signature analysis, behavior analysis, protocol analysis, and RF spectrum analysis to detect potential threats. Signature analysis looks for patterns associated with common WLAN attacks. Behavior analysis looks for 802.11 anomalies. Protocol analysis dissects the MAC layer information from 802.11 frames. Protocol analysis may also look at the Layer 3–7 information of 802.11 data frames that are not encrypted. Spectrum analysis monitors RF statistics, such as signal strength and signal-to-noise ratio (SNR). Performance analysis can be used to gauge WLAN health statistics, such as capacity and coverage.

Some solutions might additionally require a software-based *management console* that is used to communicate back to a WIDS/WIPS server from a desktop station. The management console is the software interface used for administration and configuration of the server and sensors. The management console can also be used for 24/7 monitoring of 802.11 wireless networks. However, the majority of the WIDS/WIPS solutions do not require an additional console and all security monitoring is viewed directly from the server. As shown in Figure 14.1, a WLAN administrator can monitor all of the potential WLAN security threats from the *graphical user interface (GUI)* of the WIDS/WIPS server.

FIGURE 14.1 WIDS/WIPS monitoring

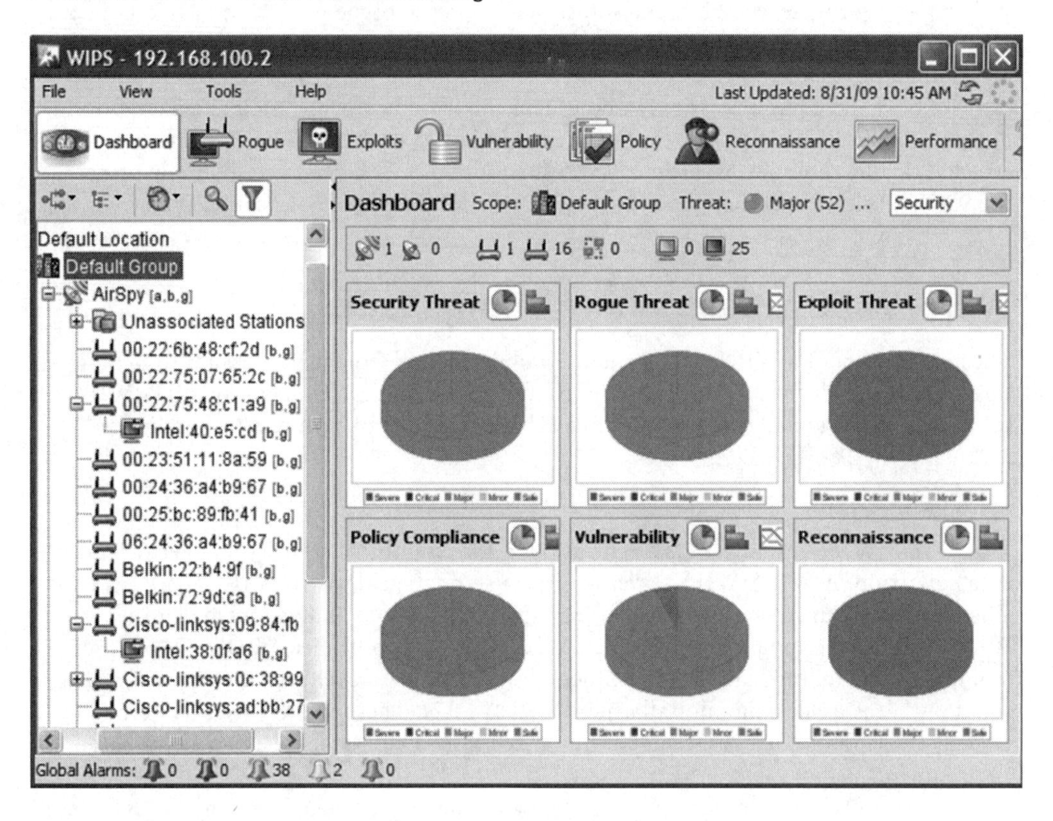

Hardware or software-based *sensors* may be placed strategically to listen to and capture all 802.11 communications. Sensors are the eyes and ears of a WIDS/WIPS monitoring solution. Sensors use 802.11 radios to collect information used in securing and analyzing WLAN traffic. As shown in Figure 14.2, WIDS/WIPS sensors use 802.11 radio chipsets and most often the same hardware as 802.11 access points. However, dedicated sensors are tasked with being a listening device rather than an AP that provides client access. Sensors constantly scan across all 14 channels of the 2.4 GHz ISM band as well as all channels of the 5 GHz UNII frequency bands. Although rarely used in WLAN deployments, some channels in the 4.9 GHz range—which is reserved for public safety in the United States but is a common channel band in Japan—may also be monitored. The channel scanning interval is usually set at a fixed rate between 100 ms and 1 second. However, the channel scanning interval can be adjusted for shorter or longer times. Usually sensors are set to continuously scan across all 802.11 channels. Sensors can also be configured to monitor only a single fixed channel. The majority of WIDS/WIPS sensors are hardware based. Some WIDS/WIPS solutions offer sensor software that can be installed on a computer. The sensor software can then use the computer's 802.11 radio for scanning. Figure 14.2 shows hardware sensors from two different WIPS vendors.

FIGURE 14.2 WIDS/WIPS hardware sensors

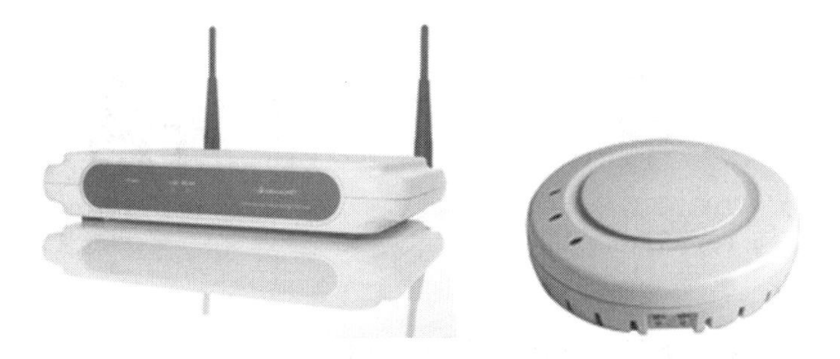

Communications from the sensors back to the server can be either be a standards-based management protocol such as *Control and Provisioning of Wireless Access Points (CAPWAP)* or a proprietary management protocol. The management protocol is normally protected by an encrypted *Secure Sockets Layer (SSL)* tunnel. Typically, a sensor also sends a continuous heartbeat message back to the WIPS server to indicate that the sensor is still functional. Sensors can usually be centrally managed from the server or may be managed individually through Telnet, SSH, or a web browser. Most WIPS use port 443 for SSL communications; port 443 will need to be open on any firewall located between a sensor and the WIDS/WIPS server. Depending on the vendor, other vendor-specific ports may also need to be open to permit communications between the sensors and the WIDS/WIPS server. Sensors may also be used for remote packet capturing.

As shown in Figure 14.3, the two components of the WIDS/WIPS distributed design can exist at a single WLAN enterprise site or can scale to monitor multiple WLAN sites across a wide area network (WAN). The WIDS/WIPS server might reside in Sydney, Australia, while the WLANs that are being monitored may reside in Austin and London. The administrator monitoring the WIPS solution can access the server across the Internet from anywhere in the world.

FIGURE 14.3 Distributed WIDS/WIPS

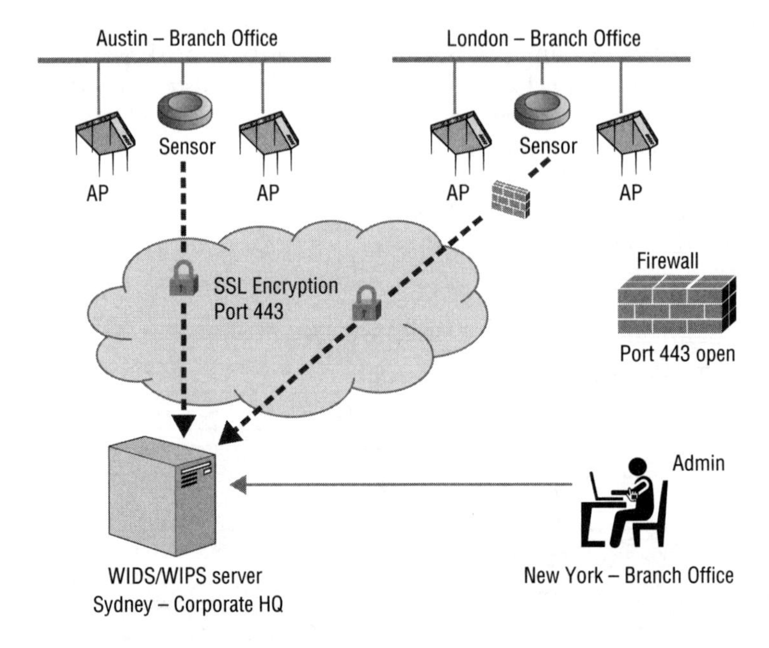

WIDS/WIPS Architecture Models

The components of a WLAN security monitoring solution are usually deployed within one of the two major WIDS/WIPS architectures:

- Overlay
- Integrated

As Figure 14.4 shows, an *overlay* WIDS/WIPS architecture is deployed on top of the existing wireless network. This model uses an independent vendor's WIDS/WIPS solution and can be deployed to monitor any preexisting or planned WLAN. The overlay systems typically have more extensive features and monitoring capabilities, but they are usually more expensive. The overlay solution consists of a WIPS server and sensors that are not part of the WLAN solution that provides access to clients.

An overlay solution uses *standalone sensors* to monitor the preexisting WLAN. Standalone sensors use dedicated radios that function only as sensors and are not used as APs. The standalone sensors, also known as dedicated sensors, require their own cable to run and communicate back to the WIDS/WIPS server over the IP network. Usually, standalone sensors can use a single 802.11 dual-band radio to monitor both the 2.4 GHz ISM band and the 5 GHz UNII bands.

FIGURE 14.4 Overlay WIDS/WIPS

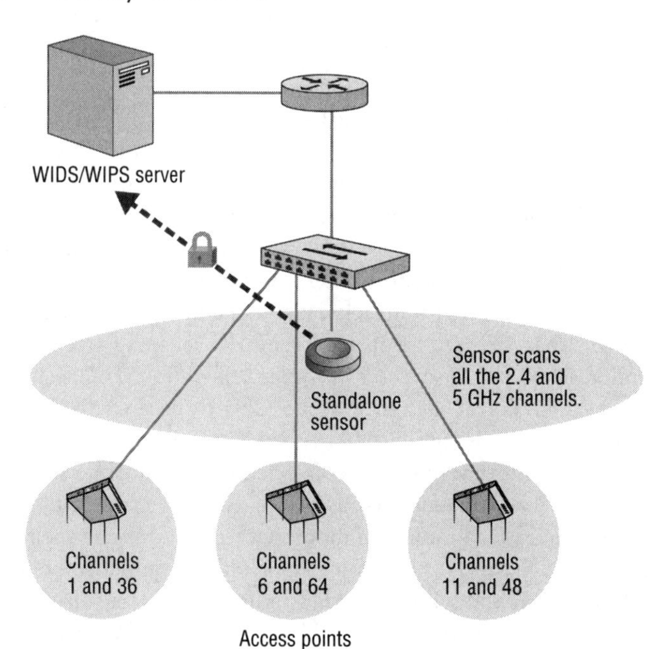

Standalone sensors may also have two radios: one dedicated to monitor the 2.4 GHz channels and another dedicated to monitor the 5 GHz channels. Sensors used in overlay WIPS architecture are normally independent of the WLAN being used to provide access. However, WLAN vendor access points can often integrate software code that can be used to convert the APs into standalone sensors that will communicate with the third-party WIDS/WIPS server. Overlay solutions increase hardware and deployment costs but generally offer more functionality and security. Overlay WIDS/WIPS servers typically use a wider range of attack signatures to recognize potential threats and collect more information for WLAN health analysis. Another major advantage of the overlay model is that if the WLAN goes down, the WIDS/WIPS monitoring continues because the overlay solution is independent of the WLAN infrastructure.

In the early days of Wi-Fi deployments, overlay WIDS/WIPS solutions were deployed with autonomous APs. Over the years, integrated WIDS/WIPS solutions have become much more commonplace mainly due to the cost savings. Overlay WIDS/WIPS solutions are now primarily used in enterprise verticals that can afford an overlay WLAN security monitoring solution. Overlay WIDS/WIPS are still often used in verticals such as federal government, military, financial industries and big-box retailers.

Most WLAN vendors now offer an *integrated* WIDS/WIPS architecture to provide both client access and security monitoring. An integrated architecture requires the APs to function as WIPS sensors while also providing 802.11 connectivity and access to WLAN clients. The device that functions as a WIDS/WIPS server within an integrated model depends on the WLAN architecture that is deployed. If a centralized WLAN architecture is deployed, typically the WLAN controller also operates as the WIDS/WIPS server. The controller-based APs, which also function as sensors, are centrally managed and monitored from the WLAN controller. If a distributed WLAN design is deployed, normally a network management server (NMS) is tasked with WIPS/WIPS server duties. The distributed APs, which also function as sensors, are centrally managed and monitored from the NMS. The NMS can be either a cloud-based solution or an on-premises solution.

As shown in Figure 14.5, APs used in an integrated WIPS architecture have the capability to be converted to *full-time sensors*. Instead of providing access to clients, the APs scan all channels, continuously listening for attacks, just like a dedicated sensor model. The form factors of the APs that are used for integrated WIPS sensors vary widely depending on the WLAN vendor. Many APs use a *software-defined radio (SDR)* that has the ability to operate either as a 2.4 GHz transceiver or as a 5 GHz transceiver, but it cannot transmit on both frequency bands at the same time. However, if the SDR uses a dual-frequency chipset, a single radio access point that has been converted into a full-time sensor can listen to both the 2.4 GHz and 5 GHz frequency bands. Many APs have two radios, both of which can be converted to a sensor. One radio can monitor the 2.4 GHz channels while the other radio monitors the 5 GHz channels. An AP with two radios could also be used to provide access on a single frequency band while the other radio is used as a full-time sensor to scan the channels of both bands. Some APs deployed in an integrated WIPS architecture have three radios. One radio is used for 2.4 GHz client access, another radio is used for 5 GHz client access, and a third SDR is used to monitor both frequency bands.

FIGURE 14.5 Integrated WIDS/WIPS

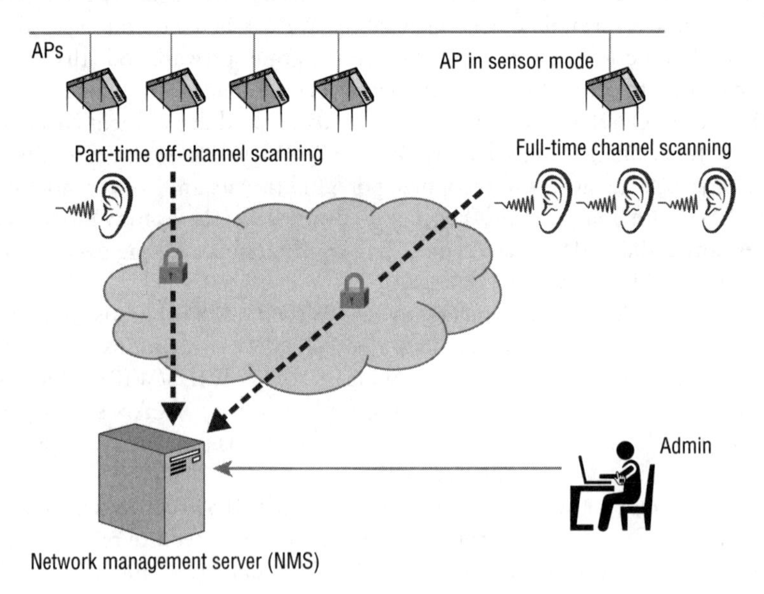

APs used in an integrated WIPS architecture can also function as part-time sensors. In this case, access points use *off-channel scanning* procedures for dynamic RF spectrum management purposes. For example, an AP that is providing client access on channel 6 will also monitor other channels where the AP does not transmit. The AP might stay on channel 6 for 10 seconds. During the 10-second interval, the AP is capable of sending transmissions to an associated client as well as receiving transmissions from an associated client. After the 10-second interval, the AP will listen off-channel on channel 7 for 110 ms. The AP will then return to channel 6 for 10 seconds and then go off-channel to monitor channel 8 for 110 ms. This round-robin method of off-channel scanning is used by the APs to listen for the beacon frame transmissions of other access points as well as to monitor for any other RF transmissions off-channel. How often an AP spends on-channel and scans off-channel is dependent on the WLAN vendor. Based on all the RF monitoring from multiple access points, there might be dynamic changes to the AP's channel and transmit power settings. Although the main purpose of off-channel scanning is to provide dynamic RF capabilities, the off-channel scanning also allows the APs to function as *part-time sensors* for the WIDS/WIPS server module in the WLAN controller or NMS. The off-channel scanning used by the APs effectively provides time slicing between AP and sensor functionality.

Time slicing between AP and sensor functionality may reduce hardware and deployment expense, but it offers limited detection and prevention. The majority of customers of WLAN vendors opt not to incur the extra expense of deploying APs as full-time sensors and only use the part-time, time-slicing capabilities of the access points. If you hired a security guard to watch the main entrance to your place of business, would you want the security guard to take a break for 55 minutes every hour and only watch the main gate for 5 minutes of each hour? An attacker might recognize that a WIPS solution is only using part-time sensors. The attacker could then launch a brief attack during the period that the APs are providing access. The attack occurs when the APs are not performing off-channel scanning and the attack is therefore not detected.

Another problem with part-time sensors is that they may suspend off-channel scanning if a VoWiFi phone is associated with the access point. Off-channel scanning is notorious for causing "choppy audio" during an active voice call from a VoWiFi device associated to an AP. Most WLAN vendors now have an option that suspends off-channel scanning based on the detection of QoS priority markings that indicate voice traffic. Effectively off-channel scanning is suspended during an active VoIP call over the WLAN. If the off-channel scanning is suspended due to VoWiFi communications, the WLAN security monitoring is also suspended.

An additional problem with part-time sensor use is wireless rogue containment, which will be discussed later in this chapter. If a time slicing AP/sensor must go off-channel for an extended period of time to contain a rogue device, the AP is not on its home channel providing access to clients. It should be noted, however, that many WLAN vendors support a configuration setting that prevents a time slicing AP/sensor from performing wireless rogue containment when clients are associated.

Although the APs act as part-time sensors for the integrated WIPS server, it is also a highly recommended practice to deploy some APs or AP radios as full-time sensors when using an integrated WIDS/WIPS server solution.

It should be noted that either an overlay or integrated WIPS architecture can very often interface with other security solutions that might be vital to the WLAN design. As shown in Figure 14.6, the WIPS solution might also interoperate with enterprise *security and management (SIEM)* platforms and *mobile device management (MDM)* platforms. SNMP and Syslog interfaces can also be used to integrate event and log management tools.

FIGURE 14.6 Integration with other security solutions

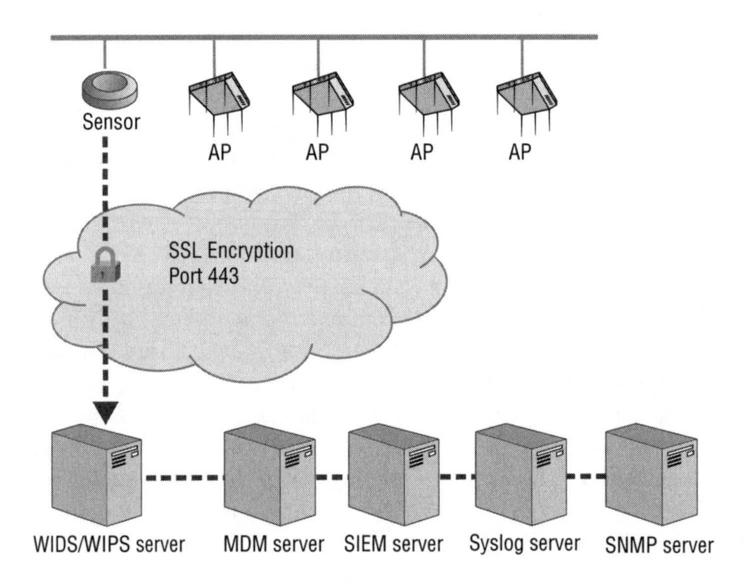

Multiple Radio Sensors

As Figure 14.7 shows, the latest innovations in WLAN security monitoring use multiple radio devices that are not locked to a frequency, protocol, or function. This innovative technique allows a single cable drop to be used, thereby reducing deployment costs; a single housing to be used, thus reducing hardware costs; and multiple radios to be used in the housing, thereby increasing functionality. A device with three radios can use one radio for 2.4 GHz coverage, one radio for 5 GHz coverage, and the third radio as a sensor to monitor and scan both bands. In a mesh deployment, one radio can be used as a 2.4 GHz AP for local coverage, one can be used for a 5 GHz mesh back-haul connection, and a third can be used as a sensor for the coverage area. Some WLAN vendors are even making 802.11 devices with three radios along with a modular spectrum analyzer sensor used for RF spectrum analysis. Multiple radio devices allow WLAN coverage and WIDS/WIPS on separate dedicated radios, increasing security and WLAN performance monitoring abilities at a reduced expense. As mentioned earlier, the form factors of the APs that are used for integrated WIPS sensors vary widely depending on the WLAN vendor.

FIGURE 14.7 Multiple 802.11 radio devices

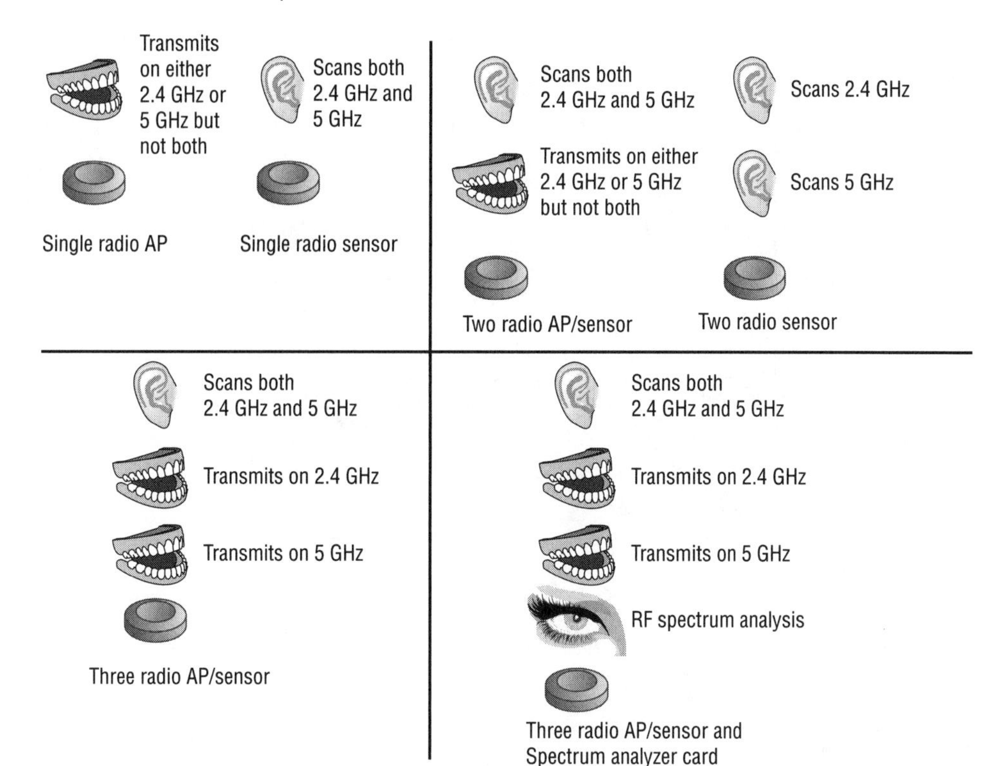

Sensor Placement

Sensor placement is an often-discussed topic when deploying a WIDS/WIPS solution. The question that is always asked is, "How many sensors do I need?" The answer often depends on the budget and the value of the network resources that are being protected by WLAN security monitoring. The best answer is that you can never have too many sensors. When WLAN security monitoring is deployed, the more ears the better.

Every WLAN vendor has its own sensor deployment recommendations and guidelines; however, a ratio of one sensor for every three-to-five access points is highly recommended. As shown in Figure 14.8, full-time sensors are often placed strategically at the intersection points of three AP coverage cells. A common mistake is placing the sensors in a straight line as opposed to staggered sensor arrangement (which will assure a wider area of monitoring). Another common sensor placement recommendation is to arrange sensors around the perimeter of the building. Perimeter placement increases the effectiveness of triangulation and also helps to detect WLAN devices that might be outside the building. Some of the better WLAN predictive modeling software solutions will also create models for recommended sensor placement.

FIGURE 14.8 Sensor placement

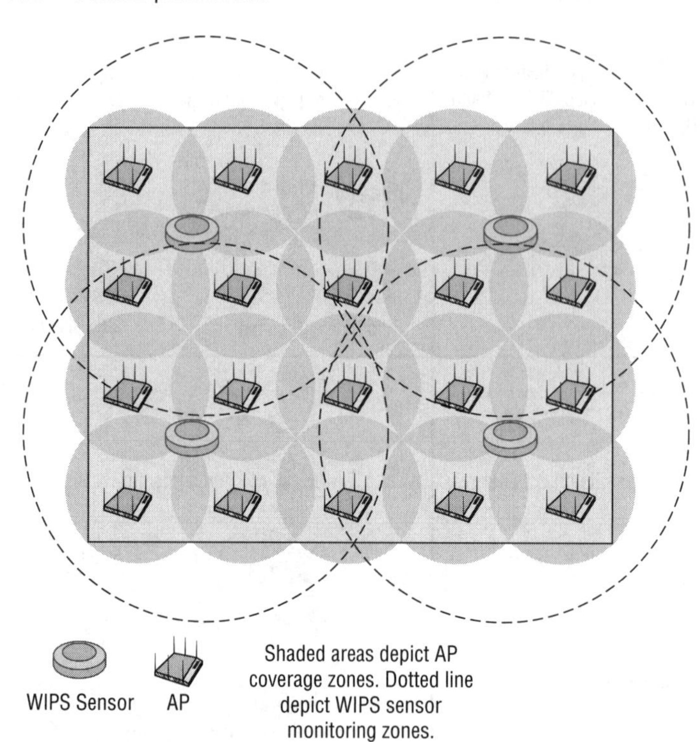

If an integrated WIDS/WIPS solution is deployed, all access points are deployed as part-time sensors. Once again, it is strongly recommended to deploy some full-time sensors with an integrated solution. When WLAN security monitoring is an extremely high priority and cost is not an issue, the more sensor devices the better. WIDS/WIPS deployments at military bases often follow a ratio of one sensor for every two APs or may even deploy sensors with a 1:1 ratio.

Device Classification

Although the upper-layer payload of 802.11 data frames is usually encrypted, the Layer 2 information remains exposed to allow for MAC layer communications to occur. All Layer 2 information is captured and analyzed by WIDS/WIPS solutions. Any 802.11-based device transmitting within the hearing range of a sensor will eventually be detected as the sensor sweeps the channels. WIDS/WIPS can also integrate with WLAN controllers, APs, and access layer switches pulling information from them for use in detecting wired devices. Information freely available in 802.11 management frames will allow the WIDS/WIPS initially to determine if an 802.11 device is an access point, client station, or ad hoc client station. If an 802.11 wireless device is transmitting within the hearing range of a sensor, it will be detected by the WIDS/WIPS and then will be classified, as shown in Figure 14.9.

FIGURE 14.9 WIDS/WIPS device classification

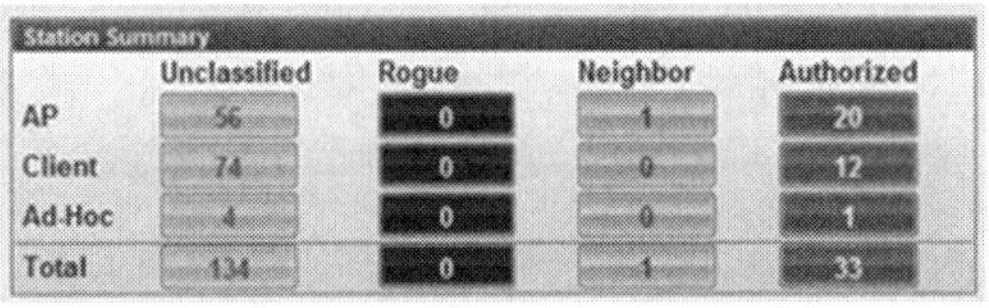

Station Summary

	Unclassified	Rogue	Neighbor	Authorized
AP	56	0	1	20
Client	74	0	0	12
Ad-Hoc	4	0	0	1
Total	134	0	1	33

Most WIDS/WIPS vendors categorize access points and client stations in four or more classifications. Wi-Fi vendors may have different names for the various classifications, but most solutions classify 802.11 radios as follows:

Authorized Device This classification refers to any client station or access point that is an authorized member of the company's wireless network. An overlay WIDS/WIPS solution usually requires initial manual input to classify radios as authorized infrastructure devices. A network administrator can manually label each radio as an infrastructure device after detection from the WIPS or can import a list of all the company's client MAC addresses into the system. Devices may also be authorized in bulk from a comma-delimited file. Integrated solutions automatically classify any APs as authorized devices. An integrated solution will also automatically classify client stations as authorized if the client stations are properly authenticated.

Unauthorized Device The unauthorized device classification is assigned automatically to any new 802.11 radios that have been detected but not classified as rogues. Unknown devices are considered to be unauthorized and are usually investigated further to determine whether they are a neighbor's device or a potential future threat. Unauthorized devices may later be manually classified as a known neighbor device.

Neighbor Device This classification refers to any client station or access point that is detected by the WIPS and whose identity is known. This type of device initially is detected as an unauthorized device. The neighbor device label is then typically assigned manually by an administrator. Devices manually classified as known are most often 802.11 access points or client radio devices of neighboring businesses that are not considered a threat.

Rogue Device The rogue classification refers to any client station or access point that is considered an interfering device and a potential threat. Most WIDS/WIPS solutions define rogue access points as devices that are actually plugged into the wired network backbone and are not known or managed by the organization. Most of the WIDS/WIPS vendors use a variety of methods to determine whether a rogue access point is actually plugged into the wired infrastructure.

Many WIDS/WIPS solutions also have the ability to conduct *auto-classification*. As shown in Figure 14.10, WLAN devices can be automatically added to any classification based on a variety of variables, including authentication method, encryption method, SSID, IP addresses, and so on. Auto-classification capabilities should be used carefully to ensure that only proper devices are classified as authorized.

FIGURE 14.10 WIDS/WIPS device auto-classification

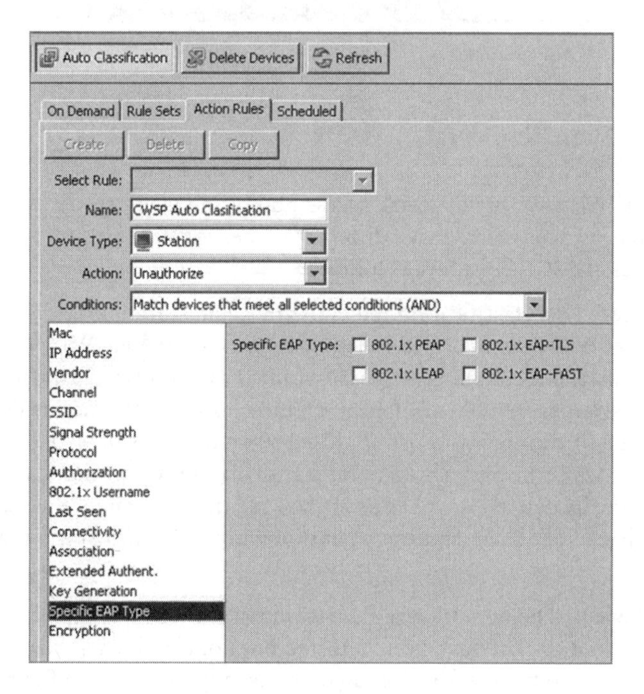

Rogue Detection

As already mentioned, most WIDS/WIPS define rogue APs as devices that are actually plugged into the wired network. Most of the WIDS/WIPS vendors use a variety of wireless and wired detection methods to determine whether a rogue access point is plugged into the wired infrastructure. Some of the rogue detection and classification methods are published whereas many remain proprietary and trade secrets. Any 802.11 device that is not already authorized will automatically be classified as an unauthorized device. However, rogue classification is a little more complex.

One effective approach for classifying rogue APs is to poll access layer switches with *Simple Network Management Protocol (SNMP)* to determine MAC addresses associated with each physical port on the switch. Given that an AP acts as a Layer 2 bridge, the WIDS/WIPS solution builds a MAC table that correlates both the wired-side MAC address and wireless-side MAC address (BSSID) of the access point. This correlated MAC table can then be compared to the database of authorized devices. Any unauthorized device that is detected by both a sensor on the wireless side and by SNMP on the wired side will then be classified as a rogue AP.

As shown in Figure 14.11, another method often used to determine whether a device is connected to the wired backbone is by looking at broadcast traffic such as an ARP request from a wired device—a core router, for example—and analyzing MAC tables. The following is an example of steps that would have to occur to classify the device as a rogue:

1. A rogue AP with a BSSID of 02:02:02:02:02:02 is plugged into the wired network.
2. A sensor detects a new BSSID on the air and initially classifies this AP as unauthorized. The new AP has not yet been classified as a rogue AP.
3. A default gateway router on the wired network with the MAC address 33:33:33:33:33:33 broadcasts an ARP request packet looking for a host on a particular subnet. Because this ARP packet is a broadcast packet, the rogue AP receives the packet and transmits it out the wireless interface.
4. If the sensor and authorized APs are on the same subnet as the rogue AP, the sensor receives the ARP broadcast on its wired interface. This MAC address is stored in the wired-side MAC table and is shared with all other sensors within the WIDS.
5. When the ARP packet is transmitted into the air by the rogue AP, the source address (SA) is the originating router's address of 33:33:33:33:33:33, and the transmitter address is the BSSID of the rogue 02:02:02:02:02:02.
6. The WIDS/WIPS solution will look at the wired/wireless MAC tables. Any unauthorized BSSID transmitting an ARP request with the source address of the wired router will now be classified as a rogue AP.

FIGURE 14.11 Rogue detection—single broadcast domain

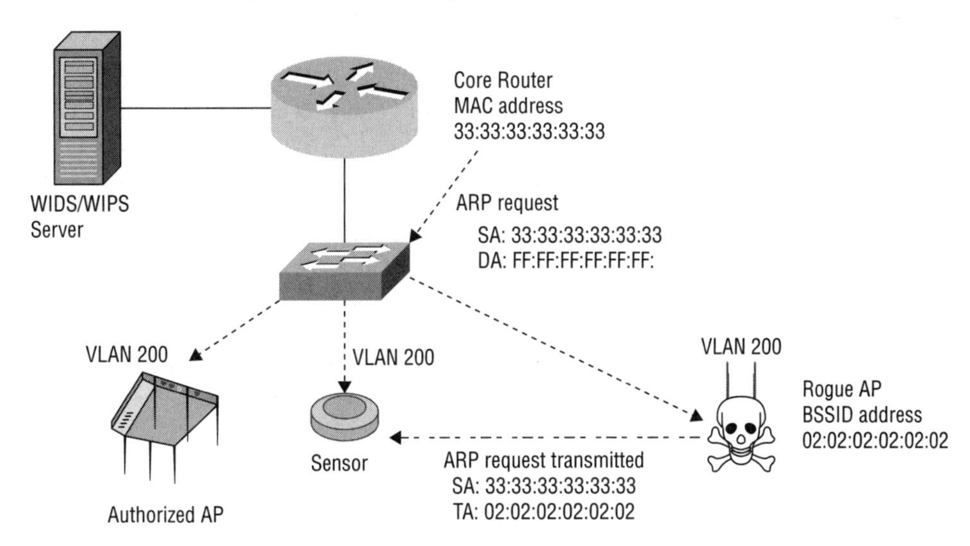

This will only work if the rogue device is plugged into the same broadcast domain as the sensor. Because many networks are designed with a large number of VLANs, a rogue AP could be plugged into a different VLAN than the sensors. As Figure 14.12 shows, the sensor would need to have an 802.1Q trunk link in order to receive the ARP request from the wired side.

FIGURE 14.12 Rogue detection—trunked sensor

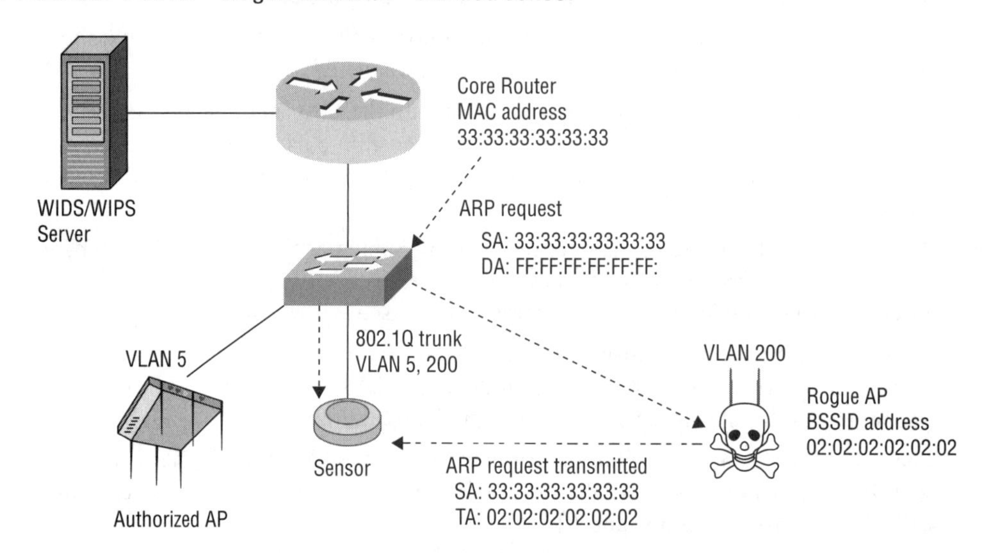

All of these rogue detection and classification methods work well if the rogue device is indeed an access point. An access point is a Layer 2 device that bridges the 802.3 Ethernet interface and the 802.11 radio interface. Unfortunately, most rogue devices that are installed are low-cost home Wi-Fi routers and are not bridged devices. Therefore, most rogue devices have separate Layer 3 interfaces and are normally configured to use network address translation (NAT). As shown in Figure 14.13, one method of classifying Layer 3 rogue devices is to have a nearby sensor associate as a client station to the unauthorized suspect rogue AP or Wi-Fi router. The sensor then sends traffic, such as a ping, back to the WIDS/WIPS server. If the traffic reaches the server on the wired side, the suspected rogue AP is confirmed to be on the internal network and then classified as a rogue.

The problem with this method is that very often a rogue AP will have configured security, such as WEP or WPA. The sensor will not be able to associate with the rogue AP and send traffic because the sensor does not know the rogue APs security settings.

Luckily, the majority of the WLAN vendors that manufacture home Wi-Fi routers use MAC addresses that are one bit apart on the wireless and wired interfaces. As shown in Figure 14.14, the methods of Layer 2 rogue detection mentioned earlier can be augmented by looking for wired MAC addresses within a range of the BSSID detected by a sensor. A simple comparison of wireless and wired MAC is then used by the WIPS solution to classify the device as a rogue device.

FIGURE 14.13 Layer 3 Rogue detection—active sensor

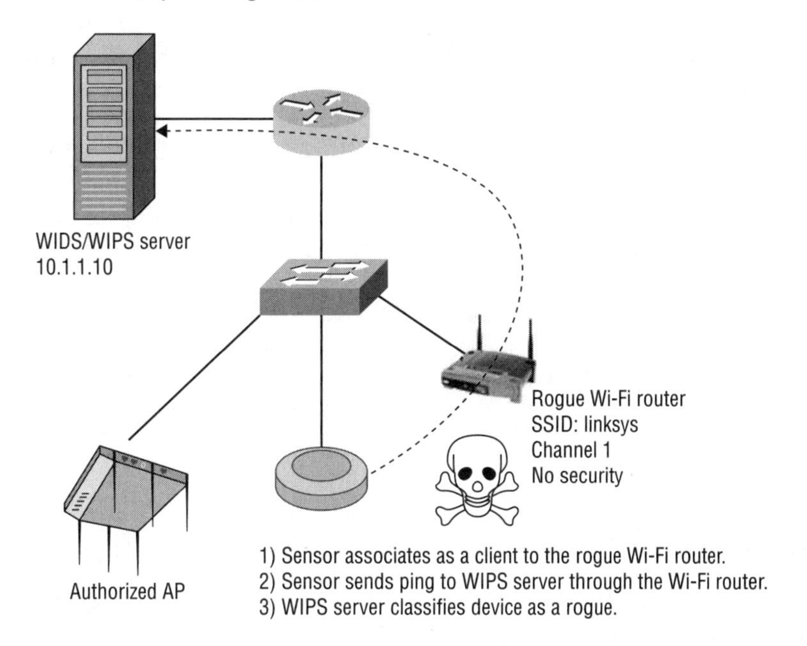

FIGURE 14.14 Layer 3 Rogue detection—MAC comparison

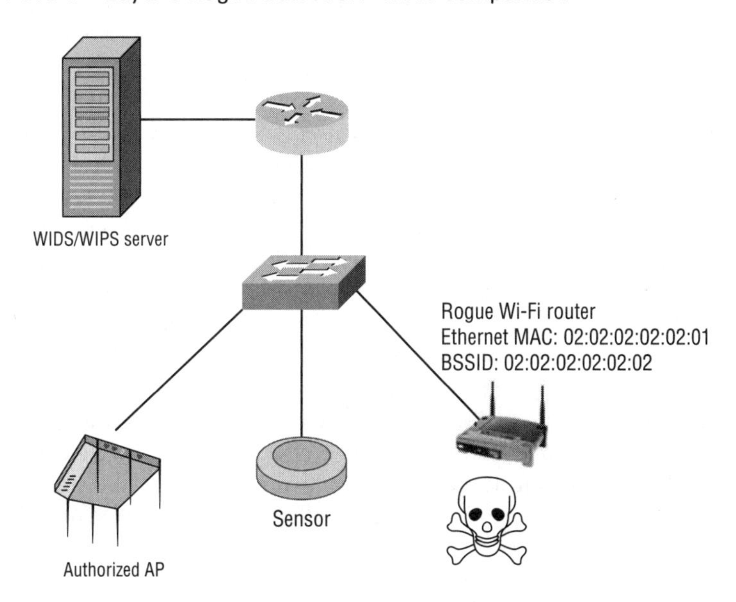

Another method for possibly determining whether there is a potential rogue device connected to the wired network is to examine the *time to live (TTL)* values of IP packets. Wi-Fi routers will lower the TTL value of a packet when it flows through the device. As mentioned earlier, WIDS/WIPS vendors may also use proprietary methods of rogue detection and classification. Some of these methods are proprietary published and some are not. One example is the use of a marker packet. The WIPs solution introduces a special marker packet on the wired side of the network. The marker packet is the transmitted into the wireless environment. If the WIPS solution detects any unauthorized BSSID transmitting the marker packet, the device is classified as rogue.

Although the art of rogue detection and classification has become quite successful, any device that is initially classified as unauthorized should also be investigated and treated as a potential rogue threat until determined otherwise. Once an unauthorized 802.11 device has been determined not to be a threat, the device can then be classified manually as a neighbor device.

Rogue Mitigation

Once a client station or access point has been classified as a rogue device, the WIPS can effectively mitigate the attack. Every WIPS vendor has several ways of accomplishing this, but the most common method is wireless *rogue containment* using spoofed deauthentication frames. Rogue containment is accomplished wirelessly when the WIPS's sensors become active and begin transmitting deauthentication frames that spoof the MAC addresses of the rogue access points, rogue ad hoc networks, and rogue clients. The WIPS is using a known Layer 2 denial-of-service attack as a countermeasure. The effect is that all communications between the rogue access point and clients are rendered useless. Any client devices trying to communicate through the rogue AP will be deauthenticated at Layer 2 and all upper-layer 3–7 communications will be disrupted. This prevents an attacker from accessing network resources through the unauthorized portal of the rogue AP. This also prevents accidental associations of legitimate clients to the rogue AP. This countermeasure can be used to disable rogue access points, individual client stations, and rogue ad hoc networks. Every WIPS vendor has its own marketing name for Layer 2 wireless rogue containment, including air termination, rogue blocking, and rogue disabling. As shown in Figure 14.15, the sensor transmits deauthentication frames spoofing the rogue APs MAC address as the transmitter address (TA). The receiver address (RA) of the spoofed frames may be a broadcast address to deauthenticate all client stations or may be a unicast address to a single, associated rogue client station. Hackers have figured out how to tinker with client radio firmware so that client stations will ignore deauthentication frames. Therefore, as shown in Figure 14.15, most WIPS sensors will also transmit deauthentication frames spoofing the client station's MAC address as the transmitter address and spoofing the rogue AP MAC address as the receiver address.

FIGURE 14.15 Wireless rogue containment

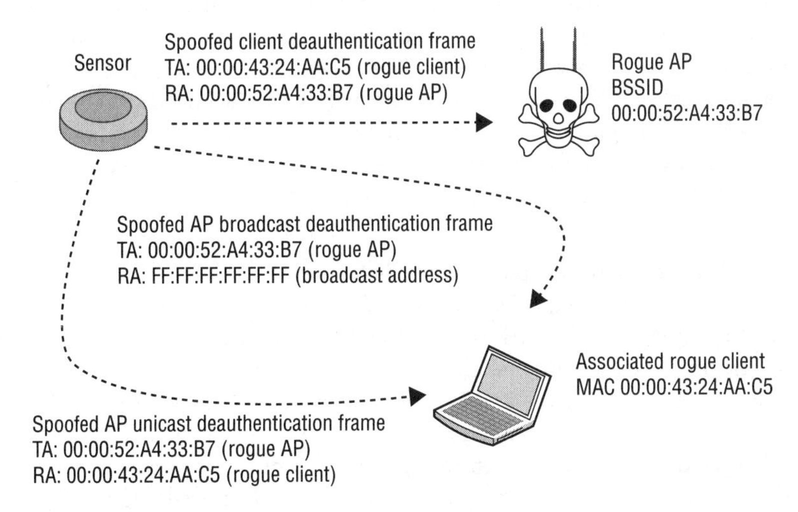

Wireless rogue containment should be used very carefully. Rogue devices can be manually terminated using wireless rogue containment, or a WIPS can be configured to automatically terminate any devices that are classified as rogue. Many WIPS can also be configured to terminate all devices except for those that are classified as authorized. Using a deauthentication countermeasure against all unauthorized devices is not a good idea because the WIPS may accidentally terminate legitimate APs and clients from neighboring businesses. Improper use of wireless rogue containment capabilities can create legal problems. Any device that has been classified as rogue is a device that has been determined by the WIPS to be connected to the corporate backbone and probably should be wirelessly contained. It is up to your organization to choose whether wireless rogue containment is a manual procedure when rogue devices are discovered or whether legitimate rogue devices are automatically contained. Often for legal reasons, manual containment might be the wiser choice. Please understand that either manual or wireless rogue mitigation can only be used for a very short period of time. As soon as mitigation begins, the WLAN administrator should locate the rogue device and disconnect it from the wired network.

Improper Use of Rogue Containment

In 2014 a major hotel chain operator admitted that employees improperly used a Wi-Fi monitoring system to block mobile hotspots. They also agreed to a three-year compliance plan with the FCC during which time they had to implement a compliance plan and report to the FCC every three months. As part of the agreement the hotel chain had to pay a civil penalty of $600,000 to resolve the Wi-Fi blocking investigation. The FCC Enforcement

Bureau's investigation determined that the hotel chain's employees used the containment features of a WLAN monitoring system to prevent individuals from connecting to the Internet via their own personal Wi-Fi networks, while at the same time charging consumers and exhibitors as much as $1,000 per device to access the hotel's Wi-Fi network. The complaint alleged that the hotel was jamming mobile hotspots to prevent their use within the hotel's grounds. More information about this incident can be found at `www.fcc.gov/document/marriott-pay-600k-resolve-wifi-blocking-investigation`.

As mentioned earlier in this chapter, using APs for rogue containment is not a recommended practice. If a time slicing AP/sensor must go off-channel for an extended period of time to contain a rogue device, the AP is not on its home channel providing access to clients. If the AP is performing rogue containment instead of providing access, the performance of the WLAN can be affected. Most WLAN vendors have a configuration setting that prevents a time slicing AP/sensor from performing rogue termination when clients are associated. If clients are associated to all the APs, then rogue containment may not occur when needed. Although the access points act as part-time sensors for the integrated IDS server, it is a highly recommended practice also to deploy some APs as full-time sensors when using an integrated WIDS/WIPS server solution. The full-time sensors would be used for rogue containment instead of the time slicing AP/sensors.

As mentioned in Chapter 12, "Wireless Security Risks," ad hoc WLANs are a huge security risk because the Ethernet connection and the Wi-Fi radio can be bridged together. An intruder might also access the ad hoc WLAN and then potentially route their way to the Ethernet connection and get onto the wired network. A WIPS solution can easily detect an ad hoc WLAN because ad hoc stations transmit 802.11 Beacon frames that indicate they are participating in an *independent basic service set (IBSS)*. As shown in Figure 14.16, after detecting the ad hoc stations, a WIPS can send spoofed deauthentication frames to disrupt communications within the IBSS. Temporary wireless termination of ad hoc WLANs may also be required.

FIGURE 14.16 Ad hoc containment

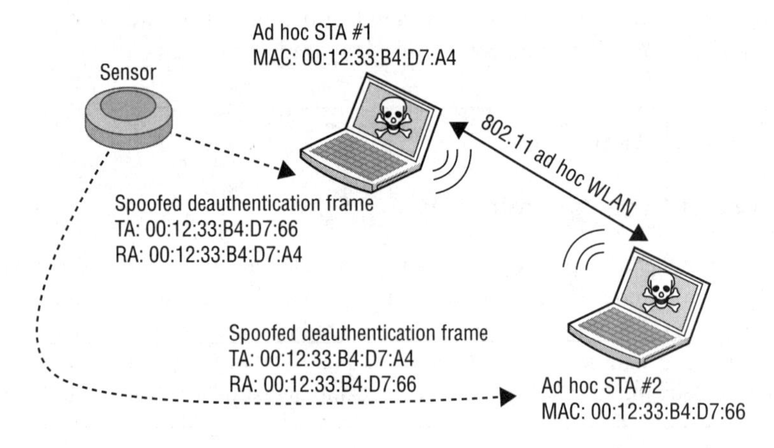

Many WIPS also use a wired-side termination process to effectively mitigate rogue devices. The wired-side termination method of rogue mitigation uses the Simple Network Management

Protocol (SNMP) for *port suppression*. Most WIPS can determine that the rogue access point is connected to the wired infrastructure and may be able to use SNMP to disable the managed switch port that is connected to the rogue access point. Port suppression uses an SNMP agent to shut down the physical port on the network switch through which a rogue device is communicating. If the physical port on the switch is disabled, the gateway to wired network is effectively closed and an attacker cannot use the rogue AP to access network resources.

WIPS vendors have other proprietary methods of disabling rogue access points and client stations and often their methods are not published. Currently, the main attack mitigated by a WIPS is an attack by rogue devices. In the future, other wireless attacks might be mitigated as well.

Device Tracking

Once a device has been detected and classified, the internal monitoring capabilities of a WIDS/WIPS server can be used to locate the device. As shown in Figure 14.17, the location of devices can be shown visually onto a graphic image of the building's floor plan that has been imported into the WIDS/WIPS server. Location tracking is often used to pinpoint the location of rogue APs, but *location tracking* can also be used to establish the vicinity of authorized APs and client stations.

FIGURE 14.17 Location tracking

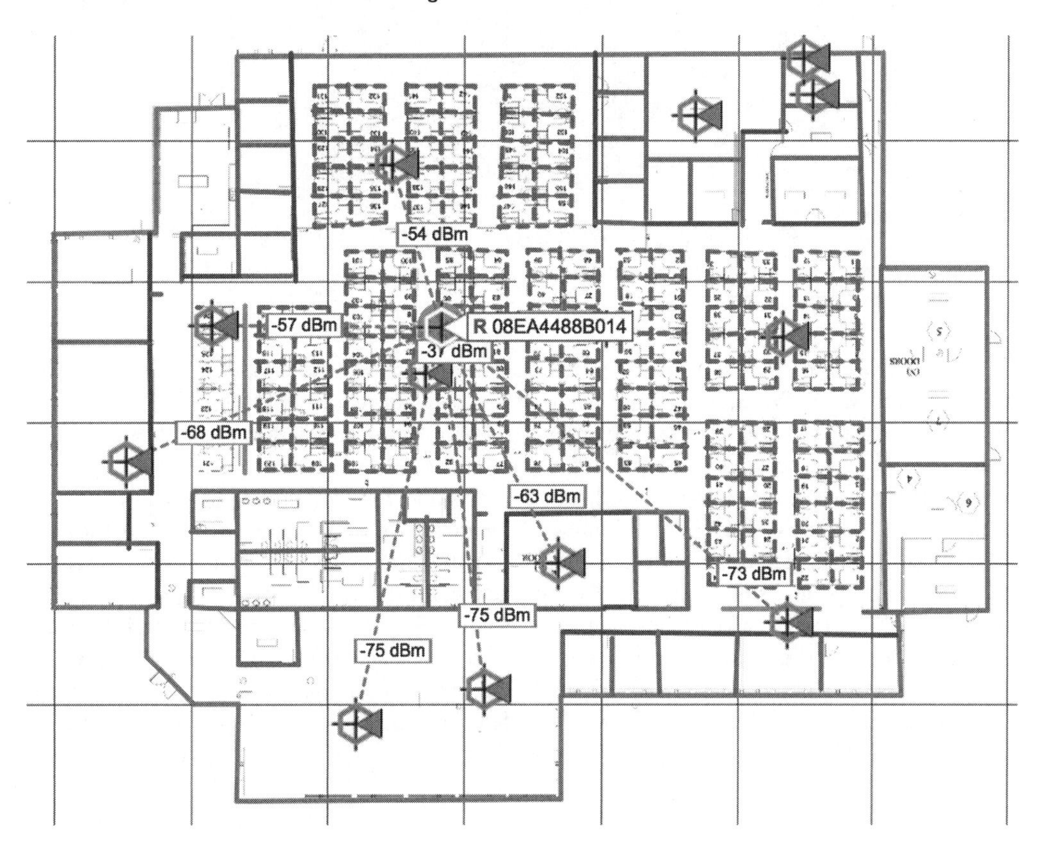

The method used to find devices will vary based on vendor implementation and optional configurations. Some methods of location tracking use *received signal strength indicator (RSSI)* values reported from sensors and authorized APs within listening range of the device being tracked. Newer features also include historical location tracking. As you can see in Figure 14.18, historical tracking allows the WIDS/WIPS to show where a device has been detected in the past, even if the device is not currently transmitting. Historical tracking capabilities are used to monitor the movements of rogue client stations. Historical tracking also provides data that can be used to find missing devices, such as handheld scanners that may have become misplaced in a retail, warehousing, or logistics environment. Furthermore, the historical data can also be used to find trends in worker movement throughout the day as required by their jobs. Whether the device you have detected is rogue, authorized, or just neighboring, it eventually becomes necessary to know the physical location of a device.

FIGURE 14.18 Historical tracking

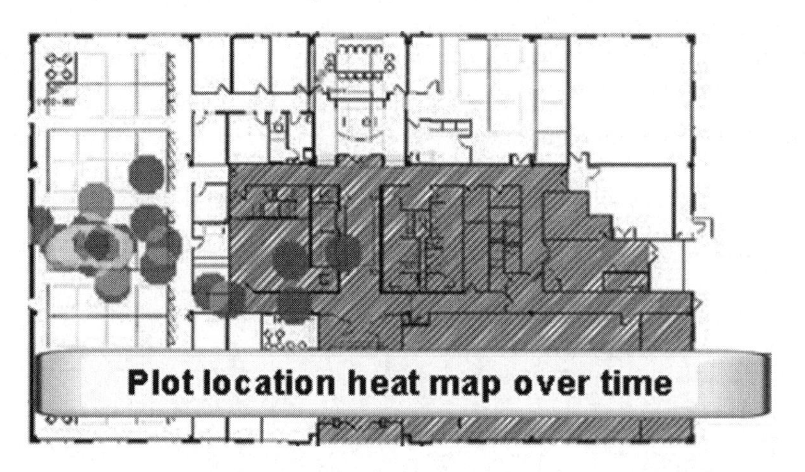

The most common location tracking method is known as *RF triangulation*. Triangulation, as it relates to WIDS/WIPS, is a way of using received signals detected by sensors that are in known locations to find devices that are in unknown locations. Within the WIDS/WIPS tracking module, sensors are positioned on a map of the area where the sensors are actually deployed. Each sensor reporting that it hears the target device provides a data point for use in finding the target based on the RSSI value of the target's signal. Sensors that detect the target with stronger signal strength (a higher RSSI value) are believed to be closer to the target. Those sensors detecting the target with weaker signal strength (a lower RSSI value) are determined to be farther away from the target. Because RSSI provides an estimate of distance but not direction, at least three sensors are needed to determine the location of a monitored device. As shown in Figure 14.19, RF triangulation provides an approximation of the target's location based solely on the varied RSSI values detected and reported by multiple sensors.

FIGURE 14.19 RF triangulation

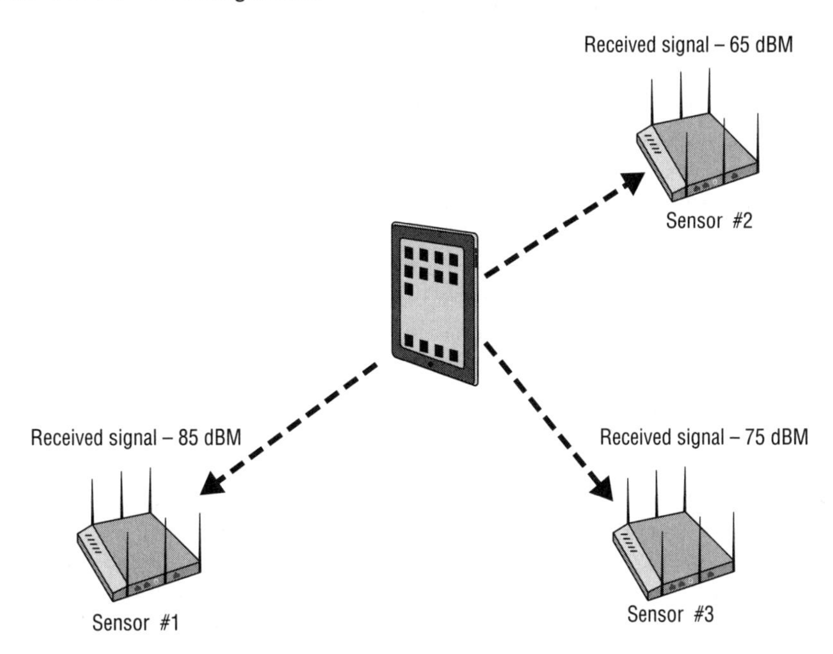

RF triangulation usually results in an estimated location within approximately 10 meters. However, this method does not take into account the RF noise in the area, nor does it account for the attenuation, reflection, absorption, scattering, or multipath that may be occurring as the target's signal propagates throughout the area. The antenna orientation of a mobile client station will also affect the RSSI measurement. The accuracy of RSSI measurement is affected by distance as well. The doubling of the distance between a sensor and a device tends to double the inaccuracy of the measurement. Therefore, RF triangulation is usually used as an approximation. In most cases, an approximation of 10 meters is accurate enough to locate a device physically.

Will Location Tracking Improve in the Future?

In the past, some vendors used proprietary software mechanisms on client stations to measure RSSI from access points continuously and to report this measurement back to a location tracking module of a WIDS/WIPS server. Because more statistical data is collected from the clients and not just APs or sensors, the accuracy of tracking authorized clients is enhanced. The ratified 802.11k-2008 amendment defines standards for *radio resource measurement (RRM)*. WLAN radio resource measurements enable STAs to understand the radio environment in which they exist by measuring RSSI, signal noise, and other statistics. Radio measurement data can be made available to upper protocol layers, where it may be used by other applications, including location tracking services of a WIPS. One optional measurement defined by 802.11k is *location configuration information (LCI)*, which can be used by an 802.11 client device that might also have GPS capabilities. LCI measurements can be used to report latitude, longitude, and altitude. Although 802.11k mechanisms still have the potential to improve location tracking accuracy of Wi-Fi clients, support for LCI mechanisms on Wi-Fi clients has yet to happen. In reality, triangulation or trilateration will never be as accurate as real-time location-tracking systems. At best, the estimated locations may become more accurate but they cannot compare to an active RFID tag on the device being tracked that helps you find it.

A more complex yet more accurate method of device location tracking is *RF fingerprinting*. RF fingerprinting also uses the RSSI values of transmitting devices as detected by sensors. However, RF fingerprinting does not rely on these values alone. RF fingerprinting also uses the RSSI values of devices whose locations are known as points of comparison. Rather than just listening to the target device's signal and estimating the target's location, this method compares the target's detected RSSI values with the RSSI values of the known reference points. The idea here is that if the target and the known reference point have similar RSSI values, they must be close to each other. RF fingerprinting relies on building up a database of actual measurements and relates them to particular locations. This method can reduce the size of the search pattern from 10 meters down to about 1 or 2 meters and may be deployed using fewer sensors. However, RF fingerprinting can also be time-consuming and expensive.

The method used to document RSSI reference points on a map is known as *RF calibration*. The calibration is often done by an administrator using an 802.11 device (usually a laptop) moving throughout the area and taking many samples of RSSI data to be imported into the RF fingerprinting engine. If anything in the area changes the RF environment—such as new walls, neighboring devices, new interference sources, or the removal of these things—the RF fingerprinting will not function correctly until the engine can be recalibrated to include these changes in the RF environment. Recalibration is often overlooked by WLAN administrators who are busy with other tasks. The result of using a poorly calibrated RF Fingerprinting engine to track devices may be less accurate than just using RSSI values and RF triangulation. Although the RF fingerprinting method is more accurate than triangulation alone, it is more costly to implement and more labor intensive to maintain. The 8 or 9 meter accuracy improvement that RF fingerprinting provides is useful for locating rogue APs and for tracking Wi-Fi RFID tags. Several companies such as AeroScout and Ekahau provide WLAN *real-time location systems (RTLSs)* that use RF fingerprinting methods. Several of the WLAN vendors and WIDS/WIPS vendors partner with RTLS vendors to take advantage of the increased accuracy provided by RF fingerprinting.

Time difference of arrival (TDoA) is another method that can be used for location tracking. As shown in Figure 14.20, TDoA uses the variation of arrival times of the same transmitted signal at three or more receivers—in our case, sensors. The transmitted signal will arrive at the sensors at different times due to the distance between the transmitter and the multiple sensors. The speed of travel of the radio frequency is a known factor, and each of the synchronized TDoA sensors reports the time of arrival of the signal from the transmitting device. In theory, if the transmitter were exactly at the midpoint between sensors, there would be no difference of arrival times, thus making the target easily located in the center of sensor coverage, equally distanced from all sensors. Ordinarily, the sensor that receives the signal first is deemed closer to the source of the transmission and the one receiving the signal last should be the farthest away from the source. The TDoA is also used in determining the *angle of arrival (AoA)* of a signal in an antenna array. TDoA does not require the calibration of RF fingerprinting, nor does it use the RSSI values as triangulation methods. TDoA simply uses the time stamps of the signal's arrival at various known locations. Some WLAN vendors are using TDoA technology to assist in tracking Wi-Fi RFID tags. Wi-Fi TDoA sensors are typically used for location tracking in line-of-sight environments, such as outdoor deployments or warehouses with high ceilings.

FIGURE 14.20 Time difference of arrival (TDoA)

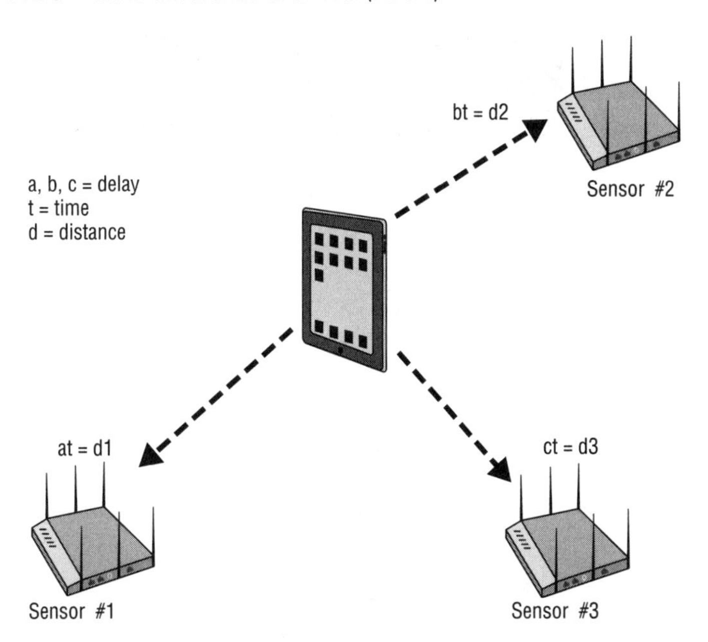

It is possible to conduct device tracking with a single mobile receiver taking multiple readings in several locations by hand using a laptop or handheld device. However, the speed at which distributed systems are able to accomplish this same task makes location tracking an important and reliable part of WLAN security monitoring. Using a single mobile device to supplement location tracking is a good practice for "last mile" discovery. As shown in Figure 14.21, mobile devices usually use a graph, needle, and/or sound to indicate proximity to the target device. Using the WIDS/WIPS to approximate the location and then following through by using the portable device is a common method of locating rogues and other devices in a WLAN enterprise. No matter which method you choose, you will most likely encounter a situation that requires finding the physical location of a device.

FIGURE 14.21 Mobile tracking tool

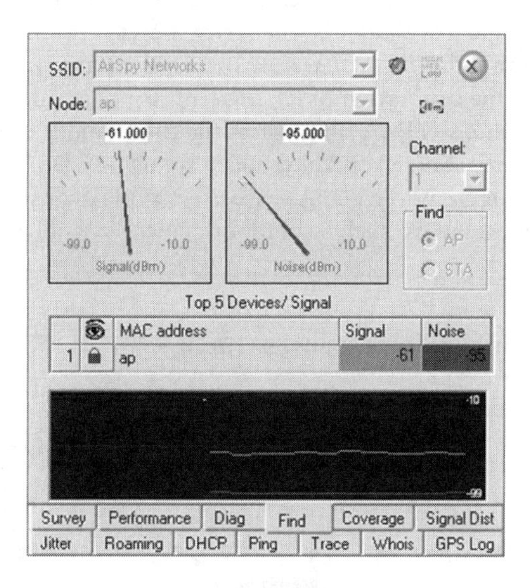

WIDS/WIPS Analysis

As you have learned, WLAN security monitoring using a distributed WIDS/WIPS solution is capable of collecting information continuously. Because the information gathered from multiple sensors can be extensive, the task of analyzing all the collected data can be overwhelming. Every WIDS/WIPS solution uses a variety of software modules or software engines to simplify the task of analyzing massive amounts of collected data.

Signature Analysis

WIDS/WIPS solutions use *signature analysis* to analyze frame patterns or "signatures" of known wireless intrusions and WLAN attacks. The WIDS/WIPS has a programmed database of hundreds of threat signatures of known WLAN attacks. As shown in Figure 14.22, threat signatures can include man-in-the-middle attacks, DoS attacks, flood attacks, and many more. WIDS/WIPS signatures are based on Layer 1 and Layer 2 attacks. The WIDS/WIPS uses some sort of signature analysis engine that processes 802.11 frames and RF data. Automatic signature learning systems require extensive logging of complex network activity and historic data mining that can impact performance. Most WIDS/WIPS solutions use manual signature detection, which is comparable to most virus protection systems, where the signature database is updated automatically as new signatures are discovered. WIDS/WIPS vendors are constantly updating their signature databases as new attacks emerge. Usually, the systems also have the capability of creating custom signatures. Custom signatures are useful to WLAN administrators who want to monitor for a behavior or attack that could be specific to their WLAN environment.

FIGURE 14.22 Attack signatures

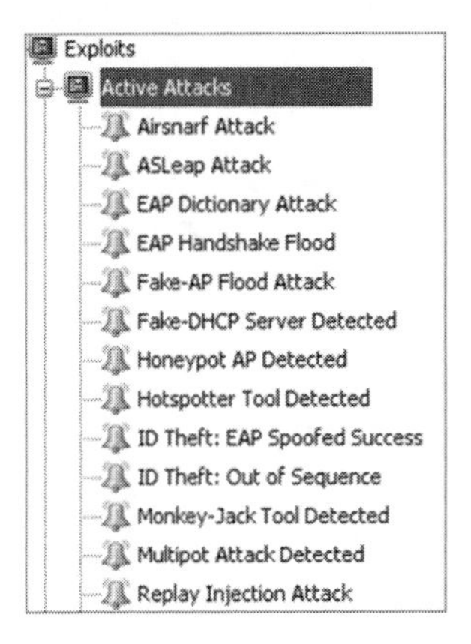

Behavioral Analysis

Many WIDS/WIPS solutions also use *behavioral analysis* to recognize any patterns that deviate from normal WLAN activity. Behavioral analysis identifies abnormal network behavior based on historical metrics. Because historical normal WLAN behavior is the baseline, anomalies can be detected that would not necessarily be discovered by other intrusion detection techniques. Whereas the signature analysis identifies known threats, the anomalous behavior analysis recognizes new, unknown attacks or threats that have no signature.

Detection of anomalies can be based on various thresholds of 802.11 management, control and data frames, fragmentation thresholds, and many other variables. Behavior analysis helps detect *protocol fuzzing*, where an attacker sends malformed input to look for bugs and programming flaws in AP code or client station firmware. Attackers transmit malformed data by tampering with bits and fields of data in 802.11 frames. Protocol fuzzing attacks often identify driver vulnerabilities and find weaknesses that result in a buffer overflow attack.

As you have learned, known attacks can be easily identified by signature analysis. However, the greatest threat to WLAN security is a new attack that is not known and cannot be detected.

An unknown threat used to exploit computer networks is referred to as a *zero day attack*. Very often, a zero day attack will create some sort of anomaly in 802.11 behavior that can be detected. As shown in Figure 14.23, behavior thresholds can be configured on the WIDS/WIPS that can trigger an alarm. When setting thresholds, you may find it difficult and time-consuming to achieve a balance in detecting possible zero day attacks versus triggering false positive alarms.

FIGURE 14.23 Behavior thresholds

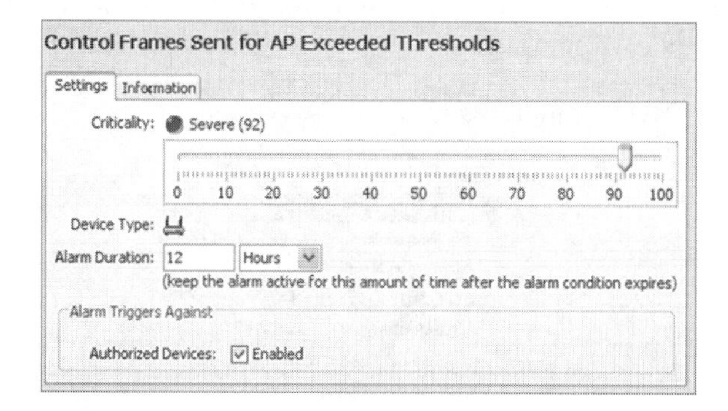

Protocol Analysis

All WIDS/WIPS vendors use *protocol analysis* to dissect the MAC layer information from 802.11 frames. Protocol analysis may also be used to analyze Layer 3–7 information of 802.11 data frames that are not encrypted. In Chapter 13, "Wireless LAN Security Auditing," you learned that standalone WLAN protocol analyzers are usually installed on a laptop and are used for security audits. Standalone WLAN protocol analyzers are also used for Layer 2 troubleshooting and WLAN performance analysis. The Wireshark software you have been using in the book's exercises is a standalone WLAN protocol analyzer. All standalone WLAN protocol analyzers have some security analysis capabilities, but some place a greater emphasis on security auditing and can effectively be used as a *mobile wireless intrusion detection system* solution. One example is Fluke Networks' AirMagnet Wi-Fi Analyzer. Although this approach is rarely used, mobile WIDS analyzers can also be integrated with a distributed WIDS/WIPS, effectively acting as software-based sensors.

An enterprise WIDS/WIPS solution provides for distributed protocol analysis, with each hardware sensor acting as a listening device. Distributed protocol analysis is mainly used to monitor all the 802.11 frame exchanges that occur at Layer 2. That information can be leveraged for both security reasons and for WLAN performance analysis. Attacks can be detected at Layer 2 by reading the headers and trailers of all of the frames captured with a distributed protocol analyzer.

An in-depth discussion of 802.11 protocol analysis is beyond the scope of this book; however, a quick discussion of the 802.11 frame format is necessary. The technical name for an 802.11 data frame is a *MAC Protocol Data Unit (MPDU)*. The 802.11 frame, as shown in Figure 14.24, contains a Layer 2 MAC header, a frame body, and a trailer, which is a 32-bit CRC known as the *frame check sequence (FCS)*. The Layer 2 header contains MAC addresses and the duration value. The frame body contains the *MAC Service Data Unit (MSDU)*, which is the Layer 3–7 payload.

FIGURE 14.24 MAC Protocol Data Unit

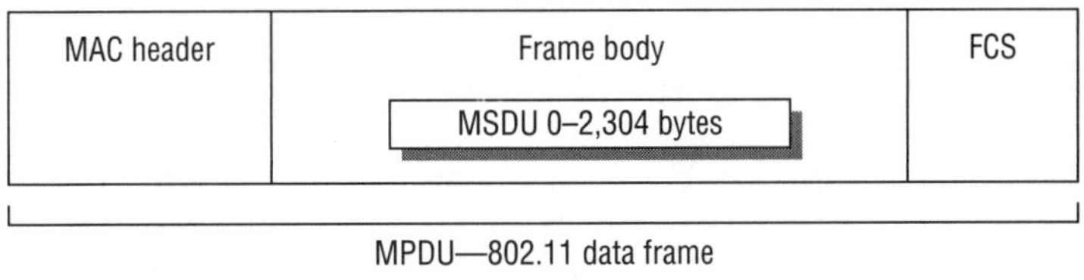

MAC header	Frame body MSDU 0–2,304 bytes	FCS

MPDU—802.11 data frame

The IEEE 802.11-2012 standard defines three major frame types:

Management 802.11 management frames are used by wireless stations to join and leave the basic service set (BSS). Another name for an 802.11 management frame is a *Management MAC Protocol Data Unit (MMPDU)*. Management frames do not carry any upper-layer information. There is no MSDU encapsulated in the MMPDU frame body, which carries only Layer 2 information fields and information elements. Information fields are fixed-length mandatory fields in the body of a management frame. Information elements are variable in length and are optional. Examples of management frames are beacons, probe requests, and association request frames.

Control 802.11 control frames assist with the delivery of the data frames. Control frames must be able to be heard by all stations; therefore, they must be transmitted at one of the basic rates. Control frames are also used to clear the channel, acquire the channel, and provide unicast frame acknowledgments. They contain only header information. Control frames do not have a frame body. Examples of control frames are acknowledgments, request-to-send, and clear-to-send frames.

Data Most 802.11 data frames carry the actual data that is passed down from the higher-layer protocols. The Layer 3–7 MSDU payload is normally encrypted for data privacy reasons.

Most WIDS/WIPS have the capability to monitor 802.11 frame exchanges in real time just like a standalone WLAN protocol analyzer. Enterprise WIDS/WIPS also usually have the ability to use sensors and APs for *remote packet capture*. As shown in Figure 14.25, an individual sensor is configured to capture on a single channel and then mirror the captured 802.11 traffic to a remote IP address.

FIGURE 14.25 Remote packet capture

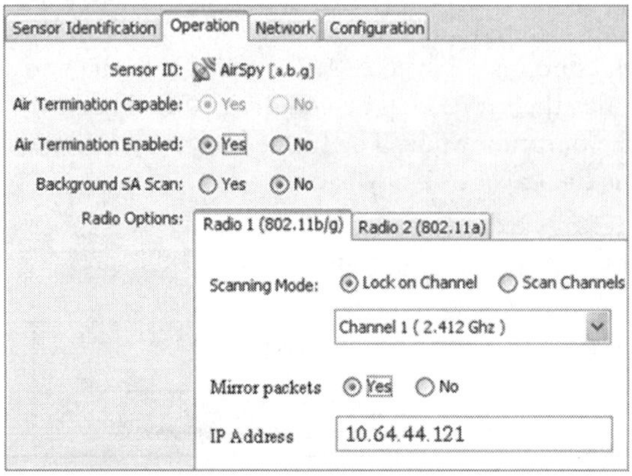

Spectrum Analysis

Always remember that 802.11 devices operate at both Layers 1 and 2. The Layer 1 physical medium is the uncontrolled, unlicensed, and unbounded RF spectrum. Traditionally, WIDS/WIPS solutions have mostly been used strictly to monitor Layer 2 communications and have mostly ignored Layer 1 for security monitoring. As you learned in Chapter 12, DoS attacks can occur at Layer 1. Any continuous transmitter will cause a DoS. RF jamming devices can be used by an attacker to cause an intentional Layer 1 DoS attack. Denial of service at Layer 1 usually occurs as an unintentional result of transmissions from non-802.11 devices. Video cameras, baby monitors, cordless phones, and microwave ovens are all potential sources of interference. Unintentional interference may cause a continuous DoS, but the disruption of service is often sporadic. This disruption of service will upset the performance of Wi-Fi networks used for data applications but can completely disrupt VoWiFi communications within a WLAN. At the very least, unintentional interference will result in retransmissions that negatively affect WLAN performance. The majority of unintentional interfering devices transmit in the 2.4 GHz ISM frequency band. The 5 GHz UNII bands are less susceptible to unintentional interference. RF interference often is undetected by traditional WIDS/WIPS sensors and cannot be properly classified.

Although wireless intrusion prevention systems are outstanding products that can mitigate most rogue attacks, some rogue devices will go undetected. The radios inside the WIPS sensors typically monitor the 2.4 GHz ISM band and the 5 GHz UNII frequencies. Older legacy wireless networking equipment exists that transmits in the 900 MHz ISM band, and these devices will not be detected. The radios inside the WIPS sensors use only *direct sequencing spread spectrum (DSSS)* and *orthogonal frequency division multiplexing (OFDM)* technologies. Wireless networking equipment exists that uses *frequency hopping spread spectrum (FHSS)* transmissions in the 2.4 GHz ISM band and will go undetected by traditional WIDS/WIPS sensors. The only tool that will detect with 100-percent certainty either a 900 MHz or a frequency hopping rogue access point is a spectrum analyzer capable of operating in those frequencies.

A spectrum analyzer is a frequency domain measurement and troubleshooting tool. A spectrum analyzer can help identify and locate an interfering transmitter. Spectrum analyzer hardware can cost upward of $40,000, thereby making them cost-prohibitive for many small and medium-sized businesses. The good news is that several companies have standalone solutions, both hardware and software based, that are designed specifically for 802.11 spectrum analysis and are drastically less expensive. As shown in Figure 14.26, WLAN spectrum analysis is most often achieved with a standalone software-based solution installed on a laptop that works with special USB adapters that use a spectrum analyzer chipset.

FIGURE 14.26 WLAN spectrum analyzer

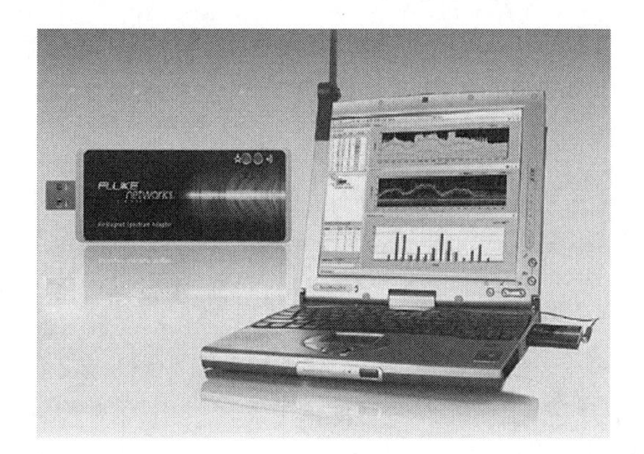

One of the most important capabilities of a spectrum analyzer is the ability not only to detect RF energy but also to classify the sources of the interference. The better spectrum analyzers use *RF signature analysis* to identify and classify interfering RF transmitters, such as Bluetooth, microwave ovens, wireless cameras, jammers, and so on.

In the past, a major oversight in many WIDS/WIPS solutions was that they were unable to detect previously discussed Layer 1 security threats. However, in recent years, enterprise WIPS have begun to operate as *distributed spectrum analysis systems (DSASs)*. The advantage of any distributed solution is that they run 24/7 and can be administered remotely. Most DSAS solutions use access points for the distributed spectrum analysis. Some vendor APs use an integrated spectrum analyzer that operates independently from the 802.11 radio. Other vendor APs use the 802.11 radio to accomplish a lower grade of spectrum analysis. A good DSAS is capable of RF signature analysis and can also physically pinpoint sources of RF interference using the location-tracking capabilities discussed earlier in this chapter.

Forensic Analysis

Enterprise WIDS/WIPS solutions may also provide *forensic analysis* that allows an administrator to retrace the actions of any single WLAN device down to the minute.

With forensic analysis, investigating an event takes minutes instead of potentially hours. Administrators can rewind and review minute-by-minute records of connectivity and communication within a WLAN. As shown in Figure 14.27, the WIPS records and stores hundreds of data points per WLAN device, per connection, per minute. This allows an organization to view months of historical data on any suspicious WLAN device as well as all authorized devices. Information such as channel activity, signal characteristics, device activity, and traffic flow and attacks can all be viewed historically.

FIGURE 14.27 Forensic analysis

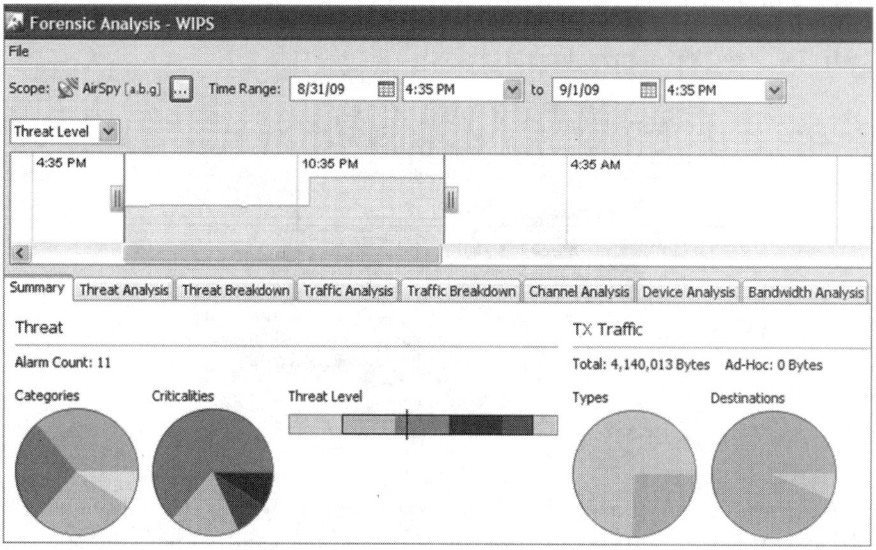

Performance Analysis

A very useful by-product of WIDS/WIPS deployment is that you are able to collect a large amount of data for analysis for performance monitoring. Although the main purpose of an enterprise WIDS/WIPS is security monitoring, information collected by the WIPS can also be used for *performance analysis*. Since everything WLAN devices transmit is visible to the sensors, the Layer 2 information gathered can be used to determine the performance level of a WLAN. The WIPS can detect hidden nodes, excessive Layer 2 retransmissions, excessive wired to wireless traffic, excessive roaming, and many other events and traffic types that lower the capacity performance of a WLAN.

Knowing how the WLAN functions on a regular basis can help reduce problem-solving time greatly. Performance analysis can be used to define performance baselines used to establish expected performance standards or levels. Baselining involves determining how the WLAN is functioning in terms of performance. An administrator must take several samples of traffic at various times. These samples should be taken at both peak and off-peak times and over a long enough period to capture adequately an idea of what is normal for the WLAN. The baseline should include normal use conditions as well as peak and off-peak captures. Using captures from longer periods will allow you to better understand normal use of the network and have a more accurate baseline. Since the WIDS/WIPS is collecting information all day, every day, the sampling for a baseline is more accurate and more inclusive. Once a baseline has been established for the WIDS/WIPS, performance thresholds can be configured to send alerts when performance drops to detrimental levels. Performance monitoring allows an administrator to act on potential network issues before the users notice any performance problems. Performance monitoring also aids in the planning for any necessary expansion of the WLAN. If an administrator knows how well the WLAN functions with the current number of APs and stations running the current applications, the impact of adding more users and devices can be more accurately predicted.

Monitoring

With the vast amount of data that can be collected, it is of great importance to have the WIDS/WIPS properly tuned for your environment. Alarm policies can be configured to define thresholds for security and performance. Custom alarm threat thresholds can be configured to match the WLAN's security and performance requirements. Alarm notifications can also be triggered to alert you about attacks. You can then evaluate the alarms and take the proper actions.

Policy Enforcement

Enterprise WIDS/WIPS solutions allow administrators to define, monitor, and enforce wireless LAN policies in the areas of security, performance, usage, and vendor types. Organizations can minimize vulnerability by ensuring that WLAN devices are using the proper security protocols. Improper configuration of WLAN devices is one of the most common causes for wireless security breaches.

Security policies must be defined to set thresholds for acceptable network operations and performance. As shown in Figure 14.28, you can define a security policy that requires that all client stations use an 802.1X/PEAP solution for authentication and CCMP/AES for encryption. If an end user configures a client station that is not using PEAP and CCMP, the WIPS will generate a policy-based alarm. Defining security policies ensures that all devices are properly configured with the mandated level of protection.

Security policies need to be set for both access point and client station configuration thresholds. Policies should be defined for authorized APs and their respective configuration parameters, such as Vendor ID, authentication modes, and allowed encryption modes. Define allowable channels of operation and normal activity hours of operation for each AP. Performance thresholds can also be defined for minimum signal strength from a client station associating with an AP to identify potential attacks from outside the building.

FIGURE 14.28 Security policy

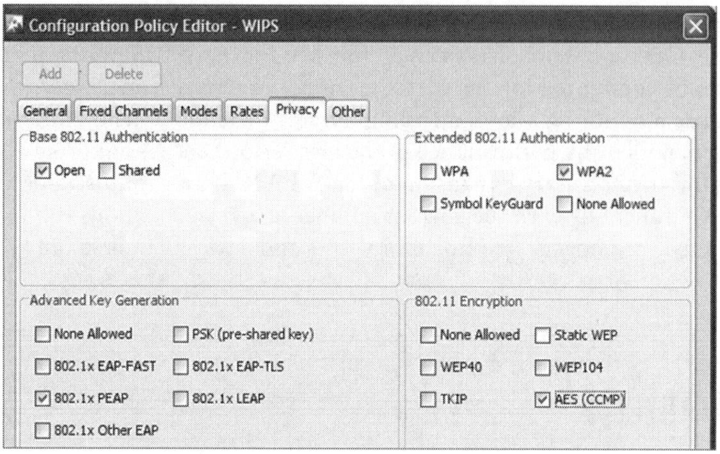

The defined security policies form the baseline for how the WLAN should operate. The thresholds and configuration parameters should be adjusted over time to tighten or loosen the security baseline to meet real-world requirements. For example, normal activity hours for a particular AP could be scaled back due to working hour changes. The security policy should also be changed to reflect the new hours of operation. No one security policy fits all environments or situations. There are always trade-offs between security and usability.

The WIPS can also be used for policy enforcement. Once again, you can define a security policy that requires all client stations use an 802.1X/PEAP solution for authentication and CCMP/AES for encryption. If an end user configures a client station that is not using PEAP and CCMP, the WIPS will generate a policy-based alarm. However, the security policy can also be set to trigger an automatic response in addition to the alarm. The WIPS can use spoofed deauthentication frames against the misconfigured client, similar to rogue containment measures. In this way, authorized devices are not able to place their traffic at risk by communicating without the use of required authentication or encryption methods. A policy is only as good as its ability to be enforced uniformly. An alarm will alert you to an unsecure environment or device but does not take steps to enforce the policy. A WIPS offers the additional protection of preventing devices from communicating outside of policy by terminating noncompliant devices connections. This approach does have the potential to disrupt business and should be properly weighed against potential security problems when making the decision to terminate noncompliant connections. Many users of WIPS in larger enterprise deployments with 24/7 staffing prefer to receive notifications of noncompliant device communications and manually remediate the problem to avoid business interruption.

Prior to implementing actions to enforce policy, it is of great importance that any written organizational security policy be consulted, followed, and or updated and required. Policy violation reporting and policy enforcement may be dictated by outside organizations based on industry and governmental regulations. You can find a more detailed discussion about policies and regulations in Chapter 15, "Wireless Security Policies."

Alarms and Notification

In a congested WLAN environment or in an area with very little traffic, any 802.11-based device that transmits a signal can be heard by the WIDS/WIPS sensors. The WIDS/WIPS will detect all 802.11 transmissions and then, if necessary, generate the appropriate alarms. Depending on the configured threshold, the alarms can be triggered by signature analysis, spectrum analysis, behavioral analysis, or performance analysis. Alarms can also be policy based, as we discussed in the previous section. Practical questions then arise once the alarms have been triggered:

- What do I do with all of this information?
- What do these alarms mean?
- Do I need to be informed about every device detected?
- Who is going to respond to the alarms?
- Am I under attack?
- Is this normal or acceptable behavior?
- Are any or all of these detected devices mine?
- Are my devices safe?

As shown in Figure 14.29, the triggered alarms will often have a detailed description of the attack or performance problem. The WIPS alarm may also have suggested mitigation actions. The detailed description and recommended actions will often help you answer the questions we just listed.

As we discussed earlier, WIDS and WIPS solutions are able to discover and classify devices as well as conduct behavioral analysis. Event alerts or alarms are used to indicate that a device or particular behavior has been detected by the system. Different behaviors will trigger different alarms. If a user turns on a new client device within hearing range of a sensor, an unauthorized device alarm will be triggered. If that user then connects to an authorized AP without their new client device first being authorized, a rogue station alarm will be triggered. What happens beyond that depends on the vendor of the WIDS/WIPS and the customization done to the system. A WIPS can proactively begin to protect the network using rogue station containment.

FIGURE 14.29 WIDS/WIPS alarm

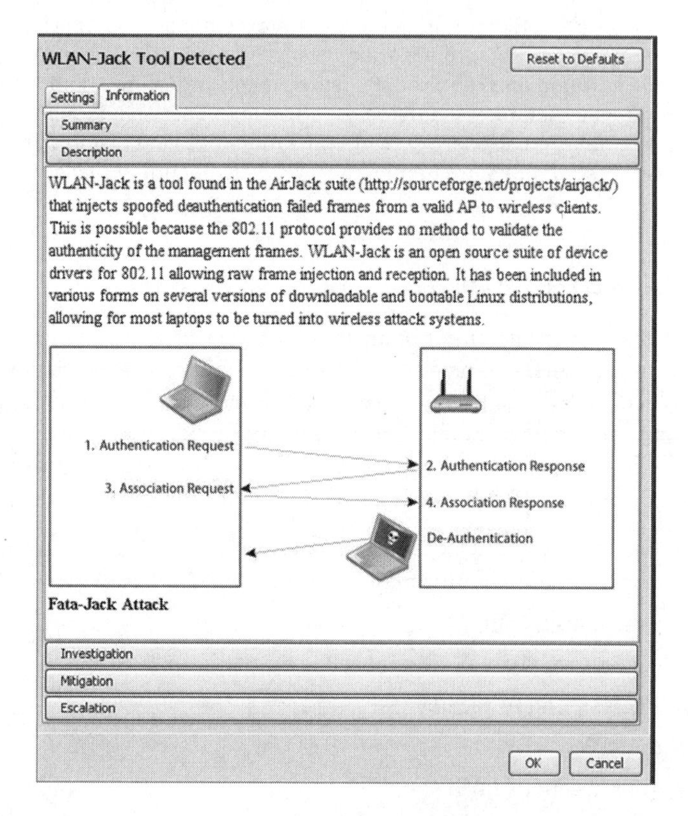

Users, contractors, and visitors often have the credentials to connect and do so using their own devices. This will trigger alarms within the WIDS/WIPS based on the classification of their device as unauthorized and the behavior analysis of the device showing it as associated with an authorized AP. Any device detected and any behavior detected can be used to trigger alarms and possible automatic responses.

Alerts or alarms can be classified just as devices can be classified. Events that trigger alarms are not always indications of security threats or vulnerabilities. Some events are normal behavior, such as protection mode in use by an 802.11g access point or including its SSID in the beacon frames it transmits. The alarms can be broken into several categories:

- Behavioral
- Exploits
- Performance
- Policy compliance
- Reconnaissance
- Rogue activity
- Performance
- Vulnerabilities

WIDS/WIPS alarms are usually set to a threat-criticality level the vendor has determined to be what most users want. However, you may wish to make some alarms more or less important than the preset threat levels. Within the categories, alarms or alerts can be given custom threat levels from "everything is fine" to "we are under attack." These levels include the following:

Safe No immediate threat

Minor Potential problem alarms that may worsen if ignored

Major Potentially serious alarms that require priority attention

Critical Serious alarms that require immediate attention

Severe Serious alarms that may have catastrophic effects

Tuning the alerts or alarms to threat levels is an important and possibly time-consuming task when deploying a WIDS/WIPS. However, spending the time up front to calibrate alarm thresholds properly will make the alerts more meaningful. The intended use of a WIDS/WIPS along with vendor-specific options will dictate how you should tune the alarm thresholds. Alarms can be disabled or have their threat levels lowered if they are not of importance.

Keeping a record of everything that is detected is a sound practice for forensic or even legal reasons. However, you may not want to receive a notification about everything the system detects. An alarm can be configured to trigger notifications, which can be sent from the system in several forms:

- Email
- SMS
- Syslog

The notifications should be configured to trigger only if an alarm is of specific importance to you. You would want to be alerted to the fact that a rogue is on the network or an attack is occurring. However, you would not want to know every time a WLAN device is detected unless a no-wireless zone is being enforced. Typically, alarms with a threat level of critical or severe are also configured for notification. It should be noted that any AP or client that is initially classified as unauthorized should still be investigated as soon as time permits.

False Positives

The physical radio frequency medium used for 802.11 communications is both harsh and unpredictable. RF behaviors, such as reflections and multipath, often create a hostile environment that results in corrupted 802.11 frames and Layer 2 retransmissions. WLAN problems, such as adjacent cell interference, low SNR, and hidden nodes, can also lead to corrupted data frames. Because the RF environment is at worst unstable and at best fluctuating, not every WIDS/WIPS alarm is going to be perfectly accurate. In other words, a certain number of false positive alarms are to be expected. A *false positive*, also known as a false detection or false alarm, is a result that is erroneously positive when the situation is actually normal. A false positive is simply another way of saying "mistake." All intrusion detection systems, both wired and wireless, will have some occurrence of false positives. A false positive WIDS/WIPS alarm indicates that a WLAN attack is occurring when in fact the threat does not exist. False positive alarms can be time consuming for you to verify or invalidate. Even worse, false positives are often ignored due to their volume, thus increasing the possibility that a real attack alarm will also be ignored.

Corrupted frames are the leading cause of false positives. However, improper configuration of the WIPS or misinterpretation of alarms by administrators can also lead to false positive alarms. Proper classification of all devices as either authorized devices or neighbor devices is important. If devices are not properly classified, many alarms will be triggered. As mentioned earlier in this chapter, setting alarm thresholds can often be difficult and it may be time consuming to achieve a balance in detecting possible attacks versus triggering false positive alarms. Some inaccurate reporting of events, seen as false positives, are unavoidable. However, properly setting alarm thresholds that are fine-tuned to the on-site RF environment can greatly reduce the number of false positives. A reduction in false positives will save time and improve the security of the WLAN.

Reports

Enterprise WIDS/WIPS solutions usually offer extensive report generation capabilities. Reports can be created manually or can be scheduled to be created automatically. Viewing and saving these reports is part of maintaining a secure and healthy wireless environment. When properly created, reports are useful tools in problem analysis and resolution. Security issues that may require disciplinary and or legal action are better supported with documentation. The reports can also be used to validate expenditure issues with upper management with regard to network security and development requirements.

The report software engine of the WIDS/WIPS will have many predefined reports that cover compliance, security, and performance. Additionally, the WIPS administrator will usually have the option to create custom reports that address the individual needs of a specific WLAN that is being monitored.

Examples of two WIPS-generated reports are available for download from the book's website: `www.sybex.com/go/cwsp2e`. The reports are titled `Security_Rogue_Detail_report.pdf` and `Vulnerability_Assessment_Report.pdf`.

802.11n/ac

The 802.11n-2009 amendment introduced the use of *High Throughput (HT)* radios that use both PHY and MAC layer enhancements to achieve these high data rates. The amendment introduced the use of multiple-input and multiple-output (MIMO) radios that use both PHY and MAC layer enhancements to achieve much higher data rates than legacy 802.11a/b/g single-input and single-output (SISO) radios. The 802.11ac *Very High Throughput (VHT)* amendment tweaked and defined more PHY and MAC enhancements so that 5 GHz MIMO radios could achieve even higher data rates than 802.11n radios. Does a WLAN administrator need to update WIPS sensors to use MIMO radios as well? The answer depends on the level of protection and monitoring that is required.

The biggest concern is always rogue access points. The good news is that rogue 802.11n and rogue 802.11ac APs will still transmit beacons at data rates that can be decoded by legacy 802.11a/b/g WIPS sensors. This means that most rogue detection methods should still work. However, 802.11n introduced an HT Greenfield mode that uses a PHY header that only 802.11n radios can understand. A legacy 802.11a/b/g sensor will not be able to decipher any transmissions using the HT Greenfield frame format. An attacker could potentially install a rogue 802.11n AP that is only transmitting using the HT Greenfield frame format. The HT Greenfield PHY header cannot be detected by a WIPS that is using legacy 802.11a/g sensors. HT Greenfield mode was never really used in the enterprise and has been dropped in 802.11ac.

802.11n HT radios also have the ability to transmit on 40 MHz OFDM channels. As shown in Figure 14.30, the 40 MHz channels used by HT radios are essentially two 20 MHz OFDM channels that are bonded together. Each 40 MHz channel consists of a primary and a secondary 20 MHz channel. 802.11ac VHT radios have the ability to transmit on 20 MHz, 40 MHz, and 80 MHz channels. The good news is that the majority of 802.11 management frames, such as beacons, are transmitted on the primary 20 MHz channel so that legacy 802.11a/b/g radios can also communicate with 802.11n/ac radios. However, the legacy radios will not be able to understand 40 MHz or 80 MHz transmissions. Any Layer 2 denial-of-service (DoS) attacks on the larger channels would go undetected by the older SISO radios. As new technologies are introduced, new attacks always follow. In an ideal world, your WIPS sensors should be updated when newer 802.11 radio technologies are introduced.

FIGURE 14.30 Channel bonding

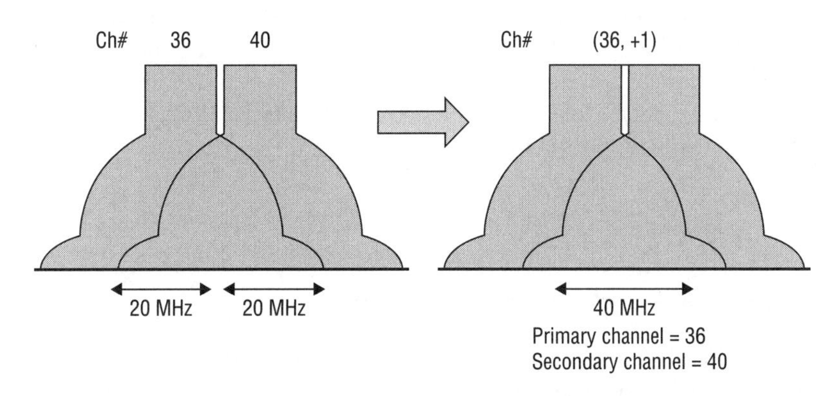

802.11w

The IEEE ratified the 802.11w-2009 amendment defines management frame protection for the prevention of Layer 2 DoS attacks.

Disassociation frames can be sent by either an AP or a client station. Disassociation is a notification, not a request. In the past, disassociation could not be refused by the receiving station. 802.11w allows the receiving STA to refuse disassociation when *management frame protection (MFP)* is negotiated and the message integrity check fails. Deauthentication frames can also be sent by either an AP or a client station, and deauthentication is a notification, not a request. In the past, deauthentication could not be refused by the receiving station. 802.11w allows the receiving STA to refuse deauthentication when management frame protection is negotiated and the message integrity check fails.

The 802.11w-2009 amendment defines a *robust management frame* as a management frame that can be protected by the management frame protection service. The robust management frames include robust action frames, disassociation frames, and deauthentication frames. The majority of action frames are considered robust, including QoS action frames, radio measurement action frames, block ACKs, and many more.

As you learned in Chapter 3, "Encryption Ciphers and Methods," TKIP/RC4 and CCMP/AES provide protection against replay attacks against 802.11 data frames. However, only CCMP/AES is supported for protection of robust management frames. Replay protection is provided for robust management frames for STAs that use CCMP and *Broadcast/Multicast Integrity Protocol (BIP)*. BIP provides message integrity and access control for group-addressed robust management frames. 802.11w should help put an end to deauthentication attacks and disassociation attacks as well as Layer 2 DoS attacks that involve action frames. Although 802.11w does address these two popular DoS attacks, 802.11w will not address many other Layer 2 DoS attacks.

Numerous Layer 2 DoS attacks exist, including association floods, EAPOL floods, PS-Poll floods, and virtual carrier attacks. 802.11w does not address these types of Layer 2 DoS attacks. Luckily, any good wireless intrusion detection system will be able to alert an administrator immediately to a Layer 2 DoS attack. Please understand that 802.11w compliance will eventually prevent some of the popular Layer 2 DoS attacks that currently exist, but it is doubtful that all Layer 2 DoS attacks will ever be circumvented.

As of this writing, there currently is very little support for 802.11w management frame protection in 802.11 clients. Most of the WLAN vendors support 802.11w mechanisms in their APs, but client-side support is still scarce.

Real World Scenario

How Does 802.11w Affect Rogue Mitigation?

As you know, most WIPS solutions use a wireless rogue containment that uses spoofed deauthentication frames. Rogue containment is accomplished wirelessly when the WIPS's sensors become active and begin transmitting deauthentication frames that spoof the MAC addresses of the rogue access points, rogue ad hoc networks, and rogue clients. The

WIPS is using a known Layer 2 DoS attack as a countermeasure. 802.11w defines mechanisms that protect deauthentication frames. Therefore, wireless rogue containment based on management frames protected by 802.11w will not work against 802.11w rogue devices. As 802.11w-compliant devices become more widely available, WIPS vendors will have to rely on other methods for rogue mitigation and prevention. Port suppression will become a more important method of rogue AP deterrence. Additionally, use of 802.1X/EAP for switch port security is becoming more important for rogue AP prevention.

Other WIPS vendors may rely more on rogue *tarpitting* methods. Tarpitting is a method used to get rogue devices to associate to a WIPS sensor. The rogue client is kept busy by the sensor and is stuck in a wireless tar pit where the rogue client cannot do any damage. WIPS vendors will most likely also develop new wireless rogue containment methods based on management frames that are not protected by 802.11w. Multiple Layer 2 DoS attacks are not protected by 802.11w, which might also be used as countermeasures against rogue devices.

Summary

As part of maintaining a secure and healthy WLAN environment, regular security monitoring and auditing is a requirement. Steps should be taken to ensure that all the devices on WLANs are legitimate and are operating within written security guidelines. Both Layer 1 and Layer 2 should be monitored as part of a complete WLAN security monitoring solution. Distributed WIDS monitoring offers numerous detection and analysis capabilities. Distributed WIPS monitoring provides protection against rogue devices as well as enforcement of predefined security policies.

Exam Essentials

Explain the difference between a WIDS and a WIPS. WIDS solutions offer monitoring and analysis capabilities, whereas WIPS solutions can also offer mitigation against some attacks.

Define the components of a WIDS/WIPS architecture. Understand the distributed architecture that includes a server, sensors, and management consoles. Mobile monitoring devices can also be integrated into a distributed architecture.

Explain the difference between an overlay, integrated, and integration-enabled WIDS/WIPS. An overall WIPS is installed on top of a preexisting WLAN. An integrated solution uses the existing WLAN infrastructure. Integration-enabled is a hybrid of the other two solutions.

Define the various types of sensors. Sensors can be standalone devices or access points that operate as either a part-time sensor or a full-time sensor. Sensors are usually hardware based and can use either a single radio or multiple radios.

Understand the importance of device classification. Explain the difference between authorized, unauthorized, rogue, and neighbor devices.

Explain rogue detection and mitigation. Be able to explain the multiple ways that a WIPS can determine whether a rogue device is connected to the wired infrastructure. Understand how rogue devices are mitigated using either wireless rogue containment or wired-side port suppression.

Describe WIDS/WIPS analysis. Outline the differences between signature, behavioral, protocol, spectrum, forensic, and performance analysis.

Understand WIDS/WIPS monitoring capabilities. Explain the importance of alarm thresholds and policy enforcement. Understand that false positives are a reality.

Explain management frame protection. Explain 802.11w management frame protection, but understand that not all Layer 2 DoS attacks will be prevented.

Review Questions

1. Which of these tools will best detect frequency hopping rogue devices? (Choose all that apply.)
   - **A.** Standalone spectrum analyzer
   - **B.** DSAS
   - **C.** Distributed Layer 2 WIPS
   - **D.** Mobile Layer 2 WIPS
   - **E.** Layer 2 WIPS
2. Name the labels that a WIPS uses to classify an 802.11 device. (Choose all that apply.)
   - **A.** Authorized
   - **B.** Unauthorized
   - **C.** Enabled
   - **D.** Disabled
   - **E.** Rogue
   - **F.** Neighbor
3. What are some of the different types and form factors of WIPS servers? (Choose all that apply.)
   - **A.** WLAN controller
   - **B.** Cloud network management server
   - **C.** Standalone appliance
   - **D.** On-premises network management server
   - **E.** Virtual appliance
   - **F.** All of the above
4. A WIDS/WIPS consists of which of the following components? (Choose all that apply.)
   - **A.** WIDS/WIPS server
   - **B.** Midspan injector
   - **C.** Sensors
   - **D.** MDM server
   - **E.** SNMP server
5. As defined by the 802.11w-2009 amendment, which 802.11 management frames will be safeguarded by management frame protection (MFP) mechanisms? (Choose all that apply.)
   - **A.** Beacon frame
   - **B.** Deauthentication frame
   - **C.** PS-Poll frame
   - **D.** Null function frame
   - **E.** QoS action frame
   - **F.** Measurement request action frame

6. What type of WLAN attacks might be detected by a distributed WIDS/WIPS solution using a behavioral analysis software engine? (Choose all that apply.)
   A. EAP flood attack
   B. Deauthentication attack
   C. Protocol fuzzing
   D. Fake AP attack
   E. CTS flood attack
   F. Zero day attack

7. Management has asked Cathy, the WLAN administrator, to devise a plan for rogue mitigation for the ACME Company corporate headquarters, which are located in a heavily populated area. What are the best recommendations Kathy can make? (Choose all that apply.)
   A. Automatic or manual wireless rogue containment for all devices classified as unauthorized
   B. Automatic or manual wireless rogue containment for all devices classified as rogue
   C. Automatic or manual wireless rogue containment for all devices classified as ad hoc
   D. Wired-side port suppression
   E. RF rogue containment using frequency jamming

8. Which method of device location tracking compares the target device's RSSI values with a database of RSSI values of the reference points?
   A. RF triangulation
   B. RF calibration
   C. RF positioning
   D. RF fingerprinting
   E. TDoA

9. Which type of triangulation method takes into account the speed of travel?
   A. RF triangulation
   B. RF calibration
   C. RF positioning
   D. RF fingerprinting
   E. TDoA

10. What type of WLAN attacks might be detected by a distributed WIDS/WIPS solution using a signature analysis software engine? (Choose all that apply.)
   A. PS-Poll flood
   B. Deauthentication attack
   C. Protocol fuzzing
   D. Virtual carrier attack
   E. CTS flood attack
   F. Zero day attack

11. After enabling an integrated WIPS solution, company employees begin to report gaps in audio communications when using VoWiFi phones over the corporate WLAN. What are the best solutions to restore audio quality over the WLAN while maintaining WLAN security monitoring? (Choose all that apply.)

    A. Replace the integrated WIPS with an overlay WIPS.

    B. Convert some of the APs to full-time sensors.

    C. Suspend rogue detection.

    D. Suspend off-channel scanning based in QoS voice priority markings.

    E. Increase off-channel scanning intervals on all APs.

12. Which of these WIDS/WIPS software modules allows an organization to view months of historical data on any suspicious WLAN device as well as all authorized devices?

    A. Spectrum analysis

    B. Protocol analysis

    C. Forensic analysis

    D. Signature analysis

13. What is the recommended ratio of WIPS sensors providing security monitoring to access points that are providing access for WLAN clients?

    A. 1:2

    B. 1:3

    C. 1:4

    D. 1:5

    E. Depends on the customer's needs

14. Bob has been investigating numerous false positive alarms from the newly installed integrated WIPS solution using APs as part-time sensors. What can Bob do to reduce the number of false positives? (Choose all that apply.)

    A. Lower alarm threat thresholds

    B. Raise alarm threat thresholds

    C. Convert some of the APs to full-time sensors

    D. Disable behavioral analysis

    E. Properly classify all authorized devices

15. Which of these radio form factors are used in 802.11 WIDS/WIPS sensors? (Choose all that apply.)

    A. Sensor with 2.4 GHz and 5 GHz radio

    B. Sensor with 5 GHz radio and software-defined radio (SDR) for both 2.4 and 5 GHz

    C. Sensor with three radios: 2.4 GHz, 5 GHz, and SDR radio

    D. Sensor with 2.4 GHz radio, 5 GHz radio, and spectrum analyzer chip

    E. Sensor with 900 MHz radio, 2.4 GHz radio, and 5 GHz radio

    F. Depends on the vendor

16. Why does RF fingerprinting offer more accurate device location information than simple triangulation?

- **A.** RF fingerprinting uses known RSSI reference points in addition to measured device RSSI values.
- **B.** RF fingerprinting uses RSSI values combined with RFID tags to find other devices.
- **C.** Triangulation uses known signal measurement points in addition to RSSI values.
- **D.** RF fingerprinting uses more sensors to locate devices by using a greater number of RSSI values.

17. Brooke is using an integrated WIDS/WIPS solution with APs as part-time sensors to protect against rogue APs. Brooke has deployed VoWiFi phones and wants to ensure maximum performance. Which WIDS/WIPS configuration setting is advisable when APs function as part-time WIPS sensors?

- **A.** Disable spectrum analysis
- **B.** Disable rouge containment
- **C.** Suspend off-channel scanning based on QoS priority markings
- **D.** Suspend rogue containment

18. What are some of the methods used by WIPS vendors to determine if a rogue device is connected to the wired network infrastructure? (Choose all that apply.)

- **A.** TTL packet analysis
- **B.** RF triangulation
- **C.** Signature analysis
- **D.** Behavioral analysis
- **E.** MAC table analysis
- **F.** Proprietary analysis

19. What are some of the methods used by WIPS vendors to determine if a Layer 3 rogue device is connected to the wired network infrastructure? (Choose all that apply.)

- **A.** Sensor associates with a suspected rogue device and sends traffic back to the WIPS.
- **B.** Sensor deauthenticates rogue clients from the suspected rogue AP and captures the association frames from the rogue clients.
- **C.** The WIPS looks for decremented MAC addresses.
- **D.** The WIPS looks for spoofed MAC addresses.

20. Which of these alarms should be configured to send an automatic notification to the WIPS administrator's phone and/or email account? (Choose all that apply.)

- **A.** Man-in-the-middle attack detected
- **B.** Unauthorized client detected
- **C.** Rogue AP detected
- **D.** Unauthorized AP detected

Chapter 15

Wireless Security Policies

IN THIS CHAPTER, YOU WILL LEARN ABOUT THE FOLLOWING:

✓ **General policy**

- Policy creation
- Policy management

✓ **Functional policy**

- Password policy
- RBAC policy
- Change control policy
- Authentication and encryption policy
- WLAN monitoring policy
- Endpoint policy
- Acceptable use policy
- Physical security policy
- Remote office policy

✓ **Government and industry regulations**

- Department of Defense (DoD) Directive 8420.1
- Federal Information Processing Standards (FIPS 140-2)
- Sarbanes Oxley (SOX)
- Graham-Leach-Bliley Act (GLBA)
- Health Insurance Portability and Accountability Act (HIPAA)
- Payment Card Industry (PCI) Standard
- Compliance reports

✓ **802.11 WLAN Policy Recommendations**

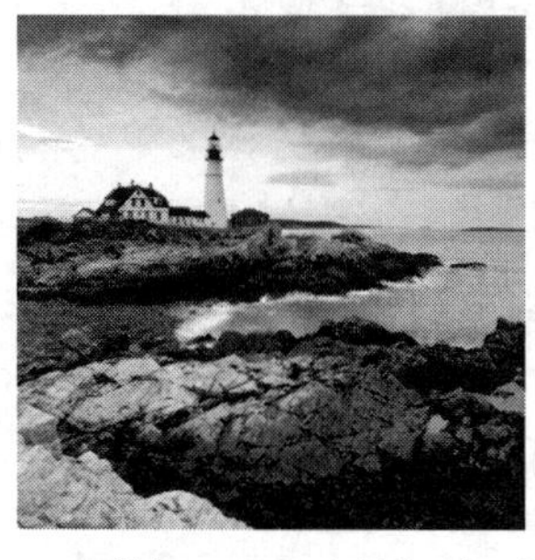

In the previous chapters of this book, you learned about many of the major components that are required for strong WLAN security. Segmentation, data privacy, AAA, and monitoring can all be accomplished with the technology solutions that were discussed. Another component is still needed to provide the foundation for a fortified WLAN solution. This additional item, and perhaps most important component, is policy. All companies should have a written WLAN security policy just as they should have a written wired network security policy. All of the technical WLAN security solutions you have learned about in this book are worthless unless proper WLAN security policies are documented and enforced. WLAN security policies can be as diverse as the variety of 802.11 wireless deployments they cover. Each organization or person deploying wireless devices must use their own set of criteria and evaluation methods in reaching what they believe to be their level of acceptable risk. WLAN security that may be safe enough in a small office/home office (SOHO) environment may not be safe enough in an enterprise environment. Likewise, WLAN security that may be secure enough in an enterprise environment could be viewed as overkill as well as an undesirable expense in a SOHO environment.

In addition to an organization's budget and internal acceptable use policies, governmental and industry regulations greatly influence security models in wireless networking. A corporate WLAN may be secure but could still fall out of compliance with an external government or industry regulation with which they must comply. WLAN policy makers must often strike a balance between productivity, budget, and security. Very often, mandated security policies hinder normal business operations. Even more common is the possibility that allowing business operations to flow without proper security can put company assets at risk. A well-crafted security policy must take into account business functions and security as well as any external regulations. Finding the proper balance is a key part of any security policy development.

After the policy is created to meet all of the business and regulatory requirements, the policy must be enforced to be effective. The policy must be enforceable and adaptable as requirements change. All WLAN policies must have the full backing of management if employees are to be expected to abide by policy. Management and employees alike must be properly trained so that all security policies are understood. The best written policy that is neither followed nor enforced is truly worthless. All policies must also be flexible and adaptable if the intended use of the WLAN changes and as WLAN technology progresses. With the introduction of the *Internet of Things (IoT)* and *bring your own device (BYOD)*, having an inclusive yet flexible security policy has never been so important. This chapter will explore some of the fundamental aspects of a wireless security policy that are needed to cement the foundation of good Wi-Fi security.

General Policy

When establishing a wireless security policy, you must first define a *general policy*. A general wireless security policy establishes why a wireless security policy is needed for an organization. Even if a company has no plans for deploying a wireless network, there should be, at minimum, a policy for dealing with rogue wireless devices. A general wireless security policy will define the following items:

Statement of Authority Defines who put the WLAN policy in place and the executive management that backs the policy

Applicable Audience Defines the audience to whom the policy applies, such as employees, visitors, and contractors

Risk Assessment and Threat Analysis Defines the potential wireless security risks and threats and what the financial impact will be on the company if a successful attack occurs

Security Auditing Defines internal auditing procedures as well as the need for independent outside audits

Violation Reporting Procedures Defines how the WLAN security policy will be enforced, including what actions should be taken and who is in charge of enforcement

Increasingly, organizations of all sizes and types have started amending their network usage policies to include a wireless policy section. If you have not done so already, a WLAN section should be added to your organization's security policy. Two good resources for learning about best practices and computer security policies are the SANS Institute and the National Institute of Standards and Technology (NIST).

Security policy templates from the SANS Institute can be downloaded from `www.sans.org/resources/policies`. You can download the NIST special publication documents 800-153 and 800-48 regarding WLAN security from `http://csrc.nist.gov/publications/nistpubs`.

General security policies covering networking are not directly focused on any one area of technology. For this reason, you will find that general security policies tend to encompass vast portions of networking if not all of it. These policies are the larger framework from which other more specific policies are formed. In this section, we will explore policy creation and policy management, as well as internal and external policy influences.

Policy Creation

Policy creation involves decisions that can impact business as a whole in both positive and negative ways. The creation of network security policy should be part of the overall network design. However, security and its related policies are often afterthoughts that develop once it is too late. WLAN security, such as a WIPS monitoring solution, is often deployed in response to a breach in WLAN security. Likewise, security policies are often applied after the WLAN has been deployed in response to an industry concern. Creating documented security policies in advance during the design phase of a WLAN is a much better strategy. If the security team is the sole developer of a security policy, management and other departments may not want to subscribe to the defined policies. If the policies are not relevant or are too restrictive, they are likely to be ignored or impede business. Therefore, the first task in policy development must be to assemble a committee that will begin the construction of a relevant and usable policy. This group should include representatives from each group of stakeholders within the organization. The *stakeholders* are part of the applicable audience that should be involved in creating the policies to which they must abide. Everyone must endorse the security policy for it to be effective. Having such an endorsement requires an executive-level champion. Often a "C"-level executive is needed to get the ball rolling. The CEO, CIO, CTO, or CSO usually champions the cause so that the policy has the energy to move forward. Here are some of the departments that should be represented in policy creation:

- Security
- Legal
- Human Resources
- Management
- Networking
- Desktop Support
- Finance
- Users
- Research and Development
- Any group using the technology covered by the policy

Once the policy creation team is assembled, the next step is to define the scope of the policy. It is important to determine and specify whether you are creating a general policy or a functional policy. Whereas a general policy establishes a framework for enforcement, a functional policy defines technical aspects of security. WLAN functional policies will be discussed in greater detail later in this chapter.

When building a general policy, you should look for existing written policies first and attempt to see if they are applicable and determine if they are being followed and enforced. If there is a written policy, you may only need to adapt it to current business and security needs or just ensure it is being adhered to and enforced. If there is no written policy, you will need to start from scratch, which, in some instances, may be an easier task than adjusting an existing policy. The WLAN may also fall under an industry or governmental network security regulation that does not apply to the wired infrastructure. When building a WLAN security policy, you should determine if any external influences such as governmental regulations or industry standards exist that would impact your decision making. Internal and external influences should also be considered when modifying an existing policy. Several government and industry regulations will also be discussed in greater detail later in this chapter. Defining the applicable audience that must abide by this policy is critical. Obviously, all employees will have to abide by policy, but contractors and temporary workers are often overlooked.

With the policy development team assembled and the scope of the policy defined, you are ready to start building and documenting your policy. A large part of general policy creation is risk assessment. Risk assessment involves determining what assets are at risk, as well as the value of the assets that need protection. Often the value of the asset being protected will determine the security measures that are needed. Assets such as corporate trade secrets and credit card databases require the strongest security available. Threat analysis is also a form of risk assessment. Determining in advance what potential WLAN security threats exist will help determine the risk due to a successful WLAN attack occurring. For example, how much would it cost the company if an intruder used a rogue AP to access the customer credit card database? What would the potential legal liability be to the company if employee healthcare records were illegally accessed? What would it cost the company if a denial-of-service attack prevented the sales team from accessing email for four hours? If a specific dollar value can be assigned to these types of potential threats, better decisions can be made when choosing the proper WLAN security solution, and better policies can be written. Furthermore, if threat analysis and risk assessment can be used to determine financial impact, it will be easier to convince management to budget for needed security solutions. Having an external audit team look for vulnerabilities in your deployment and or make recommendations about your policy based on their findings or experience can improve the finished policy immensely. Even more importantly, management is more likely to endorse the enforcement of policy if they have a better understanding of the risks and are aware of the potential financial impact of a breach in security.

The final phase of policy deployment is policy communication. This involves making sure that all of the users are aware of the policies, understand them, and realize they must comply with them. This goal can be reached using several methods. Documentation of the policy must exist to eliminate confusion and to provide reference should points of clarity be required. Users need to be trained to understand and abide by security policies. This is one of the reasons why user representation in policy creation is so important. If the users are not trained on policy conformity or do not buy into the policy, it will not be followed. A formal training class regarding policy is usually necessary and highly recommended. After training, users must be required to read the corporate security policy and sign a document stating that they have read and understand it as well as agreeing to comply with the written policy.

Many organizations make the mistake of creating policy documents that are too technical, hard to read, and hard to understand. The policy document should be written in a straightforward manner that all employees can comprehend without any unneeded technical jargon. Another mistake often made within an organization is that employees read the security policy document, sign the document, and never see it again. The policy document will often change based on changes in business needs, user requirements, industry or governmental influence, or changes in technology; therefore, it should be thought of as a living and breathing document. Many companies store the security policy document on a public share on a corporate server and then provide a link to the document for employee access when the document changes, or on a scheduled basis when the employees are required to read and agree to the policy as part of the security plan. Such a link allows the employees to read and reference the security document whenever necessary. The lengths you go to in policy communication may vary by user, department, or organization. The key aspects of building an effective policy are that the policy should be relevant, include both internal and external requirements, and be properly documented and well communicated.

Policy Management

The management of any security policy is a crucial part of an effective security solution. Policy management includes policy adaptation as well as compliance monitoring, security audits, and policy enforcement. A policy that is not enforced is not effective. Policies are also not effective when they become obsolete. As mentioned earlier, the policy document needs to be a living document that is adaptable to new business requirements and advances in technology. With the rapid growth and acceptance of BYOD, having a flexible and adaptable WLAN security policy has become extremely important. Knowing the requirements and monitoring for compliance allows security staff members to identify devices or communications that fall outside the policy. When such situations are located, they can be corrected—hopefully before a security incidence occurs. Should new technology be incorporated into your network, the general policy may cover it. However, functional policy for the newer technology may not exist and may need to be created prior to allowing such new things on your network. By monitoring policy, holes such as these can be identified, documented, and included in the next revision of the policy, or they may prompt a specific policy to be created for its use. As new threats arise, a policy can be augmented to address the threats. It is important to realize that both internal and external changes can drive the need to update the security policy.

Many organizations have a specific monetary value defined for every minute of downtime faced by the network. Downtime can be caused by hardware failure, improper configuration, firmware issues, software glitches, or security problems. Striving for 99.999 percent uptime is very common. Unmanaged security policies, or those that are too restrictive due to failure to adapt to business requirements, can cause downtime. Those that fail to comply with external requirements, industry or governmental, could result in networks being shut down until compliance is reached. This could also result in fines or a loss of assets due to noncompliance. As external mandates change, the network security policy must also change. The loss of money or time due to noncompliance is, essentially, a loss of corporate assets, although not via an attack but rather through not being vigilant in regard to changes in the industry, regulations, or technology. Policy management involves monitoring for compliance, both internal and external.

Policy management may also include security audits and penetration testing. In Chapter 13, "Wireless LAN Security Auditing," you learned about many of the tools and procedures needed to perform a successful audit. Penetration testing and audits will assist the security staff in determining compliance with corporate policy, as well as finding oversights in the policies, or identifying areas for improvements based on newer technologies or methods. Before conducting a security audit, or penetration test, consult the written policies to help ensure that your actions are covered within corporate guidelines. Internal corporate audits should be conducted on a regular basis. Consideration should be given to hiring a third-party to perform penetration tests and audits. Third-party audits often reveal security flaws that were missed by the organization.

Policy management also includes the area of enforcement. As mentioned earlier, if policies are not enforced, they become meaningless. Your general policy should outline procedures to be taken if there are policy violations. Actions taken after a policy violation usually depend on the value of what was lost or compromised. An employee who violates security policy at a nuclear power plant will probably get fired, whereas an employee who violates policy at a retail store might just be retrained. Policy enforcement and reporting of violations are often hampered by the following:

- Lack of corporate standards and documentation
- Ineffective or missing management support
- Non-uniform enforcement
- No client-side enforcement
- Lack of education about risks
- Regulation compliance
- Deployment management
- Cost (this is a big one)

All policy violations and subsequent actions must be accurately documented to meet compliance with many government regulations. Documenting any violations will also help to identify where things can be improved. Was the violation caused by ignorance, attack, or changes in business practices or technologies that went unnoticed? Knowing why the problem occurred can help prevent future occurrences of the same problem or by the same person. Documentation of policy violations is also important for liability reasons and may even be needed in case human resources or law enforcement gets involved.

Functional Policy

General policies are the framework from which functional policies are derived. A *functional policy* is required to define the technical aspects of wireless security. The functional security policy establishes how to implement and secure the wireless network, defining what solutions and actions are needed. A functional wireless security policy will at a minimum define the following items:

Policy Essentials Defines basic security procedures such as password policies, training, and proper use of the wireless network

Baseline Practices Defines minimum wireless security practices such as configuration checklists, staging and testing procedures, and so on

Design and Implementation Defines the actual authentication, encryption, and segmentation solutions that are to be put in place

Monitoring and Response Defines all wireless intrusion detection procedures and the appropriate responses to alarms

Password Policy

Functional policies should include a password policy. This policy should state the length, complexity, and age limits of passwords used in authentication, in addition to simply requiring a password. Where possible, systems should be configured to enforce the password policies and not allow users to authenticate with weak or noncompliant passwords. Users should also be forced to change their password upon first login. This prevents administrators from knowing the users' password beyond the initial account creation. The complexity and length of the password is derived from a delicate balance of difficulty to guess (or brute force the password), and ease of remembrance. If the password is too simple or too short, it is easily guessed or cracked. If it is too long or too complex, users may forget them or write them down near their work area. This could allow someone to easily find the password and use it to gain access to network resources beyond their approved access. Additional authentication measures may also be required beyond passwords to improve security, such as multifactor security measures. These measures take into account usernames and passwords (what you know), token devices or client certificates (what you have), biometrics (who you are), and RFID tags (where you are). Some of these measures include

- Certificate usage
- Smart card usage
- Fingerprint scan
- Retina scan
- Voice recognition
- Facial recognition
- Real-time location systems (RTLSs)

As basic as it seems, the policy should also include a statement prohibiting users from disclosing their passwords and or preshared keys to others. Often an employee or contractor may need to use the station of their coworker, and request the coworker's credentials to log in. This practice should be prohibited, as should the user logging in and allowing others to use their system.

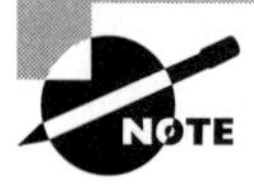

The section "Entropy" in Chapter 6, "PSK Authentication," is extremely important when considering password policy. In digital communications, *entropy* is a measure of uncertainty associated with a random variable.

A base password complexity and length that is commonly used is a password of at least eight characters in length, at least one uppercase letter, at least one lowercase letter, at least one number, and at least one special character. For example, 8H4pT@b$ would meet the criteria of an eight-character strong password. Resources within an organization may require additional complexity or length to be more secure. Some organizations require users to maintain separate credentials for each resource. As you learned in Chapter 4, "802.1X/EAP Authentication," most organizations synchronize user passwords across resources but use centralized authentication solutions, such as *Lightweight Directory Access Protocol (LDAP)* and/or *Remote Authentication Dial-In User Service (RADIUS),* in addition to complex password requirements. The length, complexity, and age of passwords should all be spelled out in a functional policy, and followed to reach the desired security level.

The same philosophies of complex passwords—not distributing them to others and age limits and so forth—that apply to a user's password should also apply to the preshared key(s) used on your network. The preshared key (PSK) should be complex enough to prevent guessing of the key without the use of some extraordinary measure such as a cracking tool and still be simple enough for legitimate users to enter the key(s) on their own. The PSK(s) should be changed on a regular interval, not repeated or too similar to the last key or keys used, just like changing a user password.

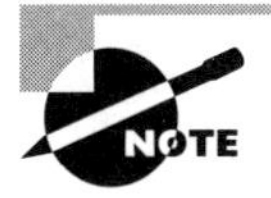

Numerous freeware tools exist that can be used to assist end users with creating strong passwords, shared secrets, and preshared keys. Additionally, free password generation websites exist such as `www.passwords-generator.net`.

RBAC Policy

In Chapter 8, "WLAN Security Infrastructure," you learned about the technical aspects of *role-based access control (RBAC)*, which is a method of providing or restricting WLAN access based on the identity of the user or device. Every user has his or her own job and should have access to the appropriate resources to do that job. To that end, users should all have their own accounts and credentials. This helps in building nonrepudiation into network activities. Granting access to resources beyond those required by users to do their job is a security hole. RBAC helps ensure that networking resources can only be accessed by users with permission to do so.

Written RBAC policies should be clearly defined, specifying what groups of users get to access which network resources. RBAC policies also define what type of devices get access to which network resources. People who need to do the same tasks usually require access to the same resources. By placing users into groups related to their job functions and assigning resource permissions to their groups, administration is an easier task and security is enhanced due to RBAC. As a user moves from job to job within an organization, permissions are easily assigned by moving the user's account into and out of the appropriate groups. If written RBAC policies are clearly defined, the RBAC technical implementations will be better planned and deployed.

Change Control Policy

Periodically, vendors release firmware and software updates to improve functionality of their products or to correct errant releases of their products or new products. A good functional policy will include a well-designed *change control* procedure governing the upgrading or downgrading of firmware and software, as well as for performing hardware maintenance or hardware replacement. Change control processes help to ensure that changes to a product or system are introduced in a controlled and coordinated manner. Additionally, they should reduce the possibility that unnecessary changes are brought into a system, which could introduce problems or undo required changes made by other users. Change control policy usually includes procedures to minimize disruption to services, reduce back-out activities, and provide for cost-effective utilization of resources, while adhering to the security policy. Changing firmware or software may open security holes that were not there before. Things such as sockets or ports that were used by a previous version may not be used by the currently installed version and may not have been checked or secured when the security baseline was created. Just because the newer version is compatible with everything else running on the device does not mean it meets the security requirements within your policies. Even when the objectives of the changes are to meet security requirements, the changes must still be covered under written policy to help protect assets and reduce your exposure to risk.

In addition to including a mandated change control process in your policies, you must include monitoring and inspection of devices to make certain that required firmware and software is installed. You should also include statements covering the installation of unapproved software, such as freeware or user-purchased software. In large environments, adhering to a change control process helps to ensure uniformity of deployment. However, some devices can still be missed. Within the auditing statements of your security policy, you should include a mandate that software and firmware be at the required version level. An auditing requirement may also be included within a general policy or any policy covering risk assessment. Auditing is often required by an external regulation or industry policy. Devices missing patches are often exposed to attackers. When a vendor releases a security patch, attackers know where holes may be in unpatched systems, if they did not already know about them. Your policies may require that security patches be installed within a given period of time from their release. Once new software or firmware is installed, a new security baseline should be established and documented.

Whenever WLAN infrastructure is deployed, or changes to configurations are made, all settings should be documented. The encryption, authentication, RBAC, and VLAN settings for APs and WLAN controllers should always be documented and validated with checklists.

Authentication and Encryption Policy

Throughout this book, you have learned about enterprise solutions that are used to protect the WLAN portal. Many of these solutions only authorize network resource access for users who have had their credentials properly validated. Whenever possible, a functional policy mandating the use of 802.1X/EAP authentication mechanisms should be used. Consideration should also be given to mandate the use of multifactor authentication solutions before access is granted. Handheld devices like VoWiFi phones may need to use weaker authentication solutions, such as WPA2-Personal, if they do not support fast secure roaming mechanisms such as opportunistic key caching (OKC) or fast BSS transition (FT). Guest networks may use captive portal authentication for liability and auditing purposes.

You have also learned about the many available enterprise encryption solutions that are used to provide data privacy for the Layer 3–7 payload of 802.11 data frames. Dynamic CCMP/AES encryption should be used. Dynamic TKIP and WEP encryption are legacy encryption methods and are not supported for 802.11n and 802.11ac data rates. Legacy encryption will not be acceptable in any modern-day WLAN deployment.

WLAN Monitoring Policy

The WLAN should be monitored by a WIDS/WIPS solution for any compromise or attack of the WLAN, including rogue devices, ad hoc clients, and DoS attacks. Functional policy should define what steps should be taken when a breach is detected by the WIDS/WIPS solution. For example, will the WIPS be used to perform automatic or manual rogue containment? What action will be taken if a Layer 1 DoS attack is detected? What thresholds will be set for WIPS alarms, and how will the WIPS administrator be alerted to alarms? As you learned in Chapter 14, “Wireless Security Monitoring,” a WIPS can also be used to help enforce certain functional policies. For example, the WIPS may detect authorized WLAN clients with improperly configured security settings. The WIPS also may be used to contain or blacklist improperly configured client stations and effectively enforce encryption and authentication functional policies.

Endpoint Policy

It is likely that end-user client devices will leave a company’s controlled environment and operate outside of the area monitored by the WIPS. Employees will use access points that do not meet the corporate encryption and authentication policies. It is almost a certainty that employees will use company-provided client devices to access the Internet at Wi-Fi hotspots and other public access WLANs. Therefore, organizations must also implement an *endpoint policy* that governs the security of all client stations. While some devices such as VoWiFi phones and barcode scanners may never leave the corporate facilities, employee laptops, tablets, smartphones, and other smaller wireless devices are coming and going every day. When end users are at off-site locations, these company-owned devices are associating with other noncompany WLANs and therefore are not able to be monitored by the company WIPS.

Throughout this book, we have discussed the reasons and solutions needed for enterprise WLAN infrastructure security, with the goal of protecting corporate assets from attackers. One of the most important facts that a WLAN security professional should always remember is that the greatest WLAN security risks will always exist off-site. The only security that Wi-Fi hotspots and public access WLANs usually have is a captive portal. There is no expectation to have encryption, and any client station is exposed to the majority of the WLAN security risks that were discussed in Chapter 12, "Wireless Security Risks." Public access WLANs are breeding grounds for peer-to-peer attacks, Wi-Fi phishing attacks, data theft or manipulation, eavesdropping attacks, and other malicious events. Always remember that WLAN hackers lurk at Wi-Fi hotspots and open public access WLANs.

> Large organizations should delegate endpoint security to a group dedicated to malware prevention, detection, and response across all technology areas.
>
> *Network World,* June 2014

Endpoint security measures within both general and functional policies should always be considered mandatory. At a minimum, the functional policy should mandate the use of a personal firewall and an IPsec VPN client solution. The personal firewall will protect client stations from peer-to-peer attacks. The IPsec VPN client will be used to establish an encrypted tunnel back to the corporate VPN server. The client's data will not only be protected across the Internet, but will also provide data privacy in the air at the Wi-Fi hotspot. Because there is no security at the Wi-Fi hotspot, it is imperative that the employee bring the security to the unprotected WLAN.

WLAN specific policies that might be enforced with an endpoint solution include the following:

- Requiring specific WLAN encryption and authentication settings on the client side
- Preventing ad hoc connectivity
- Preventing bridging between wired and wireless network interfaces
- Enforcing VPNs
- Authorizing only certain SSIDs
- Blacklisting certain SSIDS

In recent years, WLAN client endpoint security and management has changed drastically due to the influx of BYOD devices such as smartphones and mobile tablets. As discussed in detail in Chapter 10, WLAN client policy enforcement is usually handled by more robust mobile device management solutions (MDM) or network access control (NAC) solutions. MDM and NAC software agents are often used for client-side policy enforcement for both personal and corporate-owned WLAN devices.

Acceptable Use Policy

The *acceptable use policy* defines the purpose of the WLAN and how the company employees may utilize the WLAN. The WLAN is obviously not to be used for illegal purposes and usually only for company activities. Acceptable use policies will also define what applications may be used over the WLAN. For example, the use of video streaming applications may not be allowed for employees. Certain types of downloading activities may also be banned because of the effect on throughput of the WLAN. The good news is that RBAC mechanisms, such as firewall restrictions and bandwidth throttling, can often be used to limit the activities of WLAN end users and therefore enforce acceptable use policies.

Physical Security

If you do not have physical security, do you really have security at all or just the illusion of security? The physical security policy will define how WLAN infrastructure will be protected from theft and vandalism. An attacker may also access the exposed ports of an unsecured AP to extract information about WLAN configuration settings. The location of all access points will be documented, and often the APs will be secured in indoor enclosure units that can be locked. Outdoor APs will also be secured in National Electrical Manufacturers Association (NEMA) enclosure units for protection from the weather. All WLAN controllers and servers should be secured in a data closet. Inventory and documentation of all client devices is also highly recommended.

Remote Office Policy

Given the widespread small office and home use of 802.11 wireless networks, the discussion of wireless security policy for the home and small office is worth having. To date, there are no regulations or industry standards focused on how to secure wireless home networks that are used for nonbusiness reasons. However, some laws, regulations, and standards apply to businesses no matter what the size, whereas others vary the requirements based on size of the business and the amount and type of information to be protected. Although the policy formation process may be less involved and the policy created less encompassing than required by an enterprise security solution, SOHO environments should also create and follow wireless security policies. Budgets and network size do not allow the same protections in SOHO environments as found in the enterprise. A small five-person office with only five laptops and one printer is not likely to install a WIPS for protection. There is no security team conducting daily audits in the SOHO environment. However, small and medium-sized business (SMB) offices must comply with external policy influences just as large corporations must comply. A great example would be a doctor's office having the need to be Health Insurance Portability and Accountability Act (HIPAA) compliant. HIPAA requires any needed access to patient data to be accomplished in a secure manner based on maintaining patient confidentiality. If the office uses wireless communications to transfer patient data, the office must take measures to protect that data. A small collection agency may be required to meet Gramm-Leach-Bliley Act of 1999 (GLBA) compliance to continue to do business with their customers, putting a policy in place to protect the customers' information from foreseeable security and data integrity threats. The Sarbanes-Oxley Act of 2002 (SOX) has a financial privacy rule that applies to companies receiving private financial information, whether or not they are financial institutions, requiring that anyone processing or storing such information must be SOX compliant. They may also be required to meet Payment Card Industry (PCI) standards as well. The business partner of a large corporation could have a small branch office that is insecure, potentially providing an attacker with an entrance into the network of the larger firm.

Securing your wireless network should be part of the design process, whether you are an office with a single remote AP or an enterprise with thousands of APs. Just as in the enterprise, SMB deployments should develop, follow, and adapt wireless security policies to protect the assets of the organization. If the business is not required to follow industry or governmental guidelines, a simple policy covering physical security, authentication, and encryption may suffice. However, if the SMB operates in business areas where regulations exist, the policy must take into account all the external policy influences as well as simple protection policies. The smaller deployment size of SOHO wireless networks in no way negates the need for security and written, followed, enforced, and adaptable wireless security solutions and policies.

Government and Industry Regulations

In most countries, mandated regulations exist, describing how to protect and secure data communications within all government agencies. In the United States, NIST is responsible for maintaining the *Federal Information Processing Standards (FIPS)*. The FIPS PUB 200 standard specifies minimum security requirements for federal information and information systems in seventeen security-related areas. The FIPS specifications for minimum security requirements include access control, configuration management, authentication, risk assessment, auditing, and much more. Even if your organization is not required to meet FIPS PUB 200 requirements, reading FIPS PUB 200 and adopting some if not all of the requirements into your own security policy would not be a bad idea. Of special interest to wireless security is the FIPS 140-2 standard, which defines security requirements for cryptography modules. The use of validated cryptographic modules is required by the U.S. government for all unclassified communications. Other countries also recognize the FIPS 140-2 standard or have similar regulations. All the FIPS publications can be found online at `http://csrc.nist.gov/publications/PubsFIPS.html`.

From government to healthcare, banking, or even energy sectors, more and more external regulations, laws, and standards are affecting how we use wireless networking. In this section, we will explore how some of the more widely known external mandates are directly influencing wireless use and how they should be incorporated into WLAN security policies. We will specifically examine the following as they pertain to the use of 802.11-based wireless devices:

- The U.S. Department of Defense (DoD) Directive 8420.01
- The U.S. Federal Information Processing Standards (FIPS 140-2)
- Sarbanes-Oxley Act (SOX)
- Graham-Leach-Bliley Act (GLBA)
- Health Insurance Portability and Accountability Act (HIPAA)
- Payment Card Industry (PCI) Standard

Does the CWSP Exam Test the Specifics of Government and Industry Regulations?

The simple answer is no. The CWSP certification is recognized worldwide and the exam is available in many countries. All of the regulations discussed in the following sections are U.S. government and industry regulations. This information is being provided strictly as reference material. It would not be practical to test a CWSP exam candidate in New Zealand about U.S. government regulations. However, it should be noted that many countries have almost identical regulations for government and specific industries. A WLAN security professional in any country would be well advised to learn about all the applicable regulations as they apply to WLAN technology within any region.

The U.S. Department of Defense (DoD) Directive 8420.1

On April 24, 2004, the U.S. Department of Defense issued a directive covering the use of commercial wireless devices, services, and technologies within the Global Information Grid (GIG). This directive, titled the Department of Defense (DoD) Directive 8100.2, details the DoD policy for the secure use of wireless networks and devices. It also includes policy statements requiring the monitoring of wireless devices that are deployed, and monitoring for wireless devices in areas that do not have authorized wireless deployments. Furthermore, this directive states that there are areas in which the use of wireless networking is banned. The policy covers both authorized and banned wireless usage.

In November of 2009, the original directive was replaced with DoD Directive 8420.1 for Commercial Wireless Local-Area Network (WLAN) Devices, Systems, and Technologies. The directive allows for the embracing of technologies from open standards–based solutions while providing a framework from within which this can be done securely. Directive 8420.1 applies to 802.11-based communications and devices but does not cover cellular, Worldwide Interoperability for Microwave Access (WiMAX), Bluetooth, or proprietary RF communication mechanisms. Directive 8420.1 applies to all DoD components, military and civilian alike. It covers any personal electronic devices, laptops, PDAs, cell phones, and so on, that are able to process, store, or transmit information wirelessly.

All components parts of the DoD must ensure that devices are certified and validated under DoD regulations for interoperability and that there is secure end-to-end communications. The 8420.1 directive requires the following measures be taken:

- All cryptographic functionality must meet, at a minimum, NIST FIPS 140-2 Level 1 validation.
- All WLAN devices deployed must be National Information Assurance Partnership (NIAP) Common Criteria Certified.
- WLAN devices must be certified by the Wi-Fi Alliance and be WPA2 certified.
- Personal electronic devices, laptops, PDAs, and so forth must use a NIAP Common Criteria Certified Personal Firewall and Antivirus.

All portions of the DoD components were required to implement WLAN solutions that are IEEE 802.11i compliant and WPA2 Enterprise certified. The policy also mandates that they implement 802.1X access control with EAP-TLS mutual authentication and a configuration that ensures exclusive use of FIPS 140-2 (minimum overall Level 1) validated AES-CCMP encrypted transmissions. Directive 8420.1 also requires the use of wireless intrusion detection systems (WIDSs), something not included in the original 8100.2 directive. Some components of the DoD have included wireless intrusion prevention systems (WIPSs) in addition to basic WIDS functionality. Directive 8420.1 states that

- WIDSs are required for all DoD wired and wireless networks.
- WIDSs must continually scan for and detect both authorized and unauthorized devices. Continually scanning is defined as 24/7 monitoring.
- WIDSs must also have location tracking capability.
- WIDSs must be validated under NIAP Common Criteria Certification.

The DoD has well-defined policies at both the general and functional policy levels. Given that users within the DoD travel the globe and may have access to high-priority data, the policies for their network may be too restrictive for some environments to the point of enforcing no wireless zones. Directive 8420.1 addresses security beyond configuration. It mandates monitoring for wireless devices, even where none are deployed, and covers encryption, certification, and validation. Such high standards limit the hardware and software that can be used due to the level of compliance, certification, and validation required. So who must comply with DoD Directive8420.1? The answer is any DoD employee, contractor, or visitor to any DoD facility as well as any others who have access to DoD information.

DoD Directive 8420.01

The 8420.01 document can be downloaded at `www.dtic.mil/whs/directives/corres/pdf/842001p.pdf`.

Federal Information Processing Standards (FIPS) 140-2

The Federal Information Processing Standards (FIPS) are issued by NIST. They were created in the United States under the Information Technology Management Reform Act (Public Law 104-106). The Secretary of Commerce approves the standards and guidelines that are developed by NIST. These standards are created by NIST when there is a specific need for security and/or interoperability, and there are no industry standards that meet the requirements. We will focus on only FIPS 140-2 compliance requirements as they relate to wireless communication. FIPS Publication 140-2 was released on May 25, 2001. It supersedes FIPS 140-1, which was released on January 11, 1994. FIPS 140-2 covers the federal requirements for cryptographic modules. Specifically, it covers the security requirements that will be satisfied by a cryptographic module utilized within a security system protecting sensitive but unclassified information. Many businesses and individuals conducting operations with the U.S. government and/or selling communication devices to the government must comply with this publication.

FIPS 140-2 has increasing levels of requirements from Level 1 to Level 4. Each higher level must meet all of the lower-level requirements in addition to their own specifications. These mandates cover areas of both secure design and secure deployment of applications and communications. The appendices of this standard summarize the requirements and describe each of the four levels of requirements. Each of the four levels spell out their functional requirements. As FIPS 140-2 relates to cryptographic modules, the four levels are as follows:

Level 1 Security requirements are specified for a cryptographic module at this level. No specific physical security mechanisms are required in a Security Level 1 cryptographic module beyond the basic requirement for production-grade components. However, at least one Approved algorithm or Approved security function should be used.

Level 2 Level 2 requires features that show evidence of tampering, including tamper-evident coatings or seals that must be broken to attain physical access to the plaintext cryptographic keys and critical security parameters (CSPs) within the module, or pick-resistant locks on covers or doors to protect against unauthorized physical access. (Many devices use painted screws, seals, or tamper-evident tape to meet this requirement.)

Level 3 Level 3 attempts to prevent the intruder from gaining access to critical security parameters (CSPs) held within the cryptographic module. Physical security mechanisms required at Security Level 3 are intended to have a high probability of detecting and responding to attempts at compromising the cryptographic module. The practice of erasing sensitive parameters such as keys from a cryptographic module to prevent their disclosure if the equipment is captured is known as *zeroization*. Physical security mechanisms may include the use of strong enclosures and tamper detection/response circuitry that zeroizes all plaintext CSPs when the removable covers/doors of the cryptographic module are opened.

Level 4 Level 4 provides the highest level of security. Here the physical security mechanisms have the intent of detecting and responding to all unauthorized attempts at physical access. Any compromise of the cryptographic module enclosure is most likely going to be detected, resulting in the immediate zeroization of all plaintext CSPs. Level 4 cryptographic modules are important to use in physically unprotected environments. Level 4 also requires protection for a cryptographic module against compromise due to environmental conditions, such as temperature and power variations beyond normal operation. Cryptographic modules are required to include special environmental protection features designed to detect fluctuations and zeroize CSPs, or as an alternative undergo rigorous environmental failure testing providing reasonable assurance that the module will not be compromised by fluctuations outside of the normal operating range in a manner that can reduce or negate the security of the module at Level 4 certification.

So who must comply with FIPS 140-2? Just about all U.S. government agencies must comply. The applicable audience for FIPS 140-2 is all federal agencies that use cryptographic-based security systems to protect sensitive information in computer and telecommunication systems (including voice systems) as defined in Section 5131 of the Information Technology Management Reform Act of 1996, Public Law 104-106. Any devices sold for deployment in these areas must meet this standard. Any systems, applications, or devices approved for classified transmissions can be substituted, since they comply with standards that exceed those mandated in FIPS 140-2. In simpler words, within any federal agency that handles data of sensitive nature, the computer network solution must be FIPS 140-2 compliant.

The standard also includes an implementation schedule, export rules, and contact information for valid interpretation of its content. Realizing that there may be a need for some exceptions, there is even a waiver process outlined to allow for exemptions if needed and justified. This applies to wireless communications and the associated devices used by anyone required to follow this standard. Even outside areas where FIPS compliance and certification is mandated, following these measures improves an organization's security posture. With all of the security measures spelled out in FIPS 140-2, it is easy to see why your organization may want to follow its guidelines. However, the most compelling reason to follow FIPS 140-2 is that it is a federal mandate. This mandate will continue to be adapted as technology changes, as evidenced by the continued exploration of security standardization by NIST.

Information about the FIPS regulations can be found at `http://csrc.nist.gov/publications/fips`. You will find a list of FIPS-compliant vendors, devices, and software at `http://csrc.nist.gov/groups/STM/cmvp/documents/140-1/1401vend.htm`.

The Sarbanes-Oxley Act of 2002 (SOX)

The Sarbanes-Oxley Act of 2002 (SOX) defines more stringent controls on corporate accounting and auditing procedures with a goal of corporate responsibility and enhanced financial disclosure. The act is named *Sarbanes-Oxley* after Senator Paul Sarbanes and Representative Michael Oxley, the key figures who drafted it in 2002. The legislation set new or enhanced standards for all U.S. public company boards, management, and public accounting firms. Since SOX applies to publicly traded companies, it does not apply to privately held companies or those publicly traded companies that do not operate within the United States. The act contains 11 titles and multiple sections, ranging from additional corporate board responsibilities to criminal penalties, and it requires the Securities and Exchange Commission (SEC) to implement rulings on the requirements to comply with the new law. The two most relevant sections for a wireless security discussion are 302 and 404.

Section 302: Corporate Responsibility for Financial Reporting This section may be the best known one. It requires the CEO and CFO to certify that they have reviewed the financial reports, that the information is complete and accurate, and that effective disclosure controls and procedures are in place to ensure material information is made known to them. Section 302 of the Act mandates a set of internal procedures that ensure accurate financial disclosure.

Section 404: Management Assessment of Internal Controls This is a newer section with the following three basic requirements:

- Establishment of effective internal controls by corporate management for accurate and complete reporting
- Annual assessment by management of the effectiveness of internal controls supported by documented evidence
- Validation of management's assessment by a registered public accounting firm

SOX Section 404 does not specifically discuss information technology (IT) and the related security requirements. However, most financial reporting systems are heavily dependent on technology. The burden falls on the CIO and IT staff members to establish effective internal controls over the network and storage infrastructure supporting the financial reporting process. Businesses are increasingly implementing wireless networks to improve productivity and reduce costs. The introduction of wireless networking also brings new security challenges and reporting and monitoring requirements for those requiring SOX compliance. Since its passing in 2002, SOX Section 404 has been the topic of debate to either remove or replace the section. Section 404b is thought to increase investor risk.

SOX also created a quasi-public agency, the Public Company Accounting Oversight Board (PCAOB). The PCAOB has the responsibility of overseeing, regulating, inspecting, and disciplining accounting firms in their roles as auditors of public companies. Furthermore, SOX covers auditor independence, corporate governance, internal control assessment, and enhanced financial disclosure. The goal of SOX was to restore faith in publicly traded companies' reporting of earnings in the wake of several large trading scandals. The applicable audience for SOX consists of all public companies in the United States, international companies that have registered equity or debt securities with the SEC, and the accounting firms that provide auditing services to them.

Where does *information assurance* (IA) fit into SOX compliance? IA fits within SOX compliance in the following manner:

- SOX covers everything revolving around the confidentiality, availability, and (especially) integrity of financial data.
- SOX requires a recognized internal control framework.
- The Committee of Sponsoring Organizations (COSO) framework provides structured and comprehensive guidelines for implementing internal controls for SOX.
- The COSO recommended (but did not require) a framework by PCAOB.
- Five components of effective internal control are stipulated: Control Environment, Risk Assessment, Control Activities, Information & Communication, and Monitoring.
- The Control Objectives for Information and Related Technology (COBIT) framework, which was developed by the Information Systems Audit and Control Association (ISACA), bridges the gap between IT governance and SOX, and it addresses 34 IT processes that can all be mapped to COSO.

IA and IT fit within SOX under information security policies, network security, access controls, authentication, encryption, logging, monitoring and alerting, incident response and forensics, and IT audit. The secure implementation, use, and monitoring of wireless devices and networks clearly fits into these areas and therefore into SOX compliance.

Information about SOX can be found at `www.sarbanes-oxley-101.com`.

Graham-Leach-Bliley Act (GLBA)

The Financial Modernization Act of 1999, which is more commonly known as the *Gramm-Leach-Bliley Act (GLBA)*, requires banks and financial institutions to notify customers of policies and practices disclosing customer information. The goal is to protect personal information such as credit card numbers, Social Security numbers, names, addresses, and so forth. There are five core requirements for GLBA:

- Designate employees for information security.
- Identify and assess all risks and safeguards.
- Design, implement, test, and monitor safeguard programs.
- Respond to security events, and remediate and adjust based on monitoring.
- Select appropriate service providers.

To these ends, GLBA gives authority to states and eight federal agencies to administer and enforce two rules: the Financial Privacy Rule and the Safeguards Rule. Financial institutions are the applicable audience for these two regulations. This includes banks, securities firms, insurance companies, and companies providing several other types of financial products and services. These cover a wide range of services, including lending, brokering or servicing any type of consumer loan, transferring money, safeguarding money, preparing individual tax returns, providing financial advice, credit counseling, providing residential real estate settlement services, collecting consumer debts, and several other common financial services. Let's examine the two rules as well as discuss the concept of pretexting:

The Financial Privacy Rule This rule covers the collection and disclosure of any personal financial information by financial institutions. It also applies to companies receiving such information, whether or not they are financial institutions.

The Safeguards Rule This rule requires that all financial institutions design, implement, and maintain safeguards protecting customer information. This applies to financial institutions that collect information from their own customers and any other financial institutions that receive customer information from other financial institutions, such as reporting agencies.

Pretexting There is also coverage of fraudulently obtaining personal information. Obtaining this information under false pretenses is called pretexting. Social engineering is often used to obtain the private information of others. Attackers solicit the data by using what appears to be a legitimate request.

Whether or not a financial institution discloses personal information, there must be a policy in place to protect the information from foreseeable threats in security and data integrity. Institutions must take reasonable protective measures to secure client data. This is a mandatory portion of GLBA compliance and also where wireless security comes into play. Maintaining data privacy is specifically covered in Title 5 of the Act, and has the following mandates covering data privacy:

- Requires clear disclosure to clients and others by all financial institutions of their privacy policy regarding the sharing of personal information with both affiliates and third parties.
- Requires a notice to consumers and an opportunity to opt out of sharing of personal information with nonaffiliated third parties, subject to certain limited exceptions. (Clients often must notify the institution of their desire to opt out.)
- Addresses a potential imbalance between the treatment of large financial services conglomerates and small banks by including an exception, subject to strict controls, for joint marketing arrangements between financial institutions. (Consumers often see the exemptions as additional offerings from the institutions' partners.)
- Clarifies that the disclosure of a financial institution's privacy policy is required to take place at the time of establishing a customer relationship with a consumer and not less than annually during the continuation of such relationship. (This may only be done once a year or may be done as changes to the initial agreement are implemented.)
- Provides for a separate rather than joint rulemaking to protect client privacy; the relevant agencies are directed, however, to consult and coordinate with one another for purposes of assuring, to the maximum extent possible, that the regulations that each prescribes are consistent and comparable with those prescribed by the other agencies.
- Allows the functional regulators sufficient flexibility to prescribe necessary exceptions and clarifications to the prohibitions and requirements of section 502.
- Clarifies that the remedies described in section 505 are the exclusive remedies for violations of the subtitle.
- Clarifies that nothing in this title is intended to modify, limit, or supersede the operation of the Fair Credit Reporting Act.
- Extends the time period for completion of a study on financial institutions' information-sharing practices from 6 to 18 months from the date of enactment.
- Requires that rules for the disclosure of institutions' privacy policies must be issued by regulators within six months of the date of enactment. The rules will become effective six months after they are required to be prescribed unless the regulators specify a later date.
- Assigns authority for enforcing the privacy provisions to the Federal Trade Commission and the federal banking agencies, the National Credit Union Administration, and the Securities and Exchange Commission, according to their respective jurisdictions, and it provides for enforcement of the privacy provisions by the states.

Identity theft, or using the identity of another person fraudulently, is a crime that hurts society as a whole, undermining financial institutions and the economy alike. GLBA compliance helps reduce the occurrence of this crime by regulating the protection of personal information collected by financial institutions. As financial institutions, like many other conservative organizations, begin to introduce wireless networking, additional steps must be taken to protect private data. Wireless networks introduce additional entry points into financial institutions' networks that must be monitored and secured. Mandated followers of GLBA include

- Banks, thrifts, and credit unions
- Financial advisers/investment firms
- Insurance companies
- Credit card/financial transaction processing providers
- Consumer credit reporting agencies
- Frontend/backend (core) providers and check printers
- Tax information preparers/processors
- Other service providers receiving customer information from financial institutions as regulated by the GLBA's Safeguards Rule

GLBA emphasizes that threats to information security are ever changing and describes them in general terms. Financial institutions must evaluate and adjust their information assurance programs to keep up with changes in technology, such as the increased use of Wi-Fi, and new or emerging threats that indicate they are vulnerable to attack or compromise.

More information about GLBA and identity theft can be found at `www.ftc.gov`.

Health Insurance Portability and Accountability Act (HIPAA)

The *Health Insurance Portability and Accountability Act (HIPAA)* establishes national standards for electronic healthcare transactions and national standards for providers, health insurance plans, and employers. The goal is to protect patient information and maintain privacy. One major goal of the *Privacy Rule* found in HIPAA is to assure that health information is properly protected while still allowing the flow of information about an individual's health needed to provide and promote high-quality healthcare as well as protecting the public's health and well-being. The Privacy Rule seeks a balance that permits important uses of information, while protecting the individual's privacy. Healthcare is a diverse industry. The Privacy Rule is designed to be flexible and comprehensive, allowing coverage of the large variety of uses and disclosures that need to be addressed within reasonable privacy.

The use of wireless technology has proven to be a method of delivering patient care that is faster for the patients and more convenient for the caregivers. However, this more timely delivery of care may introduce situations that are outside the industry's guiding privacy standard, HIPAA. Securing WLANs within the healthcare industry is critical. Interruptions in service can lead to life-threatening situations as well as data compromise. Electronic patient records make much of the behind-the-scenes work done by doctors, nurses, and other healthcare professionals an easier task. Multidimensional bar codes on drugs and arm bands allow caregivers rapid access to life-saving information. Unlike many other enterprise environments, hospitals have many areas of physical public access that also have access to data ports such as the patient rooms. The use of wireless networking increases this risk to care and patient data privacy.

The applicable audience for the Privacy Rule, as well as all the administrative simplification rules, are health plans, healthcare clearinghouses, and any healthcare provider that transmits health information in electronic form in connection with transactions for which the Secretary of Health and Human Services (HHS) has adopted standards under HIPAA. The electronic transfer of health information includes wireless transmissions. Under section 164.530(i), HIPAA requires covered entities to develop and implement written privacy policies and procedures that are consistent with the Privacy Rule. They must also assign a person to oversee the procedure for receiving complaints about their privacy procedures and to provide individuals with information about the procedures upon request. The two largest sections that pertain to wireless as well as networking in general are as follows:

Mitigation Covered entities must reduce the effect of any loss of private data, to the extent practicable, and any harmful effect it learns was caused by use or disclosure of protected health information. This loss could have been caused by its own workforce or its business associates in violation of its privacy policies and procedures or the Privacy Rule.

Data Safeguards Covered entities must maintain reasonable and appropriate administrative, technical, and physical safeguards to prevent compromise of protected health information in violation of the Privacy Rule. Covered entities must also limit their incidental use and disclosure pursuant to otherwise permitted or required use or disclosure.

HIPAA allows each organization to determine the appropriate measures they must take to secure private information. This is largely due to the variation in healthcare areas. Both large and small organizations determine how best to meet the requirements of HIPAA.

Therefore, there is no specified format for compliance. In general, covered entities must do the following:

- Put in place administrative safeguards to manage the selection and execution of security measures chosen.
- Ensure physical safeguards are in place protecting electronic systems, buildings, and the equipment used from environmental and unauthorized intrusions that may compromise private data.
- Make certain that technical safeguards are in place, including automated processes to control access to private data and to protect private data.
- Conduct risk assessments and document security policies and procedures.
- Covered entities must have a device, such as a firewall, to screen traffic from the Internet.

Accountability is a great area to look for a wireless fit in HIPAA when trying to ensure your organization's compliance. The Privacy Rule has the foundation for accountability within an electronic health information exchange environment. It requires all covered entities involved in the exchange of protected health information, on paper or electronically, to comply with the administrative requirements of HIPAA and to extend these obligations to all business associates. The Privacy Rule promotes this accountability with established mechanisms addressing potential noncompliance. The Privacy Rule has standards, through a covered entity's voluntary compliance, a resolution agreement, and a corrective action plan, or the imposition of civil money penalties ranging from $100 to $50,000 or higher per incident.

More information about HIPAA can be found at `www.hhs.gov/ocr/privacy`.

Payment Card Industry (PCI) Standard

As more of us continue to rely on credit cards as our primary method of payment, more of us risk losing our card numbers to attackers and identity thieves through unsecure processing and/or storing of our cardholder information. The Payment Card Industry (PCI) realizes that in order to sustain continued business growth, measures must be taken to protect customer data and card numbers. The PCI Security Standards Council (SSC) has implemented regulations for those processing and storing cardholder information. This is commonly referred to as the *Payment Card Industry (PCI) Standard*. Within this standard are components governing the use of wireless devices.

As a further explanation of PCI requirements governing wireless devices, the information supplement PCI Data Security Standard (DSS) Wireless Guideline was prepared by the PCI SSC Wireless Special Interest Group (SIG) Implementation Team and released in July 2009 and updated in 2011. The PCI standards cover many aspects of credit card use, acceptance, and processing. This standard is designed to protect the cardholder data environment (CDE). The CDE is defined as the computer environment wherein cardholder data is transferred, processed, or stored, and any networks or devices directly connected to that environment.

Let's focus on the PCI implications of wireless networking and device use. The PCI DSS wireless requirements can be broken down into two main categories: (1) generally applicable wireless requirements and (2) requirements applicable for in-scope wireless networks. They cover protection against rogue devices and protection of networks storing or processing cardholder data.

What does PCI DSS call a rogue AP? A *rogue AP* is any device that adds an unauthorized (and therefore unmanaged and unsecured) WLAN to the organization's network. This rogue device need not be placed by an attacker. In this context, a rogue AP could be added by inserting a WLAN card into a back-office server, attaching an unknown WLAN router to the network, or by various other means. Most rogue devices are placed by internal sources, often just trying to be more productive.

There is also provision for legitimate wireless use on the same network as the CDE. If an organization decides to deploy a WLAN for any purpose and connects the WLAN to the CDE, that WLAN becomes part of the CDE and is therefore within the scope of the PCI

DSS and must comply with the PCI guidelines for wireless device use. If an organization deploys a WLAN that in no way touches the network that is part of the CDE, then that WLAN is out of the scope of the PCI DSS. However, should that WLAN touch the firewall that is connected to the CDE, the firewall then falls into the scope of PCI DSS. Roughly, any network that processes or stores cardholder information or any network, wired or wireless, that touches the CDE must be PCI DSS compliant. PCI DSS requires that wireless networks that do not store, process, or transmit cardholder data must be isolated from the CDE using a firewall. Any organization required to comply with PCI standards using WLANs that do not touch the CDE must also be able to prove that there is no connectivity from the WLANs to the CDE.

Specifically the two main areas of PCI covering wireless devices state the following:

Generally Applicable Wireless Requirements All organizations that wish to comply with PCI DSS should have in place methods to protect their networks from attacks via rogue or unknown wireless APs and clients. They apply to organizations regardless of their use of wireless technology and regardless of whether the wireless technology is a part of the CDE.

Requirements Applicable for In-Scope Wireless Networks All these requirements apply in addition to the universally applicable set of requirements. All organizations that wish to comply with PCI DSS that transmit payment card information over wireless technology should have in place mechanisms to protect those systems. These requirements are specific to the usage of wireless technology that is in scope for PCI DSS compliance.

Organizations that must comply with the PCI standard must also maintain an inventory of devices. It is recommended that such organizations scan for wireless devices to keep an up-to-date listing of all known WLAN devices, making rogue detection easier. PCI standard section 11.1 states that organizations must test for the presence of wireless APs by using a wireless analyzer at least quarterly or by deploying a WIDS/WIPS to identify all wireless devices in use. This means that every location, retail outlet, warehouse, office, and so on that stores or processes cardholder information be scanned each quarter or have a WIDS/WIPS deployed to scan for them. Since travel to thousands of locations of a large retailer, for example, is time consuming and expensive, many organizations have come to rely on the scanning and protection offered by WIPS. Section 12.9 states that there must also be an incidence response plan that is followed should a rogue device be detected. This is another compliance requirement that can be automated by an effective WIPS deployment.

The PCI Security Standards Council recommends the following with regard to rogue wireless devices and PCI compliance:

- Use a wireless analyzer or a WIDS/WIPS at least quarterly at all locations to detect unauthorized/rogue wireless devices that could be connected to the CDE. For large organizations having several CDE locations, a centrally managed WIDS/WIPS to detect and contain unauthorized/rogue wireless devices is recommended.
- Enable automatic alerts and containment mechanisms on the wireless IPS to eliminate rogues and unauthorized wireless connections into the CDE.
- Create an incident response plan to physically eliminate rogue devices immediately from the CDE in accordance with PCI DSS requirement 12.9.

Section 1.2.3 requires that organizations install perimeter firewalls between any wireless networks and the CDE. These firewalls must be configured to deny or control (if such traffic is necessary for business purposes) any traffic from the wireless environment going into or leaving the CDE. These firewalls are required to be audited at least every six months. This section also states that the reliance on VLAN protection of the CDE alone is not sufficient. To reduce the risk to the CDE, any protocol or traffic that is not necessary in the processing or storing of credit card transactions or cardholder data should be blocked. To this end, the PCI Security Standards Council recommends the following practices for firewalls:

- Use a stateful packet inspection firewall to block wireless traffic from entering the CDE. Augment the firewall with a WIDS/WIPS.
- Do not use VLAN-based segmentation with MAC address filters for segmenting wireless networks.
- Monitor firewall logs daily and verify firewall rules at least once every six months.

PCI DSS compliance for networks that include WLANs as a part of the CDE requires extra attention to WLAN-specific technologies and processes such as the following:

- Physical security of wireless devices
- Changing of default passwords and settings on wireless devices
- Logging of wireless access and intrusion prevention
- Strong wireless authentication and encryption
- Use of strong cryptography and security protocols
- Development and enforcement of wireless usage policies

Many older devices and WLANs touching or used by the CDE have been using WEP for security. Under PCI compliance, new wireless implementations were prohibited from implementing WEP after March 31, 2009. PCI compliance further required that WEP not be used and stronger security be in place on any network touching the CDE by the end of June 2010. This mandate meant that many wireless devices had to be upgraded or replaced. The PCI Security Standards Council recommends the following practices for WLANs and APs:

- Enable WPA or WPA2, and make sure that default PSKs are changed. Enterprise mode is recommended.
- Disable SNMP access to remote APs if possible. If not, change default SNMP passwords and use SNMPv3 with authentication and privacy enabled.
- Do not advertise organization names in the SSID broadcast.
- Synchronize the APs' clocks to be the same as other networking equipment used by the organization.
- Disable all unnecessary applications, ports, and protocols.

- WPA or WPA2 Enterprise mode with 802.1X authentication and AES encryption is recommended for WLAN networks.
- It is recommended that WPA2-Personal mode be used with a minimum 13-character random passphrase and AES encryption.
- Preshared keys should be changed on a regular basis.
- Centralized management systems that can control and configure distributed wireless networks are recommended.
- The use of WEP in the CDE is prohibited for all deployments.

The PCI requirements are much more specific than other industry standards covering wireless device use. For example, section 12.3 requires organizations to develop usage policies for critical employee-facing technologies to define proper use of these technologies for all employees and contractors. This requirement covers PDAs, remote access technologies, wireless technologies, removable electronic media (such as USB drives, laptops, and personal data/digital assistants [PDAs]), email usage, and Internet usage.

The PCI standard has many more wireless requirements than most other external influences. This is largely due to the widespread use of wireless devices in networks touching the CDE. There have been several large losses of cardholder information. One of the largest and most publicly known losses involved wireless devices. As technology continues to drive business efforts, look for even more regulation covering the use of electronic devices and communication, especially wireless networking due to its unbounded nature. Keep in mind that where there is something of value that can be stolen, such as the data in the CDE, attackers are more likely to exist. Being compliant is not always the same as being secure. However, the PCI guidelines are forcing those that process payment cards or store cardholder information to create safer environments for the data they use and store as part of meeting that compliance.

More information about the PCI standard can be found at `www.pcisecuritystandards.org`.

Detailed information about PCI wireless guidelines can be found at `www.pcisecuritystandards.org/pdfs/PCI_DSS_v2_Wireless_Guidelines.pdf`.

Compliance Reports

Many WIPS solutions and other WLAN auditing software tools have the capability of analyzing existing WLANs and generating industry-specific compliance reports. These compliance reports can be helpful when identifying areas in need of improvement. Furthermore, industry-specific compliance reports can be useful when writing WLAN policies.

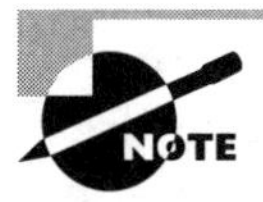

An example of a WIPS-generated compliance report can be downloaded from this book's web page, `www.sybex.com/go/cwsp2e`. The report is titled `PCI_Report.pdf`.

802.11 WLAN Policy Recommendations

Although a detailed and thorough policy document should be created for every organization's deployment of wireless technology, we highly recommend these five wireless security policies:

BYOD Policy Employees like to bring their personal Wi-Fi devices, such as tablets and smartphones, to the workplace. Employees usually expect to be able to use their personal Wi-Fi devices on the secure corporate WLAN. Each employer needs to define a bring your own device (BYOD) policy that clearly states how personal devices will be onboarded onto the secure corporate WLAN. The BYOD policy should also state how the personal devices can be used while connected to the company WLAN and which corporate network resources are accessible. BYOD is discussed in great detail in Chapter 10, "BYOD and Guest Access."

Guest Access Policy Guest users and devices will most likely need to access the corporate WLAN. Guest WLAN access is typically used only to provide a wireless gateway to the Internet. A guest access policy in regard to guest credentials, guest access times, guest SSID security, and firewall policies will need to be clearly defined.

Remote Access WLAN Policy End users will be taking their laptops and handheld devices off site and away from company grounds. Most users will likely use wireless networks at home and at wireless hotspots to access the Internet. By design, many of these remote wireless networks have absolutely no security in place, and it is imperative that a remote access WLAN policy be strictly enforced. This policy should include the required use of an IPsec VPN solution to provide device authentication, user authentication, and strong encryption of all wireless data traffic. Hotspots are prime targets for malicious eavesdropping attacks. Personal firewalls should also be installed on all remote computers. Personal firewalls will not prevent hijacking attacks or peer-to-peer attacks but will prevent attackers from accessing most critical information. As you have learned, endpoint WLAN policy enforcement software solutions exist that force end users to use VPN and firewall security when accessing any wireless network other than the corporate WLAN. The remote access policy is mandatory because the most likely and vulnerable location for an attack to occur is at a public access hotspot.

Rogue AP Policy No end users should ever be permitted to install their own wireless devices on the corporate network. This includes access points, wireless routers, wireless USB clients, and wireless cards. Any users installing their own wireless equipment could open unsecured portals into the main infrastructure network. This policy should be strictly enforced.

Ad Hoc Policy End users should not be permitted to set up ad hoc or peer-to-peer networks. Peer-to-peer networks rarely use encryption, are susceptible to peer attacks, and can serve as unsecured portals to the infrastructure network if the computer's Ethernet port is also in use.

Wireless LAN Proper Use Policy A thorough policy should outline the proper use and implementation of the main corporate wireless network. This policy should include proper installation procedures, proper security implementations, and allowed application use on the wireless LAN.

WIDS Policy Policies should be written defining how properly to respond to alerts generated by the wireless intrusion detection system. An example would be how to deal with the discovery of rogue access points and all the necessary actions that should take place.

These seven policies are simplistic but are a good starting point in writing a well-rounded WLAN security policy document. Keep in mind that a proper WLAN policy document should be much more detailed and cover many more topics than XX, and be strictly enforced.

Summary

All of the wonderful benefits and return on investment offered by wireless networking may cease to exist or cease to be allowed should a security and management policy not be in place and enforced. The security of the WLAN should be an integral part of the WLAN design and deployment. The wireless devices must be physically secure. The transmission of data wirelessly must also be secured. To reach the goal of effective and secure communications, written and enforced policy must be in place and communicated to all end users from conception forward. Data confidentiality, integrity, and nonrepudiation have long been goals of network security. The introduction of wireless networking does not change these goals but rather introduces new challenges in meeting them. As the use of wireless technology continues to grow, fueled by productivity gains, convenience, and cost savings, new security challenges are continually faced by networks. WLANs allow users untethered connectivity. This new freedom can often bypass existing security measures. As technology and business requirements change, security measures must keep up in order to avoid loss of private data and/or network functionality. Written, enforced, and adaptable policies help to ensure that when business requirements and technology move forward, security measures will be in place to protect the resources of the organization.

Exam Essentials

Explain the differences between general and functional WLAN security policies. General policy establishes a framework for enforcement; functional policy defines technical aspects of security.

Describe all aspects of general policy creation and management. General policy requires a statement of authority, defines an applicable audience, and requires risk assessment and threat analysis. Security auditing and violation procedures must also be clearly defined.

Explain the importance of functional password policy and functional RBAC policy. All end users should abide by well-defined password policies that require strong passwords and possibly two-factor authentication. Passwords should never be shared. RBAC policies ensure that certain groups of users are only allowed access to certain network resources.

Explain the importance of functional change control policy and WLAN monitoring policy. Change control procedures govern the upgrading or downgrading of firmware and software as well as hardware maintenance. Change control processes help ensure that changes to a product or system are introduced in a controlled and coordinated manner. WLAN monitoring policy defines what steps should be taken when an attack is detected by the WIDS/WIPS solution.

Explain the importance of functional authentication and encryption policies. Whenever possible, a functional policy requiring the use of 802.1X/EAP tunneled authentication should be mandated. Consideration should also be given to mandate the use of multi-factor authentication solutions that require two sets of credentials before access is granted. Dynamic CCMP/AES encryption should always be mandated for data privacy.

Define the various aspects of endpoint security policy. Minimum endpoint compliance should mandate the use of IPsec VPNs and personal firewalls. Heavy consideration should also be given to implementing an endpoint policy enforcement software solution.

Describe the functional policies of acceptable use, physical security, and remote office security. All end users should only use applications that are intended for use on the WLAN. All WLAN infrastructure should be locked and protected from theft and vandalism. Policies should also protect WLANs at corporate headquarters as well as at remote offices.

Review Questions

1. Your organization is developing a wireless device usage policy. Which group(s) should be represented in the committee that actually develops this policy?

   A. IT staff

   B. Security staff

   C. Management

   D. End users

   E. Support staff

   F. All of the above

2. A large retail store chain has decided to implement wireless registers in their gardening department to allow for seasonal reconfiguration of the sales area. Knowing that they must follow the PCI standard since they process credit cards at these stations, they have hired you to ensure that their designs are PCI compliant. Which portion of their network designs should you examine to help determine if they are compliant?

   A. The wireless registers

   B. The APs used by the wireless registers

   C. The switches to which the APs are connected

   D. The portions that touch the CDE

   E. The PCI standard does not allow wireless device use.

3. A merchant hires you to install a wireless network for processing credit cards in all of their new retail outlets as the stores are opened over the next six months. They have already purchased APs at an auction for you to install. Upon inspection of the APs, you find that they cannot be upgraded to support authentication and encryption beyond 64-bit WEP. You immediately inform the merchant that the APs they have purchased cannot be deployed in their new stores. What is the best reason you can give the merchant for not being able to use the APs they have already purchased?

   A. They were not purchased through you, so you cannot warranty them.

   B. PCI mandates that 128-bit WEP be used in all new deployments.

   C. 64-bit WEP uses a 24-bit initialization vector and 128-bit WEP uses a 48-bit initialization vector.

   D. PCI compliance does not allow new installation to use WEP.

   E. WEP encryption will slow down card processing, resulting in angry customers.

4. To protect users and their laptops as they travel, your organization has decided to implement a wireless mobile device policy. You have explained to management that you are not able to offer the traveling users the same protection they have within your facilities as they travel, because your company does not own or manage all the hotspots to which

your users may need to connect. The executives have instructed you to develop a policy that can be enforced and followed that will improve on the security users find at hotspots. What three things can you include in the policy that will help increase wireless security for traveling users while still allowing them to use the hotspots for connectivity? (Choose all that apply.)

A. All wireless connections must use PEAP.

B. All wireless connections must use AES.

C. All wireless connections must use a VPN.

D. All wireless connections must use a personal firewall.

E. All wireless connections must use antivirus software.

5. As the head of network security for your organization, you discover an unauthorized access point plugged into your network. Upon investigation, the person who placed the rogue AP on your network is determined to be one of your own users, not an attacker. You wish to proceed with appropriate disciplinary measures but are not allowed to do so by the Human Resources department. What is the most likely reason that the HR staff would not allow disciplinary measures to be taken?

A. There is no written policy preventing users from adding their own APs to the network.

B. The user who placed the AP is a member of management.

C. The user who placed the AP is a member of the union.

D. You are not the user's direct supervisor and cannot discipline the user.

E. No data was lost or altered as a result of the rogue device placement.

6. You have been working with the network staff to expand the wireless coverage within your customer's building. Halfway through the project you are asked by a member of management to stop and leave the premises. Which step in WLAN deployment did you most likely not take prior to beginning the WLAN expansion?

A. Obtaining a Scope of Work (SOW) agreement

B. Signing a mutual nondisclosure agreement (NDA)

C. Reviewing written corporate security policies

D. Requesting that a facility escort be present

7. When deploying a corporate 802.11 WLAN, what password-related items should always be included in a security policy? (Choose all that apply.)

A. The password policy should mandate a procedure on how passphrases are created for handheld devices that use WPA2-Personal.

B. End-user WPA2-Enterprise passwords should contain numbers, special characters, and upper- and lowercase letters.

C. Client-side certificates should always be used instead of passwords when securing a WLAN.

D. Machine authentication should always be mandated.

8. When developing a security policy, it is important to include many influences such as internal requirements, governmental regulations, and industry standards. When is it allowable not to include a specific external influence in your policy development?
   - **A.** When there is little to no chance of being audited for compliance
   - **B.** When your organization is not part of the applicable audience of the external policy influence
   - **C.** When implementing wireless devices without the knowledge of the governing body that developed the external policy
   - **D.** When adherence to the external regulation or standard is cost prohibitive

9. Over the years, your company has deployed various wireless devices throughout its network. Many of these devices have been in place for quite some time and can only be configured for secure transmissions using WEP. What can you do to improve network security with the least disruption of service?
   - **A.** Create a new security policy requiring WPA2 to be used.
   - **B.** Modify the existing policy requiring WPA2 to be used.
   - **C.** Update the firmware on legacy devices to support WPA2.
   - **D.** Replace legacy hardware with new devices that support WPA2.

10. After consulting your written security policy, to meet the new demands of an industry standard with which your organization must be compliant, an administrator logs into your WLAN controller and changes the authentication and encryption configurations on all your APs. The help desk becomes overwhelmed with calls from angry users stating that they can no longer access the network. One by one, the users are reconfigured to reconnect to the network, causing significant loss of time. Which portion of a well-written security policy is most likely missing from your company's wireless security policy that caused this problem?
    - **A.** External influence compliance
    - **B.** Authentication requirements
    - **C.** Encryption compliance
    - **D.** Change control process
    - **E.** User notification process

11. Your network just passed an external compliance audit. However, the same day it passed the audit there was an intrusion that compromised your company's private data. Your staff conducted an internal audit immediately after the intrusion was detected and found that your network is still compliant. The security audit files indicate that the network was compliant during the compromise as well. What is the most likely reason that this compromise was possible on your compliant network?
    - **A.** The auditor was indeed the attacker.
    - **B.** The attacker was a network administrator.
    - **C.** Being compliant is not the same as being secure.
    - **D.** Your network is not the applicable audience for this compliance.

12. One of the first things you should do when creating a wireless security policy is to conduct an impact analysis to assist you in determining the level of acceptable risk. What should be included in your impact analysis? (Choose all that apply.)

    A. Direct costs of compromise
    B. Indirect cost of compromise
    C. Legal implications
    D. Plausible deniability
    E. Implementation costs

13. You have been hired to conduct a vulnerability assessment for a chipset manufacturing company. As you conduct the physical inspection of the WLAN devices, you notice that some of their APs are in locked enclosures and others are not. Before you report the unlocked APs as being vulnerable, what should you do first?

    A. Determine if the locked and unlocked APs are on the same network.
    B. Press the reset button on the unlocked APs to prove their vulnerability.
    C. Open the locked enclosures to make sure APs are in them.
    D. Verify the physical protection requirements within the written WLAN security policy.

14. While conducting a routine security analysis of your company's network, you discover an unauthorized access point installed on the network under the vice president's desk. What should you do in dealing with the rogue device since the vice president is likely to have placed the device there to provide coverage in the office for her own wireless devices?

    A. Remove the device and take it to the IT office for forensics.
    B. Unplug the device from the network but leave it in place.
    C. Follow the procedures for rogue device management in company policy.
    D. Ask the vice president what she would like done with the device.

15. Although your organization's written policy and many external policy influences may require only periodic scanning for rogue devices, you are trying to make a case for deploying a WIPS. What are some of the benefits of using a WIPS to achieve policy compliance that make it more desirable than using periodic handheld or laptop-based scanning solutions? (Choose all that apply.)

    A. WIPSs are less expensive and easier to implement.
    B. WIPSs can provide 24-hour scanning and protection.
    C. WIPSs are a more scalable solution for security.
    D. WIPSs can correlate across multiple locations.
    E. WIPSs can provide both compliance and security.

16. Your WIPS has detected several ad hoc devices within your building. After performing location tracking in the WIPS on these devices, you and your staff physically locate them. All of the ad hoc devices detected and found have turned out to be new printers that shipped with their wireless cards enabled for ad hoc wireless networking. Ad hoc wireless

networking is specifically disallowed in written security policy. Which section of your security policy should be updated to reduce the likelihood of this happening again?

A. Printing policies

B. Deployment policies

C. Purchasing policies

D. Rogue mitigation policy

17. To provide temporary access to the Internet for a group of customers visiting your corporate headquarters, an administrator has installed an access point in your boardroom. Worried about unauthorized access, the administrator has set up the AP to use WPA2 with a very long and complex key along with AES encryption. Your WIPS sees this AP as a rogue and begins to contain it using port suppression and wireless rogue containment procedures. Your visiting customers are not unable to reach the Internet as desired. What part of your security policy may need to be updated to avoid this problem in the future?

A. WIPS usage

B. Rogue containment

C. Rogue definitions

D. Secret key length

E. Guest access

18. Your WIDS detected a rogue AP and sent an email alert to an administrator in the same building in which the rogue was detected. The administrator reads the email and does not respond to the alarm, but rather waits until after lunch and then calls you for direction. This delay has allowed the device to be on the network for over an hour and placed the organization's private information at risk. What is the most likely reason the administrator took no action?

A. The WIDS detected the rogue, and no further action was required.

B. You are the only person who knows how to deal with rogue APs.

C. The security policy lacks response procedures.

D. Only a properly configured WIPS can mitigate a rogue AP.

19. A very well-written wireless security policy is in place within your organization. However, you still find rogue devices, poorly configured APs, users who are wired leaving their wireless cards enabled, and ad hoc networks throughout your building—all of which are against policy. Policy requires that such events are remediated and documented as they occur. Your WIPS has been configured to assist you in meeting this requirement and is doing so. The only people allowed in your building are your own employees. You are not under an attack but still keep finding the same problems daily. Which portion of good security policy is most likely lacking that has led to this problem?

A. Impact analysis

B. Management approval

C. User training

D. Documentation

20. An employee has installed his own AP on your network. Each day when he leaves, he unplugs the AP and plugs it back in the morning. He has not implemented any security on the AP. After months of being on the network, this "rogue" AP finally leads to a compromise of corporate secrets. Corporate security policy prohibits the installation of APs without approval. What other requirement(s) should be added to the security policy that could have prevented this compromise? (Choose all that apply.)

A. Monitoring for rogue devices

B. Rogue device remediation

C. User training

D. Policy enforcement procedures

Appendix A

Answers to Review Questions

Chapter 1: WLAN Security Overview

1. E. The 802.11-2012 standard defines what is known as a robust security network (RSN) and robust security network associations (RSNAs). CCMP/AES encryption is the mandated encryption method, whereas TKIP/ARC4 is an optional encryption method.

2. A, C. The Wi-Fi Protected Access (WPA) certification was a snapshot of the not-yet-released 802.11i amendment, supporting only the TKIP/ARC4 dynamic encryption-key generation. 802.1X/EAP authentication was required in the enterprise, and passphrase authentication was required in a SOHO or home environment. LEAP is Cisco-proprietary and is not specifically defined by WPA. Neither dynamic WEP nor CCMP/AES were defined for encryption. CCMP/AES dynamic encryption is mandatory under the WPA2 certification.

3. B, D. The 802.11-2012 standard defines CCMP/AES encryption as the default encryption method, whereas TKIP/ARC4 is the optional encryption method. This was originally defined by the 802.11i amendment, which is now part of the 802.11-2012 standard. The Wi-Fi Alliance created the WPA2 security certification, which mirrors the robust security defined by the IEEE. WPA2 supports both CCMP/AES and TKIP/ARC4 dynamic encryption key management.

4. D, F. The required encryption method defined by an RSN wireless network is Counter Mode with Cipher Block Chaining Message Authentication Code Protocol (CCMP), which uses the Advanced Encryption Standard (AES) algorithm. An optional choice of encryption is the Temporal Key Integrity Protocol (TKIP). The 802.11-2012 standard also requires the use of an 802.1X/EAP authentication solution or the use of preshared keys for robust security. It should be noted that TKIP encryption cannot be used for 802.11n and 802,11ac data rates.

5. A, D. 802.11-2012 is the IEEE standard, and WEP (Wired Equivalent Privacy) is defined as part of the IEEE 802.11-2012 standard. PSK is not a standard; it is an authentication technique. Wi-Fi Multimedia (WMM) is a Wi-Fi Alliance certification program that enables Wi-Fi networks to prioritize traffic generated by different applications. WPA2 is a certification program that defines Wi-Fi security mechanisms.

6. C. The IEEE 802.11-2012 standard defines communication mechanisms at only the Physical layer and MAC sublayer of the Data-Link layer of the OSI model. The Logical Link Control (LLC) sublayer of the Data-Link layer is not defined by the 802.11-2012 standard. WPA is a security certification. FSK is a modulation method.

7. D, E, F. WPA-2 is a Wi-Fi Alliance certification and not an encryption method. The WPA-2 certification does mandate the use of CCMP encryption and TKIP is optional. The IEEE 802.11-2012 standard defines the use of both CCMP and TKIP dynamic encryption methods. Also defined by the IEEE is the use of static WEP encryption.

8. C. Requests for Comments are known as RFCs and are created by the Internet Engineering Task Force (IETF), which is guided and directed by the Internet Engineering Steering Group (IESG).

9. A, B, D. There is no 802.11X amendment. The x in 802.1x should be capitalized (802.1X), and the N in 802.11N should not be capitalized (802.11n). These are not trivial errors. Standards and amendments should be written and used with the proper capitalization.

10. B, C, D, E, F. By default, WPA-Personal and WPA-Enterprise use TKIP for encryption. WPA-2 Personal, WPA-2 Enterprise, and 802.11-2012 (RSN) mandate the use of CCMP, but TKIP is optional. TKIP is essentially being phased out because the IEEE and Wi-Fi Alliance mandate the use of only CCMP encryption for 802.11n and 802.11ac data rates.

11. E. A code is simply a way of representing information in a different way, such as ASCII or Morse code.

12. A, E. Wi-Fi Protected Access (WPA) version 2 mirrors the 802.11i security amendment, mandating CCMP/AES dynamic encryption key management. TKIP/ARC4 dynamic encryption key management is optional. 802.1X/EAP authentication is required in the enterprise and passphrase authentication in a SOHO environment. PEAP can be used for 802.1X/EAP authentication but is not specifically defined by WPA. Dynamic WEP is not defined for encryption.

13. C. Core devices perform high-speed switching as part of the backbone of the network. Distribution devices route traffic between VLANs and subnets. Access devices provide connectivity directly to the wireless end user. The OSI network layer and session layer are not involved with wireless communications.

14. D. Cryptography is the science of concealing the plaintext and then revealing it. Cryptanalysis is the science of decrypting the ciphertext without knowledge of the key or cipher. Cryptology is the practice or science that includes the techniques to encrypt and decrypt information. Steganography strives to hide the fact that there is a message by concealing it.

15. A. WEP was defined in the original 802.11 standard. WPA was considered a partial preview of the 802.11i amendment. WPA-2 was defined as somewhat of a mirror of 802.11i. 802.11i was incorporated into the revised 802.11-2007 standard, which was later revised as the 802.11-2012 standard.

16. C. The original 802.11 standard ratified in 1997 defined the use of a 64-bit or 128-bit static encryption solution called Wired Equivalent Privacy (WEP). Dynamic WEP was never defined under any wireless security standard. The use of 802.1X/EAP, TKIP/ARC4, and CCMP/AES is defined under the current 802.11-2012 standard.

17. E. The IEEE 802.11 task group defines the WLAN standards, and the Wi-Fi Alliance defines interoperability certification programs.

18. A, C, E. WPA-2 and 802.11i are used to allow or deny access to the network, but not to limit access to parts of it. VLANs, firewalls, and role-based access control (RBAC) can all limit or restrict access to parts or segments of the network.

19. B, D. When a wireless network is implemented in an enterprise environment, the use of 802.1X/EAP is mandatory. When used in a SOHO environment, preshared keys are used.

20. E. WIDS is a wireless intrusion detection system responsible for detecting potentially malicious wireless activity on your network. WIPS is a wireless intrusion prevention system capable of mitigating attacks on your network. FIPS is the Federal Information Processing Standards (FIPS), a U.S. standard mandating how to protect and secure data communications within government agencies.

Chapter 2: Legacy 802.11 Security

1. C, D. In order for a client to connect to the WLAN and pass data, the client must authenticate and associate to the AP. The other three choices could occur, but do not have to.

2. A, C. When a WEP encrypted frame is created, the 24-bit Initialization Vector is included in the Layer 2 header of the 802.11 data frame in cleartext format. A Key Identifier is also included, indicating to the receiving computer which of the four potential WEP keys it should use to decrypt the frame.

3. B. 128-bit WEP is known as WEP-104 in the 802.11-2012 standard. A 104-bit WEP key is provided by the user and the system adds to it the 24-bit Initialization Vector to equal 128 bits.

4. A, C. When SSID cloaking, or Hidden SSID, is enabled, beacon management frames are still transmitted, but the SSID field is set to null. The probe response frame is almost identical to the beacon frame, and it too has the SSID field set to null. In any environment, the probe request frame may or may not contain a value for the SSID field, depending on whether the client knows the SSID of the network to which it is connecting. The AP cannot stop transmitting beacons because the frames contain additional information that is critical to the functioning of the network. When an AP receives a probe request frame, by default it will respond with a probe response frame.

5. A, B, E. WEP, whether it is statically configured or whether it changes periodically, uses this cipher to encrypt and decrypt frames. MPPE (Microsoft Point-to-Point Encryption) uses the ARC4 algorithm. TKIP (Temporal Key Integrity Protocol), which is covered in later chapters, also uses ARC4.

6. B, C, D, E. WEP is the only option that is actually defined by the 802.11-2012 standard. All the other options are considered to be non-802.11 security measures.

7. C. The original 802.11 standard ratified in 1997 defined the use of a 64-bit or 128-bit static encryption solution called Wired Equivalent Privacy (WEP). WEP is considered pre-RSNA security. Dynamic WEP was never defined under any wireless security standard. The use of 802.1X/EAP, TKIP/ARC4 and CCMP/AES is defined under the current 802.11-2012 standard for robust network security.

8. A, D. Temporal Key Integrity Protocol (TKIP) is defined in the 802.11-2012 standard and is still considered to be an RSN mechanism. Point-to-Point Tunneling Protocol (PPTP) is a VPN technology that is not part of the 802.11-2012 standard. Shared Key authentication and Wired Equivalent Privacy (WEP) are the two pre-RSNA (robust security network association) security mechanisms that have been deprecated. Deprecated technologies have been superseded by new technologies and should be avoided. Open System authentication is the one pre-RSNA security mechanism that has not been deprecated.

9. A, C, E, F. Each group should be configured with a separate SSID and a separate VLAN. There will be an 802.1Q trunk connection from the standalone AP to the access layer switch. This trunk will carry the traffic from all of the VLANs that are supported on

the AP and the switch. If legacy encryption such as static WEP is being used, a different encryption key should be used for each SSID. Dynamic encryption such as CCMP/AES and 802.1X/EAP security would be preferred. The best strategy is to assign multiple VLANs and access policies for different groups of users and/or devices to a single SSID by leveraging RADIUS attributes.

10. C. Up to four WEP keys can be entered on a Wi-Fi device. In addition to four WEP keys being entered, one will be designated to be used to encrypt all transmitted data. When the encrypted frame is received, part of the frame tells the receiving system which key (1, 2, 3, or 4) was used to encrypt the frame. The receiving system then attempts to decrypt the frame using the specified key. If the value of the key is the same on the receiving system, then the frame will be decrypted. Each system can use a separate key to encrypt the data.

11. A, E. Since there are only two 802.11 authentication frames, Open System authentication is being used. Shared Key authentication would generate four 802.11 authentication frames. If 802.1X/EAP or WEP were being used, then the client would be doing Layer 2 encryption and the DHCP frames would be encrypted and not visible. Therefore, 802.1X/EAP and WEP are not being used. IPsec VPNs utilize Layer 3 encryption that would allow Laura to see the DHCP exchange and any other IP traffic.

12. A, C, D. The graphic shows an 802.11 Shared Key authentication that is made up of four authentication frames: an authentication request followed by a cleartext challenge frame, followed by a challenge response with the cleartext data encrypted, followed by an authentication response. 802.1X/EAP works together with Open System authentication but cannot be deployed when WEP is used. In order to use Shared Key authentication, WEP must be enabled. Shared Key authentication is optional with WEP, although not recommended.

13. B, E. The WLAN discovery utility displays the SSID field with a value of Unknown ten times. These are hidden or cloaked networks. Cloaking the SSID usually keeps the SSID hidden from most WLAN discovery tools that use null probe requests. Any of the SSIDs shown could be hidden; however, since hiding an SSID does not guarantee that it cannot be seen, from this screen there is no way of knowing which, if any, of the other SSIDs are configured for cloaking. Because hidden SSIDs are still transmitted in cleartext in some 802.11 management frames, a protocol analyzer can always find a hidden SSID. Even simple WLAN discovery tools may sometimes discover hidden SSIDs.

14. B. 128-bit WEP encryption uses a secret 104-bit static key that is combined with a 24-bit Initialization Vector for an effective key strength of 128 bits.

15. B. The graphic shows a two-frame Open System authentication. 802.1X/EAP works together with Open System authentication. An unencrypted session uses Open System authentication.

16. C, E. Message Digest 5 (MD5) and Secure Hash Algorithm 1 (SHA-1) are both hash algorithms. Diffie-Hellman is a protocol that allows two devices to exchange a secret key across an insecure communications channel. MS-CHAPv2 is used for authentication with Point-to-Point Tunneling Protocol (PPTP). Internet Security Association and Key Management Protocol (ISAKMP) uses IKE to set up security associations.

17. B, C. TKIP uses a two-phase mixing function to combine a 128-bit temporal key, the transmitter address (TA), and the TKIP sequence counter (TSC) as seeding material for the ARC4 algorithm. The 128-bit temporal key is generated by the 4-Way Handshake process. Therefore, the temporal keys used by TKIP/ARC4 are either a pairwise transient key (PTK) or a group temporal key (GTK).

18. D. When Open System authentication is used without WEP, all client stations are allowed to join the BSS and no data privacy is provided. WEP encryption is used as part of the Shared Key authentication process, but the WEP key could potentially be exposed and the 802.11 data frames are at risk. If static WEP is the chosen encryption solution, the WEP key is slightly safer when used with Open System authentication. Because WEP has been cracked, it should be avoided entirely. 802.1X/EAP provides the strongest authentication solution and is used together with dynamic encryption such as TKIP/ARC4 and CCMP/AES.

19. B, C, D. SSIDs can be up to 32 characters long, they are case sensitive, and they can have spaces in them, although we do not recommend putting spaces in an SSID.

20. D. Open System authentication and Shared Key authentication are both Layer 2 authentication methods defined by the 802.11-2012 standard. Open System authentication must be used in an 802.1X framework. Shared Key authentication requires WEP, which cannot be used with 802.1X. Extensible Authentication Protocol is a universal authentication framework but is not defined by 802.11. MS-CHAP is also not defined by 802.11.

Chapter 3: Encryption Ciphers and Methods

1. A. CCMP/AES encryption will add an extra 16 bytes of overhead to the body of an 802.11 data frame. Eight bytes are added by the CCMP header and 8 bytes are added by the MIC. WEP encryption will add an extra 8 bytes of overhead to the body of an 802.11 data frame. When TKIP is implemented, because of the extra overhead from the extended IV and the MIC, a total of 20 bytes of overhead is added to the body of an 802.11 data frame.

2. C. When TKIP is implemented, because of the extra overhead from the extended IV and the MIC, a total of 20 bytes of overhead is added to the body of an 802.11 MPDU. CCMP/AES encryption will add an extra 16 bytes of overhead to the body of an 802.11 MPDU. WEP encryption will add an extra 8 bytes of overhead to the body of an 802.11 MPDU.

3. D. The IEEE 802.11n amendment states that an HT station should not use WEP or TKIP when communicating with other STAs that support stronger ciphers. HT STAs should not use pre-RSNA security methods to protect unicast frames if the RA or address 1 of the

frame corresponds to an HT STA. On September 1, 2009, the Wi-Fi Alliance also began requiring that all HT radios (802.11n) not use TKIP when using HT data rates. TKIP is also not supported for VHT radios (802.11ac) when using VHT data rates.

4. A. The AES algorithm is defined in FIPS PUB 197-2001. All AES processing used within CCMP uses AES with a 128-bit key and a 128-bit block size.

5. C, D. Asymmetric algorithms use a pair of keys—a private key that is used for decryption and a public key that is used for encryption. Asymmetric algorithms generally require more computer processing power than symmetric algorithms.

6. A, D. When using asymmetric encryption, the message is encrypted with the public key and decrypted with the private key. An asymmetric encryption protocol uses different keys to encrypt the data and decrypt the data. A symmetric encryption protocol uses the same key to encrypt the data and decrypt the data.

7. A, B. The migration from TKIP to CCMP can be seen in the IEEE 802.11n amendment, the IEEE 802.11ac amendment, and the IEEE 802.11-2012 standard, which all state that *High Throughput (HT)* or *Very High Throughput (VHT)* data rates are not allowed to be used if WEP or TKIP is enabled.

8. D. Additional Authentication Data (AAD) is constructed from portions of the MPDU header. This information is used for data integrity of portions of the MAC header. Receiving stations can then validate the integrity of these MAC header fields. The MIC protects the AAD information and the frame body for data integrity.

9. C. RC4 and RC5 were never FIPS encryption standards. 3DES is a FIPS encryption standard, but it uses three keys with an effective key size of 168 bits. AES is a FIPS encryption standard with key sizes of 128, 192, and 256 bits.

10. A, C, D. CCMP is the acronym for Counter Mode with Cipher-Block Chaining Message Authentication Code Protocol. Counter Mode is often represented as CTR. Cipher-Block Chaining is CBC. CBC-MAC is the acronym for Cipher-Block Chaining Message Authentication Code.

11. E. A pseudo-random function creates a pseudo-random value; however, this is simply a process to generate a number. There are no restrictions on how many times this function or value is used. A one-time password is a password that is used once; it is not necessarily random. Single-sign on is a way of providing a single login process for accessing multiple systems or resources. A throw-away variable does not exist.

12. A, D, E. WEP, TKIP, and CCMP are defined by the IEEE 802.11-2012 standard. GCMP is an encryption method standardized in the 802.11ad-2012 amendment and is an optional method defined under the 802.11ac-2012 amendment for VHT data rates. WPA and WPA2 are certification testing standards defined by the Wi-Fi Alliance.

13. A, B, D. WEP, TKIP, and CCMP use symmetric algorithms. WEP and TKIP use the ARC4 algorithm. CCMP uses the AES cipher. Public-key cryptography is based on asymmetric communications.

14. C. AES is based on the Rijndael algorithm. CCMP is an encryption protocol that uses the AES cipher. TKIP uses the ARC4. DES and 3DES are both block ciphers unrelated to Rijndael.

15. B, C. A stronger data integrity check known as a Message Integrity Code (MIC), or by its common name, Michael, was introduced with TKIP to correct some of the weaknesses in WEP. CCMP also uses a MIC. AES and DES are encryption algorithms and are not concerned with message integrity.

16. B, D, E, F. Certain fields in the MPDU header are used to construct the additional authentication data (AAD). The MIC provides integrity protection for these fields in the MAC header as well as in the frame body. All of the MAC addresses, including the BSSID, are protected. Portions of the other fields of the MAC header are also protected. Receiving stations will validate the integrity of these protected portions of the MAC header. For example, the frame type and the distribution bits, which are subfields of the Frame Control field, are protected. The AAD does not include the header Duration field, because the Duration field value can change due to normal IEEE 802.11 operation. For similar reasons, several subfields in the Frame Control field, the Sequence Control field, and the QoS Control field are masked to 0 and therefore not protected. For example, the Retry bit and Power Management bits are also masked and not protected by CCM integrity.

17. A, F. ARC4 is used in WEP and TKIP, and AES is used in CCMP. RC5 is a symmetric block cipher design by Ron Rivest. IPsec is a protocol suite that uses other encryption protocols, and it is not defined by the 802.11-2012 standard. DES and 3DES are symmetric block ciphers, part of the NIST FIPS standard.

18. A, D. The CCMP header includes the Key ID and the packet (PN), which is divided into 6 octets. The format of the CCMP header is basically identical to the format of the 8-octet TKIP header (IV/Extended IV). The CCMP header is not encrypted.

19. A, C, E. 3DES defines three keying options:

Keying Option 1	All three keys are unique.
Keying Option 2	K1 and K2 are unique, but K3 = K1.
Keying Option 3	All three keys are identical: K1 = K2 = K3.

Keying option 1 is the strongest, because all three keys are unique, giving it an effective key size of 168 bits. Keying option 3 is the weakest, and is essentially equal to very slow DES. Remember that with a symmetric algorithm, the same key that encrypts the data also decrypts the data. With 3DES, after the first pass with K1 encrypts the data, the second pass with K2 actually decrypts the data, and the third pass with K3 encrypts the data again. Keying option 2 provides an effective key size of 112 bits.

20. B, C, D. AES uses a block size of 128 bits, which is actually a 4 × 4 array of bytes, called a state. The number of rounds performed on the block varies depending on the key sizes. AES-128 performs 10 rounds, AES-192 performs 12 rounds, and AES-256 performs 14 rounds.

Chapter 4: 802.1X/EAP Authentication

1. B, C, D, E. Tunneled authentication is used to protect the exchange of client credentials between the supplicant and the authentication server within an encrypted TLS tunnel. All flavors of EAP-PEAP use tunneled authentication. EAP-TTLS and EAP-FAST also use tunneled authentication. While EAP-TLS is highly secure, it rarely uses tunneled authentication. Although rarely supported, an optional privacy mode does exist for EAP-TLS, which can be used to establish a TLS tunnel. EAP-MD5 and EAP-LEAP do not use tunneled authentication.

2. C, E, F. EAP-TLS and EAP-PEAPv0 (EAP-TLS) require client-side certificates to be used as the supplicant credentials. Client-side certificates are optional with EAP-TTLS. EAP-FAST does not use X.509 digital certificates. It is typically recommended that you deploy EAP-TLS when using client-side certificates because of the wide support for the protocol.

3. D. EAP-PEAP and EAP-TTLS both use two phases of operation. Phase 1 is used to create an encrypted TLS tunnel, and the supplicant credentials are exchanged during Phase 2. EAP-FAST also uses Phase 1 and 2 operations to accomplish the same goals. However, EAP-FAST also defines an optional Phase 0 that is sometimes used for automatic PAC provisioning.

4. A, B, C, E. All versions of EAP-PEAP and EAP-TTLS require a server-side certificate to create an encrypted TLS tunnel. EAP-FAST uses a Protected Access Credential (PAC) to create the encrypted tunnel as opposed to a server-side certificate. EAP-LEAP and EAP-MD5 do not use a TLS tunnel. EAP-TLS requires a server certificate; however, establishing a TLS tunnel is optional.

5. B, F. EAP-MD5 uses the MD5 hash algorithm to validate the supplicant credentials during a password challenge and response exchange. EAP-LEAP uses the MS-CHAPv2 hash algorithm to validate the supplicant credentials during a password challenge and response exchange. Both hash methods can be cracked with hacker tools. EAP-MD5 and EAP-LEAP do not protect the supplicant validation exchange within a TLS tunnel and are therefore susceptible to offline dictionary attacks.

6. D. Unlike EAP-MD5 and EAP-LEAP, which have only one supplicant identity, EAP methods that use tunneled authentication have two supplicant identities. These two supplicant identities are often called the outer identity and the inner identity. The outer identity is a bogus username, and the inner identity is the actual username of the supplicant. The outer identity is seen in clear text outside the encrypted TLS tunnel, whereas the inner identity is protected within the TLS tunnel.

7. A, E. The RADIUS protocol uses UDP ports 1812 for RADIUS authentication and 1813 for RADIUS accounting. These ports were officially assigned by the Internet Assigned Number Authority (IANA). However, prior to IANA allocation of UDP ports 1812 and 1813, the UDP ports of 1645 and 1646 (authentication and accounting, respectively) were used as the default ports by many RADIUS server vendors. TCP is not used. All Layer 2 EAP traffic sent between the RADIUS server and the authenticator is encapsulated in RADIUS IP packets. The encrypted TLS tunnel communications are between the supplicant and the authentication server. IPsec is not used.

8. A. The root bridge would be the authenticator, and the nonroot bridge would be the supplicant if 802.1X/EAP security is used in a WLAN bridged network.

9. C. The supplicant, authenticator, and authentication server work together to provide the framework for 802.1X port-based access control, and an authentication protocol is needed to assist in the authentication process. The Extensible Authentication Protocol (EAP) is used to provide user authentication. The other protocols are all legacy protocols.

10. B. All of these EAP protocols create a TLS tunnel to protect the supplicant credentials. However, only EAP-TTLS offers support for legacy authentication protocols within the TLS tunnel. EAP-TTLS supports the legacy methods of PAP, CHAP, MS-CHAP, and MS-CHAPv2. EAP-TTLS also supports the use of EAP protocols as the inner authentication method. EAP-PEAP only supports EAP protocols for inner authentication, while EAP-TTLS supports just about anything for inner authentication. EAP-FAST only supports the use of EAP-GTC within the TLS tunnel.

11. D. Controller-based APs normally tunnel their 802.11 user traffic to a WLAN controller where control plane and data plane mechanisms reside. The WLAN controller usually functions as the authenticator. When an 802.1X/EAP solution is deployed in a wireless controller environment, the virtual controlled and uncontrolled ports exist on the WLAN controller.

12. A, D. An 802.1X/EAP solution requires that both the supplicant and the authentication server support the same type of EAP. The authenticator must be configured for 802.1X/EAP authentication, but it does not care which EAP type passes through. The authenticator and the supplicant must support the same type of encryption.

13. A, D, E. The purpose of 802.1X/EAP is authentication of user credentials and authorization to access network resources. Although the 802.1X framework does not require encryption, it highly suggests the use of encryption. A by-product of 802.1X/EAP is the generation and distribution of dynamic encryption keys. While the encryption process is actually a byproduct of the authentication process, the goals of authentication and encryption are very different. Authentication provides mechanisms for validating user identity while encryption provides mechanisms for data privacy or confidentiality.

14. C. EAP-TLS and EAP-PEAPv0 (EAP-TLS) both require the use of client-side certificates and therefore would be considered costly and hard to manage. EAP-FAST with manual PAC provisioning would also be difficult to administer. EAP-MD5 is cheap and simple to set up; however, it will only work with static WEP encryption and therefore would not meet the data privacy needs. EAP-PEAPv0 (MSCHAPv2) only requires the use of a server-side certificate and is easy to administer. EAP-PEAPv0 (MSCHAPv2) is the most widely supported EAP protocol available and is therefore cost-effective. CCMP/AES dynamic encryption is now widely supported and meets the data privacy objectives.

15. F. A shared secret is used between the authenticator and the authentication server for the RADIUS protocol exchange. The shared secret exists between the authenticator and the AS so that they can validate each other with the RADIUS protocol. The shared secret is only used to validate and encrypt the communication link between the authenticator and the authentication server. The shared secret is not used at all for any validation of the supplicants.

16. D. EAP-Authentication and Key Agreement (EAP-AKA) is an EAP type primarily developed for the mobile phone industry and more specifically for third-generation (3G) mobile networks. EAP-AKA defines the use of the authentication and key agreement mechanisms already being used by the two types of 3G mobile networks. The 3G mobile networks include the Universal Mobile Telecommunications System (UMTS) and CDMA2000. EAP-SIM was primarily developed for the mobile phone industry and more specifically for second-generation (2G) mobile networks.

17. B, C, D, E, F. Depending on which type of EAP protocol is used, the supplicant identity credentials can be in many different forms, including usernames and passwords, client-side certificates, PACS, security token devices, smart cards, USB devices, proximity badges, and many more.

18. D. The authentication server can maintain a user database or may proxy with an external user database to authenticate user or device credentials. In almost all cases, a RADIUS server functions as the authentication server. The RADIUS server may indeed hold a master native user database, but usually it will instead query a preexisting external database. Any Lightweight Directory Access Protocol (LDAP)-compliant database can be queried by the RADIUS authentication server. Microsoft Active Directory is the most commonly used external LDAP database, but a RADIUS server can also query LDAP-compliant databases such as Apple Open Directory and Novell eDirectory.

19. A, C. EAP-Generic Token Card (EAP-GTC) was developed to provide interoperability with existing security token device systems that use one-time passwords (OTP), such as RSA's SecurID solution. The EAP-GTC method is intended for use with security token devices, but the credentials can also be a clear-text username and password. EAP-Protected One-Time Password Protocol (EAP-POTP) is another EAP method suitable for use with one-time password (OTP) token devices. EAP-POTP and EAP-GTC are both intended to be used as inner authentication protocols. EAP-MSCHAPv2 is an inner authentication protocol that only uses usernames and passwords. LEAP, PEAP, and TTLS are not inner authentication protocols.

20. A, C, G. Many EAP protocols such as EAP-PEAPv0 (MSCHAPv2) support tunneled authentication to protect user credentials. For tunneled authentication to be successful, a server certificate must be installed in the RADIUS server and the root CA certificate must be distributed and installed in the WLAN clients that function as supplicants. Distributing and installing the root CA certificate on Windows laptops that have joined to the domain can be very easily automated with a Group Policy Object (GPO). Distributing the root certificate is much harder with devices that do not belong to the domain and use non-Windows operating systems. For this reason, mobile device management (MDM) solutions are often deployed. An MDM solution uses an encrypted over-the-air provisioning of certifications during the MDM enrollment process. Instead of a full-blown MDM solution, another option is a self-service device onboarding solution. Several WLAN vendors offer self-service solutions so employees can easily self-install security credentials such as an 802.1X /EAP root CA certificate.

Chapter 5: 802.11 Layer 2 Dynamic Encryption Key Generation

1. B, D. Open System and Shared Key authentication are legacy authentication methods that do not provide seeding material to generate dynamic encryption keys. A robust security network association requires a four-frame EAP exchange known as the 4-Way Handshake that is used to generate dynamic TKIP or CCMP keys. The handshake may occur either after an 802.1X/EAP exchange or as a result of PSK authentication.

2. A, D, E. The PTK/GTKs are generated by the 4-Way Handshake. As defined by the 802.11-2012 standard, the temporal keys generated by the 4-Way Handshake use either CCMP/AES or TKIP/ARC4 ciphers. However, the 4-Way Handshake can also be used to generate keys for proprietary encryption. Although WEP can be dynamically generated, a simpler two EAPOL-Key frame exchange is used to generate the WEP keys instead of a 4-Way Handshake.

3. C. The third EAPOL-Key frame of the 4-Way Handshake may also contain a message to the supplicant to install the temporal keys. The frame capture indicates that the temporal key is to be installed. The third EAPOL-Key frame also sends the supplicant the ANonce, the authenticator's RSN information element capabilities, and a MIC. If a group temporal key (GTK) has been generated, the GTK will be inside the third EAPOL-Key frame. The GTK confidentiality is protected because it will be encrypted with the pairwise temporal key (PTK).

4. B, D, E. The pairwise transient key (PTK) is composed of three separate keys. The Key Confirmation Key (KCK) is used to provide data integrity during the 4-Way Handshake and Group Key Handshake. The Key Encryption Key (KEK) is used by the EAPOL-Key frames to provide data privacy during the 4-Way Handshake and Group Key Handshake. The Temporal Key (TK) is the temporal encryption key used to encrypt/decrypt the MSDU payload of 802.11 data frames between the supplicant and the authenticator. The STSL transient key (STK) and STSL master key (SMK) are used during the PeerKey Handshake.

5. A, D, E. To create the pairwise transient key (PTK), the 4-Way Handshake uses a pseudo-random function that combines the pairwise master key, the authenticator nonce (ANonce), the supplicant nonce (SNonce), the authenticator's MAC address (AA), and the supplicant's MAC address (SPA). The master session key (MSK) is used to derive the pairwise master key (PMK). The group master key (GMK) is used to create the group temporal key (GTK).

6. E. The frame capture depicts a two EAPOL-Key frame exchange. When dynamic WEP is deployed, a two EAPOL-Key frame exchange always follows the EAP frame exchange. The two EAPOL-Key frames are both sent by the authenticator to the supplicant. The first EAPOL-Key frame carries the broadcast key from the access point to the client. The second EAPOL-Key frame is effectively a confirmation that the keys are installed and that the WEP encryption process can begin.

7. C. The frame capture shows an RSN information element field that can be found in a management frame. The RSN information element shows that the group cipher that is being used is WEP. A transition security network (TSN) supports RSN-defined security as well as legacy security such as WEP within the same BSS. Within a TSN, some client stations

will use RSNA security using TKIP/ARC4 or CCMP/AES for encrypting unicast traffic. However, some legacy stations are still using static WEP keys for unicast encryption. All of the clients will use WEP encryption for the broadcast and multicast traffic. Because all the stations share a single group encryption key for broadcast and multicast traffic, the lowest common denominator must be used for the group cipher.

8. A, C. The RSN information element field is found in four different 802.11 management frames: beacon management frames, probe response frames, association request frames, and reassociation request frames. Within a basic service set, an access point and client stations use the RSN information element within these four management frames to communicate with each other about their security capabilities prior to establishing association. The access point radio will use beacons and probe response frames to inform client stations of the AP security capabilities. The security capabilities include supported encryption cipher suites and supported authentication methods. Any clients that want to associate to the AP will also have to support the same minimal RSNA capabilities.

9. D. The 4-Way Handshake process creates temporal keys that are used by the client station and the access point to encrypt and decrypt 802.11 data frames. The pairwise transient key (PTK) is used to encrypt all unicast transmissions between a client station and an access point. Each PTK is unique between each individual client station and the access point. Every client station possesses a unique PTK for unicast transmissions between the client STA and the AP. PTKs are used between a single supplicant and a single authenticator. The group temporal key (GTK) is used to encrypt all broadcast and multicast transmissions between the access point and multiple client stations. Although the GTK is dynamically generated, it is shared among all client STAs for broadcast and multicast frames. The GTK is used between all supplicants and a single authenticator.

10. D, F. The RSN information element field is found in four different 802.11 management frames: beacon management frames, probe response frames, association request frames, and reassociation request frames. Within a basic service set, an access point and client stations use the RSN information element within these four management frames to communicate with each other about their security capabilities prior to establishing association. Client stations use the association request frame to inform the access point of the client station security capabilities. When stations roam from one access point to another access point, they use the reassociation request frame to inform the new access point of the roaming client station's security capabilities. The security capabilities include supported encryption cipher suites and supported authentication methods.

11. B, C, E. The access point is configured with three unique logical identifiers (SSIDs) that are also linked to specific VLANs. If multiple SSIDs are transmitted from the same AP radio, multiple BSSIDs are also needed. Multiple BSSIDs are effectively virtual MAC addresses that are incremented or derived from the actual physical MAC address of the AP radio. Each SSID also has a different type of security. Because multiple virtual BSSIDs exist with different security requirements, an RSN, pre-RSNA, and a TSN WLAN all exist within the same coverage area of an access point. Effectively, three basic service sets (BSS) exist within the same coverage cell, each with different security. BSS #1 is a robust security network (RSN) because it is only using CCMP/AES encryption. BSS #2 is a transition security network (TSN) because it is using TKIP/ARC4 and legacy WEP/ARC4 encryption. BSS #3 is a pre-RSNA security network because it is only using legacy WEP/ARC4 encryption.

12. B, C, E. The 802.11-2012 standard defines authentication and key management (AKM) services. The AKM services are a set of one or more algorithms designed to provide authentication and key management, either individually or in combination with higher layer authentication and key management algorithms. An authentication and key management protocol (AKMP) can either be a preshared key (PSK) or an EAP protocol used during 802.1X/EAP authentication. WPA2-Personal uses PSK authentication. WPA2-Enterprise uses 802.1/EAP.

13. B, C, D. Whenever a client joins a basic service set (BSS) for the first time, the client must authenticate and create new keys. Either an 802.1X or PSK authentication process is needed to produce the pairwise master key (PMK) that seeds the 4-Way Handshake. Every time a client roams, unique encryption keys must be generated using a 4-Way Handshake process between the access point and the client STA. Therefore, every time a client roams to a new BSS, the client must authenticate and create new keys. All the stations within an independent basic service set (IBSS) also have to utilize the 4-Way Handshake with each other because all unicast communications are peer-to-peer. PSK authentication is used within the IBSS to seed the 4-Way Handshake. Therefore, every time a client joins an IBSS with a peer station, the client must reauthenticate and create new keys.

14. B. The 802.11-2012 standard defines a two-frame handshake that is used to distribute a new group temporal key (GTK) to client stations that have already obtained a PTK and GTK in a previous 4-Way Handshake exchange. The Group Key Handshake is used only to issue a new group temporal key (GTK) that has previously formed security associations. Effectively, the Group Key Handshake is identical to the last two frames of the 4-Way Handshake. The purpose of the Group Key Handshake is to deliver a new GTK to all client stations that already have an original GTK generated by an earlier 4-Way Handshake.

15. E. The TDLS Peer Key Handshake consists of three messages between a TDLS initiator client station and a TDLS responder client station. The three-way handshake exchange is communicated through the AP to which the clients are associated. At the end of the exchange, each client station has a copy of the unicast *TDLS Peer Key (TPK)* and the clients can now have an encrypted direct link for sidebar communications. Keep in mind that both clients remain associated with the original AP and the clients can still communicate with other devices using normal BSS communications.

16. C. The RSN information element can also be used to indicate what authentication methods are supported. The authentication key management (AKM) suite field in the RSN information element indicates whether the station supports either 802.1X authentication or PSK authentication. If the AKM suite value is 00-0F-AC-01, authentication is negotiated over an 802.1X infrastructure using an EAP protocol. If the AKM suite value is 00-0F-AC-02, then PSK is the authentication method that is being used.

17. A. Security associations can be defined as group policies and keys used to protect information. A robust security network association (RSNA) requires two 802.11 stations (STAs) to establish procedures to authenticate and associate with each other as well as to create dynamic encryption keys. The 802.11-2012 standard defines multiple RSNA security associations. A pairwise master key security association (PMKSA) is defined as the conditions resulting from a successful 802.1X authentication exchange between the supplicant and authentication server, or from a preshared key (PSK).

18. A, C. The main advantage of using dynamic keys is that the keys are not static and are not compromised by social engineering attacks because the users have no knowledge of the keys. Another advantage of dynamic keys is that every user has a different and unique key. If a single user's encryption key is somehow compromised, none of the other users would be at risk because every user has a unique key. All client stations do share a broadcast/multicast key.

19. A, B. 802.1X/EAP authentication must first occur to achieve mutual validation of both the supplicant and authentication server credentials. 802.1X/EAP provides the PMK necessary for the 4-Way Handshake. The 4-Way Handshake then generates the temporal keys used for encryption. Once the temporal keys are created and installed, the controlled port of the authenticator is no longer blocked and the supplicant can now send encrypted 802.11 data frames through the controlled port onward to network resources. No traffic can pass through the controlled port until mutual authentication occurs and dynamic keys are created.

20. A, B, D. The purpose of the 4-Way Handshake is to confirm the existence of the PMK at the peer station and ensure that the PMK is current. A pairwise transient key (PTK) must be derived from the PMK and installed on both the supplicant and authenticator. The GTK must be transferred from the authenticator, and the GTK must be installed on the supplicant and, if necessary, the authenticator. The GMK is never transferred to the supplicant. The 4-Way Handshake is also used to confirm the selection of encryption cipher suites.

Chapter 6: PSK Authentication

1. A, D. After obtaining the passphrase, the hacker can also associate to the AP using PSK authentication and thereby access network resources. The encryption technology is not cracked, but the key can be re-created. The ANonce, SNonce, and the MAC addresses of the supplicant and authenticator are all seen in cleartext during the 4-Way Handshake frame exchange. If a hacker is able to maliciously obtain the passphrase, the hacker could use the passphrase-PSK mapping formula to create the 256-bit PSK. The PSK is also used as the pairwise master key. If the hacker has the PMK and captures the 4-Way Handshake with a protocol analyzer, the hacker has all the variables needed to duplicate the pairwise transient key. If the hacker can duplicate the PTK, the hacker can then decrypt any unicast traffic between the AP and the individual client station that performed the 4-Way Handshake. WPA/WPA2-Personal is not considered a strong security solution for the enterprise, because if the passphrase is compromised, the attacker can access network resources and decrypt traffic.

2. B, G. WPA/WPA2-Personal uses PSK authentication. A PSK used in a robust security network is 256 bits in length, or 64 characters when expressed in hex. A preshared key (PSK) is a static key that is configured on the access point and all of the clients. However, in most SOHO environments, an 8-to-63 character passphrase is instead configured on the AP and all of the clients. The passphrase-PSK mapping formula transforms the simple ASCII passphrase to the 256-bit PSK.

3. C, F. The passphrase-PSK mapping formula is defined by the 802.11-2012 standard to allow end users to use a simple ASCII passphrase that is then converted to the 256-bit PSK. The passphrase is combined with the SSID and hashed 4,096 times to produce a 256-bit (64-character) PSK. MAC addresses and nonces are inputs used during the 4-Way Handshake.

4. A, D, E. The risks involved with both WPA-Personal and WPA2-Personal are basically twofold: network resources can be placed at risk and the encryption keys can be compromised. A hacker can obtain the passphrase using either social engineering skills or using an offline brute-force dictionary attack. Once the passphrase is compromised, the hacker can use any client station to associate with the WPA/WPA2-Personal access point. The hacker is then free to access network resources. The hacker can also use the passphrase to re-create encryption keys. The hacker can then decrypt any unicast traffic between the AP and the individual client. WPA-Personal uses TKIP/ARC4 encryption and is not necessarily a security risk; however, the stronger CCMP/AES encryption mandated by WPA2-Personal would be preferable.

5. C, F. The 802.1X/EAP process is used to create a pairwise master key (PMK), which is the seeding material for the 4-Way Handshake that is used to create the final temporal encryption keys. Every time a supplicant authenticates within an 802.1X/EAP architecture, a unique PMK is created. The same cannot be said when PSK authentication is used. The 256-bit PSK is the PMK. Because every client station uses the same passphrase that is converted to a PSK, every client station has the same pairwise master key (PMK).

6. E. The simple passphrases used by most users are susceptible to brute-force offline dictionary attacks. Both the IEEE and the Wi-Fi Alliance recommend a passphrase of 20 characters or more. The strength of the passphrase increases significantly if a combination of uppercase and lowercase letters, numbers, and special symbols are used in the passphrase.

7. B, E. Because the Maxwell Corporation is using older VoWiFi phones that do not support OKC or FT, they do not support fast secure roaming (FSR) mechanisms that will work with an 802.1X/EAP solution. WPA/WPA2 Enterprise security requires an 802.1X/EAP solution. WPA2-Personal security provides the strongest available CCMP/AES encryption, and it uses simple PSK authentication that can be used with the VoWiFi phones. Roaming is considered fast with PSK authentication because clients do not need to reauthenticate and communicate with a RADIUS server every time a client roams. A proprietary PSK solution could further enhance security because each individual VoWiFi phone would have a unique PSK.

8. C, D. The 802.1X/EAP process is used to create a pairwise master key (PMK), which is the seeding material for the 4-Way Handshake that is used to create the final temporal encryption keys. Every time a supplicant authenticates within an 802.1X/EAP architecture, a unique PMK is created. The same can be said when a per-user/per-device implementation of PSK authentication is used. The 256-bit PSK is the PMK. Every client station uses a different passphrase that is converted to a PSK. Therefore every client station has a different pairwise master key (PMK).

9. B, C, D, E. WPA-Personal and WPA2-Personal both use the PSK authentication method required by the IEEE 802.11-2012 standard. However, WLAN vendors have many names for PSK authentication, including WPA/WPA2-Passphrase, WPA/WPA2-PSK, and WPA/WPA2-Preshared Key. The correct Wi-Fi Alliance terminology is WPA/WPA2-Personal. All of these terms refer to 802.11 PSK authentication.

10. A. In digital communications, entropy is a measure of uncertainty associated with a random variable. The more entropy (measured in bits) that is contained within the method used to create a passphrase, the more difficult it will be to guess the passphrase.

11. E. The image depicts an 802.11 client station being configured with a static key that is used for WEP or possibly a passphrase used for PSK authentication. Client software utilities do not always use the proper terminology. The WPA key referenced in the graphic refers to a passphrase used in WPA-Personal security. WPA-Personal and WPA2-Personal both use the PSK authentication method. WPA-Personal specifies TKIP/ARC4 encryption and WPA2-Personal specifies CCMP/AES.

12. D. Passwords or passphrases by themselves have an entropy value of zero. It is the method you use to select the passphrase that contains the entropy. A strong passphrase will have more entropy bits per character than a password created from your name or your phone number. The more entropy (measured in bits) that is contained within the method you use to create your passphrase, the more difficult it will be to guess the passphrase.

13. D. Each character in a passphrase is only worth 2.5 bits of entropy because most users use easy-to-remember words as opposed to a mix of alphanumeric and punctuation characters. Therefore, an 8-character passphrase would only contain about 20 bits of entropy before the passphrase-to-PSK mapping hash algorithm is calculated. Because the passphrase-to-PSK mapping mixes the SSID with the passphrase, approximately 12 extra bits of entropy are added. Therefore, a typical 8-character passphrase used in a WPA2-Personal configuration would have a total of about 32 bits of entropy.

14. C. The Wi-Fi Alliance views Simultaneous Authentication of Equals (SAE) as a more secure replacement for PSK authentication. The ultimate goal of SAE is to prevent dictionary attacks altogether.

15. A, B, D. Because the SAE exchange allows for only one guess of the passphrase, the brute-force offline dictionary attacks are no longer viable. Any attacker will not be able to determine the passphrase or the PMK while listening to the SAE exchange with a protocol analyzer. Additionally, the SAE exchange is also resistant to forging and replay attacks. Even if the passphrase is compromised, it could not be used to re-create any previously generated PMKs.

16. A, B, C. Numerous freeware software tools exist that can be used to assist end users with creating strong passphrases with increased entropy. These tools can be used to increase the strength of passwords, passphrases, and even the shared secrets that are used between a RADIUS server and an authenticator.

17. D. The biggest problem with using PSK authentication in the enterprise is social engineering. The PSK is the same on all WLAN devices. If the end users accidentally give the PSK to a hacker, WLAN security is compromised. If an employee leaves the company, all the devices have to be reconfigured with a new 64-bit PSK, creating a lot of work for an administrator. Several WLAN vendors offer proprietary PSK solutions where each individual client device will have its own unique PSK. These proprietary PSK solutions prevent social engineering attacks. These proprietary PSK solutions also virtually eliminate the burden for an administrator having to reconfigure each and every WLAN end-user device.

18. C. A PSK used in a robust security network is 256 bits in length, or 64 characters when expressed in hex. A preshared key (PSK) is a static key that can be configured on the access point and all of the clients. Furthermore, do not confuse entropy security bits with the number of digital bits in a PSK. The passphrase-PSK mapping formula will always convert any passphrase of 8 to 63 characters into a 256-bit PSK. A 20-character passphrase has 62 bits of entropy. Entropy bits are a measure of uncertainty associated with a random variable. The random variable is basically the strength of the passphrase. WPA uses TKIP/ARC4 encryption, which is considered to be a legacy encryption method. WPA2 uses the more secure CCMP/AES dynamic encryption.

19. E. Wi-Fi Protected Setup (WPS) defines simplified and automatic WPA and WPA2 security configurations for home and small-business users. WPS was developed by the Wi-Fi Alliance and is a protocol specification that rides over the existing IEEE 802.11-2012 standard. The IEEE does not specify WPS mechanisms.

20. E. Multiple use cases for per-user and per-device PSK credentials have gained popularity in the enterprise. However, proprietary implementations of PSK authentication are not meant to be a replacement for 802.1X/EAP.

Chapter 7: 802.11 Fast Secure Roaming

1. C. The only way to maintain upper-layer communications when crossing Layer 3 subnets is if the WLAN vendor offers a Layer 3 roaming solution based on Mobile IP.

2. A, C, E, F. A unique identifier is created for each PMKSA that has been established between the authenticator and the supplicant. The pairwise master key identifier (PMKID) is a unique identifier that refers to a PMKSA. A PMKID can reference PMKSAs derived from a PSK authentication, cached from an 802.1X/EAP authentication, or a PMKSA that has been created through preauthentication with a target AP. A PMKID is also used as an identifier in PMK-RO and PMK-R1 security associations created during fast BSS transition.

3. A, C, E. The 802.11-2012 standard states that a client station can establish a PMKSA via 802.1X/EAP, PSK, or SAE authentication methods. A PMKSA can also be established via other mechanisms used to cache the PMK.

4. A, C. The 802.11-2012 standard defines three fast secure roaming mechanisms: preauthentication, PMK caching, and fast BSS transition. Most WLAN vendors also support another method of fast secure roaming (FSR) called opportunistic key caching. Preauthentication, PMK caching, opportunistic key caching (OKC), and fast BSS transition (FT) all allow for 802.1X/EAP authentication to be skipped when roaming; however, FT does not use the traditional 4-Way Handshake and instead uses either an over-the-air fast BSS transition frame exchange or an over-the-DS fast BSS transition frame exchange.

5. E. One key component of 802.11k is the *neighbor report*, which is used by client stations to gain information from the associated AP about potential roaming neighbors. Operation information that can be within the neighbor report includes the BSSID of the neighbor AP, mobility domain, Automatic Power Save Delivery (APSD) support, PHY support, and much more.

6. B, D, E. As defined by the 802.11-2012 standard, the RSN information element found in WLAN management frames includes three authentication and key management (AKM) suites that support fast BSS transition (FT). The three AKM suites are 802.1X/EAP, PSK, and SAE.

7. A, C, E. The 802.11-2012 standard defines a PMK identifier with the following formula:

   PMKID = HMAC-SHA1-128(PMK, "PMK Name" || AA || SPA)

   The AA is the authenticator's MAC address, and the SPA is the supplicant's MAC address. A hash function combines the PMK with the access point and client station MAC addresses to create the PMKID.

8. A, C, E. RSN security can be identified by a field found in certain 802.11 management frames. This field is known as the robust security network information element (RSNIE), which is often referred to simply as the RSN information element. The pairwise master key identifier (PMKID) is found in the RSN information element in association request frames and reassociation request frames that are sent from a client station to an AP. The PMKID is also found in FT Action frames.

9. D. Fast BSS transition (FT) defines a three-level key hierarchy. The Pairwise Master Key R0 (PMK-R0) is the first-level key of the FT key hierarchy. This key is derived from the master session key (MSK). Depending on the WLAN architecture, the PMK-R0 is cached on the original AP to which a client associates or cached on a WLAN controller.

10. A, C. Voice Enterprise defines enhanced support for voice applications in the enterprise environment. Many aspects of the 802.11r, 802.11k, and 802.11v amendments are tested for Voice Enterprise certification. Effective enterprise-grade voice over Wi-Fi solutions must address two major requirements. Voice quality has to be consistently good throughout the call, in all load conditions. Wi-Fi networks are used for both voice and data applications. Therefore, voice calls must coexist and share network resources with data traffic, which often accounts for the largest portion of the network load.

11. C. Over-the-DS fast BSS transition requires the use of FT Action frames to complete the PTK creation process. A client stations sends an FT frame to the associated AP. The FT Action request frame is forwarded over the distribution system (DS), which is the wired infrastructure. The target AP responds back to the client station over the DS with an FT Action response frame. The reassociation request and response frames are then sent from the client station to the target AP over the air.

12. C. The pairwise transient key (PTK) is the third-level key of the FT key hierarchy. The PTK is the final key used to encrypt 802.11 data frames. The PTK is created during either an over-the-air fast BSS transition frame exchange or over-the-DS fast BSS transition frame exchange. In any 802.11 robust security network (RSN), the PTK is used to encrypt the MSDU payload of an 802.11 unicast data frame.

13. A, B, D, F. Performance of equipment submitted for Wi-Fi Voice-Enterprise certification has to meet four thresholds to ensure that the Wi-Fi network preserves good voice call quality. These thresholds include latency (< 50 ms), jitter (< 50 ms), packet loss (< 1 percent), and consecutive lost packets of no more than 3.

14. C. Although there is no defined standard for opportunistic key caching (OKC), the majority of WLAN infrastructure vendors that manufacture APs support OKC. WLAN clients must also support OKC for the process to work. OKC became a de facto standard for fast secure roaming because adoption of 802.11r mechanisms has been extremely slow. Although OKC is supported by all the enterprise-grade AP manufacturers, many WLAN clients do not support OKC because it is not an official roaming standard. Over-the-air fast BSS transition is part of the FT mechanisms defined by the 802.11-2012 standard. Enterprise WLAN vendors typically support FT; however, client-side support is still growing.

15. B, D. The original legacy 802.11 standard, for the most part, only defined roaming as a Layer 2 process known as the reassociation service. However, the 802.11-2012 standard does not define two very important processes for BSS transition: client roaming thresholds and AP-to-AP handoff communications. Client roaming thresholds are usually based on RSSI.

16. D. An 802.1X/EAP exchange creates a master session key (MSK), which is used to create a pairwise master key (PMK) for non-FT systems. FT also uses the 802.1X/EAP exchange to create the master session key that seeds a centralized but much more complex key management solution. The supplicant and the RADIUS server exchange credentials, and the PMK is created from the master session key and sent to the authenticator. Depending on the WLAN architecture, the authenticator can be either an AP or a WLAN controller.

17. D. When FT is configured on an access point, the AP will broadcast management frames with new information elements. For example, the mobility domain information element (MDIE) will be in all beacon and probe response frames. Unfortunately, the drivers of some older legacy client radios may not be able to process the new information in these management frames. The result is that legacy clients may have connectivity problems when an AP is configured for FT. Whenever any new 802.11 technology requires new information elements in the 802.11 management frames, there may be a possible connectivity problem with older legacy radios.

18. B, C. The second-level key PMK-R1 is derived from the PMK-R0 and sent to the APs over a secure channel from either a WLAN controller or the original associated AP. The PMK-R1 keys are cached on the APs. The access points are the key holders for the PMK-R1 keys. The client station also derives the second-level key PMK-R1 from the PMK-R0. The PMK-R1 key is cached on the client station. The supplicants are the key holders for the PMK-R1 keys.

19. C, D. Most WLAN vendors encrypt/decrypt client traffic at the edge of the network using the access points. In this model, the WLAN controller functions as the Pairwise Master Key R0 Holder (R0KH) and the AP functions as the Pairwise Master Key R1 Holder (R1KH). However, some WLAN vendors perform encryption at the WLAN controller level instead of at the AP level. End-to-end encryption provides data privacy between the client at the access layer and the WLAN controller that is typically deployed at the core. In that scenario, the WLAN controller functions as both the Pairwise Master Key R0 Holder (R0KH) and the Pairwise Master Key R1 Holder (R1KH).

20. C. During both methods of fast BSS transition, four frames carry an FT authentication algorithm (FTAA) along with nonces and other information needed to create the final dynamic keys. The PMK-R1 key is the seeding material for both over-the-air and over-the-DS fast BSS transition frame exchanges that create the final pairwise transient key (PTK). It should be noted that over-the-air fast BSS transition is the method that is supported by most WLAN vendors.

Chapter 8: WLAN Security Infrastructure

1. A, E. In the centralized WLAN architecture, autonomous APs have been replaced with controller-based APs. All three logical planes of operation reside inside a centralized networking device known as a WLAN controller. Effectively, all planes were moved out of access points and into a WLAN controller. It should be noted that an NMS could be used to manage controllers and controller-based APs.

2. D. Telecommunication networks are often defined as three logical planes of operation. The control plane consists of control or signaling information and is often defined as network intelligence or protocols.

3. A, C. All three WLAN infrastructure designs support the use of VLANs and 802.1Q tagging. However, the centralized WLAN architecture usually encapsulates user VLANs between the controller-based AP and the WLAN controllers; therefore, only a single VLAN is normally required at the edge. An 802.1Q trunk is, however, usually required between the WLAN controller and a core switch. Neither the autonomous nor the distributed WLAN architectures use a controller. Noncontroller architectures require support for 802.1Q tagging if multiple VLANs are to be supported at the edge of the network. The access point is connected to an 802.1Q trunk port on an edge switch that supports VLAN tagging.

4. C. In most cases, 802.11 PSK security is used between the mesh radios to provide encryption. The PSK is usually created automatically in most mesh WLAN solutions. A very strong passphrase of 20 characters or more should be used if the WLAN vendor does offer the option to manually define mesh backhaul security. Although SAE has yet to be implemented for 802.11 mesh networks, the Wi-Fi Alliance views SAE as an eventual replacement for PSK authentication.

5. A, B, C. Many WLAN vendors use Generic Routing Encapsulation (GRE), which is a commonly used network tunneling protocol. Although GRE is often used to encapsulate IP packets, GRE can also be used to encapsulate an 802.11 frame inside an IP tunnel. The GRE tunnel creates a virtual point-to-point link between a controller-based AP and a WLAN controller. WLAN vendors that do not use GRE use other proprietary protocols for the IP tunneling. The CAPWAP management protocol can also be used to tunnel user traffic. IPsec can also be used to securely tunnel traffic from APs across a WAN link.

6. D. One major disadvantage of using the traditional autonomous access point is that there is no central point of management. Any autonomous WLAN architecture with 25 or more access points is going to require some sort of network management system (NMS). Although a WLAN controller can be used to manage the WLAN in a centralized WLAN architecture, if multiple controllers are deployed, an NMS may be needed to manage multiple controllers. Although the control plane and management plane have moved back to the APs in a distributed WLAN architecture, the management plane remains centralized. Configuration and monitoring of all access points in the distributed model is still handled by an NMS.

7. D. VPNs are most often used for client-based security when connected to public access WLANs and hotspots that do not provide security. Because most hotspots do not provide Layer 2 security, it is imperative that end users provide their own security. Another common use of VPN technology is to provide site-to-site connectively between a remote office and a corporate office across a WAN link. When WLAN bridges are deployed for wireless backhaul communications, VPN technology can be used to provide the necessary level of data privacy.

8. E. Because a WLAN controller is the focal point of a centralized WLAN architecture, multiple security capabilities are also centralized. WLAN controllers support full 802.11 security and also often other robust security components such as internal RADIUS server, VPN server, integrated WIPS, firewall and captive web portals. WLAN controllers also offer full support for 802.11 security.

9. D. The centralized data forwarding is the traditional data forwarding method used with WLAN controllers. All 802.11 user traffic is forwarded from the AP to the WLAN controller for processing, especially when the WLAN controller manages encryption and decryption or applies security and QoS policies. Most WLAN controller solutions also now support a distributed data plane. The controller-based AP performs data forwarding locally; it may be used in situations where it is advantageous to perform forwarding at the edge and to avoid a central location in the network for all data.

10. D, E. A distributed solution using enterprise-grade WLAN routers is often deployed at company branch offices. Branch WLAN routers have the ability to connect back to corporate headquarters with VPN tunnels using an integrated VPN client. Enterprise WLAN routers also have integrated firewalls with support for port-forwarding, network address translation (NAT) and port address translation (PAT). Enterprise WLAN routers also offer full support for 802.11 security.

11. A, D, E. Encryption is needed to protect the data privacy of the backhaul communications across bridge links. IPSec VPNs are often used for bridge security which will be discussed later in this chapter. An 802.1X/EAP solution can also be used for bridge security, with the root bridge assuming the authenticator role and the nonroot bridges assuming the supplicant role. PSK authentication is also often used as an 802.11 Layer 2 security solution for bridge security. If PSK security is used, the static passphrase should be 20 strong characters or more. A weak passphrase would make the bridge link more susceptible to a brute-force offline dictionary attack.

12. G. All of these protocols can be used to configure WLAN devices such as access points and WLAN controllers. Written corporate policies should mandate the use of secure protocols such as SNMPv3, SSHv2, and HTTPS.

13. B, C. The distribution system consists of two main components. The distribution system medium (DSM) is a logical physical medium used to connect access points. Distribution system services (DSS) consist of services built inside an access point or WLAN controller usually in the form of software.

14. C, E. Communications between an NMS and an access point are management protocols. Most NMS solutions use the Simple Network Management Protocol (SNMP) to manage and monitor the WLAN. Other NMS solutions also use the Control and Provisioning of Wireless Access Points (CAPWAP) as strictly a monitoring and management protocol. CAPWAP incorporates Datagram Transport Layer Security (DTLS) to provide encryption and data privacy of the monitored management traffic.

15. E. Most of the security features found in WLAN controllers can also be found in a distributed WLAN architecture even though there is no WLAN controller. For example, a captive web portal that normally resides in a WLAN controller instead resides inside the individual APs. The stateful firewall and RBAC capabilities found in a centralized WLAN controller now exist cooperatively in the APs. An individual AP might also function as a RADIUS server with full LDAP integration capabilities. All control plane mechanisms reside in the access points at the edge of the network in a distributed WLAN architecture.

16. A, C. SMNPv3 is much more secure than older versions of SNMP. Authentication is performed using SHA or MD5. SNMPv3 uses 56-bit DES encryption to encrypt packets. SNNPv3 requires usernames and passwords instead of community strings.

17. D. Most vendor equipment defaults to SNMP being enabled with default read and write community strings. This is an enormous security threat to the configuration and operation of any network and should be one of the very first lockdown steps to securing network devices. SNMP should be disabled if it is not used for management.

18. C. IPsec VPNs use secure tunneling, which is the process of encapsulating one IP packet within another IP packet. The first packet is encapsulated inside the second or outer packet. The original destination and source IP address of the first packet is encrypted along with the data payload of the first packet. VPN tunneling, therefore, protects your original private Layer 3 addresses and also protects the data payload of the original packet. Layer 3 VPNs use Layer 3 encryption; therefore, the payload that is being encrypted is the Layer 4–7 information. The IP addresses of the second or outer packet are seen in clear text and are used for communications between the tunnel endpoints. 802.11 encryption protects Layer 3–7 information and IP headers are never seen in encrypted 802.11 data frames.

19. A, B, C. NMS solutions can be deployed at a company datacenter in the form of a hardware appliance or as a virtual appliance that runs on VMware or some other virtualization platform. A network management server that resides in a company's own datacenter is often referred to as an on-premises NMS. NMS solutions are also available in the cloud as a software subscription service.

20. B, C. The control plane mechanisms are enabled in the system with inter-AP communication via cooperative protocols in a distributed WLAN architecture. In a distributed architecture, each individual access point is responsible for local forwarding of user traffic; therefore, the data plane resides in the APs. The management plane resides in an NMS that is used to manage and monitor the distributed WLAN.

Chapter 9: RADIUS and LDAP

1. E. Role-based access control (RBAC) is an approach to restricting system access to authorized users. The majority of WLAN controller solutions have RBAC capabilities. When a user authenticates using 802.1X/EAP, RADIUS attributes can be used to assign users to specific roles. This method is used to assign users from certain Active Directory (AD) groups into predefined roles created on a WLAN controller or access point. Each role has unique access permissions and restrictions. Permissions can be defined as Layer 2 permissions (MAC filters), VLANs, Layer 3 permissions (access control lists), Layers 4–7 permissions (stateful firewall rules), and bandwidth permissions.

2. E. Although a RADIUS server is most commonly integrated with Active Directory, integration with NetQ's eDirectory is also possible. Many enterprise RADIUS servers even allow a SQL database or flat files to be queried. A RADIUS server should be able to communicate with any LDAP-compliant database.

3. C, D. Applications such as RADIUS can be used to query an LDAP server to validate user or device credentials. LDAP sessions normally use TCP or UDP port 369; however, LDAP over SSL uses TCP port 636.

4. D. The distributed sites with RADIUS, centralized LDAP model deploys one or more RADIUS servers at each remote site and proxies the LDAP queries to the central LDAP database. The RADIUS servers at each remote site may be configured for LDAP credential caching. If the WAN link goes down, the RADIUS server will first attempt to query the centralized LDAP server but eventually fall back to the locally cached LDAP credentials on the RADIUS server. When the WAN link goes back up, the RADIUS server will once again query the centralized database. Any new user who has not authenticated prior to the WAN link failure will still not be able to access the WLAN because they would not have a cached LDAP credential. The distributed autonomous sites deployment model would also meet Kenny's requirements but is much more expensive.

5. E. The RadSec protocol secures communication between RADIUS/TCP peers using TLS encryption. The default destination port number for RadSec is TCP 2083. Authentication, accounting, and dynamic authorization changes do not require separate ports. The standard RADIUS protocol uses UDP port 1812 for authentication communications and UDP port 1813 for accounting communications.

6. D. An attacker can spoof the EAP-Success, EAP-Failure, and other EAP frames. The `(80) Message-Authenticator` attribute prevents attackers from modifying EAP within a RADIUS packet using per-packet data integrity checks. The attribute is used to protect all Access-Request, Access-Challenge, Access-Accept, and Access-Reject packets containing an `EAP-Message` attribute.

7. D. This method of authentication allows a user to authenticate to a realm or sub-realm. When a RADIUS server receives an AAA request for a user name containing a realm, the server will reference a table of configured realms. If the realm is known, the server will proxy the request to the home RADIUS server for that realm. A global example of realm-based authentication is *eduroam* (education roaming).

8. D. Active Directory queries return the `memberOf` attribute, which is a list of AD groups to which a user belongs. Access policies can be configured on the WLAN device to assign roles based on AD group values from the `memberOf` attribute. The `memberOf` attribute is an LDAP attribute and not a RADIUS attribute.

9. C. UDP port 1813 is used for RADIUS accounting communications. RADIUS Accounting Request and Response packets use the `(40) Acct-Status-Type` attribute. Values for the attribute include `start`, `stop`, and `interim update`. Accounting information includes the username, IP address, NAS information, unique session identifier, data usage, and more.

10. C. The distributed sites, centralized RADIUS and LDAP model deploys one or more RADIUS servers at a central location that query a central LDAP database. The main benefit to this design is cost, because RADIUS servers and database servers are not deployed at the remote locations. The biggest concern with this type of design model is if the WAN link goes down, no wireless users can authenticate and the users would not be able to access the local wireless network.

11. A, C, E. RADIUS clients often get confused with Wi-Fi clients (supplicants). Instead, RADIUS clients are the devices that communicate directly with a RADIUS server using the RADIUS protocol. RADIUS clients are the authenticators within an 802.1X/EAP framework. To make things even more confusing, RADIUS clients are also sometimes referred to as the network access server (NAS). When discussing 802.1X/EAP, the terms *authenticator*, *NAS*, and *RADIUS client* are all synonymous.

12. B, C, D. When a WLAN supplicant sends an EAP request to the AP to gain access to network resources, the AP then forwards the EAP request encapsulated in a RADIUS Access-Request packet to the RADIUS server. The RADIUS server can respond three different ways. The RADIUS Access-Challenge is used by the RADIUS server to request more information. RADIUS Access-Accept grants access to a supplicant, while the RADIUS Access-Reject packet denies access.

13. B. The distributed autonomous sites model uses a central authentication database that is replicated to LDAP servers at each autonomous site. Each remote site has one or more RADIUS servers that proxy to the on-site LDAP database. The main benefit to this scenario is that all of the wireless users authenticate against the local replicated LDAP database and do not have to authenticate across a WAN link. If a remote WAN link goes down, the users can still successfully authenticate and access the wireless network. However, if the WAN link goes down, replication between the authentication databases cannot occur. The downside to this deployment model is that it is very expensive.

14. D. The `(79) EAP-Message` attribute is used to encapsulate EAP frames with RADIUS packets. A RADIUS client such as an access point or WLAN controller forwards the `EAP-Message` attributes within a RADIUS Access-Request packet to the RADIUS server. The RADIUS server can return `EAP-Message` attributes in Access-Challenge, Access-Accept, and Access-Reject packets.

15. C. The distributed sites, centralized RADIUS and LDAP model deploys one or more RADIUS servers at a central location that query a central LDAP database. The main benefit to this design is cost, because RADIUS servers and database servers are not deployed

at the remote locations. Performance bottlenecks can also occur that might affect latency and have a negative impact on roaming for any WLAN clients that do not support fast secure roaming mechanisms such as opportunistic key caching (OKC) or fast BSS transition (FT). A typical 802.1X/EAP frame exchange between a supplicant and RADIUS server can take 700 ms but will very often take multiple seconds across a WAN link. If a WLAN client does not have fast secure roaming capabilities, it will have to reauthenticate across the WAN link every time the client roams.

16. B, C, E. RADIUS with 802.1X/EAP is normally deployed for strong WLAN security. However, a RADIUS server can also be used to authorize WLAN users/devices with weak authentication methods such as captive web portal authentication and MAC address authentication.

17. A, E, F. When used within an 802.1X framework, a RADIUS server must be configured to communicate with the NAS, also known as the RADIUS client. The APs function as the NAS. The RADIUS server needs to be configured with the IP addresses of any APs functioning as the NAS along with a shared secret in order to communicate with the server. Both the RADIUS client and the RADIUS authentication server should be configured to use UDP port 1812 for authentication communications and UDP port 1813 for accounting communications. Digital certificates are configured on the RADIUS server to create an SSL/TSL tunnel with the WLAN client and not the AP. EAP protocol communications also do not involve the access point.

18. C. MAC authentication does not require client-side configuration and is typically used with legacy WLAN devices that might not support PSK or 802.1X/EAP. Because MAC addresses can be very easily spoofed, MAC authentication is rarely used by itself for WLAN security. However, MAC authentication is sometimes bound together with either captive web portal authentication or 802.1X/EAP. Combining MAC authentication along with 802.1X/EAP could be considered a simple method of multifactor authentication.

19. A, B, D. All EAP communications between the supplicant and the RADIUS server will only be local if a RADIUS server is physically located at the site. If the RADIUS server is located at a central location at a different site, RADIUS packets that encapsulate the EAP traffic will have to be forwarded across a WAN link.

20. A, B. WLAN vendors often use vendor-specific attributes (VSAs) for role assignment, but the standard IETF RADIUS attribute `(11) Filter-Id` is also often used for role assignment of user traffic permissions. RADIUS vendor-specific attributes (VSAs) are derived from the IETF attribute `(26) Vendor-Specific`. This attribute allows a vendor to create an additional 255 attributes however they wish. Data that is not defined in standard IETF RADIUS attributes can be encapsulated in the `(26) Vendor-Specific` attribute.

Chapter 10: Bring Your Own Device (BYOD) and Guest Access

1. B, C, E. Firewall ports that should be permitted include DHCP server UDP port 67, DNS UDP port 53, HTTP TCP port 80, and HTTPS TCP port 443. This allows the guest user's wireless device to receive an IP address, perform DNS queries, and browse the Web. Many

companies require their employees to use a secure VPN connection when they are connected to a SSID other than the company SSID. Therefore, it is recommended that IPsec IKE UDP port 500 and IPsec NAT-T UDP port 4500 also be permitted.

2. A, E. The guest firewall policy should allow for DHCP and DNS but restrict access to private networks 10.0.0.0/8, 172.16.0.0/12, and 192.168.0.0/16. Guest users are not allowed on these private networks because corporate network servers and resources usually reside on the private IP space. The guest firewall policy should simply route all guest traffic straight to an Internet gateway and away from corporate network infrastructure.

3. A, D, E. The four main components of an MDM architecture are the mobile device, an AP and/or WLAN controller, an MDM server, and a push notification service. The mobile Wi-Fi device requires access to the corporate WLAN. The AP or WLAN controller quarantines the mobile devices inside a walled garden if the devices have not been enrolled via the MDM server. The MDM server is responsible for enrolling client devices. The push notification services such as Apple Push Notification service (APNs) and Google Cloud Messaging (GCM) communicate with the mobile devices and the MDM servers for over-the-air management.

4. A, C. 802.1X/EAP requires that a root CA certificate be installed on the supplicant. Installing the root certificate onto Windows laptops can be easily automated using a Group Policy Object (GPO). An MDM uses over-the-air provisioning to onboard mobile devices and provision root CA certificates onto the mobile devices that are using 802.1X/EAP security. Self-service device onboarding applications can also be used to provision root CA certificates on mobile devices.

5. B. The MDM profiles used by Mac OS and iOS devices are Extensible Markup Language (XML) files.

6. B, D, E. An MDM server can monitor mobile device information, including device name, serial number, capacity, battery life, and applications that are installed on the device. Information that cannot be seen includes SMS messages, personal emails, calendars, and browser history.

7. D. The operating systems of some mobile devices require MDM agent application software. An MDM agent application can report back to an MDM server unique information about mobile devices that can later be used in MDM restriction and configuration policies.

8. A, B, E. A captive portal solution effectively turns a web browser into an authentication service. To authenticate, the user must launch a web browser. After the browser is launched and the user attempts to go to a website, no matter what web page the user attempts to browse, the user is redirected to a logon prompt, which is the captive portal logon web page. Captive portals can redirect unauthenticated users to a logon page using an IP redirect, DNS redirection, or redirection by HTTP.

9. B, C, D. The AP holds the mobile client device inside a walled garden. Within a network deployment, a walled garden is a closed environment that restricts access to web content and network resources while still allowing access to some resources. A walled garden is a closed platform of network services provided for devices and/or users. While inside the walled garden designated by the AP, the only services that the mobile device can access include DHCP, DNS, push notification services, and the MDM server. In order to escape from the walled garden, the mobile device must find the proper exit point, much like a real walled garden. The designated exit point for a mobile device is the MDM enrollment process.

10. C. Over-the-air provisioning differs between different device operating systems; however, using trusted certificates and SSL encryption is the norm. iOS devices use the Simple Certificate Enrollment Protocol (SCEP), which uses certificates and SSL encryption to protect the MDM profiles. The MDM server then sends a SCEP payload, which instructs the mobile device about how to download a trusted certificate from the MDM's certificate authority (CA) or a third-party CA. Once the certificate is installed on the mobile device, the encrypted MDM profile with the device configuration and restrictions payload is sent to the mobile device securely and installed.

11. A. An IP tunnel normally using Generic Routing Encapsulation (GRE) can transport guest traffic from the edge of the network back to the isolated DMZ. Depending on the WLAN vendor solution, the tunnel destination in the DMZ can be either a WLAN controller or simply a Layer 2 server appliance. The source of the GRE tunnel is the AP.

12. E. A guest management solution with employee sponsorship capabilities will integrate with an LDAP database such as Active Directory. Guest users can also be required to enter the email address of an employee, who must approve and sponsor the guest prior to allowing the guest access on the network. The sponsor typically receives an email requesting access for the guest, with a link in the email that allows the sponsor to easily accept or reject the request. Once users have registered or been sponsored, they can log on using their newly created credentials.

13. C. When enrolling their personal devices through the corporate MDM solution, typically employees will still have the ability to remove the MDM profiles because they own the device. If the employee removes the MDM profiles, the device is no longer managed by the corporate MDM solution. However, the next time the employee tries to connect to the company's WLAN with the mobile device, they will have to once again go through the MDM enrollment process.

14. D. The phrase *bring your own device (BYOD)* refers to the policy of permitting employees to bring personally owned mobile devices such as smartphones, tablets, and laptops to their workplace. A BYOD policy dictates which corporate resources can or cannot be accessed when employees access the company WLAN with their personal devices.

15. A. Social login is a method of using existing logon credentials from a social networking service such as Twitter, Facebook, or LinkedIn to register into a third-party website. Social login allows a user to forgo the process of creating new registration credentials for the third-party website. Retail and service businesses like the idea of social login because it allows the business to obtain meaningful marketing information about the guest user from the social networking service. Businesses can then build a database of the type of customers who are using the guest Wi-Fi while shopping.

16. F. A mobile device can still be managed remotely even if the mobile device is no longer connected to the corporate WLAN. The MDM servers can still manage the devices as long as the devices are connected to the Internet from any location. The communication between the MDM server and the mobile devices requires push notifications from a third-party service. Push notification services will send a message to a mobile device telling the device to contact the MDM server. The MDM server can then take remote actions over a secure connection.

17. B, D, E. Client isolation is a feature that can often be enabled on WLAN access points or controllers to block wireless clients from communicating with other wireless clients on the same wireless VLAN. Client isolation is highly recommended on guest WLANs to prevent peer-to-peer attacks. Enterprise WLAN vendors also offer the capability to throttle bandwidth of user traffic. Bandwidth throttling, which is also known as rate limiting, can be used to curb traffic at either the SSID level or the user level. Rate limiting the guest user traffic to 1024 Kbps is a common practice. A web content filtering solution can block guest users from viewing websites based on content categories. Each category contains websites or web pages that have been assigned based on their prevalent web content.

18. D. Captive portals are available as standalone software solutions, but most WLAN vendors offer integrated captive portal solutions. The captive portal may exist within a WLAN controller, or it may be deployed at the edge within an access point.

19. B. The mobile device must first establish an association with an AP. The AP holds the mobile client device inside a walled garden. Within a network deployment, a walled garden is a closed environment that restricts access to web content and network resources while still allowing access to some resources. A walled garden is a closed platform of network services provided for devices and/or users. While the mobile device is inside the walled garden designated by the AP, the only services it can access are DHCP, DNS, push notification services and the MDM server. After the mobile device completes the MDM enrollment process, the device is released from the walled garden.

20. A, B, C. A NAC server will use system health information, as reported by a posture agent, to determine whether the device is healthy. DHCP fingerprinting is used to help identify the hardware and operating system. RADIUS attributes can be used to identify whether the client is connected wirelessly or wired, along with other connection parameters. RADIUS CoA is used to disconnect or change the privileges of a client connection.

Chapter 11: Wireless Security Troubleshooting

1. A, C, E. Wi-Fi Troubleshooting 101 dictates that the end user first enable and disable the Wi-Fi network card. This ensures that the Wi-Fi NIC drivers are communicating properly with the operating system. The passphrase that is used to create a PSK can be 8–63 characters and is always case-sensitive. The problem is almost always a mismatch of the PSK credentials. If the PSK credentials do not match, a pairwise master key (PMK) seed is not properly created and therefore the 4-Way Handshake fails entirely. Another possible cause of the failure of PSK authentication could be a mismatch of the chosen encryption methods. An access point might be configured to only require WPA2 (CCMP-AES), which a legacy WPA (TKIP) client does not support. A similar failure of the 4-Way Handshake would occur.

2. C. Whenever you troubleshoot a WLAN, you should start at the Physical layer and 70 percent of the time the problem will reside on the WLAN client. Security issues usually evolve around improperly configured supplicant settings. This could be something as simple as a mistyped WPA2-Personal passphrase or as complex as 802.1X/EAP digital certificate problems. Many roaming problems are also a direct result of lack of support for fast secure roaming (FSR) mechanisms on a WLAN client.

3. D. If there is any client connectively problems, WLAN Troubleshooting 101 dictates that you disable and re-enable the WLAN network adapter. The drivers for the WLAN network interface card (NIC) are the interface between the 802.11 radio and the operating system (OS) of the client device. For whatever reason, WLAN drivers and the OS of the device may not be communicating properly. A simple disable/re-enable of the WLAN NIC will reset the drivers. Always eliminate this potential problem before investigating anything else.

4. B. The chosen Layer 2 EAP protocols must match on both the supplicant and the authentication server. For example, PEAPv0 (EAP-MSCHAPv2) might be selected on the supplicant while PEAPv1 (EAP-GTC) is configured on the RADIUS sever. Although the SSL/TLS tunnel might still be created, the inner tunnel authentication protocol does not match and authentication will fail. Although it is possible for multiple flavors of EAP to operate simultaneously over the same 802.1X framework, the EAP protocols must match on the both supplicant and the authentication server.

5. C, D. Although all of these answers could cause an IPsec VPN to fail, only two of these problems are related to certificate issues during IKE Phase 1. If IKE Phase 1 fails due to a certificate problem, ensure that you have the correct certificates installed properly on the VPN endpoints. Also remember that certificates are time based. Very often, a certificate problem during IKE Phase 1 is simply an incorrect clock setting on either VPN endpoint. If the encryption and hash settings do not match on both sides, the VPN will fail during IKE Phase 2. If the public/private IP address settings are misconfigured, the VPN will fail during IKE Phase 1.

6. C. Actually, all of these answers are correct. However, if Andrew asked his boss why he was looking at Facebook, Andrew might get fired. Options A and B are the correct technical answers, however; his boss only wants to blame the WLAN. Remember, from an end-user perspective, the WLAN is always the culprit. The correct option is C because Andrew's boss just wants it fixed.

7. D. All of these answers would cause the 4-Way Handshake to fail between the WLAN client and the AP in the reception area. Wi-Fi NIC drivers need to communicate properly with the operating system for PSK authentication to be successful. If the PSK credentials do not match, a pairwise master key (PMK) seed is not properly created and therefore the 4-Way Handshake fails entirely. The final pairwise transient key (PTK) is never created. Remember there is a symbiotic relationship between authentication and the creation of dynamic encryption keys. If PSK authentication fails, so does the 4-Way Handshake that is used to create the dynamic keys. Another possible cause of the failure of PSK authentication could be a mismatch of the chosen encryption methods. An access point might be configured to only require WPA2 (CCMP-AES), which a legacy WPA (TKIP) client does not support. A similar failure of the 4-Way Handshake would occur. Only option D is correct because the problem exists on a single AP in the reception area.

8. C, D. The graphic shows a supplicant (Wi-Fi client) trying to contact a RADIUS server. The authenticator forwards the request to a RADIUS server, but the RADIUS server never responds. This is an indication that there is a backend communication problem. If the RADIUS server never responds to the supplicant, there are four possible points of failure. Potential backend communication failure points include shared secret mismatch, authentication port mismatch, LDAP query failure, or incorrect IP settings on the AP and RADIUS server.

9. A, F. The graphic shows a supplicant and a RADIUS server trying to establish an SSL/TLS tunnel to protect the user credentials. The SSL/TLS tunnel is never created and authentication fails. This is an indication that there is a certificate problem. There is a whole range of certificate problems that could be causing the SSL/TLS tunnel not to be successfully created. The most common certificate issues are that the Root CA certificate is installed in the incorrect certificate store, the incorrect root certificate has been chosen, the server certificate has expired, the root CA certificate has expired, or the supplicant clock settings are incorrect.

10. B, D. VoWiFi requires a handoff of 150 ms or less to avoid a degradation of the quality of the call or, even worse, a loss of connection. Therefore, faster, secure roaming handoffs are required. Opportunistic key caching (OKC) and fast BSS transition (FT) produce roaming handoffs of closer to 50 ms even when 802.1X/EAP is the chosen security solution. The APs and the supplicant must both support FT; otherwise, the suppliant will reauthenticate every time the client roams.

11. A. All of these answers would cause the 4-Way Handshake to fail between the WLAN client and the AP in the reception area. Wi-Fi NIC drivers need to communicate properly with the operating system for PSK authentication to be successful. If the PSK credentials do not match, a pairwise master key (PMK) seed is not properly created and therefore the 4-Way Handshake fails entirely. The final pairwise transient key (PTK) is never created. Remember there is a symbiotic relationship between authentication and the creation of dynamic encryption keys. If PSK authentication fails, so does the 4-Way Handshake that is used to create the dynamic keys. Another possible cause of the failure of PSK authentication could be a mismatch of the chosen encryption methods. An access point might be configured to only require WPA2 (CCMP-AES), which a legacy WPA (TKIP) client does not support. A similar failure of the 4-Way Handshake would occur. Only option A is correct because iPads support CCMP encryption and all the other corporate iPads were connecting.

12. C, F. The graphic shows a supplicant and a RADIUS server trying to establish an SSL/TLS tunnel to protect the user credentials. The SSL/TLS tunnel is never created and authentication fails. This is an indication that there is a certificate problem. A whole range of certificate problems could be causing the SSL/TLS tunnel not to be successfully created. The most common certificate issues are that the root CA certificate is installed in the incorrect certificate store, the incorrect root certificate is chosen, the server certificate has expired, the root CA certificate has expired, or the supplicant clock settings are incorrect.

13. D. The graphic shows an EAP frame capture that indicates that EAP-FAST is the chosen Layer 2 authentication protocol. EAP-FAST does not use certificates and instead uses protected access credentials (PACs) to establish an SSL/TLS tunnel between the supplicant and the RADIUS server. If the PACs are not provisioned properly on the supplicants, the SSL/TLS tunnel will fail.

14. A, D, E. Troubleshooting best practices dictate that the proper information be gathered by asking relevant questions. Although options B and C might be interesting, they are not relevant to the potential problem.

15. B, D. If the backend 802.1X/EAP and RADIUS communications are working, the culprit is most likely the supplicant. Although certificates can cause numerous problems, very often the problem is a related to the supplicant credentials. You will notice in the graphic that the RADIUS server rejects the hostname credentials of the Windows laptop. Machine authentication requires the machine accounts of the laptops to be joined to the domain. Furthermore, the supplicant must be configured for both machine and user authentication.

16. A, B, C. Troubleshooting best practices dictate that the proper information be gathered by asking relevant questions. Option D is a question asked by the bridge keeper in a famous Monty Python movie.

17. A, B. All of these answers could cause an IPsec VPN to fail; however, IKE Phase 2 problems are usually a result of mismatched encryption and hash settings. If the public/private IP address settings are misconfigured, the VPN will fail during IKE Phase 1. Certificate problems will also cause IKE Phase 1 to fail.

18. A, F. The graphic clearly shows that 802.1X/EAP authentication completes and the 4-Way Handshake creates the dynamic encryption keys for the AP and client radio. At this point, Layer 2 authentication is complete and the virtual controlled port on the access point opened for the supplicant. However, the supplicant fails to obtain an IP address. This is not an 802.1X/EAP problem and is instead a networking issue. An incorrect user VLAN configuration on a switch could cause this problem. Another potential cause could be that the DHCP is offline or out of leases.

19. D, E. If you can verify that you do not have any certificate issues and the SSL/TLS tunnel is indeed established, the supplicant problems are credential failures. Notice that the RADIUS server is rejecting the supplicant credentials. Possible supplicant user credential problems include an expired user account, an expired user password, the wrong password, and that the user account does not exist.

20. A. Regardless of the networking technology, the majority of problems occur at the Physical layer of the OSI model. The majority of performance and connectivity problems in 802.11 WLANs can be traced back to the Physical layer.

Chapter 12: Wireless Security Risks

1. E. Eavesdropping is unauthorized passive capturing of data, often conducted with packet analyzing software. The other attacks listed are active attacks that would be detectable and may interrupt traffic flow. Because eavesdropping is a passive attack, it will be undetected by a WIDS/WIPS solution.

2. B, C. DoS attacks can occur at either Layer 1 or Layer 2 of the OSI model. Layer 1 attacks are known as RF jamming attacks. A wide variety of Layer 2 DoS attacks exist that are a result of tampering with 802.11 frames, including the spoofing of deauthentication frames.

3. C, D, E. The majority of unauthorized devices placed on networks, known as rogues, are placed there by people with access to the building. This means that they are more often placed by people you trust: employees, contractors, and visitors. Wardrivers and attackers are not usually allowed physical access.

4. C, D. Malicious eavesdropping is achieved with the unauthorized use of protocol analyzers to capture wireless communications. Any unencrypted 802.11 frame transmission can be reassembled at the upper layers of the OSI model.

5. F. Because rogue devices are unauthorized WLAN portals, all of your network resources are potentially exposed. Any data stored on network servers or desktop workstations is entirely at risk. Data theft is usually the most common risk associated with rogue access. Destruction of data can also occur. Databases can be erased and drives can be reformatted. Network services can also be disabled. An attacker can use the unauthorized portal for malicious data insertion and upload viruses and pornography. Remote control applications and keystroke loggers can also be uploaded to network resources and used to gather information at a later date. Attackers have been known to upload illegally copied software and set up illegal FTP servers to distribute the illegal software. Once an attacker has accomplished rogue access, a wired network can be used as a launching pad for third-party attacks against other networks across the Internet.

6. A, C. Microwave ovens operate in the 2.4 GHz ISM band and are often a source of unintentional interference. 2.4 GHz cordless phones can also cause unintentional jamming. A signal generator is typically going to be used as a jamming device, which would be considered intentional jamming. 900 MHz cordless phones will not interfere with 802.11 equipment that operates in either the 2.4 GHz ISM band or the 5 GHz UNII bands. There is no such thing as a deauthentication transmitter.

7. A, C, D, F, G. Currently, there is no such thing as a Happy AP attack or an 802.11 reverse ARP attack. Wireless users are especially vulnerable to attacks at public-use hotspots because there is no security. Because no encryption is used, the wireless users are vulnerable to malicious eavesdropping. Because no mutual authentication solution is in place, they are vulnerable to hijacking, man-in-the-middle, and phishing attacks. The hotspot access point might also be allowing peer-to-peer communications, making the users vulnerable to peer-to-peer attacks. Every company should have a remote-access wireless security policy to protect their end users when they leave company grounds.

8. A, B, D. Wired Equivalent Privacy (WEP) encryption has been cracked, and currently available tools may be able to derive the secret key within a matter of minutes. The size of the key makes no difference, and both 64-bit WEP and 128-bit WEP can be cracked. CCMP/AES encryption has not been cracked.

9. A, C, D, E. Numerous types of Layer 2 DoS attacks exist, including association floods, deauthentication spoofing, disassociation spoofing, authentication floods, PS-Poll floods, and virtual carrier attacks. RF jamming is a Layer 1 DoS attack.

10. A, B. The best way of preventing rogue access is wired port control. An 802.1X/EAP can be used to authorize access through wired ports on an access layer switch. EAP-MD5 and EAP-TLS can be used for wired 802.1 X/EAP authentication to control wired-side access. Unless

proper credentials are presented at Layer 2, upper-layer communications are not possible through the wired port. A rogue device cannot act as a wireless portal to network resources if the rogue device is plugged into a managed port that is blocking upper-layer traffic. Therefore, a wired 802.1X/EAP solution is an excellent method of preventing rogue access.

11. A, E. After obtaining the passphrase, an attacker can also associate with the WPA/WPA2 access point and thereby access network resources. The encryption technology is not cracked, but the key can be re-created. If a hacker has the passphrase and captures the 4-Way Handshake, they can re-create the dynamic encryption keys and therefore decrypt traffic. WPA/WPA2-Personal is not considered a strong security solution for the enterprise because if the passphrase is compromised, the attacker can access network resources and decrypt traffic.

12. D. An attack that often generates a lot of press is wireless hijacking, also known as the evil twin attack. The attacker hijacks wireless clients at Layer 2 and Layer 3 by using an evil twin access point and a DHCP server. The hacker may take the attack several steps further and initiate a man-in-the-middle attack and/or a Wi-Fi phishing attack.

13. B. MAC filters are configured to apply restrictions that will allow only traffic from specific client stations to pass through based on their unique MAC addresses. MAC addresses can be spoofed, or impersonated, and any amateur hacker can easily bypass any MAC filter by spoofing an allowed client station's address.

14. A, C. 802.11 uses a medium contention process called Carrier Sense Multiple Access with Collision Avoidance (CSMA/CA). To ensure that only one 802.11 radio is transmitting on the half-duplex RF medium, CSMA/CA uses four checks and balances. The four checks and balances are virtual carrier sense, physical carrier sense, the random backoff timer, and interframe spacing. Virtual carrier sense uses a timer mechanism known as the network allocation vector (NAV) timer. Physical carrier sense uses a mechanism called the clear channel assessment (CCA) to determine whether the medium is busy before transmitting. Virtual carrier sense is susceptible to a Layer 2 DoS attack when an attacker manipulates the duration value of 802.11 frames. Physical carrier sense is susceptible to a Layer 1 DoS attack when there is a continuous transmitter on the frequency channel.

15. E. Even with the best authentication and encryption in place, attackers can still see MAC address information in clear text. MAC addresses are needed to direct traffic at Layer 2.

16. C. Wired leakage often occurs when wired stations, servers, or infrastructure devices use a broadcast protocol to communicate or to find other devices with which to replicate. Access points will forward wired broadcast and multicast traffic into the air, since on a certain level APs are media converters going to and from wired and wireless mediums. Layer 2 discovery protocols such as Link Layer Discovery Protocol (LLDP) or Cisco Discovery Protocol (CDP) will reveal information about the wired network as well as what can already be seen wirelessly.

17. C, D. The only way to prevent a wireless hijacking, man-in-the-middle, and/or Wi-Fi phishing attack is to use a mutual authentication solution. 802.1X/EAP authentication solutions require that mutual authentication credentials be exchanged before a user can be authorized.

18. B. A malicious eavesdropping attack uses a protocol analyzer for unauthorized capture and viewing of 802.11 frames. A protocol analyzer uses an 802.11 radio to listen passively to

802.11 communications. A protocol analyzer does not transmit and will therefore go undetected by a WIPS solution and no alarms will be triggered.

19. A, C. The 802.11 radios inside the WIPS sensors monitor the 2.4 GHz ISM band and the 5 GHz UNII bands. Older legacy wireless networking equipment exists that transmits in the 900 MHz ISM band, and these devices will not be detected. The radios inside the WIPS sensors also use only DSSS and OFDM technologies. Wireless networking equipment exists that uses frequency hopping spread spectrum (FHSS) transmissions in the 2.4 GHz ISM band and will go undetected. The only tool that can detect either a 900 MHz or frequency hopping rogue access point is a spectrum analyzer.

20. C. A wireless intrusion prevention system (WIPS) is capable of mitigating attacks from rogue access points. A WIPS sensor can use Layer 2 DoS attacks as a countermeasure against a rogue device. SNMP may be used to shut down ports to which a rogue AP has been connected. WIPS vendors also use unpublished methods for mitigating rogue attacks.

Chapter 13: Wireless LAN Security Auditing

1. A, D, E, F. All the devices listed are known sources of RF interference in the 2.4 GHz ISM band. All-band interference is caused by frequency-hopping radio transmissions. FHSS is used by Bluetooth, legacy 802.11 FHSS access points, HomeRF equipment, and Digital Enhanced Cordless Telecommunications (DECT) telephones. DECT telephones can use multiple frequencies, including the 2.4 GHz ISM band.

2. E. The weakest link in security for any type of computer network is usually the network end users. Social engineering is the act of manipulating people into divulging confidential information for the purpose of obtaining information used to access a computer network. Therefore, social engineering techniques are often used by the security auditor to circumvent the WLAN and gain access to network resources. Social engineering performed by an auditor is a form of penetration testing and is usually the most successful method. An auditor will most likely find lapses in security due to improper employee training or employee policy enforcement. Although penetration testing using software tools is often successful, the auditor will probably have more luck with social engineering techniques used for penetration testing purposes.

3. B, D. The liability waiver, scope of work, and nondisclosure agreements should all be completed prior to conducting a WLAN audit. It is very important that any auditor get signed permission in the form of a liability waiver or hold harmless agreement to perform the security audit. An audit should never take place without authorized permission. Anyone who performs a WLAN security audit and penetration testing without permission risks arrest and prosecution.

4. D. If a distributed WIDS/WIPS monitoring solution has already been deployed, the monitoring solution should also be audited. Hacker attacks such as denial of service, MAC spoofing, rogue devices, and so forth should be simulated to see if the proper WIDS/WIPS alarms are triggered for each attack. For example, deauthentication and dissociation attacks can be simulated to verify detection by the WIPS server and sensors. If certain alarms are not triggered, alarm thresholds may need to be adjusted on the WIDS/WIPS server.

5. A, B. Once authorized onto network resources, WLAN users can be further restricted as to what resources may be accessed and where they can go. Role-based access control (RBAC) security commonly used by WLAN controllers can be used to restrict certain groups of users to certain network resources. RBAC is usually accomplished with firewall policies and access control lists. Penetration testing of the firewall used to restrict WLAN user access to certain network resources should be a mandatory procedure during the security audit. All WLAN infrastructure management interfaces should also be audited. Interfaces that are not used should be disabled. Strong passwords should be used, and encrypted login capabilities such as Hypertext Transfer Protocol Secure (HTTPS) and Secure Shell (SSH) should be utilized if available.

6. A, B, D. Stronger dynamic encryption, stringent authentication methods, role-based access control proposals, WIDS/WIPS monitoring recommendations, corporate policy suggestions, employee training, and physical security advice may all be the result of a successful WLAN audit.

7. A, B, C. The Layer 1 and Layer 2 auditing tools should be considered mandatory. WLAN protocol analyzers are typically used for Layer 2 auditing and spectrum analyzers are used for Layer 1 auditing. Many freeware Linux-based WLAN penetration testing software tools are also widely available. Penetration testing normally should be considered a mandatory part of the audit. Many other tools such as cameras, battery packs, and GPS devices may be needed but are not considered mandatory.

8. A, C. Yagi antennas are often deployed to reduce reflections in high multipath environments; however, this has nothing to do with a WLAN security audit. A high-gain unidirectional antenna can be used to hear RF signals from a great distance. Antennas are bidirectional passive amplifiers. In the case of an outside audit, proper antenna use often gives the auditor the ability to listen from a remote location. Just because the signal is not intentionally propagated beyond the property does not mean it cannot be heard with the right equipment. An outside auditor can emulate an outside attacker who can possibly launch an attack from areas that cannot be secured physically.

9. A, B. To learn which type of encryption the company has mandated, you should consult the written security policy. Any WLAN protocol analyzer can verify that all corporate WLAN devices are using the proper encryption methods. coWPAtty, Asleap, and Aircrack-ng can be used to attack devices and crack WEP and WPA-PSK keys but are not used simply to determine the encryption types that are being used and whether they meet corporate goals.

10. D. The statement of work (SOW) document outlines the audit requirements, deliverables, and timeline that the auditor will execute for a customer. Pricing and detailed terms and conditions are specified in the SOW. A liability waiver grants authorized permission to perform the security audit. A mutual nondisclosure agreement is necessary because organizations may not wish to reveal their network topology for security reasons. Additionally,

a network topology map is useful for understanding the layout of the customer's wired network infrastructure and WLAN infrastructure. If possible, obtaining a copy of the customer's written security policy document will be valuable for determining risk assessments. Very often the corporate security policy will be nonexistent or outdated.

11. A, D. Periodic auditing, in addition to the initial audit, should have revealed the flaws in security. Many post-deployment changes to a network can reduce the security posture to an unacceptable level. Devices added to the network, either as legitimate or rogue, can change the level of security of a network. Updating a written policy does not ensure that the policy is being followed. However, if the policy does not address current regulatory issues, it also needs to be updated. This should be checked regularly.

12. A, B, C, E. WPA2-Personal authentication, often referred to as WPA2-PSK, can be compromised with an offline brute-force dictionary attack. A WLAN protocol analyzer is needed to capture the 4-Way Handshake to get the seed key for cracking. Authentication cracking audit tools such as Aircrack-ng or coWPAtty use a dictionary file to crack the static passphrase used for authentication. If an auditor is not able to capture the 4-Way Handshake because the stations are already connected, a deauthentication or disassociation flood can be used to kick them off the network, forcing the client stations to reauthenticate. Asleap is an authentication cracking audit tool used to compromise LEAP. NetStumbler is a WLAN discovery tool.

13. D. If a distributed WIDS/WIPS monitoring solution has already been deployed, the monitoring solution should also be audited. Hacker attacks such as denial of service, MAC spoofing, rogue devices, and so on should be simulated to see if the proper WIDS/WIPS alarms are triggered for each attack. For example, a third-party access point can be connected to a wired port to test the rogue detection capabilities of a distributed WIPS monitoring solution. If certain alarms are not triggered, alarm thresholds may need to be adjusted on the WIDS/WIPS server.

14. C, D. Without collecting information from both Layer 1 and Layer 2, there is no certain way to determine the cause of the disconnections. There could be Layer 1 or Layer 2 DoS attack disrupting service. A rainbow table is used for cracking and the dipole antenna, regardless of gain, is not required; other antenna types could be used.

15. B. This AP should be reported to the company as quickly as possible. Often the networking team deploys APs without consulting the security team and has them added to the WIPS as authorized. Unplugging it immediately may disrupt legitimate traffic. Cracking its security and decoding the traffic may not be allowed by the scope of work agreement or corporate policy.

16. A, D. One of the main purposes of a Layer 2 audit is to detect initially any rogue devices or unauthorized 802.11 devices. Identifying rogue devices during an initial security audit is critical, especially if a distributed WIPS monitoring solution has not been deployed. A proper Layer 2 WLAN security audit will also be used to identify initially all authorized 802.11 WLAN devices, including access points and authorized clients. The audit can also be used to validate WLAN security compliance. The mandated authentication and encryption of all authorized devices can be evaluated to verify the proper security configuration. Any authorized devices that have not been properly configured will be flagged.

17. A. If there is no physical security for the AP, it could be compromised by an attacker that gains physical access. Since this one is in a common hallway of a multitenant building, there is a very high likelihood of it being compromised or stolen. Physical security of the devices being used is just as important as protecting the communications.

18. E. All of these items are of use. Part of the audit should include physical inspection of authorized devices. This may require the items listed depending on the environment and may even require other tools, such as a GPS.

19. B. Any WLAN protocol analyzer can verify that all corporate WLAN devices are using the proper authentication methods. coWPAtty, Asleap, and Aircrack-ng are used during penetration tests to compromise weaker authentication methods than EAP-TTLS. WLAN discovery tools such as NetStumbler do not capture 802.11 frames but instead find 802.11 devices using null probe request frames.

20. B. Depending on the contents of the toolkit, this is a common practice. WLAN audits and site surveys are similar and the same tools can be used to conduct both types of work. Spectrum analysis and WLAN protocol analyzer tools are widely used in both formats.

Chapter 14: Wireless Security Monitoring

1. A, B. The radios inside the WIDS/WIPS sensors currently use only DSSS and OFDM technologies. Wireless networking equipment exists that uses frequency hopping spread spectrum (FHSS) transmissions in the 2.4 GHz ISM and will go undetected by Layer 2 WIPS/WIDS sensors. The proper tool to detect a frequency hopping rogue access point is a spectrum analyzer. Some WIDS/WIPS vendors have begun to offer Layer 1 distributed spectrum analysis system (DSAS) solutions.

2. A, B, E, F. Most WIDS/WIPS solutions categorize 802.11 radios into four or more classifications. An authorized device refers to any client station or access point that is an authorized member of the company's wireless network. An unauthorized device is any new 802.11 radio that has been detected but not classified as a rogue. A neighbor device refers to any client station or access point that is detected by the WIPS and whose identity is known. This type of device initially is detected as an unauthorized device. Typically, the neighbor device label is then manually assigned by an administrator. A rogue device refers to any client station or access point that is considered a potential threat. An AP is classified as a rogue if the WIPS determines that the AP is connected to the wired network.

3. F. A WIDS/WIPS server is often a standalone device that is available as either a hardware appliance or as a software appliance that can run as a virtual machine. The WIDS/WIPS server might also be integrated in a WLAN controller. WIDS/WIPS server monitoring capabilities can also be unified with a network management server (NMS) solution that is used to monitor all aspects of a WLAN. These WLAN security servers may be deployed in a datacenter as an on-premises solution or exist in a cloud-based environment.

4. A, C. The typical wireless intrusion detection system is a client/server model that consists of two components:

   WIPS Server A software or hardware server acting as a central point of management and monitoring

   Sensors Hardware or software-based sensors placed strategically to listen to and capture all 802.11 communications

5. B, E, F. The 802.11w-2009 amendment defines protection for robust management frames. The robust management frames include robust action frames, disassociation frames, and deauthentication frames. The majority of action frames are considered robust, including QoS action frames, radio measurement action frames, block ACKs, and many more. A beacon is an 802.11 management frame that is not considered robust. A PS-Poll frame is an 802.11 control frame. A null function frame is a type of 802.11 data frame.

6. C, F. Many WIDS/WIPS solutions also use behavioral analysis to recognize any patterns that deviate from normal WLAN activity. Behavioral analysis identifies abnormal network behavior based on historical metrics. Behavior analysis helps detect protocol fuzzing, where an attacker sends malformed input to look for bugs and programming flaws in AP code or client station firmware. Very often, a zero day attack will create some sort of anomaly in 802.11 behavior that can be detected. A zero day attack is an unknown threat used to exploit computer networks. EAP floods, CTS floods, Fake AP, and deauthentication attacks are all known attacks with specific signatures that can be detected by signature analysis.

7. B, C, D. The answer to this question depends on the policy of the corporation. Defining a policy on how rogue APs are dealt with as well as a policy forbidding employees to install their own Wi-Fi devices is mandatory. Wireless rogue containment should be used very carefully. Rogue devices can be manually terminated using wireless rogue containment, or a WIPS can be configured to automatically terminate any devices that are classified as rogue. Using a deauthentication countermeasure against all unauthorized devices is usually not a good idea because the WIPS may accidentally terminate legitimate APs and clients from neighboring businesses. Improper use of wireless rogue containment capabilities can create legal problems. Most organizations choose to implement wireless rogue containment manually. Any device that has been classified as rogue is a device that has been determined by the WIPS to be connected to the corporate backbone and probably should be wirelessly contained. Temporary wireless rogue containment of ad hoc WLANs may also be necessary. The wired-side termination method of rogue mitigation uses the Simple Network Management Protocol (SNMP) for *port suppression*. Most WIPS can determine that the rogue access point is connected to the wired infrastructure and may be able to use SNMP to disable the managed switch port to which the rogue access point is connected.

8. D. RF fingerprinting is a method that compares the target's detected RSSI values with the RSSI values of the known reference points. If the target and the known reference point have similar RSSI values, they are located close to each other. RF fingerprinting relies on building up a database of actual measurements as they relate to particular locations. The method used to document RSSI reference points on a map is known as RF calibration. RF triangulation is a way of using received signals detected by sensors that are in known locations to find devices that are in unknown locations.

9. E. Time difference of arrival (TDoA) uses the variation of arrival times of the same transmitted signal at three or more receivers, in our case, sensors. The transmitted signal will arrive at the sensors at different times due to distance between the transmitter and the multiple sensors. The speed of travel of the radio frequency is a known factor, and each of the synchronized TDoA sensors reports the time of arrival of the signal from the transmitting device.

10. A, B, D, E. WIDS/WIPS solutions use signature analysis to analyze frame patterns or "signatures" of known wireless intrusions and WLAN attacks. The WIDS/WIPS has a programmed database of hundreds of threat signatures of known WLAN attacks. PS-Poll floods, CTS floods, virtual carrier attacks, and deauthentication attacks are all known attacks with specific signatures that can be detected by signature analysis. Protocol fuzzing and zero day attacks might be detected by behavioral analysis.

11. C, D. Off-channel scanning is notorious for causing "choppy audio" during an active voice call from a VoWiFi device associated to AP. WLAN vendors usually have an option that suspends off-channel scanning based on the detection of QoS priority markings that indicate voice traffic. Effectively off-channel scanning is suspended during an active VoIP call over the WLAN. If the off-channel scanning is suspended due to VoWiFi communications, the WLAN security monitoring is also suspended. Another solution would be to convert some AP to full-time sensors that scan constantly and do not allow client associations. Full-time sensors will not disrupt voice communications.

12. C. Enterprise WIDS/WIPS may provide forensic analysis that allows an administrator to retrace the actions of any single WLAN device down to the minute. You can rewind and review minute-by-minute records of connectivity and communications within a WLAN. The WIPS records and stores hundreds of data points per WLAN device, per connection, per minute. This allows an organization to view months of historical data. Information such as channel activity, signal characteristics, device activity, and traffic flow and attacks can all be viewed historically.

13. E. The answer often depends on the budget and the value of what network resources are being protected by WLAN security monitoring. The best answer is that you can never have too many sensors. When WLAN security monitoring is deployed, the more sensors the better. Every WLAN vendor has their own sensor deployment recommendations and guidelines, but a ratio of one sensor for every three-to-five access points is highly recommended. When WLAN security monitoring is an extremely high priority and cost is not an issue, the more sensor devices the better. WIDS/WIPS deployments at military bases often follow a ratio of one sensor for every two APs or may even deploy with a 1:1 ratio.

14. A, E. Corrupted frames are the leading cause of false positives, but improper configuration of the WIPS or misinterpretation of alarms by administrators can also lead to false positive alarms. Proper classification of all devices as either authorized or neighbor devices is important. If devices are not properly classified, many alarms will be triggered. Setting alarm thresholds can often be difficult, and it may be time consuming to achieve a balance in detecting possible attacks versus triggering false positive alarms. Some inaccurate reporting of events, seen as false positives, is unavoidable. However, properly setting alarm thresholds that are fine-tuned to the on-site RF environment can greatly reduce the number of false positives. When an integrated WIPS is deployed, it is always a highly recommended practice

to convert a number of the APs into full-time sensors. However, using full-time sensors does not necessarily decrease false positives.

15. A, B, D, E. The radio form factors of the APs that are used for WIPS sensors vary widely depending on the WLAN vendor.

16. A. RF fingerprinting compares RSSI values collected with measured RSSI values from known points to better pinpoint the location of the device being tracked and can do so with fewer sensors. However, RF fingerprinting solutions are more costly and require more time to set up and calibrate. Recalibration is required should the RF environment change.

17. C. APs configured as part-time sensors use off-channel scanning for security and RF monitoring purposes. Off-channel scanning can cause choppy audio during VoIP calls over the WLAN. Part-time sensors can suspend off-channel scanning based on the detection of QoS priority markings that indicate voice traffic. Effectively off-channel scanning is suspended during an active VoIP call over the WLAN.

18. A, E, F. Given that an AP acts as a Layer 2 bridge, a WIPS solution builds a MAC table that correlates both the wired-side MAC address and wireless-side MAC address (BSSID) of the access point. This correlated MAC table can then be compared to the database of authorized devices. Any unauthorized device that is detected by both a sensor on the wireless side and by SNMP on the wired side will then be classified as rogue AP. Another method often used to determine whether a device is connected to the wired backbone is to look at ARP requests from a wired device, such as a core router, and to analyze MAC tables. The WIDS/WIPS solution will look at the wired/wireless MAC tables. Any unauthorized BSSID transmitting an ARP request with the source address of the wired router will be classified as a rogue AP. Another method for possibly determining if there is a potential rogue device connected to the wired network is to examine time-to-live (TTL) values of IP packets. Wi-Fi routers will lower the TTL value of a packet when it flows through the device. WIDS/WIPS vendors may also use proprietary rogue detection and classification.

19. A, C. One method of classifying Layer 3 rogue devices is to have a nearby sensor associate with an unauthorized suspect rogue AP. The sensor then sends traffic, such as a ping, back to the WIPS server. If the traffic reaches the server on the wired side, the suspected rogue AP is confirmed to be on the internal network and then classified as a rogue. However, very often a rogue Wi-Fi router will have configured security such as WEP or WPA. A sensor will not be able to associate with the rogue Wi-Fi router because the sensor does not know the rogue Wi-Fi router security settings. The majority of the WLAN vendors that manufacture home Wi-Fi routers use MAC addresses that are one bit apart on the wireless and wired interfaces. Methods of Layer 2 rogue detection can be augmented by looking for wired MAC addresses within a small range of the BSSID detected by a sensor.

20. A, C. As the administrator, you should configure automatic notifications to trigger only if an alarm is of specific importance. You will want to be alerted to the fact that a rogue is on the network or an attack is occurring. However, you would not want to know every time a WLAN device is detected unless a no-wireless zone was being enforced. Typically, alarms with a threat level of critical or severe are also configured for notification. It should be noted that any AP or client that is initially classified as unauthorized should still be investigated as soon as time permits.

Chapter 15: Wireless Security Policies

1. F. It is important to include all stakeholders, especially management, when developing a wireless usage policy. By allowing the stakeholders to assist in policy development, you are more likely to create a usable, followed, and enforceable policy.

2. D. The PCI standard specifies how to protect the cardholder data environment (CDE) both on the wired network and wirelessly. Any organization that processes or stores cardholder information must follow PCI guidelines or risk severe penalties.

3. D. PCI compliance does not allow any deployments to include WEP. Furthermore, PCI mandates that WEP not be used within or touching the CDE.

4. C, D, E. Traveling users are obviously not going to be able to connect to your WLAN from the road. Hotspots at best only use a captive portal for security. You cannot expect a hotspot to use a WPA2-Enterprise solution that meets your needs. For traveling users, the best things you can provide for their wireless security without additional supplicant software are an up-to-date antivirus program, an up-to-date firewall, and a strong VPN solution when connecting to public networks. It is important to include security of traveling devices in your written WLAN security policies.

5. A. If there is no written policy forbidding the placement of wireless devices without authorization, it is often difficult or impossible to take appropriate disciplinary measures, because none are specified.

6. C. In addition to all of the required documentation used in network design—SOW, NDA, floor plans, and even a liability statement—you should always ask to see and examine the written security policies if they exist. Failure to do so may lead to a breach in security and an embarrassing situation, such as being escorted off the premises.

7. A, B. Using client-side certificates for end-user credentials and using machine authentication are both excellent methods of securing a WLAN. However, a policy mandating their use is not always practical due to the expense and administrative overhead. All end-user WPA-2 Enterprise passwords should contain numbers, special characters, and upper- and lowercase letters. Because a WPA2-Personal passphrase is static and shared among many legacy devices not capable of 802.1X/EAP, a policy should dictate the method of creation. Entropy is extremely important when considering password policy.

8. B. Only the applicable audience of an external policy influence is required to follow the mandates of that regulation or standard. However, following some of the guidelines from nonrequired policies is often a good practice to increase your own security levels. This is often done within organizations that do not have to meet FIPS compliance but want to have a standardization that is deemed more secure than others without developing standards internally and adopting those designed by others.

9. C. Updating the firmware of the legacy devices would be the least expensive and least disruptive way to improve the security posture of this organization. Neither creating nor changing policy improves actual security. If the policy is not followed it will be of no use. You must change the security from WEP to something more secure. Swapping out the

devices for new ones with the ability to do WPA2 would improve security, but would also be more disruptive than simply updating the firmware to do so. However, it should be noted that, in many cases, upgrading hardware is the only option because the legacy hardware does not have the processing power to support WPA2. Keeping up with security patches and making sure your devices can support the latest security should be part of a well-executed security plan.

10. D. A well-crafted security policy will include a change control process that helps to reduce the impact of making changes in security configurations. Security policies must adapt as risks and technologies change. When making changes in business practices or security measures, impact on function must be taken into consideration.

11. C. Being compliant should never be confused with being secure. For example, PCI states that WEP can still be used until June 30, 2010. If your network is using WEP to protect the CDE, you may be compliant. However, given the weakness found in WEP security and the public availability of freeware that cracks WEP and videos that teach people how to crack WEP, your network is not secure. Some of the most widely known wireless attacks happened on "compliant" networks.

12. A, B, C, E. The total impact of a compromise should be weighed when conducting an impact analysis. This includes both tangible and intangible costs. You should also consider not just the costs of implementing secure solutions but also the costs of not implementing them. You should also weigh the legal implications should there be a compromise when developing a security policy. Fines and court costs due to litigation arising from a compromise are a loss to the organization as well. Avoiding fines for noncompliance is just as much a security concern as keeping intruders off your WLAN.

13. D. Some policies mandate physical security of the APs to include locked enclosures based on the network's data privacy requirements. You should always review written security policy prior to conducting a vulnerability assessment. Hacking into the network or tampering with devices in any way is not part of an assessment but may be part of a penetration test if covered in the SOW. It is possible to have both locked and unlocked protection requirements in the same building.

14. C. When dealing with rogue device placement, or even with legitimate device placement, company security policies should always be followed. Mitigating the rogue should be covered in the policy no matter who placed the device on the network or where the device is located.

15. B, C, D, E. Compared to handheld and laptop-based scanning solutions, WIPS are more expensive but enable organizations to maintain compliance and security with less effort once deployed. WIPSs and WIDSs are mentioned in several policies and offer more than just compliance. Periodic scanning only captures information while you are scanning and only from one location. WIPS scan 24 hours a day and have the ability to use proactive measures to protect the WLAN.

16. B. Deployment policies should include how devices are configured prior to deployment. Many new printers ship with the wireless card enabled and set for default ad hoc use. A policy covering device deployment that is followed can reduce misconfigurations that lead to security policy violations.

17. E. If guests are going to be allowed the use of wireless devices in your buildings, then the use of such devices should be addressed in your security policies. Many organizations use guest networks on separate Internet connections in the DMZ or on separate VLANs within the building. The fact that a trusted employee placed the device with nothing but good intentions does not change the fact that the AP is a rogue and should be contained by the WIPS.

18. C. WIDS policies should define how to respond properly to alerts generated by the system. An example would be how to deal with the discovery of rogue access points and all the necessary actions that should take place.

19. C. Users must be trained to follow security policies. Even if the policies are well written and enforced, the users must be educated for policy enforcement to be effective.

20. A, B, C, D. Monitoring for rogue devices, rogue device remediation, user training, and policy enforcement procedures are all part of a successful security policy implementation. End users should never be permitted to install their own wireless devices on the corporate network. This includes access points, wireless routers, USB wireless clients, and wireless cards. Any users installing their own wireless equipment could open unsecured portals into the main infrastructure network. This policy should be strictly enforced.

Appendix B

Abbreviations and Acronyms

Certifications

CWAP Certified Wireless Analysis Professional

CWDP Certified Wireless Design Professional

CWNA Certified Wireless Network Administrator

CWNE Certified Wireless Network Expert

CWNT Certified Wireless Network Trainer

CWSP Certified Wireless Security Professional

CWTS Certified Wireless Technology Specialist

Organizations and Regulations

ACMA Australian Communications and Media Authority

ARIB Association of Radio Industries and Businesses (Japan)

ATU African Telecommunications Union

CEPT European Conference of Postal and Telecommunications Administrations

CITEL Inter-American Telecommunication Commission

CTIA Cellular Telecommunications and Internet Association

CWNP Certified Wireless Network Professional

DoD Department of Defense

ERC European Radiocommunications Committee

EWC Enhanced Wireless Consortium

FCC Federal Communications Commission

FIPS Federal Information Processing Standards

GLBA Gramm-Leach-Bliley Act

HIPAA Health Insurance Portability and Accountability Act

IANA Internet Assigned Number Authority

IEEE Institute of Electrical and Electronics Engineers

IETF Internet Engineering Task Force

ISO International Organization for Standardization

ITU-R International Telecommunications Union Radio Communication Sector

NEMA National Electrical Manufacturers Association

NIST National Institute of Standards and Technology

PCI Payment Card Industry

RCC Regional Commonwealth in the field of Communications

SEE-Mesh Simple, Efficient, and Extensible Mesh

SOX Sarbanes–Oxley

UTMS Universal Mobile Telecommunications System

WECA Wireless Ethernet Compatibility Alliance

WIEN Wireless Interworking with External Networks

Wi-Fi Alliance Wi-Fi Alliance

WiMA Wi-Mesh Alliance

WNN Wi-Fi Net News

Measurements

dB decibel

dBd decibels referenced to a dipole antenna

dBi decibels referenced to an isotropic radiator

dBm decibels referenced to 1 milliwatt

GHz gigahertz

Hz hertz

KHz kilohertz

mA milliampere

MHz megahertz

mW milliwatt

SNR signal-to-noise ratio

V volt

VDC voltage direct current

W watt

Technical Terms

3DES Triple DES

AA authenticator address

AAA authorization, authentication, and accounting

AAD additional authentication data

AC access category

AC access controller

AC alternating current

ACK acknowledgment

ACL access control list

AD Active Directory

AES Advanced Encryption Standard

AGL above ground level

AH Authentication Header

AID association identifier

AIFS arbitration interframe space

AKM authentication and key management

AM amplitude modulation

AMPE authenticated mesh peering exchange

A-MPDU Aggregate MAC Protocol Data Unit

A-MSDU Aggregate MAC Service Data Unit

ANonce authenticator nonce

AP access point

API application programming interface

APN Apple Push Notification

APSD automatic power save delivery

ARC4 Alleged RC4

ARP Address Resolution Protocol

ARS adaptive rate selection

ARS automatic rate selection

AS authentication server

ASCII American Standard Code for Information Interchange

ASK amplitude shift keying

ATF airtime fairness

ATIM announcement traffic indication message

AVP attribute–value pair

BA Block Acknowledgment

BER bit error rate

BIP broadcast/multicast integrity protocol

BPSK binary phase shift keying

BSA basic service area

BSS basic service set

BSSID basic service set identifier

BT Bluetooth

BVI bridged virtual interface

BYOD bring your own device

CA certificate authority

CAD computer-aided design

CAM content addressable memory

CAM continuous aware mode

CAPWAP Control and Provisioning of Wireless Access Points

CBC Cipher-Block Chaining

CBC-MAC Cipher Block Chaining Message Authentication Code

CBN cloud-based networking

CCA clear channel assessment

CC-AP cooperative control access point

CCI co-channel interference

CCK Complementary Code Keying

CCKM Cisco Centralized Key Management

CCM Counter with CBC-MAC

CCMP Counter Mode with Cipher Block Chaining Message Authentication Code Protocol

CCX Cisco Compatible Extensions

CDE cardholder data environment

CDMA2000 code division multiple access 2000

CDP Cisco Discovery Protocol

CEN cloud-enabled network

CF CompactFlash

CF contention free

CFP contention-free period

CHAP Challenge Handshake Authentication Protocol

CKIP Cisco Key Integrity Protocol

CLI command-line interface

CN common name

CoA Change of Authorization

COW computer on wheels

CP contention period

CRC cyclic redundancy check

CRM customer relationship management

CSMA/CA Carrier Sense Multiple Access with Collision Avoidance

CSMA/CD Carrier Sense Multiple Access with Collision Detection

CSR certificate-signing request

CTL Certified Trust List

CTR Counter mode

CTS clear to send

CW contention window

CWG-RF Converged Wireless Group-RF Profile

DA destination address

DBPSK differential binary phase shift keying

DC direct current

DCF Distributed Coordination Function

DDF distributed data forwarding

DDoS distributed denial of service

DECT Digital Enhanced Cordless Telecommunications

DES Data Encryption Standard

DFS dynamic frequency selection

DHCP Dynamic Host Configuration Protocol

DIFS Distributed Coordination Function interframe space

DLS direct link setup

DoS denial of service

DPI deep packet inspection

DQPSK differential quadrature phase shift keying

DRS dynamic rate switching

DS distribution system

DSAS distributed spectrum analysis system

DSCP differentiated services code point

DSM distribution system medium

DSP digital signal processing

DSRC Dedicated Short Range Communications

DSS distribution system services

DSSS direct sequence spread spectrum

DSSS-OFDM direct sequence spread spectrum-orthogonal frequency division multiplexing

DTIM delivery traffic indication message

DTLS Datagram Transport Layer Security

DMZ demilitarized zone

EAP Extensible Authentication Protocol

EAPOL Extensible Authentication Protocol over LAN

EAP-AKA Extensible Authentication Protocol-Authentication and Key Agreement

EAP-FAST Extensible Authentication Protocol-Flexible Authentication via Secure Tunneling

EAP-GTC Extensible Authentication Protocol-Generic Token Card

EAP-LEAP Extensible Authentication Protocol-Lightweight Extensible Authentication Protocol

EAP-MD5 Extensible Authentication Protocol-Message Digest5

EAP-MSCHAPv2 Extensible Authentication Protocol-Microsoft Challenge Handshake Authentication Protocol

EAP-PEAP Extensible Authentication Protocol-Protected Extensible Authentication Protocol

EAP-POTP Extensible Authentication Protocol-Protected One-Time Password Protocol

EAP-SIM Extensible Authentication Protocol-Subscriber Identity Module

EAP-TEAP Extensible Authentication Protocol-Tunneled Extensible Authentication Protocol

EAP-TLS Extensible Authentication Protocol-Transport Layer Security

EAP-TTLS Extensible Authentication Protocol-Tunneled Transport Layer Security

ECDH Elliptic Curve Diffie-Hellman

ECDSA Elliptic Curve Digital Signature Algorithm

EDCA Enhanced Distributed Channel Access

EEG enterprise encryption gateway

EIFS extended interframe space

EIRP equivalent isotropically radiated power

EM electromagnetic

EQM equal modulation

ERP Extended Rate Physical

ERP-CCK Extended Rate Physical-Complementary Code Keying

ERP-DSSS Extended Rate Physical-Direct Sequence Spread Spectrum

ERP-OFDM Extended Rate Physical-Orthogonal Frequency Division Multiplexing

ERP-PBCC Extended Rate Physical-Packet Binary Convolutional Coding

ESA extended service area

ESP Encapsulating Security Payload

ESS extended service set

ESSID extended service set identifier

EUI extended unique identifier

EWG enterprise wireless gateway

FA foreign agent

FAST Flexible Authentication via Secure Tunnel

FCS frame check sequence

FEC forward error correction

FHSS frequency-hopping spread spectrum

FILS fast initial link setup

FM frequency modulation

FMC fixed mobile convergence

FSK frequency shift keying

FSPL free space path loss

FSR fast secure roaming

FT fast BSS transition

FTAA FT authentication algorithm

FTIE fast BSS transition information element

FZ Fresnel zone

GCM Google Cloud Messaging

GCMP Glaois/Counter Mode Protocol

GFSK Gaussian frequency shift keying

GI guard interval

GMK group master key

GPO Group Policy Object

GPS Global Positioning System

GRE Generic Routing Encapsulation

GSM Global System for Mobile Communications

GTC Generic Token Card

GTK group temporal key

GUI graphical user interface

HA home agent

HC hybrid coordinator

HCCA Hybrid Coordination Function Controlled Channel Access

HCF Hybrid Coordination Function

HMAC Hashed Message Authentication Code

HR-DSSS high-rate direct sequence spread spectrum

HSRP Hot Standby Router Protocol

HT High Throughput

HTTPS Hypertext Transfer Protocol Secure

HWMP Hybrid Wireless Mesh Protocol

Hz Hertz

IAPP Inter-Access Point Protocol

IBSS independent basic service set

ICMP Internet Control Message Protocol

ICV Integrity Check Value

IdP Identity provider

IDS intrusion detection system

IE Information Element

IFS interframe space

IKE Internet Key Exchange

IoT Internet of Things

IP Internet Protocol

IPS intrusion prevention system

IPsec Internet Protocol Security

IR infrared

IR intentional radiator

IS integration service

ISAKMP Internet Security Association and Key Management Protocol

ISI intersymbol interference

ISM Industrial, Scientific, and Medical

ITS intelligent transportation systems

IV initialization vector

JDBC Java Database Connectivity

KCK Key Confirmation Key

KEK Key Encryption Key

L2TP Layer 2 Tunneling Protocol

LAN local area network

LBAC location-based access control

LCI location configuration information

LDAP Lightweight Directory Access Protocol

LEAP Lightweight Extensible Authentication Protocol

LLC Logical Link Control

LLDP Link Layer Discovery Protocol

L-LTF Legacy (non-HT) long training field

LOS line of sight

LWAPP Lightweight Access Point Protocol

MAC media access control

MAHO Mobile Assisted Hand-Over

MAN metropolitan area network

MAP mesh access point

MCA multiple-channel architecture

MCS modulation and coding scheme

MD mobility domain

MD5 Message Digest5

MDC mobility domain controller

MDID mobility domain identifier

MDIE mobility domain information element

MDI media-dependent interface

MDM mobile device management

MFP management frame protection

MIB management information base

MIC Message Integrity Code

MIMO multiple-input multiple-output

MMPDU Management MAC Protocol Data Unit

MPDU MAC Protocol Data Unit

MP mesh point

MPP mesh point collocated with a mesh portal

MPPE Microsoft Point-to-Point Encryption

MRC maximal ratio combining

MSDU MAC Service Data Unit

MSK master session key

MSSID Mesh Service Set Identifier

MTBA multiple traffic ID block acknowledgment

MTK mesh temporal key

MTU maximum transmission unit

MU-MIMO multiuser MIMO

mW milliwatt

NAC network access control

NAT Network Address Translation

NAV Network Allocation Vector

NFC Near Field Communication

NMS network management server

NOC network operations center

nQSTA Non-Quality of Service Station

OAuth open standard for authorization

ODBC Open Database Connectivity

OFDM Orthogonal Frequency Division Multiplexing

OKC opportunistic key caching

OS operating system

OSI model Open Systems Interconnection model

OTAP Over-the-air provisioning

OTP one-time password

OU organizational unit

OUI Organizationally Unique Identifier

PAC protected access credential

PAN personal area network

PAP Password Authentication Protocol

PAT Port Address Translation

PBC push-button configuration

PBCC Packet Binary Convolutional Coding

PBX private branch exchange

PCI Peripheral Component Interconnect

PCMCIA Personal Computer Memory Card International Association (PC Card)

PD powered device

PEAP Protected Extensible Authentication Protocol

PHY Physical layer

PIFS Point Coordination Function interframe space

PIN personal information number

PKI public key infrastructure

PLCP Physical Layer Convergence Procedure

PMD Physical Medium Dependent

PMK pairwise master key

PMKID pairwise master key identifier

PMK-R0 pairwise master key-R0

PMK-R1 pairwise master key-R1

PMKSA pairwise master key security association

PN packet number

PN pseudorandom number

PoE Power over Ethernet

POP Post Office Protocol

PPDU PLCP Protocol Data Unit

PPP Point-to-Point Protocol

PPTP Point-to-Point Tunneling Protocol

PSDU PLCP Service Data Unit

PRF pseudorandom function

PSE power-sourcing equipment

PSK Phase Shift Keying

PSK preshared key

PSMP Power Save Multi-Poll

PSPF Public Secure Packet Forwarding

PS-Poll power save poll

PSTN public switched telephone network

PTK pairwise transient key

PTKSA pairwise transient key security association

PTMP point-to-multipoint

PTP point-to-point

QAM quadrature amplitude modulation

QAP quality-of-service access point

QBSS quality-of-service basic service set

QoS quality of service

QSTA quality-of-service station

QPSK quadrature phase shift keying

RA receiver address

RADIUS Remote Authentication Dial-In User Service

RAP remote access point

RBAC role-based access control

RF radio frequency

RFC request for comment

RF LOS RF line of sight

RIC resource information container

RIFS reduced interframe space

R0KH R0 key holder

R1KH R1 key holder

RRM radio resource measurement

RSL received signal level

RSN robust security network

RSNA robust security network association

RSNIE robust security network information element

RSSI received signal strength indicator

RTLS real-time location system

RTS request to send

RTS/CTS request to send/clear to send

RX receive or receiver

SA security associations

SA source address

SaaS Software as a Service

SAE Simultaneous Authentication of Equals

SAML Security Assertion Markup Language

S-APSD scheduled automatic power save delivery

SCA single-channel architecture

SCEP Simple Certificate Enrollment Protocol

SD Secure Digital

SDR software-defined radio

SHA-1 Secure Hash Algorithm

SID system identifier

SIFS short interframe space

SIM Subscriber Identity Module

SIP Session Initiation Protocol

SISO single-input, single-output

SRP SpectraLink Radio Protocol

SVP SpectraLink Voice Priority

SISO single-input single-output

SM spatial multiplexing

SMB small and medium-sized business

SMK STSL master key

SMTP Simple Mail Transfer Protocol

SNMP Simple Network Management Protocol

SNR signal-to-noise ratio

SOHO small office/home office

S0KH S0 key holder

S1KH S1 key holder

SOM system operating margin

SOW statement of work

SP service provider

SPA supplicant address

SQ signal quality

SSH Secure Shell

SSID service set identifier

SSL Secure Sockets Layer

SSO single sign-on

STA Station

STBC space-time block coding

STC Space Time Coding

STK STSL transient key

STP Spanning Tree Protocol

STSL station-to-station link

SU-MIMO single user MIMO

TA transmitter address

TBTT target beacon transmission time

TCP/IP Transmission Control Protocol/Internet Protocol

TDEA Triple Data Encryption Algorithm

TDoA Time Difference of Arrival

TDLS Tunneled Direct Link Setup

TIM traffic indication map

TK Temporal Key

TKIP Temporal Key Integrity Protocol

TLS Transport Layer Security

TPC transmit power control

TPK TDLS Peer Key

TPKSA TDLS Peer Key security association

TSC TKIP sequence counter

TSN transition security network

TTAK TKIP-mixed transmit address and key

TTL time to live

TTLS Tunneled Transport Layer Security

TX transmit or transmitter

TxBF transmit beamforming

TXOP transmit opportunity

U-APSD unscheduled automatic power save delivery

UEQM unequal modulation

UNII Unlicensed National Information Infrastructure

UP user priority

USB Universal Serial Bus

USIM User Subscriber Identity Module

UTMS Universal Mobile Telecommunications System

VHT Very High Throughput

VLAN virtual local area network

VoIP Voice over IP

VoWiFi Voice over Wi-Fi

VPN virtual private network

VRRP Virtual Router Redundancy Protocol

VSA vendor specific attribute

VSWR voltage standing wave ratio

WAN wide area network

WAVE Wireless Access in Vehicular Environments

WDS wireless distribution system

WEP Wired Equivalent Privacy

WGB workgroup bridge

WIDS wireless instruction detection system

Wi-Fi Sometimes said to be an acronym for *wireless fidelity*, a term that has no formal definition; Wi-Fi is a general marketing term used to define 802.11 technologies.

WIGLE Wireless Geographic Logging Engine

WiMAX Worldwide Interoperability for Microwave Access

WIPS wireless intrusion prevention system

WISP Wireless Internet Service Provider

WLAN wireless local area network

WLSE Wireless LAN Solution Engine

WM wireless medium

WMAN wireless metropolitan area network

WMM Wi-Fi Multimedia

WMM-PS Wi-Fi Multimedia Power Save

WMM-SA Wi-Fi Multimedia Scheduled Access

WNM wireless network management

WNMS wireless network management system

WPA Wi-Fi Protected Access

WPA2 Wi-Fi Protected Access 2

WPAN wireless personal area network

WPP Wireless Performance Prediction

WPS Wi-Fi Protected Setup

WTP wireless termination point

WWAN wireless wide area network

XML Extensible Markup Language

XOR exclusive or

Index

Index

Note to the Reader: Throughout this index **boldfaced** page numbers indicate primary discussions of a topic. *Italicized* page numbers indicate illustrations.

Numbers

A

B

C

D

E

F

G

H

I

J

K

L

M

N

O

P

S

T

U

V

W

X

Z

Comprehensive Online Learning Environment

Register to gain access to the online interactive learning environment and test bank to help you study for your CWSP certification—included with your purchase of this book!

The online tool includes:

- **Assessment Test** to help you focus your study to specific objectives
- **Chapter Tests** to reinforce what you learned
- **Practice Exams** to test your knowledge of the material
- **Electronic Flashcards** to reinforce your learning and provide last-minute test prep before the exam
- **Searchable Glossary** gives you instant access to the key terms you'll need to know for the exam.

Go to `http://www.wiley.com/go/sybextestprep` **to register and gain access to this comprehensive study tool package.**